Eco Science Advances

Future Sustainability

Harper Faith

ISBN: 978-1-77961-123-9
Imprint: Telephasic Workshop
Copyright © 2024 Harper Faith.

Contents

Introduction   **1**
Overview of the book   1

Principles of Eco Science   **9**
Definitions and concepts   9
Biodiversity   26
Conservation and Restoration   79
Sustainable Development   125

Advances in Eco Technology   **179**
Clean Energy   179
Green Transportation   221
Waste Management   261
Water and Air Pollution Control   302

Eco Science in Policy and Governance   **343**
Environmental Policy   343
Corporate Sustainability   392
Environmental Governance   422

Future Directions in Eco Science   **455**
Emerging Technologies   455
Resilience and Adaptation   483
Ecosystem Services and Resilience   488
Sustainable Urban Planning   511
Sustainable Consumption and Production   540
Global Collaboration for Sustainability   572

Bibliography   **599**

Index   601

Introduction

Overview of the book

Scope and Objectives

The scope of this book, *Eco Science Advances: Future Sustainability*, is to provide a comprehensive overview of the field of eco science, focusing on its principles, advances in technology, policy and governance, and future directions. By exploring these topics, the book aims to shed light on the importance of eco science in achieving a sustainable future and provide readers with a holistic understanding of the current state and potential of eco science.

The objectives of this book are multi-fold. First, it seeks to define and explain key concepts and definitions related to eco science, such as ecology, sustainability, and environmental science. By establishing this foundation, readers will develop a solid understanding of the principles that underpin eco science.

Second, the book aims to explore the intricate relationship between biodiversity and conservation. It will delve into the importance of biodiversity, the threats it faces, and the strategies and approaches used to protect and restore it. This section will cover topics such as habitat destruction, pollution, climate change, conservation strategies, protected areas, and the role of indigenous knowledge.

Third, the book will address the concept of sustainable development and its various dimensions. It will discuss key issues such as poverty eradication, gender equality, clean energy, sustainable agriculture, sustainable cities, and waste management. By highlighting these topics, readers will gain insights into the importance of sustainable practices and their impact on society and the environment.

Fourth, the book will delve into the advances in eco technology that have the potential to revolutionize the way we address environmental challenges. It will cover topics like clean energy, green transportation, waste management, and

1

pollution control. By exploring these technological innovations, readers will gain a deeper understanding of the tools and solutions available to create a sustainable future.

Fifth, the book will examine the role of policy and governance in shaping and implementing eco science principles. It will explore international agreements and treaties, national environmental policies, environmental impact assessment, and the role of corporate sustainability. Additionally, the book will dive into topics such as stakeholder engagement, participatory decision-making, and the role of indigenous knowledge in environmental governance.

Lastly, the book will explore future directions in eco science, highlighting emerging technologies, resilience and adaptation strategies, sustainable urban planning, sustainable consumption and production, and global collaboration for sustainability. By discussing these topics, the book aims to provide readers with a glimpse into the future of eco science and inspire them to contribute to a sustainable world.

Throughout the book, real-world examples, case studies, and exercises will be included to help readers apply the principles and concepts learned. Additionally, the book will provide resources for further exploration of each topic, encouraging readers to delve deeper into areas of interest.

By the end of this book, readers will have gained a comprehensive understanding of eco science and its potential to address the urgent sustainability challenges of our time. It is hoped that this knowledge will empower and inspire readers to become change agents and contribute to the creation of a sustainable and resilient future.

Importance of Eco Science

Eco science, also known as environmental science, plays a critical role in understanding and addressing the complex challenges of today's world. It is an interdisciplinary field that encompasses various scientific disciplines and integrates their knowledge to provide insights into the interactions between the environment and society. In this section, we will explore the importance of eco science and its significance for future sustainability.

Understanding Ecosystem Dynamics

Eco science focuses on studying the relationships among living organisms and their environment. It provides us with a deep understanding of ecosystem dynamics, such as the flow of energy and matter, biogeochemical cycles, and the intricate web of interactions between species. This knowledge is crucial for identifying the underlying causes of environmental issues and developing effective conservation and management strategies.

For example, by studying the impacts of habitat destruction and fragmentation on biodiversity, eco scientists can assess the ecological consequences and propose ways to restore and protect ecosystems. By unraveling the complex relationship between human activities and climate change, eco science can help us mitigate and adapt to the challenges posed by global warming.

Conserving Biodiversity

One of the main areas of focus in eco science is the conservation of biodiversity. Biodiversity is the variety of life on Earth, including the diversity of species, genes, and ecosystems. It is essential for maintaining the healthy functioning of ecosystems and providing us with numerous ecological, economic, and cultural benefits.

Eco science provides the tools and knowledge necessary for understanding the importance of biodiversity to human well-being. It assists in identifying threatened species, assessing the impacts of human activities on ecosystems, and developing strategies for biodiversity conservation.

For instance, eco science helps identify key habitat areas that are critical for the survival of endangered species. It also guides the implementation of conservation measures, such as protected areas and sustainable land use practices. Moreover, eco science highlights the importance of considering indigenous knowledge and practices in biodiversity conservation, as indigenous communities often possess valuable insights into sustainable resource management.

Promoting Sustainable Development

Sustainable development is the key to meeting the needs of the present generation without compromising the ability of future generations to meet their own needs. Eco science plays a vital role in promoting sustainable development by providing scientific evidence, tools, and frameworks for decision-making.

Eco science contributes to the development and implementation of sustainable development goals (SDGs) at various levels, such as poverty eradication, clean energy, sustainable agriculture, and sustainable cities. It helps us understand the interactions between the social, economic, and environmental dimensions of sustainability and guides the formulation of policies and actions that balance these dimensions.

For example, eco science informs the design of sustainable transportation systems by providing insights into the environmental impacts of different modes of transportation and proposing solutions for reducing carbon emissions. It also supports the transition to a circular economy by examining the life cycle of products and identifying opportunities for resource efficiency and waste reduction.

Addressing Environmental Challenges

Eco science is essential for addressing the environmental challenges we face today, including climate change, pollution, habitat loss, and natural resource depletion. It provides the knowledge and tools necessary for understanding the root causes of these challenges and developing effective solutions.

By studying the impacts of human activities on the environment, eco scientists can help policymakers and society make informed decisions about resource management, pollution control, and conservation. They contribute to the development and implementation of environmental policies and regulations that promote sustainable practices and protect ecosystems.

For instance, eco science provides innovative solutions for clean energy generation, such as solar and wind power, which can reduce our dependence on fossil fuels and mitigate climate change. It also offers strategies for sustainable waste management, including recycling, composting, and waste-to-energy technologies, to minimize environmental pollution and promote resource recovery.

Engaging Society

Finally, eco science plays a crucial role in engaging society and raising awareness about environmental issues. It bridges the gap between scientific knowledge and public understanding, facilitating informed decision-making and promoting sustainable behaviors.

Through education, communication, and outreach efforts, eco scientists contribute to building a more environmentally literate society. They empower individuals and communities to take action towards sustainable living, conservation, and environmental stewardship.

For instance, eco science initiatives often include public participation in scientific research, such as citizen science projects, which allow individuals to contribute to data collection and analysis. This approach fosters a sense of ownership and responsibility for the environment and promotes active engagement in sustainability efforts.

In conclusion, eco science is of paramount importance for future sustainability. It provides a holistic and interdisciplinary approach to understanding and addressing environmental challenges. By studying ecosystem dynamics, conserving biodiversity, promoting sustainable development, addressing environmental challenges, and engaging society, eco science lays the foundation for a more sustainable and resilient future. It is through the integration of scientific knowledge, policy development, and societal engagement that we can achieve a harmonious relationship between human well-being and the environment.

Future Perspectives

The field of eco science is constantly evolving, driven by the pressing need to address environmental challenges and achieve sustainable development. As we look to the future, several key perspectives emerge that will shape the trajectory of eco science advancements.

Integration of Technology

One significant trend is the increasing integration of technology in eco science research and practices. Advanced technologies such as artificial intelligence, gene editing, nanotechnology, and the Internet of Things hold enormous potential for addressing environmental issues. For instance, artificial intelligence can be used for predictive modeling of ecological systems, enabling more effective decision-making in conservation efforts. Gene editing techniques have the potential to enhance the resilience of species and ecosystems in the face of climate change. Nanotechnology can contribute to the development of sustainable materials and clean energy technologies. The Internet of Things can facilitate real-time monitoring of environmental parameters and help optimize resource management. The integration of technology in eco science will undoubtedly revolutionize our approach to sustainability and conservation.

Resilience and Adaptation

Another important perspective is the increasing recognition of the need for resilience and adaptation to environmental changes. As the impacts of climate change become more evident, it is essential to develop strategies that enhance the resilience of ecosystems, communities, and infrastructure. Resilience thinking emphasizes the need to consider social and ecological systems as interconnected and dynamic, and focuses on building adaptive capacity to cope with disturbances. This perspective encourages the adoption of nature-based solutions and the integration of indigenous knowledge in conservation and adaptation efforts. By embracing resilience thinking, we can better prepare for and respond to the complex and uncertain challenges posed by a changing environment.

Sustainable Urban Planning

The rapid growth of urban areas presents both challenges and opportunities for sustainable development. Sustainable urban planning is a perspective that recognizes the importance of creating environmentally friendly and socially inclusive cities. It involves designing urban spaces that promote efficient land use, reduce pollution, and prioritize public transportation and active modes of transportation. Green building design, urban agriculture, and the mitigation of urban heat islands are some examples of sustainable urban planning practices. By adopting a holistic approach to urban development, we can create cities that are not only environmentally sustainable but also livable and resilient.

Circular Economy

The concept of a circular economy is gaining momentum as a powerful solution to the problems of resource depletion and waste generation. A circular economy is characterized by the continuous reuse and regeneration of materials and resources, aiming to minimize waste and maximize the value of products and materials. It involves designing products with a focus on longevity, promoting recycling and repurposing, and emphasizing sustainable production and consumption patterns. The circular economy perspective promotes a shift from the traditional linear "take-make-dispose" model to a more sustainable and circular approach. By fostering innovations in product design, waste management, and resource utilization, the circular economy can contribute significantly to a sustainable future.

Global Collaboration for Sustainability

The challenges we face in achieving sustainability transcend national boundaries, making global collaboration crucial. International cooperation is necessary to address complex environmental issues such as climate change and biodiversity loss. Global environmental governance frameworks, such as international agreements and treaties, play a crucial role in coordinating efforts and setting targets. Collaboration between governments, organizations, and individuals is needed to share knowledge, resources, and best practices. Initiatives like the United Nations Sustainable Development Goals provide a roadmap for global sustainability efforts. By promoting cooperation and collective action, we can work together towards a more sustainable and equitable world.

In summary, the future of eco science holds exciting prospects for leveraging technology, embracing resilience and adaptation, promoting sustainable urban planning, advancing the circular economy, and fostering global collaboration. By embracing these perspectives, we can pave the way for a more sustainable and resilient future for generations to come. Let us embark on this journey together, for the betterment of our planet and all its inhabitants.

Principles of Eco Science

Definitions and concepts

Ecology

Ecology is the scientific study of the interactions between organisms and their environment. It is a branch of biology that focuses on understanding how living organisms, including plants, animals, and microorganisms, interact with each other and with their physical surroundings.

1. Introduction to Ecology

Ecology is the study of the relationships between organisms and their environment. It involves the study of populations, communities, and ecosystems, as well as the processes and patterns that drive these systems. By understanding the intricate web of interactions between living organisms and their surroundings, ecologists can gain insights into the functioning and dynamics of ecosystems.

2. Ecological Levels of Organization

Ecology examines biological systems at different levels of organization. These levels include:

2.1 Individual Level: At this level, ecologists study the adaptations and behaviors of individual organisms and how they interact with their environment. For example, they may investigate how animals find and utilize food sources, or how plants respond to changes in light intensity.

2.2 Population Level: Ecologists analyze the characteristics and dynamics of populations, which are groups of individuals of the same species living in a specific area. They study population size, density, distribution, and growth rates. This information helps understand factors that influence the survival and reproduction of species.

2.3 Community Level: Communities consist of multiple interacting populations. Ecologists study the species composition, diversity, and interactions within communities.   They investigate how different organisms compete for resources or cooperate to survive and reproduce.

2.4 Ecosystem Level: An ecosystem encompasses both living organisms and their physical environment, including abiotic factors such as climate, soil, and water. Ecologists examine the flow of energy and nutrients within ecosystems, as well as the interactions and feedbacks among different organisms and their environment.

3. Ecological Interactions

Ecological interactions are fundamental to understanding the dynamics of ecosystems. They include:

3.1 Predator-Prey Interactions: Predators consume prey, regulating prey populations and shaping community structure.   For example, the interaction between lions and zebras in the African savanna is a classic predator-prey relationship.

3.2 Symbiotic Relationships: Symbiosis refers to close and prolonged interactions between different species. Examples include mutualism (both species benefit), commensalism (one species benefits while the other is unaffected), and parasitism (one species benefits at the expense of the other).

3.3 Competition: Organisms often compete for limited resources, such as food, water, and shelter.   Competition can influence population sizes and community structure. For instance, trees in a forest may compete for sunlight and nutrients.

3.4 Mutualistic Interactions: Some organisms engage in mutually beneficial relationships, where both species gain an advantage. For example, plants and pollinators have a mutualistic relationship as the plants provide nectar and the pollinators transfer pollen.

4. Key Concepts in Ecology

4.1 Energy Flow: Energy is the driving force of ecological systems. The sun provides energy for photosynthesis, which is then transferred through food chains or webs. Ecologists study energy flow to understand how organisms utilize and transform energy.

4.2 Nutrient Cycling: Nutrients, such as carbon, nitrogen, and phosphorus, are essential for life. Ecological processes, such as decomposition and nutrient uptake by plants, drive the cycling of these elements. Understanding nutrient cycling helps ecologists evaluate ecosystem health and stability.

4.3 Ecological Succession: Ecological succession refers to the gradual change in the species composition of a community over time. Primary succession occurs in newly formed or exposed habitats (e.g., after a volcanic eruption), while secondary succession occurs in habitats that have been disturbed but retain some soil and seed banks.

4.4 Ecological Resilience: Resilience is the ability of an ecosystem to recover from disturbances. Ecologists study how ecosystems respond and adapt to disturbances like fires, hurricanes, or human activities. They examine factors that promote or hinder ecosystem resilience.

5. Applications of Ecology

Ecological knowledge is crucial for addressing environmental challenges and promoting sustainability. Some applications of ecology include:

5.1 Conservation Biology: Conservation biologists apply ecological principles to protect and restore biodiversity. They work on issues such as habitat preservation, species conservation, and the management of protected areas.

5.2 Restoration Ecology: Restoration ecologists use ecological knowledge to restore ecosystems that have been degraded or damaged. They aim to recreate functional ecosystems and promote biodiversity recovery.

5.3 Ecosystem Services: Ecosystem services are the benefits that humans obtain from ecosystems, such as clean water, air quality regulation, and pollination. Ecological research helps quantify and protect these services.

5.4 Climate Change Adaptation: Ecologists study how ecosystems can adapt to the impacts of climate change and explore strategies for mitigating these impacts. They focus on understanding how species and communities respond to changing environmental conditions.

5.5 Sustainability: Ecology plays a vital role in promoting sustainable practices and policies. By understanding ecological principles, society can develop more sustainable approaches to agriculture, resource management, and urban planning.

In conclusion, ecology provides the foundation for understanding the intricate relationships between living organisms and their environment. It offers valuable insights into ecosystem dynamics, species interactions, and environmental processes. By studying ecology, we can develop strategies for conserving biodiversity, restoring degraded ecosystems, and promoting sustainable practices for the benefit of both humans and the natural world.

Sustainability

Sustainability is a fundamental concept in eco science that lies at the heart of efforts to achieve a more balanced and harmonious relationship between human beings and the environment. It involves meeting the needs of the present generation without compromising the ability of future generations to meet their own needs. In this section, we will explore the principles and practices of sustainability, as well as the challenges and opportunities it presents.

The Three Pillars of Sustainability

Sustainability is often depicted as having three interconnected pillars: environmental, social, and economic. These three dimensions are equally important and must be considered together in order to achieve true sustainability.

Environmental sustainability is concerned with the preservation and responsible use of natural resources and ecosystems. It involves minimizing pollution, reducing waste, protecting biodiversity, and promoting the conservation of ecosystems. Environmental sustainability aims to ensure that we can continue to enjoy clean air, fresh water, and a healthy natural environment for years to come.

Social sustainability focuses on creating equitable and inclusive societies that prioritize the well-being and quality of life for all individuals. It involves promoting social justice, addressing inequality, respecting human rights, and supporting strong communities. Social sustainability recognizes the interconnectedness of human well-being and recognizes that a sustainable future must be a just and fair one.

Economic sustainability involves fostering economic systems that are ecologically responsible, socially inclusive, and economically viable. It seeks to balance economic growth with environmental protection and social well-being. Economic sustainability aims to achieve prosperity without depleting resources or compromising the welfare of future generations.

The Triple Bottom Line

The concept of the triple bottom line is closely related to sustainability. It suggests that organizations should measure success not only by financial performance but also by their impact on people and the planet. This holistic approach recognizes that businesses and other entities have a responsibility to society and the environment beyond generating profits.

The triple bottom line consists of three interrelated dimensions: economic, social, and environmental. By considering all three aspects, decision-makers can make more informed choices that not only benefit their organization but also contribute to a more sustainable world.

Challenges and Opportunities

Achieving sustainability is not without its challenges, but it also presents numerous opportunities for innovation, collaboration, and positive change. Some of the key challenges and opportunities in sustainability include:

Resource Scarcity and Climate Change - The increasing demand for resources, coupled with the impacts of climate change, poses significant challenges to sustainability. However, addressing these challenges also provides opportunities for developing new technologies, improving efficiency, and transitioning to low-carbon and renewable energy sources.

Poverty and Inequality - Poverty and inequality are obstacles to sustainability as they limit access to healthcare, education, and basic resources. However, working towards poverty eradication and reducing inequality can lead to more resilient and sustainable communities.

Consumer Awareness and Behavior - Encouraging sustainable consumption and behavior change among individuals and communities can have a significant positive impact on sustainability. Educating consumers about the environmental and social implications of their choices can empower them to make more sustainable decisions.

Policy and Governance - Effective environmental policies and governance frameworks play a crucial role in driving sustainable practices and behaviors. Governments, businesses, and civil society must work together to create enabling environments for sustainable development.

Collaboration and Partnerships - Addressing complex sustainability challenges requires collaboration and partnerships between different stakeholders, including governments, businesses, academia, and communities. Working together can unlock innovative solutions and ensure the successful implementation of sustainable practices.

Education and Research - Education and research play a vital role in advancing sustainability. By equipping individuals with knowledge and skills, and by conducting research on sustainable solutions, we can drive positive change and empower future generations to address sustainability challenges effectively.

Case Study: Circular Economy

A compelling example of a sustainability initiative is the concept of a circular economy. The traditional linear economy follows a "take-make-dispose" model, where resources are extracted, processed into products, and eventually discarded as waste. In contrast, a circular economy aims to design out waste and keep materials in use for as long as possible.

In a circular economy, products and materials are recycled, repaired, or repurposed, creating a closed-loop system that minimizes resource extraction and waste generation. By adopting a circular economy approach, businesses can reduce their environmental impact, promote resource efficiency, and enhance economic resilience.

For instance, the fashion industry, known for its significant environmental footprint, is exploring circular economy principles by implementing clothing rental services, recycling initiatives, and innovative design practices that prioritize durability and repairability. This shift towards a circular fashion system not only reduces the industry's immense waste but also offers new business opportunities and fosters a more sustainable and ethical approach to fashion.

Key Takeaways

Sustainability is a multidimensional concept that addresses the environmental, social, and economic aspects of our actions and decisions. It requires a holistic approach that balances the needs of the present and future generations. By considering the three pillars of sustainability and adopting the principles of the triple bottom line, we can work towards a more sustainable and equitable future.

Despite the challenges, sustainability presents opportunities for innovation, collaboration, and positive change. Addressing resource scarcity, climate change, poverty, and consumer behavior are just some of the avenues through which we can drive sustainable practices. Additionally, policy and governance, collaboration and partnerships, and education and research are essential for achieving sustainability goals.

The case study of the circular economy demonstrates how a systemic approach can transform industries and promote sustainable practices. By adopting circular economy principles, businesses can reduce waste, conserve resources, and contribute to a more sustainable and resilient economy.

In summary, sustainability is not only a goal but also a mindset and a way of life. It requires a collective effort and a commitment to balancing the environmental, social, and economic dimensions of our actions. Through sustainable practices and innovative solutions, we can create a world that is better equipped to address the challenges of the present and the future.

Environmental science

Environmental science is a multidisciplinary field that seeks to understand the complex interactions between the natural world and human activities. It encompasses the study of the environment, including the physical, chemical, and biological processes that shape it, as well as the impact of human activities on ecosystems and natural resources.

Principles of Environmental Science

Environmental science is guided by several fundamental principles that shape its approach to understanding and managing the environment:

1. **Interdisciplinary approach:** Environmental science recognizes that environmental issues are complex and require knowledge from various disciplines, including biology, chemistry, physics, geology, sociology, economics, and policy. The integration of these disciplines allows for a more comprehensive understanding of environmental problems and potential solutions.

2. **Systems thinking:** Environmental science employs systems thinking, which recognizes that the environment is composed of interconnected systems, such as ecosystems and social systems. This approach emphasizes the need to study and manage the environment as a whole, rather than focusing on isolated components.

3. **Sustainability:** Central to environmental science is the concept of sustainability, which refers to the responsible use and management of resources to meet the needs of current and future generations. Environmental scientists strive to find solutions that balance environmental, social, and economic considerations to achieve sustainability.

4. **Environmental stewardship:** Environmental science emphasizes the importance of responsible environmental stewardship. This involves promoting conservation, sustainable resource use, and the protection of ecosystems and biodiversity. It also entails promoting awareness and education about environmental issues to empower individuals and communities to make informed decisions.

Key Concepts in Environmental Science

In order to understand and address environmental challenges, it is important to grasp some key concepts in environmental science:

1. **Ecology:** Ecology is the study of the relationships between organisms and their environment. It investigates how living organisms interact with each other and with their physical surroundings. Understanding ecological principles is crucial for understanding the functioning of ecosystems and the impacts of human activities on biodiversity and ecosystem services.

2. **Sustainability:** Sustainability involves meeting the needs of the present generation without compromising the ability of future generations to meet their own needs. It requires the responsible use of resources, the protection of biodiversity, and the promotion of social equity. Sustainability is a guiding principle in environmental science, providing a framework for addressing environmental challenges and striving for long-term viability.

3. **Environmental Science:** Environmental science is the study of the environment and the impact of human activities on it. It combines scientific research, data analysis, and interdisciplinary approaches to understand environmental processes and develop strategies for sustainable resource management.

4. **Environmental Impact:** Environmental impact refers to the consequences of human activities on the environment. It includes both direct and indirect effects, such as pollution, habitat destruction, climate change, and resource depletion. Understanding environmental impact is essential for identifying and mitigating the negative effects of human activities.

5. **Environmental Policy and Governance:** Environmental policy and governance are crucial for addressing environmental challenges at local, national, and global levels. They involve the development and implementation of regulations, laws, and management strategies to protect the environment and promote sustainability. Environmental science provides the scientific basis for environmental policy decisions and contributes to effective environmental governance.

Challenges and Solutions in Environmental Science

Environmental science faces numerous challenges in addressing environmental issues. Some of the key challenges include:

1. **Climate Change:** Climate change is one of the most pressing environmental challenges of our time. It is caused by human activities, primarily the burning of fossil fuels, which release greenhouse gases into the atmosphere. Climate change leads to rising temperatures, sea-level rise, altered precipitation patterns, and increased frequency and intensity of extreme weather events. Addressing climate change requires reducing greenhouse gas emissions, transitioning to renewable energy sources, and implementing adaptation measures to minimize its impacts.

2. **Biodiversity Loss:** Biodiversity loss is a result of habitat destruction, pollution, invasive species, climate change, and overexploitation of natural resources. It has significant ecological, economic, and social consequences. Protecting and restoring biodiversity requires the establishment of protected areas, sustainable land use practices, conservation strategies, and the integration of indigenous knowledge in conservation efforts.

3. **Pollution:** Pollution, including air, water, and soil pollution, poses a threat to human health and the environment. It is caused by industrial activities, agricultural practices, improper waste management, and other human activities. To combat pollution, environmental science focuses on developing and implementing pollution prevention and control measures, promoting sustainable waste management practices, and advocating for cleaner technologies.

4. **Resource Depletion:** The unsustainable use of natural resources, such as minerals, water, and fossil fuels, leads to resource depletion and environmental degradation. Environmental science aims to promote sustainable resource management practices, including resource conservation, efficient use of resources, and the development of renewable energy sources.

Case Study: Plastic Pollution

Plastic pollution is a significant environmental issue that highlights the challenges faced by environmental science. Plastics are widely used due to their durability and low cost, but their improper disposal and slow decomposition have resulted in widespread pollution of marine and terrestrial ecosystems.

Plastic pollution has detrimental effects on wildlife, including entanglement and ingestion by marine animals. It also poses risks to human health, as plastic debris can enter the food chain through the consumption of contaminated seafood.

Addressing plastic pollution requires a multifaceted approach. Environmental science plays a crucial role in analyzing the impacts of plastic pollution, developing effective waste management strategies, and advocating for the reduction of plastic use. This includes promoting recycling, implementing policies to reduce single-use plastics, and supporting the development of biodegradable alternatives.

Furthermore, education and public awareness are essential in addressing plastic pollution. Environmental science can contribute by conducting research on the sources and impacts of plastic pollution, engaging with communities, and raising awareness about the importance of reducing plastic waste.

Conclusion

Environmental science is a multidisciplinary field that seeks to understand and address environmental challenges. By applying principles of interdisciplinary collaboration, systems thinking, and sustainability, environmental science provides insights and solutions to protect and manage the environment. Through research, policy development, and environmental stewardship, environmental science plays a vital role in achieving a sustainable future for our planet.

Interdisciplinary approaches

Interdisciplinary approaches are vital in understanding and addressing complex environmental challenges. By bringing together different disciplines, such as biology, chemistry, physics, mathematics, social sciences, and engineering, we can develop comprehensive and innovative solutions for sustainable development. These approaches promote collaboration, integration of knowledge, and the application of diverse perspectives to tackle environmental issues from multiple angles.

Why interdisciplinary approaches?

Environmental problems are often interconnected and require a holistic understanding to effectively address them. Interdisciplinary approaches enable us to bridge the gaps between disciplines and explore the relationships and interactions between various components of the environment. By combining expertise from different fields, we can gain a more complete understanding of complex systems, identify potential trade-offs, and develop innovative solutions that consider social, economic, and ecological factors.

Application areas of interdisciplinary approaches

Interdisciplinary approaches can be applied to various areas within environmental science and sustainability. Some key application areas include:

+ **Ecosystem management and conservation:** Understanding the interactions between species, ecosystems, and human activities requires inputs from ecologists, sociologists, economists, and policymakers. By integrating their knowledge, we can develop effective strategies for conservation and sustainable management of natural resources.

+ **Climate change adaptation and mitigation:** Addressing the challenges of climate change requires a multidisciplinary approach. Climate scientists, engineers, social scientists, and policymakers collaborate to develop mitigation strategies, assess the impacts of climate change, and design adaptation measures to reduce vulnerabilities and enhance resilience.

+ **Sustainable agriculture and food systems:** Creating a sustainable and resilient food system requires the integration of knowledge from agronomists, nutritionists, economists, sociologists, and policymakers. By considering the ecological, social, and economic dimensions, interdisciplinary approaches can help optimize resource use, reduce waste, and ensure food security.

+ **Water resource management:** Managing water resources sustainably involves understanding hydrological processes, water quality, social dynamics, and policy frameworks. Collaboration between hydrologists, biologists, chemists, sociologists, and policymakers can lead to effective water management strategies that balance human needs and environmental conservation.

+ **Urban sustainability:** Developing sustainable cities requires interdisciplinary collaboration between urban planners, architects, engineers, social scientists, and policymakers. By considering factors such as energy efficiency, transportation, waste management, and social equity, interdisciplinary approaches can help create livable, resilient, and environmentally friendly urban environments.

Challenges and solutions

While interdisciplinary approaches offer great potential, they also come with challenges. Some of these challenges include:

- **Communication and language barriers**: Each discipline has its own jargon and ways of communicating, which can make it difficult for experts to understand each other. This can be addressed through effective communication and the development of a common language that facilitates knowledge exchange among disciplines.

- **Differences in methodologies and research practices**: Different disciplines often have different methodologies and research approaches. This can hinder collaboration and integration of knowledge. To overcome this challenge, interdisciplinary teams can adopt a transdisciplinary approach, where there is a shared understanding and co-creation of knowledge across disciplines.

- **Power dynamics and disciplinary hierarchy**: Some disciplines may hold more influence or power in decision-making processes, which can limit the contributions of other disciplines. It is important to promote equal participation and recognition of the contributions from all disciplines involved.

Example: Interdisciplinary approach in landscape restoration

Landscape restoration is a complex environmental challenge that requires an interdisciplinary approach. Let's consider the example of a degraded forest ecosystem. An interdisciplinary team focusing on landscape restoration might include ecologists, foresters, soil scientists, hydrologists, social scientists, and policymakers.

The ecologists would study the biodiversity and ecological processes, while foresters would assess the tree species composition and identify suitable species for reforestation. Soil scientists would analyze the soil health and nutrient content, ensuring that the right soil amendments are used during restoration efforts. Hydrologists would study the water dynamics and suggest measures to improve water availability and quality.

Social scientists would engage with local communities to understand their needs and aspirations, ensuring that restoration efforts align with their goals. Policymakers would provide the necessary legal and regulatory frameworks to support the restoration process and ensure long-term sustainability.

By integrating the knowledge and expertise of these different disciplines, the interdisciplinary team can develop a comprehensive restoration plan that addresses ecological, social, and economic aspects. They can plan the restoration activities, monitor the progress, and evaluate the ecological and social outcomes. This collaborative approach increases the chances of success and ensures that the restored ecosystem is resilient and able to provide multiple benefits to the environment and local communities.

Further resources

Interdisciplinary approaches are a rapidly evolving field, and there are several resources available for further exploration. Some recommended resources include:

- **Interdisciplinary Environmental Studies:** A journal dedicated to publishing interdisciplinary research on environmental topics, providing insights into the latest research and methodologies in the field.

- **Bridging Boundaries:** A book by Monica Berger and Paul H. Thistle, which explores the challenges and benefits of interdisciplinary collaboration in environmental science. It offers practical tips and strategies for successful interdisciplinary research.

- **Interdisciplinary Research: Process and Theory:** A book by Allen F. Repko, which provides a comprehensive overview of interdisciplinary research approaches. It discusses the theoretical foundations and practical applications of interdisciplinary research in various disciplines, including environmental science.

Takeaways

Interdisciplinary approaches are essential for addressing complex environmental challenges. By bringing together expertise from various disciplines, we can develop comprehensive and innovative solutions for sustainable development. These approaches enable us to understand the interconnectedness of environmental issues and consider the social, economic, and ecological dimensions. While interdisciplinary collaboration comes with challenges, effective communication, shared methodologies, and equal participation can help overcome these obstacles. Ultimately, interdisciplinary collaboration is key to achieving a sustainable future.

Systems Thinking

Systems thinking is a fundamental concept in eco science that provides a holistic understanding of complex environmental issues. It is a way of thinking that analyzes the relationships between different components of a system and how they interact to influence the overall behavior and dynamics of the system.

Definition and Concepts

At its core, systems thinking recognizes that the behavior of a system is not solely determined by the individual components, but also by their interactions and feedback loops. It considers the system as a whole and its connections to the larger context in which it exists. This approach is particularly relevant in eco science, as many environmental problems are inherently complex and interconnected.

Systems thinking emphasizes the following key concepts:

- **Emergence:** Systems thinking recognizes that properties and behaviors emerge from the interactions of components within a system. These emergent properties cannot be fully understood by examining the individual parts in isolation.

- **Interconnectedness:** Systems thinking emphasizes the interdependencies and relationships between different elements within a system. It recognizes that changes in one part of the system can have far-reaching effects on other parts.

- **Feedback loops:** Systems thinking recognizes that systems often have feedback loops, where the outputs of a system can affect its inputs. Feedback loops can be either positive (reinforcing) or negative (balancing) and play a crucial role in shaping the behavior and stability of the system.

- **Non-linearity:** Systems thinking acknowledges that the relationships within a system can be non-linear, meaning that small changes in one component can lead to disproportionately large effects or unexpected outcomes in other components. Non-linearity makes predicting the behavior of a system challenging.

- **Boundaries:** Systems thinking defines boundaries to demarcate the system being studied. These boundaries help identify and understand the relevant components and interactions within the system, while also recognizing the system's interactions with its external environment. Boundaries can be flexible and allow for different levels of granularity in analysis.

Applications in Eco Science

Systems thinking has numerous applications in eco science, enabling researchers and practitioners to tackle complex environmental issues. Here are some key examples:

- **Ecosystem analysis:** Systems thinking provides a framework for understanding the dynamics of ecosystems, including the relationships between species, the flows of energy and matter, and the feedback mechanisms that maintain ecosystem stability. By considering the various components and their interconnectedness, systems thinking aids in predicting and managing ecosystem responses to disturbances such as climate change or invasive species.

- **Sustainability assessment:** Systems thinking is integral to assessing the sustainability of human activities and policies. It helps identify the interconnections between different sectors and dimensions of sustainability (e.g., environmental, social, economic), allowing for a more comprehensive evaluation of actions and their potential impacts across different scales. This approach is critical for promoting integrated and balanced decision-making that supports long-term sustainability.

- **Climate change modeling:** Systems thinking is indispensable in modeling and predicting the impacts of climate change. It enables scientists to analyze complex climate systems, including feedback loops and tipping points, and understand the potential consequences of different climate scenarios. Systems thinking also highlights the interconnected nature of climate change with other environmental, social, and economic systems, guiding the development of integrated adaptation and mitigation strategies.

- **Sustainable development planning:** Systems thinking plays a central role in sustainable development planning at various scales. By considering the interactions and trade-offs between different aspects of development (e.g., economic growth, social welfare, environmental conservation), it helps identify synergies and identify potential conflicts. Systems thinking guides the design of policies and interventions that maximize positive impacts and minimize unintended consequences, promoting sustainable and equitable development.

Example: Systems Thinking in Forest Management

To illustrate the importance of systems thinking, let's consider an example in forest management. A systems thinking approach would involve analyzing the forest ecosystem as a complex system with various components, including trees, soil, wildlife, and human activities.

By understanding the interrelationships between these components, such as the impact of human logging practices on wildlife habitat or soil erosion, forest managers can develop strategies that aim to maintain ecosystem health and resilience. This includes considering the feedback loops within the system, such as the potential for forest degradation leading to reduced rainfall patterns, which can exacerbate the negative impacts on the ecosystem.

Moreover, systems thinking in forest management requires recognizing the larger context in which the forest exists, such as its connection to adjacent ecosystems and the potential for cross-boundary effects. This holistic perspective allows for the consideration of long-term goals, such as biodiversity conservation, sustainable timber production, and the well-being of local communities dependent on the forest resources.

In practice, systems thinking in forest management involves stakeholder engagement, multidisciplinary collaboration, and the integration of scientific knowledge with local ecological knowledge. By considering the complex interactions within the ecosystem and the wider socio-ecological context, forest managers can make informed decisions that promote sustainability and balance the needs of both humans and the environment.

Key Resources and Further Reading

To delve deeper into systems thinking and its applications in eco science, here are some key resources:

+ *Thinking in Systems: A Primer* by Donella H. Meadows: This book offers a comprehensive introduction to systems thinking principles and provides practical guidance on understanding complex systems and making effective interventions.

+ *Resilience Thinking: Sustaining Ecosystems and People in a Changing World* by Brian Walker and David Salt: This book explores the concept of resilience in socio-ecological systems and offers valuable insights into managing and adapting to change in complex systems.

+ *Systems Thinking, Systems Practice: Includes a 30-year Retrospective* by Peter Checkland: This book provides a practical guide to systems thinking and its application in problem-solving and decision-making, with examples from a range of disciplines, including environmental management.

+ *Introduction to Systems Thinking* by Daniel H. Kim: This introductory book provides a clear explanation of systems thinking concepts and presents real-world examples of how systems thinking can be applied to analyze and address complex issues.

Remember, systems thinking is an integral part of eco science, enabling a holistic understanding of complex environmental challenges. By embracing this approach, we can develop sustainable solutions that consider the interdependencies and feedback loops within natural and human systems, ultimately contributing to a more resilient and harmonious world.

Biodiversity

Importance of Biodiversity

Biodiversity is the variety of life on Earth, including the genetic, species, and ecosystem diversity. It is the foundation of a healthy and functioning planet. Biodiversity plays a crucial role in sustaining various ecological processes and services that are essential for human well-being. In this section, we will explore the importance of biodiversity and its various benefits.

Ecological Stability

Biodiversity ensures the stability and resilience of ecosystems. A diverse range of species within an ecosystem provides functional redundancy, which means that if one species is lost or its population declines, other species can compensate for its role in the ecosystem. This redundancy helps maintain ecosystem functions, such as nutrient cycling, pollination, and pest control. For example, bees and other pollinators play a vital role in the pollination of flowering plants, resulting in the production of fruits, seeds, and other plant products. The loss of these pollinators could have a severe impact on crop yields and food production.

Ecosystem Services

Biodiversity is the foundation of numerous ecosystem services that directly or indirectly support human well-being. These services include provisioning services (e.g., food, water, fuel), regulating services (e.g., climate regulation, water purification), supporting services (e.g., soil formation, nutrient cycling), and cultural services (e.g., recreational and spiritual benefits). For instance, forests provide timber for construction, clean air through photosynthesis, and habitat for a diverse range of species. Wetlands act as natural filters, improving water quality by trapping sediments and removing pollutants. Biodiversity-rich habitats also attract tourists, providing economic benefits to local communities.

Medicinal Resources

Biodiversity is a valuable source of medicinal resources. Many species of plants, animals, and microorganisms produce unique bioactive compounds that have potential therapeutic properties. These compounds are used to develop medicines, including antibiotics, anticancer drugs, and pain relievers. For example, the bark of the Pacific yew tree contains a compound called taxol, which is used in the treatment of ovarian, breast, and lung cancers. The loss of biodiversity could result in the loss of potential future medicines and restrict our ability to combat diseases.

Genetic Resources

Biodiversity provides a vast array of genetic resources that are essential for the development of crops, livestock, and industrial products. Wild relatives of cultivated plants often contain valuable genetic traits, such as disease resistance or tolerance to environmental stresses. By preserving biodiversity, we ensure the availability of genetic resources for future genetic improvements and crop breeding programs. This genetic diversity also enhances the adaptability of species to changing environmental conditions, such as climate change.

Intrinsic Value

Every species has its intrinsic value, regardless of its direct value to humans. Each species is a unique product of millions of years of evolution and has its role in the complex web of life. The ethical and aesthetic value of biodiversity cannot be understated. Species inspire art, literature, and spiritual beliefs, providing cultural and emotional enrichment. Conservation efforts are not only about preserving ecosystems for human benefit but also about respecting the intrinsic value and right to exist for all species.

The importance of biodiversity cannot be overstated. It is crucial for maintaining ecological stability, providing essential ecosystem services, offering medicinal and genetic resources, and serving intrinsic and cultural values. However, biodiversity is increasingly threatened by various human activities, including habitat destruction, pollution, climate change, and the introduction of invasive species. It is our responsibility to protect and conserve biodiversity for the benefit of current and future generations. By promoting sustainable practices and raising awareness about the importance of biodiversity, we can strive towards a more sustainable and ecologically balanced future.

Key Points to Remember:

- Biodiversity ensures ecological stability and resilience by providing functional redundancy within ecosystems.

- Biodiversity is the foundation of various ecosystem services, including provisioning, regulating, supporting, and cultural services.

- Biodiversity is a valuable source of medicinal resources, providing compounds for the development of medicines and treatments.

+ Genetic resources from biodiversity are crucial for genetic improvements in crops and adaptability to changing environmental conditions.

+ Each species has intrinsic value and contributes to cultural, aesthetic, and emotional enrichment.

Discussion Questions:

1. Discuss an example of how the loss of biodiversity can impact a specific ecosystem and its services.

2. What are some potential solutions or strategies for conserving biodiversity in the face of increasing threats?

3. How can awareness and education about the importance of biodiversity be promoted among communities and policymakers?

Further Reading:

1. Wilson, E. O. (1992). The Diversity of Life. Harvard University Press.

2. Daily, G. C. (1997). Nature's Services: Societal Dependence on Natural Ecosystems. Island Press.

3. Millennium Ecosystem Assessment. (2005). Ecosystems and Human Well-being: Synthesis. Island Press.

4. Maffi, L. (Ed.). (2001). On Biocultural Diversity: Linking Language, Knowledge, and the Environment. Smithsonian Institution Press.

Overall, understanding and appreciating the importance of biodiversity is crucial for promoting sustainable practices and ensuring the long-term health and well-being of both ecosystems and human societies. By recognizing the intrinsic value of all species and the services they provide, we can work towards preserving and conserving biodiversity for future generations.

Threats to Biodiversity

Biodiversity, the variety of life on Earth, is facing numerous threats that are jeopardizing the delicate balance of ecosystems and the survival of many species. These threats arise from human activities and have accelerated in recent decades due to factors such as population growth, industrialization, and unsustainable consumption patterns. Understanding these threats is crucial in designing effective strategies for biodiversity conservation. In this section, we will explore some of the key threats to biodiversity and their impacts on ecosystems.

Habitat Destruction

One of the major threats to biodiversity is habitat destruction. This occurs when natural habitats, such as forests, wetlands, and coral reefs, are cleared, fragmented, or degraded to make way for human activities such as agriculture, urbanization, and infrastructure development. Habitat destruction directly affects biodiversity by reducing the availability of suitable habitats for species to thrive. It leads to the loss of specialized habitats that support unique and endemic species, disrupts ecosystem processes, and reduces overall species richness.

For example, deforestation in the Amazon rainforest not only results in the loss of countless plant and animal species but also has far-reaching impacts on global climate patterns. The destruction of coral reefs due to pollution and climate change not only reduces marine biodiversity but also affects the livelihoods of communities dependent on reef ecosystems for food and tourism.

Pollution

Pollution, both in the form of chemical contaminants and waste accumulation, poses a significant threat to biodiversity. Chemical pollutants, such as pesticides, heavy metals, and industrial chemicals, can have detrimental impacts on sensitive organisms, including disruption of reproductive systems, impaired growth and development, and even death. These pollutants can enter ecosystems through various pathways, including air, water, and soil, and can accumulate in the food chain, causing widespread impacts.

Water pollution, for example, caused by industrial discharge, agricultural runoff, and improper waste management, can lead to the contamination of freshwater ecosystems, affecting fish populations, amphibians, and other aquatic organisms. Air pollution, resulting from industrial emissions, vehicle exhaust, and burning of fossil fuels, not only affects human health but also damages ecosystems by depositing harmful substances on vegetation, reducing plant productivity, and altering the composition of plant and animal communities.

Climate Change

Climate change is one of the most pressing threats to biodiversity worldwide. Rising global temperatures, changing precipitation patterns, and extreme weather events are disrupting ecosystems and pushing many species to their limits. Climate change affects biodiversity at various levels, from individual species to entire ecosystems.

One of the primary impacts of climate change is habitat loss and fragmentation. As temperatures rise, many species are forced to move to higher latitudes or elevations to find suitable conditions. However, this is not always possible, especially for species with limited dispersal abilities or those confined to specific habitats. Consequently, some species may become locally extinct or may be unable to find suitable habitats, leading to a loss of biodiversity.

Climate change also affects the timing of biological events, such as flowering, migration, and reproduction, disrupting the synchronized interactions between species. For example, if flowering plants and their pollinators, such as bees or butterflies, are not in synchrony, it can lead to reduced pollination success and decreased plant reproduction, ultimately impacting the entire ecosystem.

Overexploitation

Overexploitation of natural resources, including hunting, fishing, and logging, poses a significant threat to biodiversity, particularly for species with slow reproductive rates or those targeted for high-value products. Unsustainable harvesting practices can lead to population declines, ecosystem imbalances, and even species extinction.

For example, overfishing has led to the depletion of many commercially valuable fish species, disrupting marine food webs and affecting the livelihoods of communities dependent on fishing. Illegal wildlife trade, driven by the demand for exotic pets, traditional medicine, and luxury products, poses a major threat to many species, including elephants, rhinos, and pangolins, pushing them closer to extinction.

Invasive Species

Invasive species, introduced by human activities outside their natural range, can have devastating impacts on native biodiversity. These species often lack natural predators or competitors in their new environment, allowing them to outcompete native species for resources and space. Invasive species can disrupt ecosystem functions, degrade habitats, and threaten the survival of native plants and animals.

The brown tree snake, introduced to the island of Guam, has caused the extinction of several bird species and led to the decline of many others. Invasive plants, such as the kudzu vine in the United States, can quickly dominate landscapes, suppressing native vegetation and reducing biodiversity.

Population Growth and Urbanization

The ever-increasing human population, coupled with rapid urbanization, is placing immense pressure on ecosystems and contributing to the decline of biodiversity. As more land is converted for agriculture, infrastructure, and settlements, natural habitats are lost, fragmented, or degraded, leading to the displacement of wildlife and loss of biodiversity.

Urbanization also creates artificial habitats, such as parks and gardens, which may support some species but not others. This selective favoring of certain species, known as urban biotic homogenization, reduces overall biodiversity by promoting the dominance of a few adaptable species at the expense of others.

Solutions and Conservation Strategies

Addressing the threats to biodiversity requires a combination of policy interventions, conservation efforts, and sustainable practices. Some key strategies include:

- Protected Areas: Establishing and effectively managing protected areas, such as national parks and wildlife reserves, can help safeguard critical habitats and provide sanctuary for vulnerable species.

- Sustainable Land Use: Promoting sustainable land use practices, such as agroforestry, organic farming, and reforestation, can help protect ecosystems and support biodiversity.

- Conservation Genetics: Using genetic techniques and tools to identify and preserve unique genetic diversity within and among species can enhance their resilience and long-term survival.

- Community-based Conservation: Engaging local communities in conservation efforts, promoting sustainable livelihoods, and recognizing their traditional knowledge and practices can foster effective conservation outcomes.

- Policy and Governance: Strengthening environmental policies, legislations, and governance frameworks at national and international levels, along with effective enforcement mechanisms, can help mitigate threats to biodiversity.

In conclusion, the threats to biodiversity are multifaceted and interconnected, requiring integrated approaches and collective action to address them effectively. By understanding these threats and implementing conservation strategies, we can strive to protect and preserve the incredible diversity of life on Earth for future generations.

Note to the Reader: Take a moment to reflect on how your own actions and choices can contribute positively towards biodiversity conservation. Small changes, such as reducing your ecological footprint, supporting sustainable practices, and advocating for policy changes, can collectively make a significant impact in preserving our planet's biodiversity.

Habitat Destruction

Habitat destruction is a major threat to biodiversity and is one of the leading causes of species extinction worldwide. It refers to the process by which natural habitats are altered, degraded, or completely destroyed, leading to the loss of species and disruption of ecological balance. In this section, we will explore the various causes and consequences of habitat destruction, as well as discuss strategies for mitigating its impact.

Causes of Habitat Destruction

There are several key factors that contribute to habitat destruction, including:

- **Deforestation:** The clearing of forests for agriculture, urban development, and logging activities is one of the primary drivers of habitat destruction. Forests are home to a wide range of plant and animal species, and their destruction not only leads to the loss of biodiversity but also contributes to climate change by reducing the capacity of forests to absorb carbon dioxide.

- **Urbanization:** The expansion of cities and towns results in the conversion of natural habitats into built environments. Urban development often leads to habitat fragmentation, where natural habitats are divided into smaller, isolated patches. This fragmentation can disrupt the movement, breeding, and feeding patterns of many species.

- **Infrastructure Development:** The construction of roads, dams, and other infrastructure projects often requires the clearing of natural habitats. These projects can lead to the destruction of important habitats, such as wetlands and river systems, and can also result in the displacement of local communities and disruptions to ecosystems.

- **Agricultural Activities:** The conversion of natural habitats into agricultural land is another major cause of habitat destruction. Large-scale monoculture farming practices, such as those used in the production of soy, palm oil, and cattle, often result in the clearing of vast areas of forests and grasslands.

+ **Mining:** Mining activities, including surface mining and underground mining, can have significant impacts on habitats. The extraction of minerals often involves the removal of large amounts of soil and vegetation, resulting in the destruction of habitats and the pollution of water sources.

+ **Pollution:** Pollution, particularly from industrial and agricultural activities, can have detrimental effects on habitats and the species that rely on them. Chemical pollutants, such as pesticides and fertilizers, can contaminate waterways and soil, leading to the loss of biodiversity and the disruption of ecosystem functioning.

Consequences of Habitat Destruction

Habitat destruction has far-reaching consequences for both species and ecosystems. Some of the key impacts include:

+ **Loss of Biodiversity:** Habitat destruction is one of the main drivers of species extinction. When habitats are destroyed, the plants and animals that depend on them are forced to adapt, migrate, or perish. This loss of biodiversity not only reduces the aesthetic and ecological value of ecosystems but also threatens the delicate balance of food webs and ecological interactions.

+ **Disruption of Ecosystem Services:** Habitats provide a wide range of services to humans, such as water purification, climate regulation, and pollination. When habitats are destroyed, these services are compromised, leading to negative impacts on human well-being, including decreased water quality, increased vulnerability to natural disasters, and reduced agricultural productivity.

+ **Loss of Indigenous Knowledge:** Many indigenous communities have deep knowledge and understanding of their local ecosystems. Habitat destruction can result in the loss of this valuable knowledge, which has been passed down through generations. This loss of traditional ecological knowledge not only undermines indigenous cultures but also hinders our ability to manage and conserve ecosystems effectively.

+ **Climate Change:** Habitat destruction, particularly deforestation, contributes to climate change by reducing the Earth's capacity to absorb carbon dioxide. Forests act as carbon sinks, absorbing and storing large amounts of carbon dioxide. When forests are cleared, the carbon stored in trees is released back into the atmosphere, contributing to greenhouse gas emissions and exacerbating climate change.

+ **Loss of Cultural and Recreational Values:** Natural habitats and landscapes have important cultural and recreational values. They provide opportunities for recreation, tourism, and spiritual practices, and are often deeply intertwined with local cultures and identities. When habitats are destroyed, these cultural and recreational values are also lost.

Mitigation Strategies

Addressing habitat destruction requires a multifaceted approach that involves both individual and collective actions. Some key strategies for mitigating the impact of habitat destruction include:

+ **Habitat Conservation:** Protecting and managing existing habitats is crucial for maintaining biodiversity and ensuring the continued provision of ecosystem services. This includes establishing protected areas, such as national parks and wildlife reserves, as well as implementing sustainable land-use practices that minimize the impact on habitats.

+ **Forest Restoration:** Restoring degraded forests and establishing new forested areas can help reverse the impacts of deforestation. Reforestation initiatives, such as planting native tree species, can help create habitat corridors, enhance biodiversity, and mitigate climate change by sequestering carbon dioxide.

+ **Sustainable Agriculture:** Promoting sustainable agricultural practices, such as agroforestry and organic farming, can help reduce the need for further habitat conversion. These practices prioritize the conservation of natural habitats and biodiversity while ensuring food security and livelihoods for local communities.

+ **Integrated Landscape Management:** Taking a holistic approach to land management is essential for mitigating habitat destruction. Integrated landscape management involves coordinating actions across different land uses, such as agriculture, forestry, and urban development, to minimize the impact on habitats and maintain ecological connectivity.

+ **Education and Awareness:** Increasing public awareness about the importance of habitats and biodiversity is crucial for fostering a sense of stewardship and driving individual actions. Education initiatives, community-based conservation projects, and public outreach programs can help raise awareness and encourage sustainable behavior.

- **Policy and Regulations**: Adopting and enforcing strong environmental policies and regulations is vital for preventing habitat destruction. Governments can play a central role in setting targets, providing incentives for sustainable practices, and creating frameworks that promote the protection and restoration of habitats.

Case Study: Deforestation in the Amazon Rainforest

The Amazon rainforest is one of the most biodiverse regions on Earth, housing millions of plant and animal species. However, it is also experiencing significant habitat destruction due to deforestation. Illegal logging, agricultural expansion, and infrastructure development have led to the loss of vast areas of the Amazon rainforest.

The consequences of deforestation in the Amazon are far-reaching. The loss of habitat has resulted in the extinction of numerous plant and animal species, disrupted traditional ways of life for indigenous communities, and contributed to climate change by releasing large amounts of carbon dioxide.

Efforts to address deforestation in the Amazon have focused on a combination of strategies, including establishing protected areas, implementing sustainable land-use practices, and promoting responsible business practices. International cooperation and financial incentives have also played a crucial role in supporting conservation efforts in the region.

The case of the Amazon rainforest highlights the complexity of addressing habitat destruction and the need for a coordinated and multidimensional approach. By recognizing the value of habitats and taking action to protect and restore them, we can work towards a sustainable future that balances human development with the preservation of biodiversity and ecosystem services.

Summary

In this section, we discussed the causes and consequences of habitat destruction, focusing on its impact on biodiversity, ecosystem services, and human well-being. We explored various strategies for mitigating habitat destruction, including habitat conservation, forest restoration, sustainable agriculture, integrated landscape management, education and awareness, and policy and regulations. We also examined a case study on deforestation in the Amazon rainforest to illustrate the challenges and potential solutions in addressing habitat destruction. By understanding the drivers of habitat destruction and implementing effective conservation measures, we can work towards a future where ecosystems are protected, species are preserved, and sustainable development is achieved.

Pollution

Pollution is a significant environmental issue that poses threats to the well-being of ecosystems, human health, and the planet as a whole. It refers to the introduction of harmful substances or pollutants into the environment, leading to adverse effects on living organisms and the overall ecosystem balance. In this section, we will explore different types of pollution, their sources, impacts, and potential strategies to mitigate them.

Types of Pollution

There are several types of pollution, each with its own distinct characteristics and sources. Understanding these types is crucial for developing effective solutions to prevent and manage pollution. Let's explore the most common forms of pollution:

1. **Air Pollution:** This type of pollution is caused by the release of harmful gases, particulate matter, and other pollutants into the atmosphere. Common sources of air pollution include industrial emissions, vehicle exhaust, agricultural activities, and the burning of fossil fuels. Air pollution can have detrimental effects on human health, such as respiratory diseases, as well as on ecosystems, leading to the depletion of ozone layer and climate change.

2. **Water Pollution:** Water pollution occurs when pollutants contaminate water bodies such as rivers, lakes, oceans, and groundwater. Sources of water pollution include industrial waste, agricultural runoff, improper disposal of chemicals and sewage, oil spills, and littering. Water pollution not only affects aquatic life but also poses risks to human health, as contaminated water can cause waterborne diseases.

3. **Soil Pollution:** Soil pollution refers to the contamination of soil with harmful substances, such as heavy metals, chemicals, and toxic pollutants. It often occurs due to improper disposal of industrial waste, use of pesticides and fertilizers in agriculture, as well as mining activities. Soil pollution can have detrimental effects on soil fertility, agricultural productivity, and overall ecosystem health.

4. **Noise Pollution:** Noise pollution is the excessive or disturbing noise that disrupts the natural environment and human activities. Sources of noise pollution include traffic noise, industrial machinery, construction activities, and loud music. Prolonged exposure to high levels of noise can lead to various physical and psychological health problems, including hearing loss, stress, and sleep disturbances.

5. **Light Pollution:** Light pollution is the excessive or misdirected artificial light that interferes with the natural darkness of the night sky. It mainly arises from outdoor lighting, such as streetlights, advertising signs, and illuminated buildings. Light pollution disrupts ecosystems, interferes with animal behavior and migration patterns, and affects human health by disrupting sleep patterns and circadian rhythms.

6. **Thermal Pollution:** Thermal pollution refers to the release of heated water into natural water bodies, such as rivers and oceans, causing a sudden increase in water temperature. It occurs primarily due to industrial activities, power plants, and nuclear reactors. Thermal pollution can disrupt aquatic ecosystems, adversely affect aquatic organisms, and deplete oxygen levels in water bodies, leading to reduced biodiversity and fish kills.

Impacts of Pollution

Pollution has far-reaching impacts on both the environment and human society. Let's take a closer look at the consequences of pollution:

+ **Environmental Impact:** Pollution can degrade and disrupt ecosystems, leading to habitat destruction, loss of biodiversity, and ecological imbalance. For example, water pollution can contaminate aquatic habitats, impair water quality, and harm marine life. Similarly, air pollution can damage plant and animal life, reduce crop yields, and contribute to climate change.

+ **Health Impact:** Pollution, particularly air and water pollution, poses significant risks to human health. Exposure to pollutants can cause respiratory problems, cardiovascular diseases, neurological disorders, and even cancer. For instance, prolonged exposure to air pollutants such as particulate matter and nitrogen dioxide can lead to respiratory diseases like asthma and bronchitis. Contaminated water can result in waterborne diseases like cholera and dysentery.

+ **Economic Impact:** Pollution has substantial economic consequences, including the costs associated with healthcare, environmental cleanup, and loss of productivity. For example, healthcare expenses related to pollution-induced illnesses can burden individuals, communities, and healthcare systems. Moreover, pollution-related damage to ecosystems and natural resources can impact industries such as agriculture, tourism, and fisheries.

+ **Social Impact:** Pollution can have adverse social implications, particularly affecting marginalized communities and vulnerable populations. Often, these communities are disproportionately exposed to pollution sources due to factors such as proximity to industrial areas or lack of resources to address the issue. Environmental justice is an important aspect of pollution management, emphasizing equal rights and access to a clean and healthy environment for all.

Mitigation Strategies

Addressing pollution requires comprehensive strategies that focus on prevention, reduction, and remediation. Here are some effective mitigation strategies:

+ **Regulatory Measures:** Implementation of strict environmental regulations, standards, and laws is essential to control pollution. Governments and regulatory bodies can enforce emission standards for industries, establish waste management protocols, and impose penalties for non-compliance. Such measures can encourage industries to adopt cleaner technologies and practices.

+ **Technological Innovations:** Advancements in technology play a crucial role in pollution control. Developing and implementing cleaner technologies, such as emission control devices in vehicles and industrial plants, can significantly reduce pollution. For example, the use of catalytic converters in cars helps to reduce harmful emissions. Similarly, wastewater treatment technologies can help remove pollutants before discharging water into the environment.

+ **Transition to Renewable Energy:** Shifting from fossil fuels to renewable energy sources like solar, wind, and hydropower can significantly reduce air pollution and greenhouse gas emissions. Governments and policymakers can provide incentives and support for renewable energy adoption, promoting a sustainable and clean energy transition.

+ **Waste Management:** Effective waste management practices, including recycling, proper disposal techniques, and waste reduction strategies, can minimize pollution. Encouraging the use of biodegradable and recyclable materials, implementing recycling programs, and promoting the concept of a circular economy can help reduce pollution associated with waste.

+ **Education and Awareness:** Educating individuals and communities about the impacts of pollution and the importance of sustainable practices is crucial. Creating awareness about pollution-related issues, promoting responsible consumption, and encouraging environmentally friendly behaviors can help bring about behavioral changes necessary for pollution prevention.

Example: Plastic Pollution

Plastic pollution serves as a pertinent example of how pollution can have severe environmental and health impacts. The widespread use of single-use plastics, such as plastic bags and water bottles, has led to immense plastic waste accumulation in oceans and landfills. This pollution not only harms marine life but also affects human well-being, as microplastics enter the food chain.

To address plastic pollution, various strategies can be implemented, including:

+ **Plastic Waste Reduction:** Implementing policies and initiatives to reduce the usage of single-use plastics, promoting reusable alternatives, and encouraging proper waste management practices can help minimize plastic pollution.

+ **Recycling and Upcycling:** Establishing efficient recycling systems and encouraging the recycling and upcycling of plastic materials can divert plastic waste from landfills and reduce the demand for virgin plastic production.

+ **Innovation in Packaging:** Encouraging the development and use of sustainable packaging materials, such as bioplastics and compostable materials, can significantly reduce plastic pollution.

+ **Public Awareness and Education:** Creating awareness campaigns and educational programs focusing on the detrimental impacts of plastic pollution can mobilize public support and encourage behavioral changes.

Conclusion

Pollution is a complex and pressing environmental issue that requires collective action and sustained efforts to mitigate its detrimental effects. By understanding the different types of pollution, their impacts, and implementing appropriate mitigation strategies, we can work towards a cleaner and more sustainable future. It is crucial to emphasize the importance of individual responsibility, government regulations, technological advancements, and education in addressing pollution for the long-term well-being of the planet and all its inhabitants.

Climate change

Climate change refers to long-term shifts in weather patterns and temperatures on Earth. It is primarily caused by the increase in greenhouse gases (GHGs) in the atmosphere, resulting from human activities such as burning fossil fuels, deforestation, and industrial processes. The release of these GHGs, such as carbon dioxide (CO_2), methane (CH_4), and nitrous oxide (N_2O), traps heat from the sun and leads to a rise in global temperatures.

The greenhouse effect

To understand climate change, it is important to first grasp the concept of the greenhouse effect. The Earth's atmosphere acts like a greenhouse, allowing sunlight to enter and keeping the planet warm. However, excessive GHGs in the atmosphere intensify the greenhouse effect, leading to increased surface temperatures.

The greenhouse effect occurs in the following steps:

1. Sunlight enters the Earth's atmosphere. 2. Some of the sunlight is absorbed by the Earth's surface and warms it. 3. The heated surface emits thermal radiation in the form of infrared (IR) radiation. 4. GHGs in the atmosphere absorb a portion of the IR radiation and re-emit it in all directions, including back to the Earth's surface. 5. This re-emitted IR radiation further heats the Earth's surface, causing a rise in temperature.

Evidences of climate change

Multiple lines of evidence support the existence and impact of climate change:

1. Temperature records: Long-term temperature records indicate a consistent increase in global temperatures over the past century. Each of the past three decades has been successively warmer than any preceding decade since the mid-1800s.

2. Retreating glaciers: Glaciers and ice caps worldwide have been melting at an accelerated rate, leading to rising sea levels. This melting is typically a result of increased air temperatures.

3. Sea level rise: As water temperatures increase, seawater expands, resulting in rising sea levels. Coastal areas are particularly vulnerable to the impacts of sea-level rise, including increased coastal erosion and flooding during storms.

4. Changes in precipitation patterns: Climate change affects rainfall patterns, leading to more frequent and intense extreme weather events. Some regions experience increased rainfall and flooding, while others face prolonged droughts.

5. Ocean acidification: The increase in atmospheric CO_2 levels has led to increased absorption of CO_2 by the oceans. This, in turn, causes ocean acidification, which poses a threat to marine ecosystems.

Impacts of climate change

Climate change has far-reaching implications for the environment, ecosystems, and human societies. Some of the significant impacts include:

1. Biodiversity loss: As habitats change due to shifting temperatures and altered precipitation patterns, many species struggle to adapt, leading to a loss of biodiversity. This disruption can have cascading effects on entire ecosystems.

2. Extreme weather events: Climate change intensifies extreme weather events, such as hurricanes, heatwaves, and heavy rainfall. These events can result in significant property damage, loss of life, and disruption of vital infrastructure.

3. Food security: Changes in temperature and rainfall patterns affect agricultural productivity, leading to crop failures and reduced food production. This poses a threat to global food security, particularly in vulnerable regions.

4. Public health: Climate change contributes to the spread of disease vectors, such as mosquitoes carrying diseases like dengue fever and malaria. Additionally, heatwaves and extreme temperatures can lead to heat-related illnesses and deaths.

5. Economic impacts: The costs associated with climate change are substantial, including damage to infrastructure, increased healthcare expenses, and loss of livelihoods in affected industries such as agriculture, fisheries, and tourism.

Mitigation and adaptation strategies

Addressing climate change requires both mitigation and adaptation strategies:

1. Mitigation: Mitigation involves reducing greenhouse gas emissions to limit the extent of climate change. Strategies include transitioning to clean and renewable energy sources, improving energy efficiency, and promoting sustainable transportation and land use practices.

2. Adaptation: Adaptation focuses on minimizing the impacts of climate change through preparedness and resilience. Examples of adaptation measures include implementing early warning systems for extreme weather events, improving water management strategies, and developing climate-resilient infrastructure.

It is essential to involve various stakeholders, including governments, businesses, communities, and individuals, in implementing these strategies to effectively combat climate change.

Case study: The Paris Agreement

The Paris Agreement, adopted in 2015, is a landmark international climate agreement aiming to combat climate change and keep global temperature rise well below 2 degrees Celsius above pre-industrial levels. Key elements of the agreement include:

1. Nationally Determined Contributions (NDCs): Each participating country sets its own emissions reduction targets and outlines its strategies to achieve them.

2. Transparency and accountability: The agreement emphasizes the importance of reporting and monitoring progress towards achieving NDCs, as well as providing support to developing countries in their mitigation and adaptation efforts.

3. Finance and technology transfer: Developed countries commit to providing financial resources and technology transfer to help developing countries transition towards low-carbon and climate-resilient pathways.

The Paris Agreement represents a global commitment to addressing climate change, although challenges remain in implementing and maintaining the necessary actions to limit global warming.

Conclusion

Climate change is a pressing global challenge with wide-ranging consequences. Understanding the science behind climate change, its impacts on the environment and society, and the strategies for mitigation and adaptation is crucial for shaping a sustainable future. By taking collective action and adopting sustainable practices, we can work towards a more resilient and climate-friendly world.

Conservation Strategies

Conservation strategies play a crucial role in safeguarding biodiversity and promoting sustainable practices. These strategies aim to mitigate the threats that ecosystems and species face, ensuring their long-term survival. In this section, we will discuss some of the key conservation strategies that are commonly employed in the field of eco science.

Protected Areas

Protected areas are designated regions that are specifically managed to preserve their ecological values and promote biodiversity conservation. National parks, wildlife sanctuaries, and nature reserves are some examples of protected areas. These areas serve as refuges for wildlife, allowing them to thrive without human interference. Protected areas also help in maintaining essential ecological processes and preventing habitat loss.

One example of a successful protected area is the Serengeti National Park in Tanzania. This park has been instrumental in conserving the vast savannahs, which are home to a diverse range of species, including the iconic wildebeest migration. By restricting human activities and implementing effective management strategies, the Serengeti National Park has been able to protect critical habitats and maintain the region's ecological balance.

Sustainable Land Use

Sustainable land use is another important conservation strategy that focuses on utilizing land resources in a way that minimizes negative impacts on the environment while ensuring long-term productivity. It involves adopting practices that are ecologically sound, socially just, and economically viable.

One widely recognized sustainable land-use approach is agroforestry. Agroforestry involves integrating trees into agricultural systems, providing multiple benefits such as improved soil fertility, enhanced water retention, and biodiversity conservation. By diversifying crop production and incorporating trees, agroforestry systems create resilient and sustainable landscapes.

For example, in parts of sub-Saharan Africa, farmers have implemented agroforestry techniques like planting nitrogen-fixing trees alongside crops. These trees not only improve soil fertility but also offer shade and contribute to climate change mitigation by sequestering carbon dioxide. This practice promotes sustainable agriculture while conserving biodiversity and supporting local livelihoods.

Restoration Ecology

Restoration ecology focuses on repairing and restoring damaged ecosystems to their original or an ecologically functional state. It involves removing or mitigating the factors that caused the degradation and implementing interventions to facilitate the recovery of the ecosystem.

One restoration strategy widely used is habitat restoration. This involves repairing or recreating habitats that have been lost or degraded due to human activities, such as habitat destruction or pollution. Restoration activities can include removing invasive species, replanting native vegetation, and creating suitable conditions for the return of native wildlife.

An excellent example of habitat restoration is the restoration of the Everglades in Florida, USA. Due to extensive drainage and water diversion for agriculture and urban development, the Everglades experienced severe habitat loss and degradation. Restoration efforts have focused on re-establishing natural water flows, removing invasive species, and restoring wetland habitats. These initiatives have resulted in the recovery of several endangered species and water quality improvement.

Community-Based Conservation

Community-based conservation involves collaboration with local communities in the management and conservation of natural resources. It recognizes the role of indigenous people and local communities as guardians of biodiversity and incorporates their knowledge and practices into conservation strategies.

By involving local communities in decision-making processes and recognizing their rights and responsibilities over natural resources, community-based conservation promotes sustainable and equitable outcomes. It also helps address social and economic development issues while conserving biodiversity.

An example of community-based conservation is the establishment of community conserved areas (CCAs) in Nepal. CCAs are community-managed areas that conserve biodiversity and sustainably utilize natural resources. These areas are governed by local communities and involve them in decision-making, monitoring, and enforcement of conservation efforts. This approach has been successful in protecting critical habitats, conserving endangered species, and improving livelihoods.

Conservation Genetics

Conservation genetics is an interdisciplinary field that applies genetic techniques to address conservation challenges. It provides insights into the genetic diversity, population structure, and evolutionary processes of species, which are essential for effective conservation strategies.

One application of conservation genetics is the identification and management of endangered species. By studying the genetic diversity and relatedness of individuals within a population, conservationists can design strategies to minimize inbreeding and maintain genetic variability. Techniques such as captive breeding, translocation, and genetic rescue can be employed to ensure the survival and recovery of endangered species.

Conservation genetics also plays a crucial role in combating illegal wildlife trade. Genetic tools, such as DNA profiling and forensic analysis, can assist in identifying the origin of illegally traded specimens and detecting wildlife trafficking networks.

Indigenous Knowledge and Practices

Indigenous knowledge and practices have been instrumental in biodiversity conservation for centuries. Indigenous communities possess a wealth of traditional knowledge about their local ecosystems, including sustainable resource management practices and ecosystem stewardship.

Integrating indigenous knowledge with scientific approaches can lead to holistic and culturally appropriate conservation strategies. Indigenous communities' deep understanding of ecosystems and their traditional practices can contribute to ecosystem resilience and the sustainable use of natural resources.

For instance, indigenous fire management practices, such as controlled burning, have been used for generations in Australia. These practices help prevent catastrophic wildfires, promote biodiversity, and support the regeneration of native vegetation. Incorporating such practices into modern fire management strategies can enhance ecological resilience and reduce the risk of uncontrolled fires.

In conclusion, conservation strategies are essential for protecting biodiversity and promoting sustainable practices. Protected areas, sustainable land use, restoration ecology, community-based conservation, conservation genetics, and indigenous knowledge are all valuable approaches in the field of conservation. By implementing these strategies and integrating them with interdisciplinary approaches, we can work towards a future that is more sustainable and ecologically balanced.

Protected Areas

Protected areas play a crucial role in the conservation of biodiversity and the preservation of ecosystems. These designated areas are managed to safeguard natural resources, protect vulnerable species, and maintain critical habitats. In this section, we will explore the importance of protected areas, the various types of protected areas, their management strategies, and the challenges they face.

Importance of Protected Areas

Protected areas are vital for the conservation of biodiversity due to the following reasons:

1. **Habitat preservation:** Protected areas provide a refuge for a wide range of ecosystems, habitats, and species. They ensure the preservation of essential breeding grounds, feeding areas, and migration routes.

2. **Biodiversity conservation:** Protected areas support high species diversity and maintain important ecological interactions. They serve as a sanctuary for endangered and threatened species, helping to prevent their extinction.

3. **Ecosystem services:** Protected areas contribute to the provision of clean water, climate regulation, soil protection, and nutrient cycling. They serve as natural buffers against environmental disasters such as floods and wildfires.

4. **Scientific research and education:** Protected areas offer opportunities for scientific research, enabling the study of ecological processes, species behavior, and the impacts of environmental changes. They also provide educational resources for visitors and local communities.

Types of Protected Areas

Protected areas can be classified into several types, depending on their objectives and management strategies. Some common types of protected areas include:

1. **National Parks:** National parks are large areas of land designated for the protection of outstanding natural and cultural features. They often include unique ecosystems, iconic landscapes, and important historical sites. National parks focus on conservation, recreation, and education, and generally have strict regulations to minimize human impact.

2. **Wildlife Sanctuaries:** Wildlife sanctuaries primarily focus on the protection of specific species or groups of species. They provide ideal habitats for breeding, nesting, and foraging. Wildlife sanctuaries often have restrictions on human activities to minimize disturbance to the resident wildlife.

3. **Biosphere Reserves:** Biosphere reserves aim to reconcile biodiversity conservation with sustainable development. They incorporate core protected areas, buffer zones, and transition areas that allow for various human activities such as sustainable agriculture, forestry, and tourism. Biosphere reserves emphasize the conservation of natural and cultural diversity.

4. **Marine Protected Areas (MPAs):** MPAs are designated areas in oceans, seas, or estuaries that are managed to conserve marine ecosystems, habitats,

and species. They help to maintain fish populations, protect vulnerable marine species, and preserve important breeding and feeding grounds. MPAs also promote sustainable fisheries and ecotourism.

5. **Community Conserved Areas:** Community conserved areas are managed by local communities with the aim of protecting natural resources and sustaining traditional livelihoods. These areas rely on community knowledge, practices, and customary laws to regulate resource use and maintain ecosystem health.

Management Strategies

Effective management is crucial for the success of protected areas. Some key management strategies include:

1. **Zoning and regulations:** Protected areas are often divided into different zones with varying levels of protection and permitted activities. Strict regulations and guidelines are put in place to minimize human impact and maintain ecological integrity.

2. **Monitoring and enforcement:** Regular monitoring of protected areas helps assess the health of ecosystems, detect changes in biodiversity, and detect and prevent illegal activities such as poaching and illegal logging. Enforcement of rules and regulations is essential to deter harmful practices.

3. **Stakeholder engagement:** Involving local communities, indigenous groups, and other stakeholders is vital for effective management. Their knowledge, perspectives, and traditional practices contribute to the conservation and sustainable use of protected areas.

4. **Collaboration and partnerships:** Collaboration between different organizations, government agencies, and international bodies enhances the management of protected areas. Partnerships can involve research institutions, non-governmental organizations, local communities, and private sector entities.

5. **Ecotourism and sustainable use:** Generating income through ecotourism and sustainable resource use can provide financial support for the management and maintenance of protected areas. This approach promotes economic development while minimizing negative impacts on the environment.

Challenges and Solutions

Protected areas face numerous challenges that threaten their effectiveness and long-term sustainability. Some key challenges include:

1. **Illegal activities:** Poaching, illegal logging, and wildlife trafficking undermine the conservation efforts of protected areas. Enhanced law enforcement, community engagement, and international collaborations are essential for combating these illegal activities.

2. **Human-wildlife conflicts:** Encroachment of human settlements into protected areas can lead to conflicts between local communities and wildlife. Implementing strategies for sustainable land use, promoting alternative livelihoods, and improving community awareness and education can reduce these conflicts.

3. **Climate change:** The impacts of climate change, including rising temperatures, changing rainfall patterns, and extreme weather events, pose significant threats to protected areas. Implementing climate change adaptation and mitigation strategies, such as habitat restoration and carbon sequestration, can help build resilience.

4. **Lack of financial resources:** Limited funding often hinders the effective management and conservation of protected areas. Identifying innovative financing mechanisms, such as public-private partnerships, philanthropy, and eco-tourism revenue, can help address this challenge.

5. **Lack of public awareness and support:** Engaging and educating the public about the importance of protected areas is critical for their long-term sustainability. Developing outreach programs, environmental education initiatives, and involving local communities in decision-making processes can foster public support.

Case Study: Serengeti National Park

One notable example of a protected area is the Serengeti National Park in Tanzania. It is renowned for its vast savannahs, wildebeest migration, and diverse wildlife. The park, covering an area of approximately 30,000 square kilometers, is home to numerous species, including elephants, lions, giraffes, and zebras.

The management of Serengeti National Park faces various challenges, such as poaching, habitat degradation, and human-wildlife conflicts. To address these challenges, the government of Tanzania, along with conservation organizations and local communities, has implemented several strategies:

- **Anti-poaching measures:** The park authorities work tirelessly to combat poaching through the deployment of well-trained anti-poaching units. Strict penalties have been implemented to deter illegal hunting and trading of wildlife.

- **Community engagement:** Local communities living adjacent to the park are actively involved in its management through community-based conservation initiatives. This engagement provides economic opportunities for the communities and encourages their active participation in wildlife conservation programs.

- **Tourism and revenue sharing:** Serengeti National Park attracts a large number of tourists each year, providing significant revenue for its management. Revenue sharing programs ensure that a portion of the tourism income is directed towards local communities, promoting their support for conservation efforts.

- **Ecosystem monitoring:** Regular monitoring of the park's ecosystems helps assess changes in habitats, species populations, and ecological processes. This data guides management decisions and informs conservation strategies.

- **Research and collaboration:** Collaboration between international research institutions, national parks agencies, and local stakeholders facilitates scientific research, knowledge sharing, and capacity building. This collaboration strengthens the management and protection of Serengeti National Park.

The success of Serengeti National Park in preserving its unique biodiversity and facilitating sustainable development serves as a model for other protected areas around the world.

Conclusion

Protected areas are essential for conserving biodiversity, protecting ecosystems, and promoting sustainable development. Through effective management strategies, these areas can safeguard critical habitats, support endangered species, and provide ecological services. However, they face various challenges, requiring collaborative efforts from governments, communities, researchers, and conservation organizations. By understanding the importance of protected areas, implementing innovative management approaches, and addressing key challenges, we can build a more sustainable future for our planet.

Sustainable Land Use

Sustainable land use is a crucial aspect of eco science that aims to ensure the long-term viability and productivity of land resources while minimizing negative environmental impacts. It involves making informed decisions and implementing practices that balance the needs of various stakeholders, including agriculture, forestry, urban development, and conservation. Sustainable land use is based on the principles of ecological integrity, economic viability, and social equity.

Challenges in Land Use

The challenge of sustainable land use arises due to various factors such as population growth, urbanization, industrialization, and changing climatic conditions. These factors put pressure on land resources and can lead to land degradation, loss of biodiversity, soil erosion, and deforestation. The efficient management of land resources becomes crucial to address these challenges and ensure sustainable development.

Principles of Sustainable Land Use

To achieve sustainable land use, several principles need to be considered:

1. **Land Zoning and Planning:** Effective land zoning and planning are essential to allocate land for different uses based on its ecological and socio-economic characteristics. It involves considering factors such as soil fertility, water availability, biodiversity, and land productivity to determine suitable land uses.

2. **Conservation Agriculture:** Conservation agriculture promotes sustainable farming practices that minimize soil disturbance, maintain soil cover, and enhance crop diversity. It aims to improve soil health, increase water holding capacity, and reduce erosion. Techniques like crop rotation, minimum tillage, and the use of cover crops are integral to conservation agriculture.

3. **Preservation of Ecosystems:** Protecting and restoring ecosystems is crucial for sustainable land use. Ecosystems provide essential services, such as nutrient cycling, water regulation, and habitat provision. Conservation strategies, such as creating protected areas and implementing wildlife corridors, help maintain biodiversity and ecological balance.

4. **Wise Forest Management:** Forests play a vital role in sustainable land use by mitigating climate change, conserving biodiversity, and providing essential resources. Sustainable forest management practices, such as selective logging, reforestation, and protecting old-growth forests, ensure the long-term sustainability and productivity of forests.

5. **Urban Planning and Design:** Sustainable land use in urban areas involves planning and designing cities that minimize ecological footprint, promote compact development, and prioritize green spaces. Concepts like mixed land use, pedestrian-friendly neighborhoods, and the use of renewable energy contribute to sustainable urban development.

6. **Stakeholder Engagement:** Inclusive and participatory decision-making processes involving stakeholders from diverse backgrounds are essential for sustainable land use. Engaging local communities, indigenous peoples, and other relevant groups ensures that their knowledge, values, and needs are considered in land use planning and decision-making.

Land Use Strategies

Several land use strategies can help achieve sustainable land use:

1. **Integrated Land Management:** Integrated land management approaches combine various land uses, such as agriculture, forestry, and conservation, in a coordinated manner. This approach ensures that land resources are utilized efficiently, and conflicts among different land uses are minimized.

2. **Agroforestry:** Agroforestry is a land use system that combines trees with agricultural crops or livestock. It provides multiple benefits, including improved soil fertility, increased biodiversity, and enhanced carbon sequestration. Agroforestry practices can be tailored to different agro-climatic conditions and provide sustainable livelihood options for rural communities.

3. **Land Reclamation and Rehabilitation:** Land reclamation and rehabilitation involve restoring degraded or contaminated lands to their productive and functional state. Techniques such as soil remediation, re-vegetation, and land contouring help restore ecosystem services and biodiversity in previously disturbed areas.

4. **Land Use Policy and Regulation:** Effective land use policies and regulations are instrumental in guiding sustainable land use practices. These policies can include regulations on land conversion, protection of critical habitats, and incentives for adopting sustainable land management practices.

Case Study: Sustainable Agriculture in the Netherlands

The Netherlands offers an excellent example of sustainable land use in the agricultural sector. Despite being a densely populated country, the Netherlands has managed to achieve high agricultural production while minimizing environmental impacts. This success is attributed to innovative practices such as precision farming, efficient water management, and the use of advanced technologies. The Dutch agricultural sector focuses on sustainable intensification, which maximizes production while minimizing resource use. The integration of livestock and crop farming, efficient nutrient management, and the adoption of circular economy principles contribute to sustainable land use in the Netherlands.

Conclusion

Sustainable land use is critical for ensuring the long-term viability of our land resources, maintaining biodiversity, and addressing the challenges of population growth and climate change. By embracing the principles of sustainable land use and implementing appropriate strategies, we can achieve a balance between economic development and environmental conservation. Through collaborative efforts and informed decision-making, we can create a sustainable future where land resources are managed efficiently and responsibly.

Restoration ecology

Restoration ecology is a multidisciplinary field that focuses on the recovery and renewal of degraded ecosystems. It involves the application of ecological principles and practices to restore the structure, function, and biodiversity of ecosystems that have been impacted by human activities. Restoration ecology plays a crucial role in reversing the negative impacts of habitat destruction, pollution, and climate change, and in conserving and enhancing biodiversity.

Understanding Ecosystem Degradation

To effectively restore ecosystems, it is essential to understand the processes and factors contributing to their degradation. Ecosystem degradation can be caused by various human activities, such as deforestation, intensive agriculture, urbanization, and mining. These activities can result in habitat loss, fragmentation, soil erosion, water pollution, and the loss of biodiversity.

Principles of Restoration Ecology

Restoration ecology is guided by several key principles that help ensure the success of restoration efforts:

1. Clear goals: Restoration projects should have clear objectives, such as the reestablishment of specific plant or animal species, improvement of water quality, or the creation of functional habitats.

2. Reference ecosystems: Reference ecosystems serve as models of what the restored ecosystem should resemble. By studying reference ecosystems, restoration practitioners can determine appropriate species composition, functional attributes, and ecological processes for restoration.

3. Ecological succession: Restoration often involves facilitating ecological succession, which is the natural process of changing species composition and community structure over time. Understanding the stages of succession helps guide restoration efforts.

4. Monitoring and adaptive management: Monitoring the progress of restoration efforts is crucial to evaluate the effectiveness of different strategies and make necessary adjustments. Adaptive management allows for continuous learning and improvement throughout the restoration process.

Approaches to Ecosystem Restoration

Ecosystem restoration can be approached in various ways depending on the specific goals and conditions of the degraded ecosystem. Some common approaches include:

1. Passive restoration: This approach involves allowing natural processes to restore ecosystems without direct human intervention. It is often used in areas with intact seed banks and natural regeneration potential.

2. Active restoration: Active restoration involves direct intervention to accelerate the recovery process. It may include activities such as reforestation, reintroduction of keystone species, invasive species control, and soil erosion prevention.

3. Assisted natural regeneration: This approach combines elements of both passive and active restoration. It involves promoting the natural regeneration of native species by removing barriers and providing favorable conditions.

4. Structural restoration: Structural restoration focuses on reestablishing the physical and biological components of ecosystems, such as wetlands, coral reefs, or riparian zones. It aims to reconstruct the habitat to support diverse plant and animal communities.

Challenges and Considerations

Restoring ecosystems is a complex and challenging task that requires careful planning and consideration of various factors. Some of the key challenges and considerations in restoration ecology are:

1. Genetic diversity: Restored ecosystems should aim to preserve or enhance genetic diversity to ensure the long-term resilience of plant and animal populations.

2. Invasive species: The control and management of invasive species is crucial for successful restoration. Invasive species can outcompete native species and disrupt ecological processes.

3. Connectivity: Restored ecosystems should be connected to other habitats to facilitate the movement of species and maintain ecological processes. Corridors and stepping-stone habitats can help enhance connectivity.

4. Stakeholder involvement: The engagement and involvement of local communities, landowners, and stakeholders are essential for the success and long-term sustainability of restoration projects.

Case Study: Everglades Restoration

An example of a large-scale ecosystem restoration project is the restoration of the Florida Everglades. The Everglades are a unique wetland ecosystem that has been heavily impacted by water diversion, urban development, and agricultural activities.

The restoration efforts for the Everglades focus on improving water flow, water quality, and restoring natural habitat patterns. This involves the construction of reservoirs and water storage areas, reestablishment of natural water flow patterns, and removal of invasive species.

The restoration of the Everglades not only benefits the unique plants and animals that call this ecosystem home, but also provides important ecosystem services such as water purification, flood control, and habitat for numerous species.

Conclusion

Restoration ecology is a vital discipline in promoting the recovery and resilience of degraded ecosystems. By applying ecological principles and practices, restoration efforts can reverse the damage caused by human activities and create sustainable habitats for future generations. However, successful restoration requires careful planning, monitoring, and adaptive management to overcome the challenges inherent in restoring complex ecosystems. Through continued research and collaboration, restoration ecology can contribute to the long-term sustainability of our planet's biodiversity and natural resources.

Species reintroduction

Species reintroduction is a conservation strategy that aims to reestablish a population of a specific species in its historical range where it has become extirpated or locally extinct. This process involves capturing individuals from a healthy population or breeding them in captivity, and then releasing them into suitable habitats in the wild. Species reintroduction can be an effective tool for restoring ecosystems, enhancing biodiversity, and conserving endangered species.

Importance of species reintroduction

Species reintroduction plays a crucial role in conservation efforts for several reasons. Firstly, it helps to restore and maintain ecosystem balance. Many species have important ecological roles, such as controlling pests, dispersing seeds, and regulating prey populations. When these species are lost from an ecosystem, it can have cascading effects on other species and the overall functioning of the ecosystem. Reintroducing them can help to restore these important ecological processes.

Secondly, species reintroduction contributes to the conservation of endangered species. This strategy can help prevent the extinction of species by establishing new populations in areas where they have disappeared. By increasing the number of individuals and populations, the risk of catastrophic events, such as disease outbreaks or habitat loss, can be reduced. It also enhances the genetic diversity of the species, which is essential for their long-term survival and adaptation to changing environmental conditions.

Thirdly, species reintroduction can have broader ecological and socio-economic benefits. Restored populations can attract visitors and ecotourism, which can provide economic opportunities for local communities. It can also help to promote public awareness and engagement in conservation efforts, fostering a sense of stewardship towards the environment.

Challenges of species reintroduction

While species reintroduction is a valuable conservation strategy, it comes with challenges that need to be addressed for successful outcomes. One of the main challenges is the selection of suitable release sites. These sites should have appropriate habitat quality, resources, and low risks of predation or competition from other species. Conducting thorough habitat assessments, understanding species requirements, and considering potential threats are essential for choosing suitable release sites.

Another challenge is the integration of reintroduced species with the existing ecosystem. Reintroduced individuals may face difficulties in adapting to their new environment, finding food, and avoiding predators. Monitoring their behavior, habitat use, and survival rates is crucial to determine the success of reintroduction programs and identify potential issues that need to be addressed.

Furthermore, genetic considerations are important in species reintroduction. Genetic diversity is essential for the long-term viability of populations. Reintroduced individuals should be genetically diverse and representative of the source population to ensure their ability to adapt to changing environmental conditions. Genetic monitoring and management are necessary to maintain genetic diversity and prevent inbreeding depression.

Social acceptance and community involvement are also crucial for the success of species reintroduction. Engaging local communities, stakeholders, and landowners in the planning and implementation of reintroduction programs can help build support, address concerns, and ensure long-term commitment to the conservation efforts.

Case Study: Gray Wolf Reintroduction in Yellowstone National Park

A notable example of successful species reintroduction is the reintroduction of gray wolves (Canis lupus) in Yellowstone National Park in the United States. Gray wolves were extirpated from the park in the 1920s due to coordinated predator control programs. However, their absence led to an ecological imbalance, including overgrazing by elk populations and changes in vegetation.

In 1995 and 1996, gray wolves were reintroduced to Yellowstone National Park. The reintroduction involved capturing individuals from healthy wolf populations in Canada and releasing them into the park. The reintroduced wolves thrived in the park and their presence had a significant impact on the ecosystem.

The return of wolves helped regulate elk populations, reducing their overgrazing on vegetation and allowing for the recovery of streamside vegetation, such as willows and aspen trees. This, in turn, restored habitat for a variety of species, including beavers, songbirds, and fish. The cascading effects of wolf reintroduction on the ecosystem have been extensively studied and documented, highlighting the importance of top predators in maintaining ecosystem balance.

The Yellowstone wolf reintroduction serves as a powerful example of how species reintroduction can have positive ecological outcomes, restore natural processes, and contribute to the overall health and diversity of ecosystems.

Conclusion

Species reintroduction is a critical tool for conserving endangered species, restoring ecosystems, and promoting biodiversity. It requires careful planning, habitat assessments, and monitoring to ensure the success of reintroduction programs. By addressing challenges such as suitable release sites, adaptation in new environments, genetic considerations, and social acceptance, species reintroduction can contribute to the long-term sustainability of ecosystems and the conservation of threatened species.

It is worth noting that the success of species reintroduction depends not only on scientific knowledge and technical expertise but also on the collaboration and support of local communities, stakeholders, and governments. By working together, we can make a significant positive impact on the conservation of species and the preservation of the natural world.

Ecosystem Resilience

Ecosystem resilience is a fundamental concept in eco science, which refers to the ability of an ecosystem to withstand disturbances and maintain its structure and functions. Resilience is crucial for maintaining the health and sustainability of ecosystems in the face of various stressors such as climate change, pollution, habitat loss, and invasive species. In this section, we will explore the principles, factors, and strategies related to ecosystem resilience.

Principles of Ecosystem Resilience

Ecosystem resilience is guided by several key principles:

1. **Resistance:** The ability of an ecosystem to resist changes caused by disturbances. Highly resistant ecosystems are less affected by disturbances and can recover quickly.

2. **Recovery:** The ability of an ecosystem to recover and return to its original state after a disturbance. It depends on the presence of resilient species, the availability of resources, and the absence of additional stressors.

3. **Resilient Species:** Species that are adapted to disturbances and can survive in adverse conditions. Resilient species play a vital role in maintaining ecosystem functions and promoting recovery.

4. **Biodiversity:** High biodiversity enhances ecosystem resilience by providing a wider range of species with varying ecological traits and functional roles. Biodiverse ecosystems are more likely to recover from disturbances.

5. **Feedback Mechanisms:** Positive feedback loops can amplify the effects of disturbances and reduce resilience, while negative feedback loops can enhance resilience by stabilizing ecosystem processes.

6. **Adaptive Capacity:** The ability of ecosystems to adapt to changing conditions through evolutionary processes or ecological adjustments.

Factors Influencing Ecosystem Resilience

Ecosystem resilience is influenced by a variety of factors:

1. **Natural Disturbances:** Ecosystems have evolved with natural disturbances such as fire, floods, and droughts. These disturbances play a crucial role in shaping the structure and dynamics of ecosystems and can enhance their resilience.

2. **Anthropogenic Disturbances:** Human activities, such as land-use change, pollution, and climate change, can disrupt ecosystem processes and reduce resilience. It is crucial to minimize these disturbances and promote sustainable practices.

3. **Habitat Connectivity:** The degree to which habitats are interconnected influences the movement of species and the flow of energy and resources. Well-connected habitats promote biodiversity and enhance ecosystem resilience.

4. **Resource Availability:** The availability of essential resources, such as water, nutrients, and organic matter, is crucial for the functioning of ecosystems and their ability to recover from disturbances.

5. **Species Interactions:** Interactions between species, such as predation, competition, and mutualism, can affect ecosystem resilience. Some species may provide key ecological services that enhance resilience, while others may have negative impacts.

6. **Climate Change Adaptation**: Ecosystems need to adapt to changing climatic conditions to maintain their resilience. This may involve shifts in species composition, changes in phenology, and adjustments in ecosystem processes.

Strategies for Enhancing Ecosystem Resilience

To enhance ecosystem resilience, proactive measures can be taken:

1. **Conservation and Restoration**: Protecting and restoring ecosystems help maintain their resilience by preserving biodiversity, promoting natural processes, and reducing disturbances. Conservation strategies include the establishment of protected areas, sustainable land-use practices, and habitat restoration.

2. **Building Ecological Networks**: Creating interconnected networks of habitats improves the movement of species and increases their resilience to environmental changes. This can be achieved through landscape planning, ecological corridors, and restoration of degraded habitats.

3. **Diversification of Species and Genetic Resources**: Enhancing the diversity of species and genetic resources within ecosystems increases their adaptive capacity. This can be achieved through reintroduction programs, maintaining gene banks, and promoting the conservation of native species.

4. **Strengthening Ecosystem Services**: Ecosystem services, such as water purification, carbon sequestration, and pest control, contribute to ecosystem resilience. Protecting and restoring these services through sustainable management practices is essential.

5. **Promoting Adaptive Governance**: Adapting governance frameworks and decision-making processes to include scientific knowledge, stakeholder engagement, and adaptive management approaches can enhance ecosystem resilience.

Case Study: Coral Reef Resilience

Coral reefs are highly diverse and productive ecosystems that provide critical ecological and socio-economic services. However, they are facing significant threats from climate change, pollution, and overfishing. Understanding and enhancing the resilience of coral reefs is crucial for their long-term survival.

One approach to enhancing coral reef resilience is the implementation of marine protected areas (MPAs). MPAs restrict activities such as fishing and tourism, allowing coral reefs to recover from disturbances. Additionally, reducing pollution inputs, such as sedimentation and nutrient runoff, can improve water quality and enhance coral resilience.

Another strategy is the promotion of coral reef restoration. This involves interventions such as coral transplantation, artificial structures, and the removal of coral predators. Restoration efforts help to recover damaged reefs and enhance their resilience by increasing coral cover and diversity.

Furthermore, promoting community-based conservation and sustainable livelihoods can enhance coral reef resilience. Engaging local communities in reef conservation efforts can support the sustainable use of resources and the protection of reefs.

By implementing a combination of conservation, restoration, and adaptive management strategies, the resilience of coral reefs can be enhanced, improving their chances of survival and the preservation of their valuable services.

Key Takeaways

- Ecosystem resilience is the ability of ecosystems to withstand disturbances and maintain their functions and structure. - Ecosystem resilience is influenced by natural and anthropogenic factors such as disturbances, habitat connectivity, resource availability, and climate change. - Strategies for enhancing ecosystem resilience include conservation and restoration, building ecological networks, diversifying species and genetic resources, strengthening ecosystem services, and promoting adaptive governance. - Case studies, such as coral reef resilience, provide real-world examples of enhancing ecosystem resilience through various interventions and management strategies.

Community-based conservation

Community-based conservation is an approach to conservation that involves the active participation of local communities in the management and protection of natural resources. It recognizes the importance of local knowledge, values, and traditions in achieving sustainable conservation outcomes. This section will delve into the principles, methods, and benefits of community-based conservation, as well as some challenges and potential solutions.

Principles of community-based conservation

Community-based conservation is based on several key principles that guide its implementation:

1. **Local empowerment:** Local communities are empowered to take an active role in decision-making processes regarding the management of natural resources. They are given the necessary knowledge, skills, and resources to participate effectively in conservation efforts.

2. **Collaboration and partnerships:** Community-based conservation encourages collaboration among different stakeholders, including local communities, government agencies, non-governmental organizations (NGOs), and researchers. Effective partnerships enhance the collective efforts towards conservation and promote the sharing of resources and knowledge.

3. **Respect for local knowledge:** Traditional ecological knowledge held by local communities is recognized and respected. This knowledge, accumulated over generations, provides valuable insights into the dynamics of ecosystems and can help inform conservation strategies.

4. **Sustainable resource use:** Community-based conservation aims to promote sustainable resource use practices that ensure the long-term health and productivity of ecosystems. This includes setting limits on resource extraction, promoting alternative livelihoods, and implementing measures to reduce overexploitation.

Methods of community-based conservation

Several methods and approaches are commonly employed in community-based conservation:

1. **Community-led conservation planning:** Local communities actively participate in the planning and decision-making processes related to conservation initiatives. They contribute their knowledge, priorities, and preferences to develop management plans that align with their cultural values and aspirations.

2. **Community-based natural resource management (CBNRM):** CBNRM involves the delegation of resource management rights and responsibilities to local communities. This approach promotes sustainable resource use practices by giving communities a direct stake in the management and conservation of natural resources.

3. **Community-based monitoring and surveillance:** Local communities play an essential role in monitoring and surveillance efforts. They help collect data on species abundance, habitat condition, and the presence of threats. This

information is crucial for assessing conservation progress and implementing adaptive management strategies.

4. **Capacity building and education:** Community-based conservation programs often focus on building the capacity of local communities by providing training on sustainable resource management, conservation techniques, and income-generating activities. Education plays a vital role in raising awareness about the value of biodiversity and the benefits of conservation.

Benefits of community-based conservation

Community-based conservation offers numerous benefits, both for biodiversity and local communities:

1. **Enhanced conservation outcomes:** Active community involvement leads to increased compliance and support for conservation efforts. Local communities become stewards of their natural resources and are more motivated to protect and restore biodiversity.

2. **Sustainable livelihoods:** Community-based conservation promotes alternative livelihood options that are compatible with conservation goals. This helps alleviate rural poverty and reduces dependence on unsustainable practices such as illegal logging or overfishing.

3. **Preserving cultural heritage:** Community-based conservation ensures the preservation of indigenous knowledge, cultural practices, and traditions that are often closely tied to the local environment. Conservation efforts become intertwined with the cultural identity and well-being of communities.

4. **Conflict resolution:** By involving local communities in decision-making processes, community-based conservation helps address conflicts between conservation goals and local resource-use needs. Through dialogue and negotiation, solutions can be found that balance conservation objectives with the socio-economic needs of communities.

Challenges and solutions

While community-based conservation offers significant benefits, it also faces several challenges:

1. **Lack of resources:** Limited financial and technical resources can hinder the effective implementation of community-based conservation initiatives. To address this, partnerships with NGOs, research institutions, or government agencies can provide additional support.

2. **Power dynamics and governance issues:** Unequal power relations and insufficient representation of marginalized groups can negatively impact community-based conservation efforts. Transparent and inclusive governance structures are necessary to ensure the equal participation of all community members.

3. **External threats and pressures:** Local communities often face external threats, such as illegal wildlife trade, encroachment of protected areas, or climate change. Building resilience and providing support to communities in dealing with these challenges is crucial.

4. **Sustainability of community engagement:** Ensuring the long-term engagement and commitment of communities can be challenging. Providing tangible benefits, such as improved livelihoods or access to healthcare and education, can help maintain community motivation and support.

Case study: Community-based conservation in the Sinharaja Forest Reserve, Sri Lanka

One example of successful community-based conservation is found in the Sinharaja Forest Reserve in Sri Lanka. The Sinharaja Forest is a UNESCO World Heritage Site known for its high biodiversity and endemic species. Local communities in the surrounding villages actively participate in the conservation and management of the reserve through the formation of Community Based Organisations (CBOs).

These CBOs engage in various activities, including forest patrols to deter illegal logging and hunting, eco-tourism initiatives, and sustainable agriculture practices. The communities also benefit from income-generating projects like handicraft production and the sale of organic products. As a result of their efforts, the communities have experienced improved livelihoods and have contributed significantly to the conservation of the Sinharaja Forest.

Conclusion

Community-based conservation is a powerful approach to biodiversity conservation that recognizes the importance of local communities as key stakeholders. By empowering communities and involving them in decision-making processes, we can achieve more sustainable and equitable conservation outcomes. Community-based conservation not only helps protect biodiversity but also enhances the well-being of local communities, preserves cultural heritage, and promotes sustainable livelihoods.

Indigenous Knowledge and Practices

Indigenous knowledge refers to the knowledge and practices that are developed, accumulated, and passed down through generations by indigenous people. It encompasses a deep understanding of the natural world, including ecosystems, biodiversity, and sustainable resource management. Indigenous communities have lived in harmony with nature for centuries, relying on traditional knowledge to maintain a balance between human needs and the environment.

1. Importance of Indigenous Knowledge

Indigenous knowledge is crucial for the field of eco science because it offers unique insights and approaches to environmental conservation and sustainability. It provides a holistic understanding of ecosystems and the interconnections between humans and nature. Indigenous communities have developed sustainable practices that promote the long-term well-being of both the environment and the people who rely on it for their livelihoods.

Indigenous knowledge is often derived from direct observations and experiences over many generations. It incorporates knowledge of local biodiversity, weather patterns, natural resources, and ecosystems. This deep understanding helps indigenous communities make informed decisions about resource management, land use, and conservation strategies.

2. Principles of Indigenous Knowledge

2.1 Holistic View of the Environment

Indigenous knowledge acknowledges the interconnectedness of all living beings and the environment. It recognizes that actions taken in one part of an ecosystem can have far-reaching effects on other components. This holistic perspective encourages sustainable practices that consider the long-term consequences of human actions.

For example, indigenous communities have developed traditional agricultural systems that leverage the benefits of biodiversity and promote soil fertility. By planting a variety of crops, integrating livestock, and practicing crop rotation, they maintain the health of both the land and the people who rely on it for sustenance.

2.2 Intergenerational Knowledge Transmission

Indigenous knowledge is transmitted from one generation to another through stories, rituals, ceremonies, and practical experiences. Elders play a crucial role in sharing their wisdom and teachings with younger members of the community. This intergenerational knowledge transmission ensures the preservation and continuity of traditional practices and sustainable resource management strategies.

2.3 Respect for Nature and Spiritual Connections

Indigenous knowledge recognizes nature as a living entity deserving of respect and reverence. Many indigenous cultures have spiritual beliefs and rituals centered around living in harmony with the natural world. These spiritual connections foster a deep appreciation for the Earth's resources and emphasize the importance of sustainable resource use.

3. Examples of Indigenous Practices

3.1 Traditional Ecological Knowledge (TEK)

Traditional Ecological Knowledge (TEK) is a subset of indigenous knowledge that focuses on understanding ecosystems and their dynamics. TEK incorporates observations, traditional practices, and cultural beliefs to guide resource management decisions. It often combines scientific understanding with traditional practices to achieve sustainable outcomes.

For example, indigenous communities in Australia use "firestick farming" to manage their landscapes. By intentionally setting controlled fires, they maintain open grasslands, prevent wildfires, and promote the growth of native plants. This practice not only supports biodiversity but also allows for effective hunting and gathering.

3.2 Sustainable Fishing Practices

Indigenous communities around the world possess detailed knowledge of local fish populations and their seasonal migrations. This knowledge is used to develop sustainable fishing practices that ensure the long-term viability of fish stocks. For instance, indigenous communities in the Pacific Northwest of North America have sustainable salmon fishing practices that include the use of selective fishing methods and the protection of critical spawning habitats.

4. Challenges and Opportunities

Despite the importance of indigenous knowledge, it is often marginalized and undervalued in mainstream scientific and policy discussions. This marginalization can lead to the loss of traditional practices and the erosion of cultural diversity.

To address this issue, collaboration and knowledge-sharing between indigenous communities and scientists are essential. By integrating indigenous knowledge and scientific expertise, we can develop more effective and culturally appropriate approaches to environmental conservation and sustainable development.

5. Conclusion

Indigenous knowledge and practices offer valuable insights into sustainable resource management and environmental conservation. Embracing this knowledge can contribute to the development of holistic and culturally inclusive approaches to eco science. By recognizing the wisdom of indigenous communities, we can work towards a more sustainable and interconnected future.

Gene conservation

Gene conservation is a crucial aspect of eco science that focuses on protecting the genetic diversity of different species to ensure their long-term survival and adaptability.   It involves the preservation and management of genes within populations and across species, aiming to maintain the variability necessary for the resilience of ecosystems. In this section, we will explore the importance of gene conservation, the threats that endanger genetic diversity, and the strategies and methods that can be implemented to safeguard genes.

Importance of gene conservation

Genetic diversity is the foundation for the adaptation and evolution of all living organisms. It allows populations to respond to changing environmental conditions, such as climate change, disease outbreaks, or habitat degradation. By conserving genes, we maintain the potential for future generations to adapt and thrive in the face of these challenges.

Gene conservation also plays a critical role in ecosystem functioning. Genetic variability within species contributes to the stability and resilience of ecosystems. For example, diverse plant populations can enhance ecosystem productivity, nutrient cycling, and pest resistance. Moreover, many species provide essential ecosystem services, such as pollination or seed dispersal, which rely on genetic variability.

Threats to gene conservation

Despite its significance, genetic diversity is increasingly at risk due to a range of threats. Human activities, such as habitat destruction, pollution, overexploitation, and the introduction of invasive species, can lead to the loss of gene pools within populations and the extinction of species. Climate change, in particular, poses a severe threat to gene conservation as it disrupts the natural habitats and migration patterns of many species.

Fragmentation of habitats also poses a risk to gene conservation. Isolated populations may become genetically isolated, leading to reduced gene flow, inbreeding, and decreased genetic diversity. Inbreeding can result in the expression of harmful genetic traits, reduced reproductive success, and reduced adaptability to changing environments.

Conservation strategies

To address the threats to genetic diversity, various conservation strategies have been developed. These strategies aim to preserve and manage gene pools, promote gene flow, and prevent the loss of genetic diversity. Let's explore some of the key conservation strategies:

1. **In situ conservation:** This strategy focuses on protecting species within their natural habitats. It includes the establishment and management of protected areas, national parks, and wildlife reserves. In these protected areas, conservation efforts can be directed towards preserving the entire ecosystem, maintaining the interactions between species, and ensuring the preservation of genetic diversity.

2. **Ex situ conservation:** Ex situ conservation involves the conservation of species outside of their natural habitats. This includes the establishment of gene banks, botanical gardens, and zoos. In ex situ conservation, individuals or populations are maintained in captivity or in controlled environments to protect them from threats and facilitate breeding programs.

3. **Captive breeding programs:** Captive breeding programs are aimed at breeding endangered species in captivity to increase their population sizes and genetic diversity. These programs can help prevent the extinction of species with small population sizes and facilitate the reintroduction of individuals into the wild once suitable habitats are restored.

4. **Artificial selection:** Artificial selection involves selectively breeding individuals with specific genetic traits to enhance desirable characteristics or promote genetic diversity within populations. This can be done to maintain important traits for crop plants, livestock, or endangered species.

5. **Genetic monitoring and management:** Genetic monitoring allows us to assess the genetic health and diversity of populations over time. It involves collecting and analyzing DNA samples to track changes in gene frequencies and detect potential threats of inbreeding or loss of genetic diversity. Genetic management interventions, such as translocation of individuals or establishing corridors to promote gene flow, can be implemented based on the findings of genetic monitoring.

Example: Conservation of the Black-footed Ferret

The Black-footed Ferret (*Mustela nigripes*) is a species that has benefited from gene conservation efforts. It is one of the most endangered mammals in North America, primarily due to the decline of its primary prey species, the prairie dog. The Black-footed Ferret population faced severe genetic bottlenecks, leading to reduced genetic diversity and increased susceptibility to diseases.

To conserve the Black-footed Ferret, a captive breeding program was initiated, involving the collection of individuals from the wild to establish an assurance colony. This captive breeding program aimed to increase the population size and improve genetic diversity. Through careful genetic management, such as selective breeding based on genetic compatibility, the genetic health of the population was restored.

Once the captive breeding program was successful, individuals were reintroduced to specific locations in the wild, where prairie dog populations were also recovering. Genetic monitoring continued to ensure that the reintroduced ferrets were establishing genetically diverse populations while maintaining their overall genetic distinctiveness.

The role of indigenous knowledge

Indigenous communities have a wealth of traditional knowledge and practices that promote the conservation of genetic diversity. Their sustainable resource management systems and traditional seed-saving practices contribute to the maintenance of genetic diversity in agroecosystems. Including indigenous communities in decision-making processes and recognizing their rights and contributions is crucial for effective gene conservation.

Conclusion

Gene conservation is a vital component of eco science, aiming to preserve the genetic diversity necessary for the long-term survival of species and the functioning of ecosystems. By implementing various conservation strategies and engaging indigenous knowledge and practices, we can protect and manage genes to ensure the adaptability and resilience of biodiversity in the face of environmental challenges. Through collaborative efforts and global cooperation, we can work towards a sustainable future that values and conserves genetic diversity.

Conservation genetics

Conservation genetics is a field of study that focuses on understanding and preserving genetic diversity in order to ensure the long-term survival of endangered species and populations. It combines principles from genetics, ecology, and conservation biology to provide valuable insights into the genetic health and viability of species and the impact of human activities on their genetic diversity.

Genetic diversity and its importance

Genetic diversity refers to the variety of genetic traits within a species or population. It is crucial for the adaptation and survival of species in changing environments, as it provides the raw material for natural selection. Genetic diversity allows populations to better withstand environmental challenges, such as disease outbreaks or climate change, and increases their resilience.

Loss of genetic diversity can have serious consequences for species and populations. Reduced genetic diversity can result in decreased reproductive success, increased susceptibility to diseases, and reduced ability to adapt to changing environments. In extreme cases, it can lead to population decline and even extinction. Therefore, preserving and managing genetic diversity is a critical component of conservation efforts.

Genetic techniques in conservation

Conservation genetics employs several techniques to assess and manage genetic diversity in populations. These techniques provide valuable information for conservation planning and the development of effective conservation strategies. Some of the commonly used techniques include:

1. **Genetic markers**: Genetic markers are specific regions of the genome that can be used to identify and track genetic variation within a species. DNA microsatellites and single nucleotide polymorphisms (SNPs) are commonly used genetic markers. They can be used to study population structure, estimate genetic diversity, and assess gene flow between populations.

2. **Population genetics**: Population genetics is the study of genetic variation and its distribution within populations. It provides insights into the demographic history, migration patterns, and genetic structure of populations. Population genetic analyses, such as estimating effective population size and studying patterns of gene flow, can help identify populations at risk of inbreeding or genetic isolation.

3. **Parentage analysis**: Parentage analysis uses genetic markers to determine the parent-offspring relationships within a population. It can be used to track the spread of genes, identify kinship patterns, and understand mating systems and reproductive strategies. This information is crucial for guiding captive breeding programs and avoiding inbreeding.

4. **Genetic rescue**: Genetic rescue involves the introduction of individuals from genetically diverse populations into small or isolated populations to increase their genetic variation. This helps to restore genetic diversity and improve population viability. Genetic rescue can be carried out through translocation or the exchange of genetic material between populations.

5. **Assisted reproductive technologies**: Assisted reproductive technologies, such as in vitro fertilization and cryopreservation of gametes, can be used to enhance genetic management and reproductive success in endangered species. These technologies allow for the preservation and storage of genetic material, which can then be used in breeding programs or for reintroduction efforts.

6. **Conservation genomics**: Conservation genomics combines high-throughput DNA sequencing technologies with bioinformatics and statistical analyses to study the entire genome of a species. It provides a comprehensive understanding of the genetic basis of adaptation, population structure, and demographic history. Conservation genomics can help identify genes associated with traits important for species survival and inform conservation strategies.

Challenges and considerations

Conservation genetics faces several challenges and considerations that need to be addressed for effective conservation management. These include:

1. **Small population size**: Small populations are more prone to the loss of genetic diversity due to genetic drift and inbreeding. Conservation efforts need to focus on preventing further population decline and maintaining genetic diversity through genetic rescue and captive breeding programs.

2. **Hybridization**: Hybridization, the interbreeding between different species or populations, can impact genetic integrity and conservation efforts. It is important to identify and manage hybridization events to preserve the genetic distinctiveness of endangered species.

3. **Invasive species:** Invasive species can disrupt native populations and negatively impact genetic diversity. Genetic techniques can be used to identify and manage the genetic impacts of invasive species on native populations.

4. **Ethical considerations:** Genetic management strategies, such as genetic rescue and translocations, raise ethical considerations. It is important to carefully consider the potential impacts on both target and source populations and ensure that the benefits outweigh the risks.

5. **Data analysis and interpretation:** Genetic data analysis and interpretation require specialized skills and expertise. Effective conservation genetics requires collaboration between geneticists, ecologists, and conservation biologists to ensure the accurate interpretation and application of genetic information.

6. **Long-term monitoring:** Long-term monitoring of populations is essential to assess the effectiveness of conservation actions and detect changes in genetic diversity over time. This requires ongoing funding and commitment to ensure the long-term success of conservation efforts.

Case study: Genetic rescue of the Florida panther

The Florida panther (Puma concolor coryi) is a critically endangered subspecies of the cougar found in southern Florida. By the 1990s, the population had declined to fewer than 30 individuals, resulting in severe inbreeding and reduced genetic diversity. This genetic bottleneck left the population highly susceptible to diseases and reduced reproductive success.

To address this issue, a genetic rescue program was initiated in the late 1990s. Eight female pumas from a closely related subspecies, the Texas cougar (Puma concolor stanleyana), were introduced into the Florida panther population. This infusion of genetic diversity helped to restore the genetic health of the population.

Population genetic analyses have shown that the genetic rescue has increased the genetic diversity and reduced the levels of inbreeding in the Florida panther population. The population has now increased to around 200 individuals, and the genetic health has significantly improved.

The genetic rescue of the Florida panther serves as a successful example of how conservation genetics can be used to save endangered species facing genetic decline. It demonstrates the importance of genetic management strategies in ensuring the long-term viability of endangered populations.

Summary

Conservation genetics plays a crucial role in preserving and managing genetic diversity in endangered species and populations. It provides insights into the genetic health and viability of species, helps identify populations at risk, and informs conservation strategies. By utilizing various genetic techniques, such as genetic markers, population genetics, and assisted reproductive technologies, conservation genetics contributes to the preservation of species and the restoration of their genetic diversity. However, it also faces challenges, such as small population size, hybridization, and invasive species, which need to be carefully managed. Through ongoing research, monitoring, and collaboration, conservation genetics continues to be an invaluable tool in the quest for future sustainability.

Wildlife Management

Wildlife management is an essential component of eco science that focuses on the conservation and sustainable use of wild animal populations and their habitats. It involves the application of scientific knowledge and principles to ensure the long-term viability of wildlife populations while also considering the needs and interests of human societies. In this section, we will explore the key concepts, principles, and strategies employed in wildlife management.

Importance of Wildlife Management

Wildlife management plays a crucial role in promoting biodiversity conservation, ecosystem stability, and human well-being. It recognizes the intrinsic value of wildlife and aims to ensure their continued existence for future generations. Here are some key reasons why wildlife management is important:

+ **Conservation of biodiversity:** Many wildlife species are endangered or at risk due to various factors such as habitat loss, fragmentation, and poaching. Wildlife management helps protect and restore these species and their habitats, thereby preserving biodiversity.

+ **Ecosystem balance:** Wildlife species often play vital roles in maintaining ecosystem balance. By managing wildlife populations, we can ensure the stability and functioning of ecosystems, including regulating prey populations, dispersing seeds, and enhancing nutrient cycling.

+ **Economic value:** Wildlife management recognizes the economic benefits associated with wildlife, such as wildlife watching, hunting, and ecotourism.

By sustainably managing wildlife, we can support local economies, create employment opportunities, and generate revenue for conservation efforts.

+ **Cultural and recreational value:** Wildlife holds significant cultural and recreational value for communities around the world. Wildlife management ensures that these values are preserved, allowing people to connect with nature and maintain traditional practices.

Principles of Wildlife Management

Effective wildlife management is based on a set of guiding principles that help in making informed decisions and implementing strategies. Here are some key principles of wildlife management:

+ **Science-based approach:** Wildlife management relies on scientific research and data to inform decision-making. It involves studying the ecology, behavior, and population dynamics of wildlife species to understand their needs and develop appropriate management strategies.

+ **Multiple-use management:** Wildlife management recognizes the various uses and values associated with wildlife. It aims to balance conservation objectives with sustainable use, including activities like hunting, fishing, and wildlife viewing, ensuring the long-term viability of wildlife populations while also meeting human needs.

+ **Collaborative and participatory approach:** Wildlife management involves collaboration among various stakeholders, including government agencies, local communities, NGOs, and scientists. It encourages the participation of these stakeholders in decision-making, promoting shared responsibility and effective implementation of management plans.

+ **Adaptive management:** Wildlife management is an iterative and adaptive process that involves monitoring, evaluation, and continuous learning. It allows for adjustments in management strategies based on new information and changing conditions, ensuring flexibility and resilience in the face of uncertainty.

Challenges in Wildlife Management

While wildlife management plays a critical role in conserving and managing wildlife populations, it also faces numerous challenges. These challenges often arise due to the complex interactions between wildlife, ecosystems, and human activities. Here are some key challenges in wildlife management:

+ **Habitat loss and fragmentation:** One of the primary threats to wildlife is habitat loss and fragmentation due to activities such as urbanization, agriculture, and infrastructure development. Managing and restoring habitats is crucial for sustaining wildlife populations.

+ **Invasive species:** Invasive species can have detrimental impacts on native wildlife species and their habitats. Wildlife management involves controlling and eradicating invasive species to protect native biodiversity.

+ **Climate change:** Climate change poses significant challenges to wildlife management by altering habitats, disrupting migration patterns, and affecting species' survival and reproduction. Managing the impacts of climate change on wildlife requires adaptive strategies.

+ **Human-wildlife conflicts:** As human populations expand and encroach into wildlife habitats, conflicts between humans and wildlife can arise. Wildlife management seeks to minimize these conflicts through strategies such as habitat modification, fencing, and community education.

+ **Illegal wildlife trade:** The illegal trade in wildlife is a major threat to many endangered species. Wildlife management involves combating wildlife trafficking through law enforcement, international collaborations, and public awareness campaigns.

Management Strategies

Wildlife management employs a range of strategies to achieve its conservation objectives. These strategies may include:

+ **Habitat conservation and restoration:** Protecting and restoring habitats is crucial for maintaining viable wildlife populations. This involves identifying critical habitats, establishing protected areas, and implementing habitat restoration initiatives.

+ **Population monitoring and research:** Wildlife management relies on population monitoring techniques such as wildlife surveys, camera trapping, and radio telemetry to assess population trends, population size, and demographic parameters. Research helps in understanding species' ecology and behavior, guiding management decisions.

- **Sustainable hunting and fishing regulations:** When properly regulated, sustainable hunting and fishing can contribute to wildlife management by controlling populations and generating revenue for conservation efforts. Regulations are developed based on scientific data and incorporate principles of sustainable use.

- **Endangered species recovery programs:** For species on the brink of extinction, wildlife management may involve captive breeding programs, reintroduction efforts, and habitat restoration to recover their populations.

- **Conservation education and outreach:** Educating and engaging the public is essential for fostering conservation values and promoting responsible behavior towards wildlife. Conservation education initiatives may include awareness campaigns, community-based conservation projects, and environmental education programs.

- **Collaboration and partnerships:** Effective wildlife management requires collaboration among various stakeholders, including government agencies, NGOs, local communities, and private landowners. Partnerships help pool resources, share knowledge, and implement coordinated management actions.

Case Study: Conservation of African Elephants

The conservation of African elephants provides an illustrative case study for wildlife management. African elephants are an iconic, keystone species whose populations have declined due to poaching, habitat loss, and human-elephant conflicts. Effective wildlife management strategies have been crucial in protecting and conserving this species.

One key approach in elephant conservation is the establishment of protected areas and wildlife reserves to provide safe habitats. These areas are managed to ensure suitable resources for elephants, protect against poaching, and mitigate conflicts with human activities. Collaboration between governments, local communities, and NGOs has been instrumental in creating and managing these protected areas.

Population monitoring plays a vital role in assessing elephant populations and informing management decisions. Techniques such as aerial surveys, camera traps, and individual identification methods (using ear patterns or DNA analysis) help monitor population size, demographic trends, and movement patterns. This information guides the development of effective conservation strategies, such as anti-poaching measures and habitat management.

To address human-elephant conflicts, wildlife management initiatives have implemented various strategies, such as the development of community-based conservation programs. These programs involve local communities in decision-making processes, provide alternative livelihood options, and educate communities about elephant behavior and conservation benefits.

In conclusion, wildlife management is a critical discipline within eco science. It aims to conserve and sustainably manage wildlife populations, ensuring their long-term viability while addressing the needs of human societies. By employing scientific knowledge, multiple-use management approaches, and collaborative strategies, wildlife management contributes to the conservation of biodiversity, ecosystem stability, and the well-being of both wildlife and human communities.

Conservation and Restoration

Conservation biology

Conservation biology is a multidisciplinary field of study that aims to understand and protect biodiversity, ecosystem integrity, and ecological processes. It provides the scientific basis for conservation efforts and strategies to mitigate the impact of human activities on the environment. This section will explore the principles and strategies of conservation biology, as well as the challenges and opportunities in conserving biodiversity.

Principles of Conservation Biology

Conservation biology is guided by several key principles that form the foundation for understanding and managing biodiversity:

1. **Biodiversity is essential for ecosystem health**: Biodiversity refers to the variety and variability of living organisms, including genetic, species, and ecosystem diversity. It plays a crucial role in maintaining ecological processes, such as nutrient cycling, pollination, and pest control. Conservation efforts aim to preserve biodiversity at all levels.

2. **Habitat loss and fragmentation are major threats to biodiversity**: Habitat destruction and fragmentation due to human activities, such as deforestation, urbanization, and agriculture, are among the primary drivers of biodiversity loss. Conservation biology focuses on identifying and protecting important habitats to ensure the survival of species and ecological communities.

3. **Invasive species pose a significant threat**: Invasive species, introduced by human activities to ecosystems where they are not native, can outcompete and displace native species, disrupt ecosystem processes, and cause economic and ecological damage. Conservation biology aims to prevent the spread of invasive species and manage their impacts.

4. **Climate change exacerbates biodiversity loss**: Climate change and associated phenomena, such as rising temperatures, altered precipitation patterns, and sea-level rise, pose significant threats to biodiversity. Conservation biology aims to understand and mitigate the impacts of climate change on ecosystems and species.

5. **Conservation actions should be evidence-based and adaptive**: Conservation biology integrates scientific research, monitoring, and evaluation to inform conservation strategies. It recognizes the importance of adaptive

management, learning from past experiences, and adjusting conservation approaches based on new knowledge and emerging challenges.

Strategies in Conservation Biology

Conservation biologists employ a range of strategies and approaches to protect biodiversity and promote sustainable management of natural resources. Some key strategies include:

1. **Protected areas**: Protected areas, such as national parks, wildlife sanctuaries, and marine reserves, are established to conserve habitats, ecosystems, and species. They provide legal protection and management for critical biodiversity hotspots.

2. **Conservation genetics**: Conservation genetics uses genetic methods to understand population dynamics, genetic diversity, and connectivity among populations. It helps in identifying vulnerable populations, designing effective breeding programs, and implementing genetic rescue efforts.

3. **Wildlife management**: Wildlife management aims to ensure the survival and well-being of wildlife populations while considering human needs and interests. It includes strategies such as habitat restoration, population monitoring, and the regulation of hunting and fishing activities.

4. **Restoration ecology**: Restoration ecology focuses on restoring degraded ecosystems to their natural state. It involves activities such as reforestation, wetland restoration, and habitat rehabilitation. Restoration efforts help enhance biodiversity, ecosystem services, and resilience.

5. **Community-based conservation**: Community-based conservation involves engaging local communities in conservation efforts. It recognizes the importance of traditional knowledge, local values, and community participation in the long-term success of conservation initiatives.

6. **Conservation education and outreach**: Conservation biology emphasizes the importance of public awareness and engagement in conservation efforts. Education and outreach programs aim to raise awareness about biodiversity, foster a sense of stewardship, and encourage sustainable practices.

Challenges and Opportunities

Conservation biology faces various challenges in its mission to protect biodiversity and promote sustainability. Some of the key challenges include:

1. **Loss of habitat and biodiversity**: Habitat loss, driven by human activities such as deforestation, urbanization, and agricultural expansion, continues to

threaten biodiversity worldwide. Conservation biology seeks innovative approaches to address habitat loss and protect critical ecosystems.

2. **Climate change impacts:** Climate change poses significant challenges to conservation efforts. Rising temperatures, changing precipitation patterns, and extreme weather events can negatively impact species' distribution, disrupt ecological interactions, and increase the risk of extinction. Conservation biology seeks to identify climate change adaptation strategies and promote resilience in ecosystems.

3. **Invasive species:** Invasive species can have severe impacts on native ecosystems and biodiversity. The globalization of trade and travel has facilitated the spread of invasive species worldwide. Conservation biology focuses on preventing and managing invasive species through early detection, rapid response, and effective control measures.

4. **Lack of funding and resources:** Conservation efforts often struggle with limited funding and resources. Conservation biology seeks to address this challenge by promoting sustainable financing mechanisms, public-private partnerships, and innovative fundraising approaches.

5. **Human-wildlife conflicts:** As human activities encroach upon natural habitats, conflicts between humans and wildlife become more prevalent. Conservation biology aims to mitigate these conflicts through measures such as habitat management, community involvement, and the development of sustainable livelihood options.

Despite these challenges, conservation biology also presents numerous opportunities for positive change:

1. **Technological advancements:** Advances in technology, such as remote sensing, genetic tools, and data analytics, provide new opportunities for understanding and managing biodiversity. These tools can support evidence-based decision-making, enhance monitoring efforts, and facilitate targeted conservation actions.

2. **Collaborative approaches:** Conservation biology recognizes the importance of collaboration among scientists, policymakers, local communities, and various stakeholders. Collaborative approaches can lead to innovative ideas, shared knowledge, and effective conservation strategies.

3. **Sustainable development integration:** Integrating conservation goals into broader sustainable development frameworks can create synergies between conservation efforts and socioeconomic development. Conservation biology promotes the integration of biodiversity conservation into sectors such as agriculture, energy, and infrastructure planning.

4. **Policy and governance support:** Conservation biology contributes to the development of evidence-based policies, legal frameworks, and governance structures that promote biodiversity conservation. It advocates for the inclusion of conservation objectives in national and international agendas.

In conclusion, conservation biology plays a crucial role in understanding and protecting biodiversity. Its principles and strategies provide a scientific foundation for conserving ecosystems, species, and ecological processes. By addressing challenges and embracing opportunities, conservation biology can contribute to a sustainable future for both humans and the natural world.

Exercises

1. Research and discuss an example of a successful conservation project that utilized community-based conservation approaches. 2. Investigate and explain the impacts of climate change on a specific ecosystem or species of your choice. 3. Design a hypothetical restoration plan for a degraded ecosystem in your local area. Consider the necessary steps, resources, and potential challenges involved. 4. Analyze a case study of an invasive species and discuss the ecological and economic impacts it has caused. Also, propose potential strategies for managing the invasive species. 5. Evaluate the role of technology, such as citizen science initiatives or remote sensing, in monitoring and managing biodiversity. Provide examples to illustrate your points.

Further Reading

1. Soule, M. E., & Terborgh, J. (Eds.). (1999). *The Science of Conservation Biology.* Sinauer Associates. 2. Primack, R. B. (2012). *Essentials of Conservation Biology.* Sinauer Associates. 3. Kareiva, P., & Marvier, M. (2013). *Conservation Science: Balancing the Needs of People and Nature.* Roberts. 4. Sodhi, N. S. (2010). *Conservation Biology: Voices from the Tropics.* Wiley-Blackwell. 5. Wilson, E. O. (2016). *Half-Earth: Our Planet's Fight for Life.* Liveright.

Endangered Species

Endangered species are an important aspect of eco science and conservation biology. These species are at risk of extinction due to various factors such as habitat loss, climate change, poaching, and pollution. The protection and conservation of endangered species is crucial for maintaining biodiversity and ensuring the overall health of ecosystems. In this section, we will explore the key issues related to endangered species and discuss the strategies and initiatives aimed at their preservation.

Causes of Endangerment

The primary cause of endangerment for many species is habitat destruction. Human activities such as deforestation, urbanization, and industrial development have resulted in the loss and fragmentation of natural habitats. This destruction directly affects the survival and reproduction of species, leading to their decline. Climate change is another significant factor contributing to the endangerment of species. Rising temperatures, altered precipitation patterns, and extreme weather events disrupt ecosystems, making it challenging for some species to survive.

Poaching and illegal wildlife trade pose a severe threat to many endangered species. For example, elephants are targeted for their ivory tusks, and rhinos are hunted for their horns. These activities decimate populations and disrupt the natural balance within ecosystems. Pollution, both on land and in water, also endangers species by contaminating their habitats and affecting their health and reproduction. Pesticides, industrial waste, and plastic pollution all have detrimental effects on endangered species.

Conservation Efforts

Conservation biology is the field dedicated to the study and preservation of endangered species. Scientists, researchers, conservation organizations, and governments collaborate to develop strategies and initiatives to protect these species. One of the fundamental approaches is the establishment of protected areas, which serve as sanctuaries for endangered species. These areas, such as national parks and wildlife reserves, provide a safe habitat where species can thrive without human interference.

Sustainable land use practices are crucial for conserving endangered species. This involves promoting responsible forestry, agricultural practices, and urban planning that minimize habitat destruction and prioritize biodiversity conservation. Restoration ecology is another important conservation approach. It focuses on restoring degraded ecosystems to their natural state, allowing endangered species to recolonize these areas. Techniques such as habitat restoration, species reintroduction, and ecosystem management play a vital role in this process.

Conservation genetics is a specialized field that uses genetic data to guide conservation efforts. Genetic analyses help identify individuals with unique traits or adaptations that are important for the survival of a species. By preserving genetic diversity, conservationists can improve the resilience of endangered populations. Wildlife management techniques, such as population monitoring and control of invasive species, are also essential for the conservation of endangered species.

Furthermore, community-based conservation initiatives engage local communities in the protection and management of endangered species. These initiatives recognize the importance of indigenous knowledge and practices in ecosystem management. Indigenous communities have historically played a role in protecting and conserving biodiversity through sustainable practices. By involving local communities, conservation efforts become more sustainable and successful.

Case Study: The Endangered Black Rhinoceros

The black rhinoceros (Diceros bicornis) is a critically endangered species native to Africa. It has been severely affected by poaching, habitat loss, and habitat fragmentation. Poachers target black rhinos for their valuable horns, which are used primarily in traditional medicine practices. The demand for rhino horn has led to a significant decline in their population, making them one of the most endangered species in the world.

Conservation efforts have been implemented to protect the black rhino. One successful strategy is the establishment of protected areas and anti-poaching units. These measures have led to an increase in the population of black rhinos in certain areas. Additionally, community-based conservation initiatives have involved local communities in the protection of black rhinos, emphasizing the role of indigenous knowledge and practices in conservation.

Despite these efforts, the black rhino population continues to face challenges. The illegal wildlife trade and demand for rhino horn persist, creating an ongoing threat to their survival. Therefore, a multi-faceted approach involving strict law enforcement, public awareness campaigns, and international collaboration is necessary to ensure the long-term survival of the black rhinoceros.

Discussion Questions

1. Why is the protection of endangered species important for maintaining biodiversity?

2. What are the major causes of endangerment for many species?

3. How does habitat destruction contribute to the decline of endangered species?

4. Discuss the role of protected areas in the conservation of endangered species.

5. What are the different approaches used in wildlife management for the conservation of endangered species?

6. How can community-based conservation initiatives enhance the protection of endangered species?

7. Why is it essential to involve local communities in conservation efforts?

8. Describe the conservation efforts aimed at protecting the black rhinoceros.

9. What are the major challenges faced in the conservation of the black rhinoceros?

10. How can international collaboration contribute to the preservation of endangered species?

Further Reading

1. Primack, R. B. (2010). Essentials of Conservation Biology. Sinauer Associates, Inc.

2. Soule, M. E., & Wilcox, B. A. (Eds.). (1980). Conservation Biology: An Evolutionary-Ecological Perspective. Sinauer Associates, Inc.

3. Ehrlich, P. R., & Ehrlich, A. H. (1981). Extinction: The Causes and Consequences of the Disappearance of Species. Random House.

4. Noss, R. F. (1999). The Science of Conservation Planning: Habitat Conservation Under the Endangered Species Act. Oxford University Press.

5. Terborgh, J., & van Schaik, C. P. (2002). Making Parks Work: Strategies for Preserving Tropical Nature. Island Press.

Fragmentation

Fragmentation refers to the process of breaking large habitat areas into smaller, isolated patches. It is one of the key threats to biodiversity and ecosystem function. This section will discuss the causes and consequences of fragmentation, as well as the strategies and techniques used to mitigate its negative effects.

Causes of Fragmentation

Fragmentation is primarily caused by human activities such as urban development, agriculture, logging, and infrastructure expansion. These activities result in the conversion of natural habitats into smaller and isolated patches. Some specific causes of fragmentation include:

1. **Urbanization:** The rapid expansion of cities and towns leads to the conversion of natural landscapes into urban areas, resulting in the fragmentation of habitats.

2. **Agricultural expansion:** The conversion of forests and grasslands into agricultural fields and plantations results in the destruction and fragmentation of natural habitats.

3. **Road construction:** The construction of roads and highways cuts through natural habitats, creating barriers for wildlife movement and fragmenting ecosystems.

4. **Logging:** Timber extraction can fragment forests, as logging roads and clear-cut areas disrupt the connectivity of the remaining forest patches.

5. **Mining and extraction:** Mining activities can lead to habitat destruction and fragmentation, particularly in ecologically sensitive areas.

Consequences of Fragmentation

The fragmentation of habitats has numerous negative consequences for biodiversity and ecosystem functioning. Some of the key consequences include:

1. **Loss of biodiversity:** Fragmentation leads to the loss of species diversity and the decline of populations, as small and isolated habitat patches are unable to support viable populations.

2. **Reduced genetic diversity:** Fragmented populations often experience decreased gene flow, which can result in reduced genetic diversity and increased risk of inbreeding.

3. **Altered ecological processes:** Fragmentation disrupts ecological processes such as seed dispersal, pollination, and predator-prey interactions, leading to changes in ecosystem structure and function.

4. **Increased edge effects:** The creation of habitat edges through fragmentation increases the exposure of interior habitats to edge effects, such as increased predation, invasive species colonization, and microclimatic changes.

5. **Loss of ecosystem services:** Fragmentation can lead to the loss of important ecosystem services such as water purification, carbon sequestration, and climate regulation.

Strategies for Mitigating Fragmentation

To address the negative impacts of fragmentation, various strategies and techniques have been developed. These include:

1. **Habitat corridor creation:** Establishing corridors or connections between fragmented habitats allows for the movement of plants and animals, promoting gene flow and supporting population viability.

2. **Habitat restoration:** Restoring degraded habitats or creating new habitats can help to reconnect fragmented landscapes and promote biodiversity conservation.

3. **Conservation planning:** Incorporating landscape-level conservation planning approaches, such as identifying and protecting critical habitat areas and connectivity corridors, can help minimize fragmentation.

4. **Land-use planning:** Implementing effective land-use planning strategies, such as zoning and protected area designations, can help prevent further fragmentation and promote sustainable development.

5. **Ecosystem-based approaches:** Adopting ecosystem-based approaches in resource management and development projects can help minimize habitat fragmentation and maintain ecosystem integrity.

Case Study: Fragmentation and the Amazon Rainforest

The Amazon rainforest, the largest tropical rainforest in the world, is facing severe fragmentation due to various human activities, including agriculture expansion, logging, and infrastructure development. The consequences of fragmentation in the Amazon include the loss of biodiversity, loss of ecosystem services, and increased vulnerability to climate change. However, several initiatives have been implemented to mitigate fragmentation in the region. For example, the establishment of protected areas, such as national parks and indigenous territories, helps to maintain connectivity and protect key biodiversity areas. Additionally, the creation of sustainable land-use practices, such as agroforestry systems and community-based conservation initiatives, promotes landscape connectivity and supports local livelihoods.

Key Takeaways

* Fragmentation refers to the process of breaking large habitat areas into smaller, isolated patches due to human activities.

* Fragmentation leads to the loss of biodiversity, reduced genetic diversity, altered ecological processes, increased edge effects, and loss of ecosystem services.

* Strategies for mitigating fragmentation include habitat corridor creation, habitat restoration, conservation planning, land-use planning, and ecosystem-based approaches.

* The Amazon rainforest serves as a significant case study highlighting the impacts of fragmentation and the importance of conservation efforts in mitigating its effects.

Summary

Fragmentation is a significant threat to biodiversity and ecosystem function. It is caused by human activities such as urbanization, agriculture, and infrastructure development. Fragmentation leads to the loss of biodiversity, disrupted ecological processes, and reduced ecosystem services. To mitigate fragmentation, strategies such as habitat corridor creation, habitat restoration, and conservation planning are employed. The Amazon rainforest serves as an important case study illustrating the impacts of fragmentation and the need for conservation efforts. By understanding and effectively addressing fragmentation, we can contribute to the preservation of biodiversity and the sustainable management of ecosystems.

Invasive Species

Invasive species are non-native organisms that are introduced into an ecosystem and have the potential to cause harm to native species and their habitats. They can include plants, animals, and microorganisms, and they often have the ability to rapidly reproduce and spread, outcompeting native species for resources and altering the balance of the ecosystem.

Introduction and Spread

The introduction and spread of invasive species can occur through various means, including accidental transportation, intentional release, and natural dispersal.

Human activities such as international trade, travel, and agriculture play a significant role in facilitating their introduction.

Once introduced, invasive species can spread quickly and establish themselves in new areas. They often have traits that allow them to outcompete native species, such as rapid growth rates, high reproductive capacities, and the ability to adapt to different environmental conditions. As a result, they can dominate ecosystems, displacing native species and causing ecological imbalances.

Impacts

Invasive species can have significant negative impacts on ecosystems, economies, and human health. They can disrupt natural food chains, decrease biodiversity, degrade habitats, and alter ecosystem functions. This can lead to the decline or extinction of native species, including endangered ones.

Economically, invasive species can cause substantial damage to agriculture, forestry, fisheries, and infrastructure. Invasive pests, for example, can devastate crops and forests, leading to substantial economic losses. Invasive species can also impact human health directly, through the transmission of diseases or allergic reactions.

Management and Control

The management and control of invasive species require a comprehensive and integrated approach. This includes prevention, early detection, eradication, and control measures. Prevention is the most cost-effective strategy and involves reducing the introduction and spread of invasive species through measures such as biosecurity, quarantine, and public awareness campaigns.

Early detection and rapid response are crucial in minimizing the impacts of invasive species. Timely detection allows for the implementation of effective eradication and control measures before populations become established. This can involve methods such as physical removal, chemical treatment, biological control using natural enemies, and the use of barriers or mechanical devices.

However, complete eradication of invasive species is often challenging and may not always be feasible. In such cases, management strategies aim to control and minimize the impacts of invasive species, often through a combination of methods. This can include the restoration of native habitats, the promotion of biodiversity, and the development of integrated pest management approaches.

Case Study: Asian Carp

One notable example of an invasive species is the Asian carp, specifically the silver carp and bighead carp, which were introduced into North America in the 1970s for use in aquaculture. They quickly escaped into the Mississippi River system and have since spread into many waterways across the United States.

Asian carp pose a significant threat to native fish populations and aquatic ecosystems. They are voracious filter feeders, consuming large quantities of plankton that are essential for other fish species. This results in competition for resources and can lead to the decline of native fish populations.

Efforts to control Asian carp include the construction of underwater electric barriers in the Chicago Area Waterway System to prevent their entry into the Great Lakes, as well as targeted removal and monitoring programs. However, the complete eradication of Asian carp from these waterways remains a significant challenge.

Conclusions

Invasive species pose a significant threat to ecosystems and require proactive management and control measures. Preventive actions, early detection, and rapid response are essential in minimizing the impacts of invasive species. Ongoing research, monitoring, and collaboration between scientists, policymakers, and stakeholders are necessary to effectively manage and control invasive species and preserve the integrity of ecosystems.

By implementing these strategies and raising public awareness about the impacts of invasive species, we can work towards minimizing their negative effects and preserving the biodiversity and functioning of our ecosystems for future generations.

Ecosystem Restoration

Ecosystem restoration is a crucial aspect of eco science that aims to reverse the damage caused by human activities and restore the health and functioning of ecosystems. It involves a range of practices and techniques aimed at recovering and rebuilding natural habitats, improving biodiversity, and enhancing ecosystem services. Ecosystem restoration plays a vital role in ensuring the long-term sustainability of our planet and is critical for conserving biodiversity, mitigating climate change, and promoting sustainable development.

Importance of Ecosystem Restoration

Ecosystems provide us with a wide range of invaluable services, such as clean air and water, nutrient cycling, soil fertility, climate regulation, pollination, and natural pest control. However, due to various human-induced activities such as deforestation, pollution, and habitat destruction, many ecosystems have been degraded or completely destroyed. This loss of biodiversity and ecosystem services has severe implications for human well-being and the health of the planet.

Ecosystem restoration aims to restore the functionality and resilience of degraded ecosystems, enhancing their capacity to provide essential services and support biodiversity. By restoring ecosystems, we can conserve and protect endangered species, restore natural landscapes, improve water quality, prevent soil erosion, sequester carbon dioxide, and enhance the overall health and sustainability of ecosystems.

Principles of Ecosystem Restoration

Successful ecosystem restoration projects are based on a set of principles that guide the planning, implementation, and monitoring of restoration activities. These principles include:

1. **Ecological Integrity**: Restoration efforts should aim to reestablish the ecological processes and functions that were present in the original, undisturbed ecosystem. This involves restoring the structure, composition, and diversity of plant and animal communities, as well as the natural flow of energy and matter.

2. **Adaptive Management**: Restoration projects should be based on a flexible and adaptive management approach that allows for ongoing monitoring, evaluation, and adjustment of restoration activities. This ensures that the chosen strategies and techniques are effective and can be modified if necessary to achieve the desired outcomes.

3. **Stakeholder Engagement**: Successful ecosystem restoration requires collaboration and engagement with local communities, landowners, indigenous peoples, and other stakeholders. Involving stakeholders in the decision-making and implementation processes helps to ensure a sense of ownership, encourages knowledge sharing, and enhances the long-term success of restoration efforts.

4. **Holistic Approach:** Ecosystem restoration should consider the broader social, economic, and cultural contexts in addition to ecological factors. By addressing the needs and priorities of local communities, incorporating traditional knowledge, and considering the economic viability of restoration projects, the outcomes are more likely to be sustainable and successful in the long term.

5. **Monitoring and Evaluation:** Regular monitoring and evaluation of restoration projects are essential for assessing progress, identifying challenges, and adapting strategies accordingly. Monitoring helps to ensure that restoration goals are being met and provides valuable insights into the effectiveness of different techniques and approaches.

Techniques for Ecosystem Restoration

Ecosystem restoration involves a variety of techniques and approaches that can be tailored to specific ecosystem types, geographical locations, and restoration objectives. Some common techniques include:

1. **Revegetation:** This technique involves planting native vegetation in areas where it has been removed or degraded. It helps to stabilize soils, control erosion, enhance biodiversity, and provide habitat for wildlife. Native plant species are preferred as they are better adapted to local conditions and provide ecological functionality.

2. **Habitat Creation:** Restoration projects often involve creating new habitats, such as wetlands, forests, or grasslands, where they have been lost or altered. This may include constructing ponds, creating artificial reefs, or establishing new vegetation assemblages. These habitats provide valuable resources for a wide range of species and contribute to overall ecosystem health.

3. **Invasive Species Control:** Invasive species pose a significant threat to ecosystem functioning and biodiversity. Restoration efforts often involve the removal or control of invasive species to reduce competition with native plants and animals. This may involve manual removal, chemical treatments, or the introduction of natural predators or biocontrol agents.

4. **Stream and Riparian Restoration:** Stream and riparian (water-adjacent) ecosystems are critical habitats that support a wide range of plant and animal species. Restoration techniques in these areas may include bank stabilization, riparian buffer establishment, and stream channel restoration.

These measures help to improve water quality, enhance habitat connectivity, and mitigate the impacts of erosion and flooding.

5. **Ecosystem Engineering:** In some cases, restoration efforts require active engineering interventions to recreate or enhance specific ecological functions. For example, building beaver dams to restore water flow in wetlands, constructing fish ladders to facilitate fish migration, or creating artificial structures to provide nesting sites for birds or bats.

Challenges and Considerations

Despite the importance of ecosystem restoration, there are several challenges that need to be addressed in order to achieve successful outcomes. Some key challenges include:

1. **Limited Resources:** Ecosystem restoration requires significant financial and human resources, which are often limited. Securing adequate funding and support for restoration projects can be a major challenge, especially in developing countries or regions with competing priorities.

2. **Long-Term Monitoring and Maintenance:** Restoration projects need to be monitored and maintained over the long term to ensure that the desired outcomes are achieved and sustained. This requires ongoing funding, dedicated personnel, and a commitment to long-term stewardship.

3. **Lack of Awareness and Knowledge:** Many people are not aware of the importance of ecosystem restoration or the techniques and approaches available. Promoting awareness and providing education and training opportunities are vital for building capacity and garnering support for restoration efforts.

4. **Complexity and Uncertainty:** Ecosystems are complex, dynamic systems, and their restoration is rarely straightforward. Uncertainty about the best techniques, potential outcomes, and the timescales involved can make restoration challenging. Embracing adaptive management and continuous learning is essential for effective restoration.

5. **Legal and Policy Barriers:** In some cases, legal and policy barriers may impede restoration efforts. These barriers can include inadequate legislation, conflicting land use policies, or lack of enforcement mechanisms. Addressing these barriers requires collaboration among stakeholders and advocating for supportive policy frameworks.

Case Study: Everglades Restoration

The Everglades in Southern Florida, USA, is one of the largest wetlands in the world and an emblematic example of ecosystem restoration efforts. For decades, human activities, such as drainage for agriculture and urban development, disrupted the natural hydrological patterns of the region, leading to habitat loss, water pollution, and the decline of wildlife populations.

To restore the Everglades, a comprehensive restoration plan known as the Comprehensive Everglades Restoration Plan (CERP) was initiated in 2000. This plan involves a combination of strategies, including constructing reservoirs and water storage facilities, improving water quality through the reduction of nutrient runoff, and restoring natural flow patterns.

The Everglades restoration project faces numerous challenges, such as securing sufficient water resources, managing invasive species, and balancing the needs of different stakeholders. However, it serves as a model for large-scale ecosystem restoration and highlights the importance of integrated, multidisciplinary approaches, stakeholder engagement, and adaptive management in achieving restoration goals.

Conclusion

Ecosystem restoration is a critical component of eco science that seeks to reverse the damage caused by human activities and restore the health and functioning of ecosystems. By employing a variety of techniques and approaches, ecosystem restoration aims to recover and rebuild natural habitats, enhance biodiversity, and restore ecosystem services. The principles of ecosystem restoration guide the planning, implementation, and monitoring of restoration projects, ensuring their long-term success. Despite the challenges, ecosystem restoration plays a vital role in achieving sustainability and is crucial for the conservation of biodiversity, climate change mitigation, and the promotion of sustainable development.

Ecological Restoration Methods

Ecological restoration is the process of assisting the recovery of an ecosystem that has been degraded, damaged, or destroyed. It involves various methods and techniques aimed at revitalizing the natural processes and functions of the ecosystem. Ecological restoration is essential for preserving biodiversity, ensuring the provision of ecosystem services, and promoting the long-term sustainability of our planet.

Assessment and Planning

Before initiating any ecological restoration project, a thorough assessment of the site is crucial. This assessment involves evaluating the extent and causes of ecosystem degradation, identifying key ecological components, and determining the desired future state of the ecosystem. The planning phase includes setting restoration goals, developing a detailed restoration plan, and identifying the appropriate restoration methods and techniques.

Habitat Restoration

One of the fundamental approaches in ecological restoration is habitat restoration. This method focuses on recreating or enhancing habitat conditions to support the recovery of specific species or ecological communities. It involves:

- Removing alien species: Invasive alien species can outcompete native species and disrupt ecosystem functions. Removing them helps restore the balance of the ecosystem.

- Replanting native vegetation: Restoring vegetation cover with native plants improves habitat quality and provides food and shelter for wildlife. It also enhances ecosystem functions like nutrient cycling and soil stability.

- Restoring hydrological conditions: Many ecosystems, such as wetlands and riparian areas, rely on specific water regimes. Restoring appropriate hydrological conditions is crucial for their recovery.

- Creating artificial structures: In some cases, the construction of artificial structures such as nesting boxes or artificial reefs can provide essential habitat components for target species.

Reintroduction of Species

The reintroduction of species is another important method of ecological restoration, particularly for endangered or locally extirpated species. Reintroduction aims to reestablish viable populations of these species in their former habitats. The process involves:

- Selecting appropriate species: Extensive research is conducted to identify suitable candidate species for reintroduction based on their ecological requirements, genetic diversity, and potential interactions with other species.

- Sourcing individuals: Individuals for reintroduction can be sourced from captive breeding programs, rescued individuals, or other populations with similar genetic and ecological characteristics.

- Preparing for release: Prior to release, individuals undergo various treatments and conditioning to ensure their survival and adaptation to the environment. This may include quarantine, veterinary care, and acclimation to local conditions.

- Monitoring and post-release management: After release, ongoing monitoring is essential to assess the success of the reintroduction and make necessary adjustments. Post-release management, such as predator control and habitat management, may be required to support the establishment of the reintroduced population.

Ecosystem Function Restoration

Apart from habitat and species-focused restoration, restoring ecosystem functions is critical for the overall health and resilience of the ecosystem. It involves:

- Soil restoration: Soil degradation due to factors like erosion, compaction, or pollution can severely impact ecosystem functions. Techniques such as soil erosion control, soil amendment, and bio-remediation help restore soil health.

- Nutrient cycling: Restoring natural nutrient cycling processes through techniques like composting, organic farming, and cover cropping improves soil fertility and enhances ecosystem productivity.

- Reinstituting ecological processes: Ecosystems rely on various processes like pollination, seed dispersal, and natural disturbance regimes. Restoring the functioning of these processes ensures the ecological resilience of the restored ecosystem. For example, reintroducing native pollinators or using prescribed fires can mimic natural processes.

Monitoring and Adaptive Management

Monitoring the effectiveness of ecological restoration efforts is essential to evaluate progress, identify potential issues, and make adaptive management decisions. Regular monitoring involves:

- Ecological data collection: Collecting data on key ecological indicators, such as species abundance and diversity, habitat quality, and ecosystem functions, helps assess the trajectory of the restoration project.

- Data analysis and interpretation: Analyzing the collected data provides insights into the success of the restoration efforts and allows for any necessary adjustments to the restoration plan.

- Adaptive management: Based on the monitoring results, adaptive management strategies can be implemented to optimize the restoration outcomes. This may involve modifying restoration methods, intensifying management actions, or revising restoration goals.

Case Study: Everglades Restoration

The restoration of the Everglades in Florida, USA, is one of the most significant and complex ecological restoration projects in the world. The Everglades, a unique wetland ecosystem, has been severely degraded due to water diversion, pollution, and habitat loss.

The restoration efforts involve several ecological restoration methods, such as:

- Reconstructing hydrological conditions: The ecosystem's water flow has been altered by drainage canals, disrupting natural water patterns. Restoration efforts involve modifying water management infrastructure to mimic historic hydrological conditions.

- Removing invasive species: Invasive plants, like the melaleuca and Brazilian pepper, have proliferated in the Everglades, outcompeting native vegetation. Extensive efforts are underway to control and remove these species.

- Restoring wildlife corridors: Creating and enhancing wildlife corridors and connecting fragmented habitats allows for the movement of species and promotes genetic diversity within the Everglades.

- Water quality improvement: Addressing water pollution from agricultural runoff and urban sources is crucial for improving the health of the Everglades. Constructed wetlands and treatment areas are employed to filter and treat water before it reaches the ecosystem.

The Everglades restoration project showcases the interdisciplinary nature of ecological restoration, requiring collaboration between scientists, engineers, policymakers, and local communities. It highlights the importance of considering social, economic, and political factors in addition to ecological principles for successful restoration outcomes.

Conclusion

Ecological restoration methods play a vital role in reversing ecosystem degradation and ensuring the long-term sustainability of our planet. By employing a combination of habitat restoration, species reintroduction, ecosystem function restoration, and careful monitoring, we can achieve successful ecological restoration outcomes. However, it is important to recognize that each restoration project is unique and requires careful planning, adaptive management, and stakeholder involvement to achieve the desired ecological and socio-economic goals. Through continued research and innovation, we can enhance our understanding and capabilities in ecological restoration, contributing to a more sustainable and resilient future for our planet.

Exercises

1. Choose an ecosystem in your local area that has been degraded. Conduct an assessment to identify the causes and extent of the degradation. Develop a restoration plan outlining specific methods and techniques that could be employed to restore the ecosystem.

2. Research a reintroduction project for an endangered species. Identify the challenges associated with the reintroduction process and propose strategies to overcome these challenges.

3. Investigate a successful habitat restoration project in your region. Analyze the techniques used and the outcomes achieved. Discuss the importance of community engagement in the restoration process.

4. Choose a degraded soil site and propose restoration techniques to improve soil health and fertility. Explain the potential benefits of these techniques for the overall ecosystem.

5. Study a real-life case of a failed ecological restoration project. Identify the factors that led to the failure and suggest alternative approaches that could have been more successful.

Additional Resources

1. Society for Ecological Restoration: https://www.ser.org/

2. International Union for Conservation of Nature (IUCN) Restoration Resources: https://www.iucn.org/theme/valuing-restoration-funded-mde

3. Society for Ecological Restoration Science and Policy Working Group: https://ercoalition.org/

Reintroduction of Species

Reintroduction of species is a vital component of conservation efforts aimed at restoring populations and ecosystems. It involves the deliberate release of organisms into their historical range, from which they have been extirpated or become locally extinct. Species reintroduction has gained significant attention and importance due to the increasing recognition of biodiversity loss and the urgent need for conservation actions to restore ecosystems and prevent further extinctions.

Why Reintroduce Species?

The reasons for reintroducing species are multifaceted. Some key objectives of species reintroduction include:

1. Restoring ecological processes: Reintroduced species often play critical ecological roles, such as pollination, seed dispersal, and predator-prey interactions. By reintroducing these species, ecological processes can be restored, leading to healthier and more functional ecosystems.

2. Preventing extinction: Reintroducing endangered or threatened species into their natural habitats can help prevent their extinction. By boosting population sizes and genetic diversity, reintroduction programs aim to improve the survival prospects of species on the brink of extinction.

3. Ecosystem restoration: Reintroducing species is a crucial step in ecosystem restoration efforts, particularly when keystone species or those with specific ecological functions are involved. The presence of these species can help restore natural balance and promote the recovery of degraded ecosystems.

4. Enhancing biodiversity: Reintroducing species can increase biodiversity levels in areas where species have been lost. This is especially important in habitats that have been significantly altered by human activities and where biodiversity has been severely impacted.

5. Educating and engaging the public: Species reintroduction programs provide opportunities for public education and engagement. By involving local communities and raising awareness about conservation issues, reintroduction projects can foster support for broader conservation efforts.

Challenges and Considerations

While species reintroduction can be an effective conservation strategy, it is not without challenges and considerations. Several factors need to be carefully assessed and addressed to maximize the success of reintroduction efforts:

1. Habitat suitability: Before reintroducing a species, it is crucial to ensure that the habitat can support the species' needs. Factors such as food availability, shelter, water, and suitable breeding conditions must be considered.

2. Genetic considerations: Genetic diversity is vital for the long-term survival of reintroduced populations. Genetic assessments, such as determining the genetic makeup of captive populations, should be conducted to maintain healthy genetic diversity in reintroduced populations.

3. Disease transmission: Reintroduced species can be vulnerable to diseases to which they have lost natural immunity. Pre-release health screening and monitoring are essential to prevent the introduction and spread of diseases within reintroduced populations.

4. Predation and competition: Reintroduced species may face predation from introduced or invasive species. Assessing and managing predation risks, including the presence of predators and competition for resources, is crucial for successful reintroduction.

5. Reintroduction methods: Various techniques can be utilized for reintroducing species, such as soft-release (gradual acclimation to the wild), hard-release (direct release into the wild), or captive breeding followed by reintroduction. The choice of method depends on the species, its specific requirements, and the conservation objectives.

6. Monitoring and adaptive management: Regular monitoring and evaluation of reintroduced populations are essential to assess their progress and address any challenges. Adaptive management strategies allow for adjustments to be made based on ongoing assessments, ensuring the best chance of success for reintroduced species.

Case Study: Gray Wolf Reintroduction in Yellowstone National Park

A notable example of successful species reintroduction is the reintroduction of gray wolves (Canis lupus) to Yellowstone National Park in the United States. Gray wolves were extirpated from the park in the early 1900s due to predator control programs. However, their absence led to ecological imbalances, including overpopulation of elk and a decline in biodiversity.

In 1995 and 1996, a total of 31 gray wolves from Canada were reintroduced to Yellowstone. The reintroduction program aimed to restore the natural predator-prey dynamics and enhance biodiversity. The reintroduced wolves played a crucial role in controlling elk populations, reducing browsing pressure on vegetation, and indirectly benefitting other species like beavers, birds, and fish.

Since their reintroduction, the gray wolf population in Yellowstone has thrived, reaching a sustainable level. The Yellowstone wolf reintroduction case study illustrates the potential of species reintroduction to restore ecosystems and highlight the importance of top predators for maintaining ecological balance.

Conclusion

The reintroduction of species is a valuable tool in conservation efforts, aiming to restore populations, ecological processes, and ecosystems. It is a complex undertaking that requires careful planning, assessment of habitat suitability, genetic considerations, disease management, and monitoring. Successful reintroduction programs can contribute to preventing extinctions, restoring ecological balance, enhancing biodiversity, and engaging the public in conservation efforts. The case study of gray wolf reintroduction in Yellowstone National Park demonstrates the positive impacts that can be achieved through thoughtful and well-executed reintroduction initiatives.

Succession and Disturbance Ecology

Succession is a process of ecological change in which a community of organisms gradually replaces another community over time. It occurs in response to disturbance events, such as natural disasters or human activities, that disrupt the existing community structure. Disturbances can range from small-scale events, like a fallen tree, to large-scale disturbances, like a wildfire or volcanic eruption. Understanding succession and disturbance ecology is crucial for managing and restoring ecosystems, as well as predicting the long-term impacts of disturbances.

Principles of Succession

Succession can be divided into two main types: primary succession and secondary succession. Primary succession occurs in areas that are devoid of life or have a bare substrate, such as a newly formed volcanic island or a retreating glacier. The process starts with the colonization of pioneer species, which are well-adapted to harsh environmental conditions. These species, such as lichens and mosses, gradually pave the way for the establishment of more complex and diverse communities. Over time, the pioneer species are replaced by more competitive species, leading to an increase in biodiversity and the formation of a climax community.

Secondary succession, on the other hand, occurs in areas that have been previously occupied by a community but have experienced a disturbance that removes or significantly alters the existing vegetation. Examples of disturbances that can trigger secondary succession include forest fires, hurricanes, logging, or agricultural activities. Unlike primary succession, secondary succession begins with remnants of the previous community, such as seeds or roots, which allow the new community to establish more rapidly. The process of secondary succession often leads to the restoration of the original community, although with some changes in species composition.

Disturbance Regimes

Understanding the frequency, intensity, and spatial extent of disturbances is crucial for predicting the trajectory of succession and its impact on ecosystems. Disturbance regimes can be classified as either stochastic or non-stochastic. Stochastic disturbances occur randomly in space and time, such as lightning strikes or disease outbreaks. Non-stochastic disturbances, on the other hand, are predictable and often influenced by human activities, such as logging or land conversion for agriculture. The characteristics of a disturbance, such as its intensity and severity, also play a critical role in shaping the trajectory of succession. High-intensity disturbances, like severe wildfires, can remove an entire community and reset the successional process, while low-intensity disturbances, like small-scale windthrow events, may only affect a subset of the community.

Facilitation, Inhibition, and Tolerance

The trajectory of succession is influenced by the interactions among species in a community. Three key mechanisms that govern these interactions are facilitation, inhibition, and tolerance. Facilitation occurs when pioneer species modify the environment in a way that benefits subsequent species. For example, nitrogen-fixing plants can enrich the soil with nitrogen, making it more favorable for other plant species. Inhibition, on the other hand, occurs when the presence of one species inhibits the establishment of others. This can happen if a dominant species competes for resources or suppresses the growth of neighboring plants through chemical interactions. Tolerance refers to the ability of a species to tolerate and persist in a range of environmental conditions. Tolerant species can establish and grow under a wide range of disturbances and are often the final species to colonize a community.

Applications and Challenges

Succession and disturbance ecology have important applications in ecosystem management and restoration. By understanding the processes of succession and the factors that influence it, managers can develop strategies to restore ecosystems after disturbances or prevent undesirable successional trajectories. For example, in areas affected by severe wildfires, managers can promote the growth of fire-resistant species to ensure a more resilient and diverse community.

However, applying succession principles in a real-world context often poses challenges. One major challenge is the unpredictability of disturbances, especially in the face of climate change. Climate change can alter disturbance regimes, leading to more frequent and intense disturbances, which can disrupt successional processes. Additionally, the presence of invasive species can also complicate successional trajectories by outcompeting native species and altering the overall community structure. These challenges highlight the need for ongoing research and adaptive management strategies to address the complex dynamics of succession and disturbance ecology.

Case Study: Succession after a Volcanic Eruption

A volcanic eruption is a dramatic natural disturbance that can have a profound impact on the surrounding ecosystem. After a volcanic eruption, the land is covered with a layer of ash and lava, which creates a harsh and inhospitable environment for most organisms. However, over time, the ecosystem undergoes succession as pioneer species colonize the barren landscape and pave the way for the establishment of more diverse communities.

One such case study is the eruption of Mount St. Helens in Washington state, USA, in 1980. The eruption transformed the landscape into a barren wasteland, devoid of any vegetation. However, within weeks, pioneer species such as lichens and mosses were able to colonize the site, followed by grasses and herbaceous plants. These early colonizers gradually accumulated organic matter and nutrients, creating more favorable conditions for shrubs and small trees to establish. Within a few decades, a mature forest had developed, resembling the pre-eruption ecosystem.

This case study exemplifies the resilience of ecosystems and their ability to recover after major disturbances. It also highlights the importance of pioneer species in kick-starting the successional process and the importance of time in restoring ecosystems to their original state.

Exercises

1. Imagine a forest that has been clear-cut due to logging activities. Describe the process of secondary succession that would occur in this forest. Discuss how the

presence of remnant vegetation and adjacent habitats might influence the successional trajectory.

2. Research and identify a specific example of a disturbance that has occurred in an ecosystem of your choice. Describe the subsequent successional changes that have taken place and discuss the current state of the ecosystem. What challenges and opportunities are associated with managing this ecosystem in the context of succession and disturbance ecology?

3. Investigate a case study where the presence of invasive species has altered successional trajectories. Discuss the mechanisms by which invasive species influence succession and the potential ecological and economic impacts of these changes.

Further Reading

- Pickett, S. T. A., & White, P. S. (eds.) (1985). The Ecology of Natural Disturbance and Patch Dynamics. Academic Press. - Walker, L. (ed.) (2002). Ecosystems of Disturbed Ground. Elsevier Science. - Temperton, V. M., Hobbs, R. J., Nuttle, T., & Halle, S. (2004). Assembly Rules and Restoration Ecology: Bridging the Gap Between Theory and Practice. Island Press.

Ecological connectivity

Ecological connectivity refers to the degree to which ecosystems and their components are connected, allowing for the flow of energy, materials, and species between different habitats. It is a fundamental concept in ecology and conservation biology as it plays a crucial role in maintaining biodiversity, supporting ecosystem functioning, and promoting resilience in the face of environmental change.

1. Importance of ecological connectivity: Ecological connectivity is essential for the survival and persistence of species and populations. It allows for the movement of individuals between habitats, facilitating gene flow, colonization of new areas, and access to resources. By connecting different habitats, ecological connectivity promotes species interactions such as pollination, seed dispersal, and predator-prey relationships. It also enhances the ability of species to adapt to changing environmental conditions by providing opportunities for range shifts and population expansion.

2. Types of ecological connectivity: Ecological connectivity can occur at various spatial scales and through different pathways. Some common types of ecological connectivity include: - Landscape connectivity: This refers to the connectivity of habitats within a landscape, allowing for the movement of organisms across

different patches of habitat. - Corridor connectivity: Corridors are linear features, such as rivers, forest strips, or hedgerows, that connect larger patches of habitat and facilitate movement between them. - Matrix connectivity: The matrix refers to the surrounding habitat that separates patches of habitat. Matrix connectivity focuses on the permeability of the matrix for movement between patches. - Metapopulation connectivity: Metapopulations are a network of populations occupying different patches of habitat. Connectivity between these patches influences the dynamics and viability of metapopulations.

3. Challenges and threats to ecological connectivity: Ecological connectivity faces numerous challenges and threats, mainly due to human activities. Habitat fragmentation, caused by land-use changes such as urbanization and agriculture, disrupts ecological connectivity by creating barriers and isolating populations. Linear infrastructures, such as roads and railways, can act as barriers that impede the movement of individuals and fragment habitats. Climate change also poses a threat to ecological connectivity as it can alter the suitability of habitats and disrupt species' migration patterns.

4. Conservation strategies for enhancing ecological connectivity: Conservation efforts aim to enhance ecological connectivity and mitigate its fragmentation. Several strategies can be employed: - Creating and enhancing wildlife corridors: By establishing corridors that connect patches of habitat, wildlife can freely move across landscapes, maintaining gene flow and promoting biodiversity. - Implementing landscape planning: This involves considering ecological connectivity in land-use planning, ensuring that the design and location of human activities do not disrupt connectivity. - Restoring and protecting critical habitats: Conserving and restoring habitats of high ecological value, such as wetlands or riparian zones, can enhance connectivity by providing stepping stones for movement. - Managing and mitigating infrastructure impacts: Sensible infrastructure planning can reduce the negative impacts of linear infrastructure on ecological connectivity. Measures like wildlife bridges or underpasses can facilitate the movement of wildlife across roads. - Promoting transboundary collaboration: Many ecosystems span multiple jurisdictions. Collaborative efforts between different countries or regions are necessary to ensure the conservation of ecological connectivity at a broader scale.

5. Case study: Ecological connectivity in the Amazon rainforest: The Amazon rainforest is a globally significant ecosystem known for its high biodiversity. Maintaining ecological connectivity within this region is crucial for the survival of many species. The construction of roads and expansion of agriculture pose significant threats to connectivity in the Amazon. Efforts are underway to establish protected areas, create ecological corridors, and regulate land-use practices to

enhance connectivity. By promoting landscape connectivity and protecting critical habitats like river corridors, the conservation of ecological connectivity in the Amazon contributes to the long-term sustainability of this unique ecosystem.

In conclusion, ecological connectivity is a vital aspect of ecosystem health and resilience. By understanding and promoting connectivity, we can ensure the survival of species, maintain biodiversity, and safeguard the functioning of ecosystems. Conservation strategies that enhance ecological connectivity are crucial for mitigating the threats posed by habitat fragmentation and climate change, and for fostering sustainable environments for future generations.

Landscape Restoration

Landscape restoration is a crucial aspect of eco science and plays a vital role in preserving and enhancing the health and resilience of ecosystems. It involves the deliberate and planned process of restoring degraded landscapes to their original state or improving their ecological functionality. This section will delve into the principles, methods, and benefits of landscape restoration.

Principles of Landscape Restoration

To effectively restore a landscape, several principles need to be considered. These principles guide restoration practitioners in making informed decisions and developing strategies that promote the recovery of degraded ecosystems. Here are some key principles of landscape restoration:

1. **Ecological integrity**: Restoration efforts should aim to restore the ecological integrity of a landscape, ensuring that its natural functions are reinstated. This includes restoring biodiversity, ecological processes, and ecosystem services.

2. **Adaptive management**: Landscape restoration is a complex and dynamic process. Therefore, adaptive management approaches should be applied to account for uncertainties and changes during the restoration process. Regular monitoring and evaluation are necessary to adjust management actions accordingly.

3. **Holistic approach**: Restoration should consider the broader landscape context, including factors such as land use, hydrology, and connectivity. Taking a holistic approach promotes the integration of restoration efforts into larger ecological systems.

4. **Stakeholder engagement**: Involving local communities, landowners, and other stakeholders is essential for successful landscape restoration. Collaboration and engagement foster a sense of ownership, promote knowledge sharing, and enhance the long-term sustainability of restoration projects.

5. **Long-term perspective**: Landscape restoration is a time-intensive process that requires long-term commitment. Success is often measured over decades rather than short periods, highlighting the importance of sustained investment and monitoring.

Methods of Landscape Restoration

Landscape restoration involves a variety of methods and techniques, depending on the specific goals and conditions of the degraded landscape. Here are some commonly employed methods:

1. **Revegetation**: Restoring vegetation cover is crucial for the recovery of degraded landscapes. This can involve planting native tree and plant species, establishing meadows or grasslands, or encouraging natural regeneration through seed dispersal.

2. **Soil restoration**: Degraded soils often require restoration to improve fertility and nutrient cycling. Techniques such as soil amendment with organic matter, erosion control measures, and soil erosion prevention are employed to restore soil health.

3. **Water management**: Restoring hydrological processes is critical for landscape restoration. This can involve implementing measures to control erosion, restore natural water flows, and enhance water retention in the landscape.

4. **Habitat creation**: Restoration projects often aim to create or enhance specific habitats that have been lost or severely degraded. This can involve constructing wetlands, ponds, or other habitat features to support biodiversity and ecological functions.

5. **Invasive species management**: Invasive plant and animal species can significantly disrupt ecosystem dynamics and hinder the success of restoration efforts. Effective management strategies, including removal and control measures, are necessary to minimize their impact.

6. **Landscape connectivity**: Establishing ecological corridors and enhancing landscape connectivity are vital for biodiversity conservation and the movement of wildlife. Restoration efforts should consider the connectivity between different ecosystems to promote species dispersal and gene flow.

Benefits of Landscape Restoration

Landscape restoration brings about numerous benefits, both for ecosystems and human communities. Here are some key benefits:

1. **Biodiversity conservation:** Restoring degraded landscapes helps conserve biodiversity by providing habitat and food resources for a wide range of species. Restored landscapes can support the recovery of threatened and endangered species, contributing to their long-term survival.

2. **Ecosystem services:** Restored landscapes offer various ecosystem services such as clean air, water purification, carbon sequestration, and soil stabilization. These services are essential for human well-being and contribute to the overall resilience of ecosystems.

3. **Climate change mitigation:** Restored landscapes promote carbon sequestration and help mitigate climate change. By restoring vegetation cover and improving soil health, landscapes can act as carbon sinks, reducing greenhouse gas emissions and enhancing climate resilience.

4. **Water management:** Landscape restoration can improve water quality and quantity by reducing erosion, enhancing water infiltration, and restoring natural hydrological processes. Restored landscapes act as natural water filters, improving water supplies and reducing downstream flooding.

5. **Recreational and cultural values:** Restored landscapes provide opportunities for outdoor recreation, including hiking, birdwatching, and nature photography. Moreover, these landscapes often hold cultural and historical significance, preserving heritage and promoting tourism.

6. **Community engagement and well-being:** Landscape restoration engages local communities and promotes a sense of connection to the natural environment. Participating in restoration activities can enhance community well-being, mental health, and social cohesion.

Case Study: The Loess Plateau Restoration Project

The Loess Plateau Restoration Project in China is a notable example of landscape restoration on a large scale. The Loess Plateau, characterized by severe soil erosion and degradation, underwent restoration efforts starting in the late 1990s. The project incorporated various restoration techniques and approaches, including terracing, vegetation restoration, and soil erosion control measures.

The restoration project aimed to address the issues of soil erosion, water scarcity, and poverty in the region. Through the implementation of contour terracing, reforestation, and sustainable land management practices, the project achieved significant ecological and socio-economic outcomes.

The restoration efforts resulted in reduced soil erosion, increased vegetation cover, and improved water retention in the landscape. This led to increased agricultural productivity, improved water availability, and enhanced ecosystem services. Furthermore, the project lifted many local communities out of poverty and improved their livelihoods through the adoption of sustainable land management practices.

The success of the Loess Plateau Restoration Project highlights the importance of large-scale landscape restoration and its potential to address multiple environmental and social challenges. It serves as an inspiring example of how landscape restoration can lead to sustainable development and positive transformations.

Summary

Landscape restoration plays a crucial role in reversing the degradation of ecosystems and promoting sustainable development. By adhering to the principles of ecological integrity, adaptive management, holistic approaches, stakeholder engagement, and long-term perspectives, restoration efforts can effectively restore degraded landscapes.

Through methods such as revegetation, soil restoration, water management, habitat creation, invasive species management, and landscape connectivity, degraded landscapes can be transformed into vibrant and resilient ecosystems. The benefits of landscape restoration include biodiversity conservation, ecosystem services, climate change mitigation, improved water management, recreational and cultural values, and community well-being.

The case study of the Loess Plateau Restoration Project demonstrates the potential of landscape restoration to address environmental challenges while benefiting local communities. By valuing and implementing landscape restoration, we can work towards a sustainable and ecologically healthy future.

Marine Restoration

Marine restoration is a crucial aspect of eco science and plays a significant role in the conservation and protection of our oceans. It focuses on repairing, rehabilitating, and revitalizing marine ecosystems that have been damaged or degraded due to human activities, including pollution, overfishing, habitat destruction, and climate change impacts.

Importance of Marine Restoration

Marine ecosystems, such as coral reefs, seagrass beds, and mangrove forests, provide essential benefits to both marine species and humans. They support high biodiversity, serve as nurseries for many commercially important fish species, protect coastlines from erosion, and contribute to carbon storage. However, these ecosystems are under threat, and their decline has far-reaching ecological and socioeconomic consequences.

Marine restoration aims to reverse the negative impacts of human activities and restore the health and resilience of these ecosystems. By creating suitable conditions for the recovery of degraded areas, marine restoration contributes to the overall conservation and sustainable use of marine resources.

Approaches to Marine Restoration

Several approaches and techniques are employed in marine restoration, depending on the type and extent of the damage. Here are some commonly used methods:

1. Coral Reef Restoration: Coral reefs are highly sensitive to environmental changes and are facing widespread degradation globally. Restoration efforts involve coral propagation techniques, such as coral nurseries and transplantation, to enhance coral cover and diversity. Additionally, artificial reef structures can be deployed to provide substrate for new coral growth.

2. Seagrass Restoration: Seagrass meadows are crucial habitats for marine species and help maintain water quality. Restoration involves transplanting seagrass seeds or propagules into degraded areas and monitoring their growth and survival. Restored seagrass meadows can provide habitats for fish and other marine organisms, improve coastal water quality, and enhance carbon sequestration.

3. Mangrove Restoration: Mangroves are important coastal ecosystems that provide numerous ecological services, including coastal protection, carbon sequestration, and nursery grounds for fish and crustaceans. Restoration efforts focus on replanting mangrove seedlings in areas where they have been cleared or destroyed. It is essential to consider the hydrology and soil conditions to ensure successful mangrove restoration.

4. Shellfish Reef Restoration: Shellfish reefs, such as oyster and mussel beds, serve as important biodiversity hotspots and provide various ecosystem services. Restoration involves the creation of artificial structures using recycled or discarded shells to provide a substrate for shellfish settlement. As shellfish filter feed, they improve water quality by removing excess nutrients.

Challenges and Solutions

Marine restoration faces several challenges that can hinder its effectiveness. These challenges include limited funding and resources, inadequate monitoring and evaluation, and complex ecological interactions within marine ecosystems. However, innovative solutions and approaches can help overcome these challenges:

1.   Collaboration and Partnerships: By fostering collaboration among scientists, government agencies, local communities, and non-profit organizations, marine restoration efforts can benefit from shared expertise, resources, and funding.   Public-private partnerships and community engagement also play a crucial role in the success of restoration projects.

2. Adaptive Management: Adopting an adaptive management approach allows for flexibility in the design and implementation of restoration projects. It involves regularly monitoring and evaluating the progress of restoration efforts and adjusting strategies based on the collected data. This iterative process helps in identifying effective techniques and adapting to changing environmental conditions.

3.   Incorporating Climate Resilience: Climate change poses significant challenges to marine ecosystems, including rising sea temperatures, ocean acidification, and increased storm intensity.   Incorporating climate resilience considerations in restoration projects, such as selecting resilient species and designs, can enhance the long-term success of restoration efforts.

4.   Public Awareness and Education: Raising public awareness about the importance of marine ecosystems and the benefits of restoration is crucial for garnering support and volunteers.   Education programs can help communities understand the role they play in protecting marine resources and inspire them to take action.

Real-World Example: The Great Barrier Reef Restoration

The Great Barrier Reef in Australia is one of the world's most iconic and biodiverse coral reef ecosystems.   However, it is facing multiple threats, including coral bleaching events associated with climate change.   The Great Barrier Reef Restoration and Adaptation Program (GBRRAP) is a collaborative initiative aimed at restoring and enhancing the resilience of this World Heritage-listed area.

GBRRAP utilizes a range of innovative techniques, including coral larval restoration, underwater fans to re-suspend sediments, and marine cloud brightening to reduce water temperature. These methods focus on promoting coral recovery, improving water quality, and increasing the reef's resilience to climate change impacts.

This program emphasizes adaptive management, continual monitoring, and stakeholder engagement to ensure the effectiveness of restoration efforts.   By addressing multiple stressors and involving various stakeholders, GBRRAP serves as a model for large-scale marine restoration projects worldwide.

Conclusion

Marine restoration is a vital component of eco science and contributes to the conservation and sustainable use of marine ecosystems. By employing various approaches and innovative techniques, it is possible to restore and revitalize degraded marine areas. However, addressing the challenges and incorporating adaptive management and climate resilience considerations are crucial for the long-term success of restoration efforts. Through collaboration, public awareness, and education, we can work towards a future where our oceans are healthy, thriving, and resilient.

Wetland Restoration

Wetlands are valuable ecosystems that provide numerous ecological functions and services. They play a crucial role in water purification, flood control, carbon sequestration, and provide habitat for a wide range of plant and animal species. However, wetlands are under threat due to human activities such as drainage for agriculture, urbanization, and pollution. Wetland restoration aims to reverse the degradation and loss of wetland ecosystems by implementing various strategies and techniques.

Importance of Wetlands

Wetlands are one of the most productive ecosystems on Earth, supporting a diverse array of plant and animal life. They serve as breeding grounds, feeding areas, and nurseries for many species, including migratory birds, fish, amphibians, and reptiles. Wetlands act as natural filters, removing pollutants and excess nutrients from water bodies, thereby improving water quality. They also provide essential habitat for endangered and threatened species. Furthermore, wetlands have the ability to trap and store large amounts of carbon, contributing to climate change mitigation.

Threats to Wetlands

Wetlands face several threats that have resulted in their rapid decline worldwide. These threats include:

- **Drainage and Conversion** - Wetlands are often drained or converted for agriculture, infrastructure development, and urbanization. This alteration of hydrology and land use leads to loss of wetland habitat.

- **Destruction of Riparian Zones** - Riparian zones, the areas of transition between land and water, are critical for maintaining the health and integrity of wetlands. Their destruction, through activities such as deforestation and river channelization, disrupts the natural functioning of wetland ecosystems.

- **Pollution** - Wetlands are vulnerable to pollution from various sources, including industrial discharges, agricultural runoff, and domestic sewage. High levels of pollutants can negatively impact water quality and the health of wetland organisms.

- **Invasive Species** - Invasive plant and animal species can outcompete native species in wetlands, altering the balance of the ecosystem and reducing biodiversity.

The degradation of wetlands has led to a loss of ecosystem services and a decline in biodiversity. Recognizing these challenges, wetland restoration efforts have gained prominence in recent years.

Wetland Restoration Techniques

Wetland restoration involves the application of various techniques to recreate or enhance wetland ecosystems. These techniques can be broadly categorized into physical, chemical, and biological methods. Here are some commonly used approaches:

- **Re-establishing Hydrology** - Restoring the natural hydrological regime is crucial for wetland restoration. This may involve reintroducing surface water flow, controlling water levels, or redirecting water through channels and ditches. Restoring the hydrology helps recreate the specific wetland conditions required for different plant and animal communities.

- **Revegetation** - Planting native vegetation is essential for restoring wetland habitats. Native plants are adapted to local conditions and provide food and shelter for a variety of organisms. Care must be taken to select appropriate plant species that can thrive in the restored wetland environment.

+ **Removing Invasive Species** - Invasive plant species can outcompete native vegetation and disrupt ecological processes in wetlands. Removing and controlling invasive species is necessary to restore the balance of the ecosystem. This can be done through manual removal, herbicide application, or biological control methods.

+ **Creating or Enhancing Wildlife Habitat** - To encourage the return of wildlife to restored wetlands, creating or enhancing specific habitat features is important. This can include constructing nesting sites for birds, creating fish spawning areas, or building artificial structures such as beaver dams to promote biodiversity.

+ **Monitoring and Adaptive Management** - Monitoring the progress of wetland restoration projects is crucial to assess their success and make necessary adjustments. Key indicators, such as water quality, vegetation growth, and species diversity, can be monitored to evaluate the effectiveness of restoration activities. Adaptive management allows for ongoing learning and refinement of restoration approaches based on observed outcomes.

Case Study: Everglades Restoration

The restoration of the Everglades in Florida, USA, provides an excellent example of a large-scale wetland restoration effort. The Everglades is a unique wetland ecosystem that has been significantly altered over the past century due to drainage for agricultural purposes and water diversion for urban and industrial use.

The restoration plan for the Everglades focuses on restoring the natural hydrological patterns of the wetland system. This involves modifying canals and water control structures to mimic the original flow of water. By redirecting water and reestablishing the natural water depths, the restoration aims to revive the wetland's unique mix of freshwater and brackish habitats.

In addition to hydrological restoration, revegetation efforts are underway in the Everglades. Native plant species are being reintroduced to restore the ecosystem's balance and provide suitable habitat for the diverse array of wildlife that rely on the wetland.

The long-term success of the Everglades restoration project relies on collaboration between federal, state, tribal, and local stakeholders. Monitoring and research are integral components of the project, helping to evaluate the effectiveness of restoration actions and guide adaptive management.

Challenges and Considerations

Wetland restoration projects face several challenges and considerations that need to be taken into account:

- **Hydrological Complexity** - Restoring the natural hydrological regime of a wetland can be challenging due to the complexity of hydrological processes and dependencies on external factors such as rainfall patterns and upstream activities.

- **Land Ownership and Regulation** - Wetland restoration often involves multiple landowners and regulatory agencies. Coordinating efforts and ensuring compliance with regulations can be a complex task.

- **Ecosystem Connectivity** - Wetlands are interconnected with other ecosystems, such as rivers and coastal areas. Restoration efforts need to consider these ecological connections to ensure the broader health and functionality of the landscape.

- **Climate Change** - The impacts of climate change, such as changing precipitation patterns and sea-level rise, can affect the success of wetland restoration projects. Adaptive management strategies that account for climate change scenarios are essential.

In conclusion, wetland restoration is a critical endeavor to conserve and restore the ecological functions and services provided by wetlands. It requires a multidisciplinary approach, involving hydrology, ecology, and collaboration between stakeholders. Successful restoration projects have the potential to revive damaged wetland ecosystems, promote biodiversity, and enhance the resilience of landscapes in the face of environmental change.

Forest Restoration

Forest restoration is the process of restoring degraded or deforested areas to their natural state or to a state that closely resembles the original forest ecosystem. It involves the implementation of various techniques and strategies to enhance biodiversity, improve ecosystem services, and promote sustainability. Forest restoration plays a crucial role in mitigating climate change, conserving biodiversity, and improving the overall health of ecosystems.

Importance of Forest Restoration

Forests are vital components of the Earth's ecosystems, providing numerous benefits to both humans and the environment. They regulate the climate by absorbing carbon dioxide and releasing oxygen, thus helping to mitigate climate change. Forests also play a key role in the water cycle, by capturing and storing rainwater, preventing soil erosion, and maintaining healthy watersheds. Additionally, forests support biodiversity by providing habitats for a wide range of plant and animal species.

However, forests worldwide are facing significant threats, including deforestation, habitat degradation, and the spread of invasive species. These activities lead to the loss of biodiversity, increased carbon emissions, and negative impacts on local communities that depend on forests for their livelihoods. Forest restoration is a proactive approach to address these challenges and restore the ecological functions and benefits provided by forests.

Forest Restoration Techniques

1. Reforestation: Reforestation involves planting trees in areas that have been deforested or degraded. The selection of tree species is crucial and should consider the site's ecological conditions and the desired forest structure. Reforestation can be done through natural regeneration or by planting saplings.

2. Assisted Natural Regeneration: This technique promotes the natural regrowth of forests by removing competing vegetation and providing favorable conditions for native tree species to emerge. It is a cost-effective and environmentally friendly approach, particularly in areas with a viable seed bank.

3. Agroforestry: Agroforestry is a land use system that combines tree cultivation with agricultural practices. It promotes sustainable land management by diversifying income sources, improving soil fertility, and providing shade and windbreaks for crops. This approach enhances ecosystem services while ensuring the economic viability of land use.

4. Silviculture: Silviculture involves the management and cultivation of forests to achieve specific objectives, such as timber production or biodiversity conservation. It includes practices such as selective logging, thinning, and tree planting, with the goal of improving forest health and productivity.

5. Ecological Restoration: Ecological restoration aims to rehabilitate degraded ecosystems by mimicking natural ecological processes. It involves the removal of invasive species, active soil restoration, and the reestablishment of native

vegetation. This approach focuses on restoring the entire ecosystem, including soil health, hydrological processes, and wildlife habitats.

Challenges and Solutions

1. Fragmentation: The fragmentation of forests due to human activities poses a significant challenge to forest restoration efforts. Fragmentation reduces habitat connectivity, leading to decreased biodiversity and ecosystem resilience. To address this, landscape-scale restoration plans should be implemented, integrating large patches of forest and creating corridors for species movement.

2. Invasive Species: Invasive species can outcompete native vegetation, disrupt natural ecosystem processes, and hinder the success of forest restoration projects. Effective management strategies, such as early detection and rapid response, are essential to control invasive species and prevent their spread. This may include manual removal, chemical treatments, or biological control methods.

3. Climate Change: Climate change presents challenges for forest restoration, as changing environmental conditions may affect the survival and growth of tree species. To enhance the resilience of restored forests, adaptive management practices should be employed, such as planting climate-resilient tree species and promoting genetic diversity within tree populations.

4. Socioeconomic Considerations: Successful forest restoration requires the involvement and support of local communities and stakeholders. Collaborative decision-making processes, capacity building, and equitable benefit-sharing mechanisms should be established to ensure the long-term sustainability of restoration initiatives.

Case Study: The Atlantic Forest Restoration Program

The Atlantic Forest Restoration Program in Brazil is one of the largest and most successful forest restoration initiatives in the world. The Atlantic Forest, a biodiversity hotspot, has been greatly impacted by deforestation and habitat degradation. The program aims to restore and conserve the forest by implementing a combination of restoration techniques.

The program focuses on native tree species selection, ensuring the ecological integrity of the restored forests. It also emphasizes socioeconomic aspects, working closely with local communities to support their participation in restoration activities. The program has successfully restored thousands of hectares of degraded land, contributing to the conservation of biodiversity, carbon sequestration, and sustainable development in the region.

Conclusion

Forest restoration is a vital component of eco science and plays a crucial role in achieving future sustainability. By restoring degraded forests, we can mitigate climate change, conserve biodiversity, and enhance ecosystem services. The implementation of various restoration techniques, along with addressing challenges such as fragmentation and invasive species, is essential for successful restoration outcomes. Forest restoration initiatives, such as the Atlantic Forest Restoration Program, demonstrate the significance of collaborative efforts and integrated approaches in achieving effective and lasting results.

Grassland Restoration

Grasslands are important ecosystems that provide a variety of ecological services, such as carbon sequestration, nutrient cycling, and habitat for a wide range of plant and animal species. However, they are also one of the most threatened ecosystems worldwide, with vast areas being converted to agriculture, urban development, and other land uses. Grassland restoration aims to reverse these negative impacts and restore degraded grasslands back to their functioning state. In this section, we will explore the principles, methods, and challenges associated with grassland restoration.

Principles of Grassland Restoration

Restoring grasslands requires a comprehensive understanding of their ecological dynamics and the factors that contribute to their degradation. Some of the key principles of grassland restoration include:

1. **Understanding ecosystem processes:** Restoring grasslands involves understanding the key ecosystem processes that drive their functioning, such as nutrient cycling, seed dispersal, and plant-animal interactions. This understanding helps guide restoration efforts to ensure the establishment of resilient and self-sustaining grassland communities.

2. **Native species selection:** Native plant species are the building blocks of grassland ecosystems. Restoring grasslands involves selecting appropriate native species that are adapted to local environmental conditions and can provide the necessary ecological functions. Native species also support the survival of native fauna, including pollinators and herbivores.

3. **Seed collection and propagation**: Collecting and propagating seeds of native grassland plants is a critical step in grassland restoration. Seeds can be collected from remnant grassland patches or from seed banks that store genetically diverse and locally adapted seed collections. The propagation of native plant species ensures a diverse and resilient plant community in the restored grassland ecosystem.

4. **Soil preparation and management**: Soil preparation is often required to create favorable conditions for plant establishment. This may involve removing invasive species, controlling erosion, and improving soil fertility. The use of sustainable soil management practices, such as adding organic matter and minimizing soil disturbance, is crucial for long-term grassland restoration success.

5. **Monitoring and adaptive management**: Regular monitoring of restored grasslands is essential to assess the success of restoration efforts and make necessary adjustments. Adaptive management involves learning from restoration outcomes and using that information to refine restoration approaches over time. Monitoring also helps detect potential threats, such as invasive species or climate change impacts, allowing for timely intervention.

Methods of Grassland Restoration

Grassland restoration can be achieved through a variety of methods. The choice of method depends on factors such as the extent of degradation, available resources, and the specific objectives of the restoration project. Some common methods include:

1. **Direct seeding**: Direct seeding involves sowing native grassland seeds directly onto the restoration site. This method is cost-effective and can result in the establishment of a diverse plant community. Successful direct seeding requires proper seed collection, seed dormancy breaking techniques, and careful consideration of seed dispersal mechanisms.

2. **Transplanting**: Transplanting involves moving intact sod or individual plants from healthy grassland areas to degraded sites. This method can be useful for establishing native grassland species that are difficult to establish using seed-based methods. Transplanting requires careful site preparation, proper handling of transplants, and monitoring to ensure their survival and integration into the restored ecosystem.

3. **Grazing management**: Grazing management plays a crucial role in grassland restoration, especially in areas where grazing is an important ecological process. Proper grazing management, such as rotational grazing or rest periods, can help control invasive species, promote the establishment of native plants, and enhance biodiversity in restored grasslands. This method requires close collaboration with landowners and stakeholders.

4. **Fire management**: Fire has long been a natural process shaping grassland ecosystems. Controlled burning can be used as a restoration tool to remove invasive species, stimulate the germination of native seeds, and promote the growth of fire-adapted grassland species. However, fire management must be carefully planned and implemented to ensure the safety of surrounding areas and minimize negative impacts on sensitive species.

Challenges and Future Directions

Grassland restoration faces several challenges that need to be addressed for successful ecosystem recovery. Some of these challenges include:

1. **Invasive species**: Invasive plant species can pose significant challenges to grassland restoration efforts, outcompeting native species and impeding ecosystem recovery. Effective invasive species control methods, such as herbicide application, manual removal, or biological control, need to be implemented as part of restoration strategies.

2. **Fragmentation**: Fragmentation of grassland habitats due to land-use change can hinder the success of restoration efforts. Restoring ecological connectivity through the creation of wildlife corridors or the establishment of buffer zones can help facilitate the movement of species and enhance the long-term viability of restored grasslands.

3. **Climate change**: Climate change poses risks to grassland ecosystems, including altered precipitation patterns, increased frequency of extreme weather events, and shifts in plant community composition. Incorporating climate change adaptation strategies, such as selecting climate-resilient plant species and implementing appropriate water management practices, is crucial for the success of grassland restoration.

4. **Community engagement**: Grassland restoration is most successful when local communities are actively involved in decision-making and implementation processes. Engaging local stakeholders, landowners, and

indigenous communities can foster a sense of ownership and ensure the long-term sustainability of restoration initiatives.

In conclusion, grassland restoration plays a vital role in conserving biodiversity, mitigating climate change, and preserving ecosystem services. By understanding the principles, employing suitable restoration methods, and addressing challenges, we can work towards the recovery and resilience of degraded grassland ecosystems. Efforts in grassland restoration not only contribute to the conservation of a threatened ecosystem but also provide opportunities for scientific research, education, and ecotourism, fostering a deeper connection between humans and the natural world.

Socio-economic benefits of restoration

Restoration ecology focuses not only on the ecological aspects of repairing degraded ecosystems but also on the socio-economic benefits that can be derived from restoration efforts. Restoration projects have the potential to create numerous opportunities that can benefit local communities, economies, and societies as a whole. In this section, we will explore some of the socio-economic benefits that can be achieved through ecosystem restoration.

Enhanced Ecosystem Services

One of the primary socio-economic benefits of ecosystem restoration is the enhancement of various ecosystem services. Ecosystem services are the benefits that people obtain from ecosystems, including provisioning services (e.g., food, water, timber), regulating services (e.g., climate regulation, water purification), cultural services (e.g., recreation, spiritual value), and supporting services (e.g., soil formation, nutrient cycling).

Restoration efforts can lead to the recovery of ecosystem services that have been degraded due to habitat destruction, pollution, or other human activities. For example, restoring wetlands can improve water quality, reduce flood risk, and provide habitat for wildlife, thereby enhancing the provision of clean water, flood control, and biodiversity conservation services.

Job Creation

Ecosystem restoration projects have the potential to generate employment opportunities, thereby contributing to local economies and reducing unemployment rates. Restoration activities often require a wide range of skills and expertise, including ecological restoration, landscape design, construction, and monitoring.

For instance, the restoration of degraded forests can create jobs in activities such as tree planting, forest management, and forest product manufacturing. Similarly, the restoration of degraded river systems can provide employment opportunities in river restoration, sediment management, and fisheries enhancement. These jobs not only improve the livelihoods of local communities but also promote economic growth in the region.

Tourism and Recreation

Restored ecosystems can attract tourists and nature enthusiasts, which can have significant economic benefits for local communities. Natural areas that have been successfully restored often provide opportunities for outdoor recreational activities such as hiking, bird-watching, camping, and fishing.

For example, the restoration of coral reefs can create opportunities for snorkeling and scuba diving, attracting tourists interested in experiencing marine biodiversity. Similarly, the restoration of degraded urban spaces, such as abandoned industrial sites, into green parks and gardens can enhance the attractiveness of cities and promote eco-tourism.

Property Value Enhancement

Ecosystem restoration can positively impact the value of nearby properties and real estate. Restored natural areas, such as wetlands, forests, and parks, often enhance the aesthetic appeal of the surrounding landscapes and can increase property values in the area.

Studies have shown that properties located near green spaces or natural areas command higher prices compared to those in degraded or urbanized areas. This increase in property value can contribute to the local economy and provide additional income for property owners.

Climate Change Mitigation and Adaptation

Ecosystem restoration can play a crucial role in climate change mitigation and adaptation. Restored ecosystems, such as forests and wetlands, have the potential to sequester carbon dioxide and mitigate greenhouse gas emissions. They also contribute to climate change adaptation by reducing the vulnerability of communities to extreme weather events such as floods and storms.

Restoration activities that focus on planting trees, restoring mangroves, or creating green infrastructure can provide multiple benefits in terms of climate change mitigation, water regulation, and disaster risk reduction. These benefits can, in turn, reduce the economic burden associated with climate change impacts.

Community Engagement and Education

Restoration projects offer opportunities for community engagement and education, fostering a sense of ownership and stewardship among local residents. Getting communities involved in restoration activities not only creates a sense of pride but also increases awareness and understanding of the importance of conserving and restoring ecosystems.

Education programs and workshops related to restoration can empower community members with new skills and knowledge, providing them with additional economic and employment opportunities. Moreover, restoring ecosystems in urban areas can contribute to community identity, social cohesion, and overall well-being.

In conclusion, ecosystem restoration holds immense potential for socio-economic benefits. By enhancing ecosystem services, creating employment, attracting tourists, increasing property values, mitigating and adapting to climate change, and engaging local communities, restoration projects can contribute to sustainable development and improve the quality of life for both present and future generations. Such projects exemplify the interconnectedness between ecological health and socio-economic well-being, highlighting the importance of integrating restoration into policy and governance frameworks.

Sustainable Development

Sustainable Development Goals

The concept of sustainable development recognizes the need to balance economic, social, and environmental concerns in order to meet the needs of the present without compromising the ability of future generations to meet their own needs. In 2015, the United Nations adopted a set of 17 Sustainable Development Goals (SDGs) as part of the 2030 Agenda for Sustainable Development. These goals provide a holistic framework for addressing global challenges and promoting sustainable development worldwide.

Overview of the Sustainable Development Goals

The Sustainable Development Goals cover a wide range of interconnected issues, including poverty eradication, climate action, gender equality, clean energy, sustainable cities, responsible consumption and production, and biodiversity conservation, to name just a few. Each goal is important in its own right, but they are also deeply interconnected, and progress in one goal can positively impact others.

Key Sustainable Development Goals

Goal 1: No Poverty – End poverty in all its forms everywhere. This includes eradicating extreme poverty and reducing the proportion of people living in poverty.

Goal 2: Zero Hunger – Achieve food security, improve nutrition, and promote sustainable agriculture. This goal aims to end hunger, achieve food security and improved nutrition, and promote sustainable agriculture practices.

Goal 3: Good Health and Well-being – Ensure healthy lives and promote well-being for all at all ages. This goal focuses on reducing mortality rates, preventing diseases, and improving access to essential healthcare services.

Goal 4: Quality Education – Ensure inclusive and equitable quality education and promote lifelong learning opportunities for all. This goal aims to provide quality education for all, promote literacy and numeracy, and increase access to vocational training and higher education.

Goal 5: Gender Equality – Achieve gender equality and empower all women and girls. This goal aims to eliminate gender-based discrimination and violence, ensure equal opportunities and rights for women and girls, and promote women's empowerment.

Goal 6: Clean Water and Sanitation – Ensure availability and sustainable management of water and sanitation for all. This goal focuses on providing access to clean drinking water and basic sanitation facilities, promoting water efficiency, and improving water quality.

Goal 7: Affordable and Clean Energy – Ensure access to affordable, reliable, sustainable, and modern energy for all. This goal aims to promote the use of renewable energy sources, improve energy efficiency, and increase access to affordable and clean energy.

Goal 8: Decent Work and Economic Growth – Promote sustained, inclusive, and sustainable economic growth, full and productive employment, and decent work for all. This goal focuses on creating decent jobs, promoting entrepreneurship, and ensuring fair and equal opportunities for all.

Goal 9: Industry, Innovation, and Infrastructure – Build resilient infrastructure, promote inclusive and sustainable industrialization, and foster innovation. This goal aims to promote sustainable industries, improve infrastructure, and enhance technological innovation.

Goal 10: Reduced Inequalities – Reduce inequality within and among countries. This goal focuses on reducing income inequality, promoting social inclusion, and ensuring equal opportunities for all.

Goal 11: Sustainable Cities and Communities – Make cities and human settlements inclusive, safe, resilient, and sustainable. This goal aims to ensure access to affordable housing, promote sustainable urbanization, and improve urban planning and management.

Goal 12: Responsible Consumption and Production – Ensure sustainable consumption and production patterns. This goal focuses on promoting sustainable lifestyles, reducing waste generation, and increasing the efficiency of resource use.

Goal 13: Climate Action – Take urgent action to combat climate change and its impacts. This goal aims to promote climate mitigation and adaptation measures, raise awareness, and strengthen resilience to climate-related hazards.

Goal 14: Life Below Water – Conserve and sustainably use the oceans, seas, and marine resources for sustainable development. This goal focuses on protecting marine ecosystems, reducing marine pollution, and promoting sustainable fishing practices.

Goal 15: Life on Land – Protect, restore, and promote sustainable use of terrestrial ecosystems, sustainably manage forests, combat desertification, halt and reverse land degradation, and halt biodiversity loss. This goal aims to preserve and restore terrestrial ecosystems, promote sustainable land use practices, and conserve biodiversity.

Goal 16: Peace, Justice, and Strong Institutions – Promote peaceful and inclusive societies for sustainable development, provide access to justice for all, and build effective, accountable, and inclusive institutions at all levels. This goal focuses on promoting peace, ensuring access to justice, and strengthening institutions for sustainable development.

Goal 17: Partnerships for the Goals – Strengthen the means of implementation and revitalize the global partnership for sustainable development. This goal aims to enhance international cooperation, promote technology transfer, and strengthen the global partnership for sustainable development.

Examples and Case Studies

To illustrate the practical application of the Sustainable Development Goals, let's consider a few examples:

Example 1: Goal 7 - Affordable and Clean Energy In a developing country where many rural communities lack access to electricity, the government invests in renewable energy solutions such as solar power systems and mini-grids. By providing affordable and clean energy to these communities, the government not only addresses Goal 7 but also contributes to poverty reduction, improved health and education outcomes, and sustainable economic development.

Example 2: Goal 11 - Sustainable Cities and Communities In a rapidly urbanizing city, the local government recognizes the importance of sustainable urban planning. They prioritize the development of public transportation systems, invest in green infrastructure, and promote mixed-use development to reduce commuting distances and promote walkability. By focusing on Goal 11, the city improves air quality, reduces traffic congestion, and enhances the quality of life for its residents.

Example 3: Goal 15 - Life on Land In a country facing significant deforestation and biodiversity loss, a conservation organization partners with local communities to establish community-based conservation areas. Through sustainable land use practices, reforestation efforts, and wildlife conservation initiatives, they contribute to Goal 15 while also empowering local communities and safeguarding their cultural heritage.

Challenges and Future Directions

Implementing the Sustainable Development Goals is not without its challenges. Some of the key challenges include:

- Funding: Adequate financial resources are crucial for achieving the Goals, but financing sustainable development initiatives remains a major challenge.

- Monitoring and Evaluation: Measuring progress towards the Goals and ensuring accountability requires effective monitoring and evaluation systems.

- Policy Integration: Coordinating policies across different sectors and levels of government is essential for achieving the Goals, but it can be complex and challenging.

- Stakeholder Engagement: Engaging diverse stakeholders, including governments, civil society organizations, and the private sector, is crucial for successful implementation but requires effective collaboration and cooperation.

- Data Availability: Access to reliable and disaggregated data is essential for evidence-based decision-making, but data gaps and limitations can hinder progress.

Looking ahead, it is important to continue promoting awareness and advocacy for the Sustainable Development Goals. Governments, businesses, civil society organizations, and individuals all have a role to play in implementing the Goals and working towards a more sustainable future.

Conclusion

The Sustainable Development Goals provide a comprehensive framework for addressing the world's most pressing challenges and promoting sustainable development. Achieving these Goals requires collective action, political will, and innovative solutions. By focusing on interconnected issues such as poverty, inequality, climate change, and biodiversity conservation, we can create a more inclusive, resilient, and sustainable future for all. Let us embrace the Sustainable Development Goals as a roadmap towards a better tomorrow.

Poverty Eradication

Poverty eradication is a significant goal within sustainable development. It entails addressing the root causes of poverty and ensuring that all individuals have access to basic needs and opportunities for economic and social advancement. In this section, we will explore the principles, challenges, and strategies related to poverty eradication.

Understanding Poverty

Poverty is a multifaceted issue that encompasses not only the lack of income but also limited access to education, healthcare, clean water, sanitation, and other essential resources and services. It is a complex web that traps individuals and communities in a cycle of deprivation and marginalization.

Causes of Poverty

Poverty arises from a combination of economic, social, and political factors. Some common causes include unequal distribution of wealth, lack of access to education and employment opportunities, gender inequality, discrimination, conflict and violence, and inadequate social protection systems. These factors often interact and reinforce each other, perpetuating poverty.

Goals for Poverty Eradication

To combat poverty effectively, it is essential to set clear and measurable goals. The United Nations has established the Sustainable Development Goal 1 (SDG 1) to end poverty in all its forms by 2030. This goal encompasses eradicating extreme poverty, reducing poverty rates, and ensuring social protection systems for the most vulnerable.

Strategies for Poverty Eradication

1. Promoting Inclusive Economic Growth   Inclusive economic growth plays a critical role in poverty eradication. It involves fostering an enabling environment for sustainable and equitable economic development that provides opportunities for all. This can be achieved through:
- Creating decent jobs and ensuring fair wages. - Enhancing access to financial services and credit for the poor.   - Facilitating economic diversification and entrepreneurship development. - Promoting equitable access to markets and trade opportunities.

2. Investing in Human Capital Development   Investing in human capital is crucial to break the cycle of poverty. Key areas of focus are:
- Access to quality education, including early childhood education, primary, secondary, and tertiary education.   - Improving healthcare systems to ensure universal access to healthcare services. - Enhancing skills training and vocational education to improve employability.   - Empowering women and girls through gender-responsive policies and programs.

3. Strengthening Social Protection Systems   Social protection systems provide a safety net for the most vulnerable populations and help break the cycle of poverty. Key components include:
- Establishing comprehensive social protection programs, such as cash transfers, social pensions, and health insurance. - Strengthening social safety nets to provide support during economic shocks and disasters. - Promoting access to affordable housing, clean water, sanitation, and energy.

4. Fostering Inclusive Governance and Institutions   Efficient and inclusive governance and institutions are crucial to ensuring effective poverty eradication efforts. Strategies include:
- Promoting transparency, accountability, and participation in decision-making processes. - Implementing anti-corruption measures to prevent misappropriation of resources. - Strengthening the rule of law and access to justice for marginalized populations.   - Empowering marginalized communities and ensuring their representation in decision-making bodies.

Case Study: Conditional Cash Transfer Programs

Conditional Cash Transfer (CCT) programs are social protection initiatives that have been successful in poverty eradication efforts. They provide cash transfers to low-income families on the condition that they meet certain criteria, such as sending their children to school, ensuring regular health check-ups, and attending nutrition workshops.

One example is the Bolsa Familia program in Brazil, which has lifted millions of families out of poverty. By combining cash transfers with education and health incentives, the program has improved school enrollment rates, reduced child labor, and improved access to healthcare services for vulnerable populations.

Challenges in Poverty Eradication

Despite the progress made in poverty eradication efforts, numerous challenges persist, including:

1. Inequality: Economic disparities and unequal distribution of resources hinder poverty eradication efforts. Addressing income inequality is crucial to creating an equitable society.

2. Climate Change: Climate change disproportionately affects vulnerable populations. Disasters, such as droughts and floods, can exacerbate poverty by destroying livelihoods and infrastructure.

3. Conflict and Fragility: Poverty is often concentrated in conflict-affected and fragile states. Establishing peace, stability, and good governance are essential for sustainable poverty eradication.

4. Gender Inequality: Women and girls are disproportionately affected by poverty due to discrimination and limited access to resources and opportunities. Gender-responsive policies and targeted interventions are necessary to address this inequality.

Unconventional Approach: Universal Basic Income

One unconventional approach to poverty eradication is the concept of Universal Basic Income (UBI). UBI proposes giving every individual, regardless of their socioeconomic status, a guaranteed income sufficient to meet their basic needs. It has the potential to reduce poverty, ensure social protection, and empower individuals to pursue their goals.

While UBI has both proponents and critics, pilot programs and experiments are being conducted in various countries to test its feasibility, impact, and implications for social and economic development.

Conclusion

Poverty eradication is a critical component of sustainable development. It requires addressing the multifaceted nature of poverty, implementing comprehensive strategies, and promoting inclusive growth, human capital development, and social protection systems. Overcoming challenges and exploring unconventional approaches will be key to achieving the goal of ending poverty and building a more equitable and sustainable world.

Resources for Further Reading

1. The World Bank: *World Development Report 2020: Trading for Development in the Age of Global Value Chains*

2. United Nations Development Programme: *Human Development Report 2020: The Next Frontier - Human Development and the Anthropocene*

3. International Labour Organization: *The World of Work Report 2019: Transforming Jobs to End Poverty*

4. Banerjee, A., Duflo, E. (2011). *Poor Economics: A Radical Rethinking of the Way to Fight Global Poverty*

5. Moyo, D. (2009). *Dead Aid: Why Aid Is Not Working and How There Is a Better Way for Africa*

Gender Equality

Gender equality is a fundamental principle that underpins sustainable development. It refers to the equal rights, opportunities, and responsibilities of both men and women, without any discrimination based on gender. Achieving gender equality is not only a moral imperative but also crucial for social, economic, and environmental progress.

The Importance of Gender Equality

Gender equality is crucial for creating a sustainable world. It promotes social justice, reduces poverty and inequality, and enhances economic growth. When women and girls have equal access to education, healthcare, employment, and leadership opportunities, they can contribute more effectively to society, drive innovation, and participate in decision-making processes.

Research shows that gender equality leads to positive outcomes in various sectors. For example, increasing women's participation in the labor market can boost productivity and economic growth. Moreover, gender equality is linked to improved health outcomes, lower rates of violence, and better environmental stewardship.

Barriers to Gender Equality

Despite progress in recent years, gender inequality persists in many societies. Discrimination, stereotypes, cultural norms, and unequal power relations continue to hinder women's equal participation and decision-making.

One major barrier to gender equality is the persistence of gender stereotypes and traditional gender roles. Preconceived notions about gender limit opportunities for women and men, confining them to certain roles and expectations. This can affect education, employment, and leadership opportunities, perpetuating inequality.

Violence against women is another significant barrier. Gender-based violence, including domestic violence, sexual harassment, and femicide, undermines women's safety and well-being. It also reinforces power imbalances and perpetuates gender inequality.

Despite progress, women continue to face challenges in accessing and controlling resources, such as land, credit, and technology. This limits their economic empowerment and exacerbates gender inequality.

Promoting Gender Equality

Promoting gender equality requires a multi-faceted approach, addressing both structural and cultural barriers. Here are some key strategies and initiatives in achieving gender equality:

Policy and Legal Reforms Governments should enact and enforce laws and policies that protect women's rights and promote gender equality. These measures can include legislation against gender-based violence, equal pay laws, and policies supporting work-life balance and parental leave.

Education and Empowerment Investing in girls' education is crucial for breaking the cycle of gender inequality. Providing quality education that promotes gender equality helps girls develop the skills and knowledge necessary for economic empowerment and leadership roles.

Women's Economic Empowerment Enhancing women's access to economic opportunities and financial resources is vital for achieving gender equality. This can be done through initiatives such as microfinance programs, vocational training, and support for women-led businesses.

Ending Violence Against Women   Efforts to eliminate gender-based violence and ensure women's safety are essential. This includes providing support services for survivors, raising awareness, and promoting gender-responsive justice systems.

Promoting Women's Leadership and Political Participation   Increasing women's representation in decision-making roles is crucial for promoting gender equality. This can be achieved through quotas, affirmative action, and targeted leadership training programs.

Challenging Gender Stereotypes and Social Norms   Promoting gender equality requires challenging harmful stereotypes and norms that perpetuate inequality. This can be done through awareness campaigns, media engagement, and educational initiatives that promote gender equality and challenge traditional gender roles.

Case Study: Closing the Gender Gap in Renewable Energy

The renewable energy sector offers significant potential for promoting gender equality. However, women continue to be underrepresented in this sector, particularly in technical and leadership roles.

To address this gap, organizations and governments are implementing initiatives to empower women in renewable energy. For example, the "Women in Wind" program, launched by the Global Wind Energy Council, aims to increase the participation of women in the wind industry through mentorship programs, networking events, and advocacy.

In addition, the "Clean Energy Education and Empowerment (C3E)" initiative, led by the International Energy Agency, highlights the importance of gender diversity in clean energy and promotes women's leadership in this field.

By closing the gender gap in renewable energy, not only can we achieve gender equality but also tap into a wider pool of talent and perspectives, fostering innovation and driving sustainable energy transitions.

Conclusion

Gender equality is a fundamental principle for building a sustainable future. It is essential for social progress, economic growth, and environmental stewardship.

Overcoming barriers to gender equality requires comprehensive strategies, including policy reforms, education and empowerment initiatives, and challenging gender stereotypes. By promoting gender equality, we can create a more inclusive and sustainable world for all.

Clean energy

Clean energy is a vital aspect of eco science and sustainable development. As the world continues to face challenges related to climate change, pollution, and the depletion of fossil fuels, finding clean and renewable sources of energy is of utmost importance. In this section, we will explore various forms of clean energy and the technologies associated with them.

Renewable energy sources

Renewable energy sources are those that can be regenerated or replenished naturally. These sources have a minimal impact on the environment and contribute to reducing greenhouse gas emissions. Some of the major renewable energy sources include solar power, wind energy, hydroelectric power, geothermal energy, and biomass.

Solar energy

Solar energy is derived from the sun's radiation and can be harnessed in various ways. One of the most common methods is through photovoltaic (PV) cells, which convert sunlight directly into electricity. Solar thermal systems, on the other hand, use mirrors or lenses to concentrate solar energy to generate heat, which can be used for heating or electricity production.

The use of solar energy has several advantages. It is a clean and abundant source of energy, and the technology is becoming more cost-effective. Moreover, solar panels can be installed on both residential and commercial buildings, reducing dependency on the grid and allowing for decentralized energy production.

However, there are challenges to widespread solar adoption. One challenge is the intermittency of solar energy due to variations in sunlight availability. Energy storage technologies such as batteries can help address this issue by storing excess energy for use when the sun is not shining. Additionally, the manufacturing and disposal of solar panels can have environmental impacts, highlighting the need for sustainable practices within the solar industry.

Wind energy

Wind energy is another significant source of clean power. Wind turbines convert the kinetic energy of the wind into electrical energy. As the wind blows, it causes the turbine blades to rotate, which drives a generator. Wind farms, consisting of multiple turbines, are often set up in areas with consistent and strong winds, such as coastal regions or open plains.

One of the advantages of wind energy is its scalability. Wind farms can range from small-scale installations to large offshore projects, contributing to the diversification of the energy mix. Moreover, wind energy does not consume water, making it an environmentally friendly alternative to conventional power generation methods.

However, wind energy also poses challenges. Firstly, wind is an intermittent resource, and energy storage systems are required to address fluctuations in power generation. Secondly, the visual and noise impacts of wind turbines can be a concern for local communities, emphasizing the importance of proper planning and stakeholder engagement in the development of wind projects.

Hydroelectric power

Hydroelectric power harnesses the energy of flowing or falling water to generate electricity. It is one of the oldest and most widely used sources of renewable energy. Hydroelectric power plants typically consist of a dam, which creates a reservoir, and turbines, which convert the energy of the flowing water into electrical energy.

Hydroelectric power offers several advantages. It is a reliable and predictable source of energy since it is not dependent on weather conditions like solar or wind energy. It also provides a way of storing water, which can be utilized for irrigation, flood control, and recreational activities.

On the other hand, hydroelectric power can have significant environmental and social impacts. The construction of dams can lead to the displacement of communities and the loss of ecosystems. It can also alter the natural flow of rivers, affecting fish migration and water quality. Therefore, it is crucial to consider these factors when evaluating hydroelectric projects.

Energy storage

Energy storage technologies play a crucial role in the adoption of clean energy sources. They allow renewable energy to be stored and used when the demand is high or when the primary sources of energy, such as the sun or wind, are not available.

One of the most common energy storage technologies is batteries. Batteries store electricity in the form of chemical energy and can be used for various applications, from powering electric vehicles to providing backup power for homes and businesses. Advancements in battery technology, such as lithium-ion batteries, have significantly improved energy storage capacity and efficiency.

Pumped hydro storage is another form of energy storage widely used globally. It involves pumping water from a lower reservoir to an upper reservoir when excess energy is available, and then releasing the water downhill through turbines to generate electricity when needed.

Energy storage is essential for stabilizing the grid and integrating a higher share of intermittent renewable energy sources. However, there are challenges associated with energy storage, including cost, limited storage capacity, and environmental considerations related to battery production and disposal.

Energy efficiency

Energy efficiency plays a fundamental role in achieving clean and sustainable energy systems. By reducing energy waste and optimizing the use of energy, we can minimize the need for additional energy generation and lower greenhouse gas emissions.

Improving energy efficiency can be achieved through various measures. These include using energy-efficient appliances and equipment, implementing building insulation and energy-saving lighting systems, adopting smart energy management systems, and promoting behavioral changes to reduce energy consumption.

Apart from environmental benefits, energy efficiency also brings economic advantages. It reduces energy costs for households and businesses, stimulates innovation and job creation, and enhances energy security by reducing dependence on energy imports.

In conclusion, clean energy sources are crucial for achieving a sustainable and low-carbon future. Solar power, wind energy, hydroelectric power, and other forms of renewable energy offer significant potential in reducing greenhouse gas emissions and mitigating climate change. Energy storage and energy efficiency technologies complement the adoption of clean energy sources and promote a more sustainable energy system. By embracing clean energy technologies and practices, we can create a greener and more resilient planet for future generations.

Sustainable Infrastructure

In the quest for future sustainability, it is essential to focus on developing sustainable infrastructure. Infrastructure plays a crucial role in economic development and societal well-being. However, the traditional approach to infrastructure development often comes at a significant environmental cost, leading to resource depletion, increased greenhouse gas emissions, and ecosystem degradation. To address these challenges, sustainable infrastructure aims to integrate environmental, social, and economic considerations into the design, construction, and operation of infrastructure systems.

Principles of Sustainable Infrastructure

Sustainable infrastructure is guided by several key principles:

1. **Integration of Sustainable Design:** Sustainable infrastructure promotes the use of design strategies that minimize environmental impacts, conserve resources, and enhance overall system efficiency. This includes considering factors such as energy efficiency, water conservation, waste reduction, and the use of environmentally friendly materials.

2. **Lifecycle Approach:** A lifecycle approach considers the entire lifespan of infrastructure, including construction, operation, maintenance, and eventual decommissioning. Sustainable infrastructure aims to optimize resource use, reducing environmental impacts at every stage of the infrastructure's lifecycle.

3. **Resilience and Adaptation:** Sustainable infrastructure design takes into account the challenges posed by climate change, natural disasters, and other uncertainties. It aims to enhance the resilience of infrastructure systems to cope with changing conditions and minimize disruptions.

4. **Multi-functionality:** Sustainable infrastructure seeks to maximize the benefits derived from infrastructure systems by considering their multiple functions. For example, a road can also serve as a stormwater management system by incorporating green infrastructure elements.

5. **Social Equity:** Sustainable infrastructure ensures equitable access to its benefits, taking into account the needs and perspectives of different communities and vulnerable groups. It aims to minimize negative social impacts and enhance overall social well-being.

6. **Stakeholder Engagement**: Engaging stakeholders and involving local communities in the decision-making process is fundamental to sustainable infrastructure development. Including diverse perspectives helps identify and address potential social, economic, and environmental concerns.

Examples of Sustainable Infrastructure

1. **Green Buildings**: Sustainable infrastructure includes green building design principles. This involves energy-efficient construction techniques, the use of sustainable materials, and the integration of renewable energy systems. For example, buildings can incorporate solar panels, rainwater harvesting systems, and waste reduction measures.

2. **Smart Grids**: Smart grids are an example of sustainable infrastructure in the energy sector. They utilize advanced technologies to optimize the generation, distribution, and consumption of electricity. Smart grids enable the integration of renewable energy sources, demand-response mechanisms, and real-time monitoring, leading to increased energy efficiency and reduced emissions.

3. **Transportation Systems**: Sustainable transportation infrastructure focuses on providing efficient and eco-friendly mobility options. This includes the development of public transit systems, bike lanes, and pedestrian-friendly infrastructure. It also promotes the use of electric vehicles, car-sharing programs, and intelligent transportation systems that optimize traffic flow.

4. **Water Management**: Sustainable water infrastructure involves the efficient management of water resources, aiming to reduce water wastage and ensure equitable distribution. This includes the development of water recycling and reclamation systems, green stormwater management, and the restoration of natural water bodies.

5. **Waste Management**: Sustainable infrastructure in waste management focuses on the reduction, recycling, and proper disposal of waste. It includes the development of waste-to-energy facilities, recycling centers, and the implementation of composting programs to divert organic waste from landfills.

Challenges and Opportunities

While sustainable infrastructure offers numerous benefits, its implementation faces certain challenges. Some of the challenges include financial constraints, limited technical expertise, and the need for policy and regulatory frameworks.

To overcome these challenges, collaboration between governments, private sectors, and academic institutions is crucial. The opportunities lie in technological advancements, innovative financing mechanisms, and capacity-building initiatives. Governments can provide incentives and create supportive policies to encourage sustainable infrastructure development, while businesses can seize the growing market for sustainable solutions.

Case Study: The High Line Park

The High Line Park in New York City serves as an inspiring example of sustainable infrastructure development. This project transformed an abandoned elevated railway into a vibrant public park, integrating green space, pedestrian walkways, and cultural programming.

The High Line Park incorporates sustainable design principles by maximizing green spaces, utilizing native plantings, and implementing stormwater management techniques. It serves as a model for sustainable urban development, promoting community engagement and offering economic and social benefits to the surrounding neighborhoods.

Conclusion

Sustainable infrastructure is the foundation for future sustainability. By integrating environmental, social, and economic considerations, sustainable infrastructure development can address the challenges of resource depletion, ecosystem degradation, and climate change. Through the principles of sustainable design, lifecycle thinking, and stakeholder engagement, we can create infrastructure that supports the well-being of both present and future generations.

Sustainable consumption and production

Sustainable consumption and production is a key aspect of achieving long-term environmental sustainability. It refers to the use of resources and the production of goods and services in a manner that minimizes environmental impacts, conserves resources, and promotes social and economic well-being. This section will explore the principles, challenges, and strategies associated with sustainable consumption and production.

Principles of Sustainable Consumption and Production

Sustainable consumption and production is guided by several principles:

1. **Resource efficiency**: Using resources efficiently helps minimize waste and reduce environmental impacts. It involves optimizing production processes, reducing material consumption, and promoting recycling and reuse.

2. **Pollution prevention**: Preventing pollution is crucial for sustainable consumption and production. It involves the use of cleaner technologies, the reduction or elimination of hazardous substances, and the proper management and treatment of waste.

3. **Lifecycle thinking**: Adopting a lifecycle perspective helps assess the environmental and social impacts of a product or service throughout its entire lifespan, from raw material extraction to disposal. It promotes a holistic approach to decision-making.

4. **Social equity**: Sustainable consumption and production should promote social well-being and ensure fair distribution of resources and benefits. It involves providing decent working conditions, ensuring fair wages, and respecting human rights.

5. **Consumer awareness and empowerment**: Educating and empowering consumers is essential for sustainable consumption. Consumers need access to information about the environmental and social impacts of products and services, as well as the ability to make informed choices.

6. **Collaboration and multi-stakeholder engagement**: Achieving sustainable consumption and production requires collaboration among various stakeholders, including governments, businesses, civil society organizations, and consumers. Engaging all relevant parties helps foster innovation and find integrated solutions.

Challenges and Solutions

Achieving sustainable consumption and production faces several challenges. Some of the key challenges include:

1. **Overconsumption**: The increasing demand for goods and services puts a strain on natural resources and contributes to environmental degradation. It is essential to promote responsible consumption patterns and reduce overall consumption levels.

2. **Waste generation:** The production and disposal of waste pose significant environmental challenges. To address this, strategies such as waste prevention, recycling, and the development of circular economy models need to be implemented.

3. **Unsustainable production practices:** Some production processes are resource-intensive and contribute to pollution and environmental degradation. Shifting towards cleaner and more sustainable production methods, such as adopting renewable energy sources and reducing emissions, is necessary.

4. **Lack of awareness:** Many consumers lack information about the environmental and social impacts of the products they purchase. Raising awareness through educational campaigns and providing labeling systems can help consumers make more sustainable choices.

5. **Lack of policy support:** Without supportive policies and regulations, it is challenging to promote sustainable consumption and production. Governments need to implement measures such as eco-labeling, financial incentives, and regulations that encourage sustainable practices.

Addressing these challenges requires a combination of strategies:

1. **Product design:** Designing products with sustainability in mind can reduce their environmental impacts. This includes using eco-friendly materials, designing for durability and recyclability, and minimizing resource consumption.

2. **Extended producer responsibility (EPR):** EPR is a policy approach that holds manufacturers responsible for managing the entire lifecycle of their products, including their disposal. Implementing EPR can incentivize producers to design products that are easier to recycle or reuse.

3. **Sharing economy:** The sharing economy promotes the shared use of resources, reducing the need for individual ownership. Platforms that facilitate resource sharing, such as ride-sharing and co-working spaces, can contribute to sustainable consumption patterns.

4. **Circular economy:** A circular economy aims to maximize resource use, minimize waste generation, and keep products and materials in use for as long as possible. It involves strategies such as recycling, reusing, and remanufacturing.

5. **Consumer behavior change**: Encouraging consumers to adopt more sustainable behaviors, such as reducing food waste, choosing energy-efficient appliances, and opting for sustainable transportation, is crucial. This can be achieved through education, awareness campaigns, and financial incentives.

6. **Business sustainability practices**: Businesses play a vital role in promoting sustainable consumption and production. Adopting sustainability strategies, such as energy and water efficiency, waste reduction, and responsible sourcing, can help minimize environmental impacts.

Examples of Sustainable Consumption and Production

Several real-world examples showcase the principles and strategies of sustainable consumption and production:

1. **Patagonia**: Patagonia, an outdoor clothing company, has implemented a number of sustainable practices. They promote repair and reuse by offering repair services for their products, encourage customers to buy used items through their Worn Wear program, and use recycled and organic materials in their products.

2. **Toyota Prius**: The Toyota Prius, a hybrid electric vehicle, demonstrates sustainable consumption and production. It combines a gasoline engine with an electric motor, reducing fuel consumption and emissions. The Prius also employs lightweight materials and eco-friendly production processes.

3. **Zero Waste initiatives**: Cities like San Francisco and Kamikatsu in Japan have implemented Zero Waste initiatives. Through comprehensive recycling programs, composting, and public education campaigns, these cities aim to divert waste from landfills and move towards a circular economy model.

4. **Cradle to Cradle**: The Cradle to Cradle (C2C) framework is a design concept that promotes the creation of products that can be fully recycled or composted at the end of their useful life. It encourages the use of safe materials and aims to eliminate the concept of waste.

Conclusion

Sustainable consumption and production are essential for achieving environmental, social, and economic sustainability. By adopting resource-efficient practices, preventing pollution, considering the lifecycle of products, promoting social equity, and empowering consumers, we can create a more sustainable future. Overcoming challenges such as overconsumption, waste generation, and unsustainable production practices requires collaboration, innovative solutions, and supportive policies. Through strategies like product design, extended producer responsibility, the sharing economy, circular economy models, and behavior change, we can transition towards a more sustainable society. Let us embrace these principles and work together towards a greener and more equitable future.

Sustainable Agriculture

Sustainable agriculture is an approach to farming and food production that aims to meet the needs of the present without compromising the ability of future generations to meet their own needs. It emphasizes the use of environmentally friendly practices, promotes economic viability for farmers, and prioritizes the well-being and health of communities.

Importance of Sustainable Agriculture

Sustainable agriculture is crucial for several reasons. First and foremost, it helps to protect and preserve natural resources such as soil, water, and air. By implementing sustainable practices, farmers can minimize soil erosion, reduce water usage, and prevent pollution from agricultural chemicals.

Additionally, sustainable agriculture promotes biodiversity by using cropping systems that support a wide variety of plants, insects, and wildlife. This biodiversity plays a key role in maintaining ecosystem balance and resilience.

Furthermore, sustainable agriculture enhances food security by ensuring the long-term productivity of agricultural systems. By managing resources efficiently, farmers can maintain high yields and produce nutritious food without depleting the land.

Principles of Sustainable Agriculture

Sustainable agriculture is guided by several key principles:

1. **Soil health:** Maintaining and improving the fertility, structure, and nutrient content of the soil is essential for sustainable agriculture. This can be achieved through practices such as crop rotation, cover cropping, and the use of organic fertilizers.

2. **Water management:** Sustainable agriculture aims to minimize water consumption by utilizing efficient irrigation methods, such as drip irrigation systems, and by capturing and storing rainwater. It also focuses on preventing water pollution by reducing the use of agrochemicals that can leach into water bodies.

3. **Biodiversity conservation:** Sustainable agriculture promotes the preservation of biodiversity by providing habitat for beneficial organisms through the use of diversified cropping systems and the preservation of natural areas within farms. This helps to control pests, improve pollination, and maintain ecosystem balance.

4. **Integrated pest management:** Instead of relying solely on chemical pesticides, sustainable agriculture encourages the use of integrated pest management (IPM) strategies. These include biological control methods, crop rotation, and the use of resistant crop varieties to minimize pest damage while reducing reliance on synthetic chemicals.

5. **Resource efficiency:** Sustainable agriculture aims to minimize waste and optimize resource use. This can be achieved through practices such as precision agriculture, which uses technology to apply inputs only where and when needed, reducing the use of fertilizers and pesticides.

6. **Animal welfare:** Sustainable agriculture recognizes the importance of animal welfare in agricultural systems. It promotes the use of humane practices that provide animals with adequate space, access to pasture, and a balanced diet.

Challenges and Solutions

Sustainable agriculture faces various challenges, but innovative solutions are being developed to address them:

1. **Climate change:** Changing weather patterns and extreme weather events pose significant challenges to agriculture. Sustainable agriculture practices, such as agroforestry and conservation agriculture, help to mitigate and adapt to climate change by enhancing soil moisture retention, reducing greenhouse gas emissions, and improving carbon sequestration.

2. **Food waste:** Food waste is a major issue, contributing to both economic losses and environmental degradation. Sustainable agriculture promotes the reduction of food waste through efficient post-harvest handling, improved storage facilities, and better distribution systems.

3. **Small-scale farming:** Small-scale farmers often face significant barriers to adopting sustainable practices, such as limited access to resources, knowledge, and markets. Supporting small-scale farmers through training programs, access to credit, and market opportunities can help to overcome these barriers and promote sustainable agriculture in diverse farming systems.

4. **Consumer awareness:** Consumer demand plays a crucial role in driving the adoption of sustainable agriculture practices. Increasing awareness about the environmental and social benefits of sustainable farming through education, labeling initiatives, and consumer campaigns can encourage consumers to support sustainable food systems.

Agroecology: An Unconventional Approach

Agroecology is an unconventional but highly relevant approach to sustainable agriculture. It combines ecological principles with social and economic aspects to create resilient and sustainable farming systems.

In agroecology, farmers work with the natural environment rather than against it. They integrate diverse crops and livestock, use natural pest control methods, and prioritize soil health through organic matter management. Agroecological systems are designed to mimic natural ecosystems, promoting biodiversity and enhancing ecosystem services such as nutrient cycling and natural pest control.

Agroecology also emphasizes local and indigenous knowledge, recognizing that traditional farming practices often hold valuable insights and adaptive strategies that can contribute to sustainable agriculture.

Conclusion

Sustainable agriculture is essential for addressing the complex challenges facing our food systems and ensuring long-term environmental, social, and economic sustainability. By applying principles such as soil health, water management, biodiversity conservation, and integrated pest management, farmers can produce food in a way that protects natural resources, enhances ecosystem resilience, and promotes the well-being of communities.

Through the innovative application of technologies and the integration of diverse knowledge systems, sustainable agriculture offers a promising path towards a more sustainable and resilient future. By supporting and promoting sustainable agriculture, we can create a brighter and more sustainable future for generations to come.

Exercises

1. Research and describe a successful example of sustainable agriculture in your local region, highlighting the practices and strategies implemented to achieve sustainability.

2. Discuss the role of women in sustainable agriculture and the challenges they face in accessing resources and decision-making processes. Provide examples of initiatives or programs that aim to empower women in agriculture.

3. Evaluate the economic benefits of sustainable agriculture for farmers. How can sustainable practices improve profitability and market access for small-scale farmers?

4. Imagine you are a farmer interested in transitioning to sustainable agriculture. Develop a plan that outlines the steps you would take to implement sustainable practices on your farm, considering the specific characteristics of your farming system.

Agroecology

Agroecology is an interdisciplinary field that combines principles from agronomy and ecology to create sustainable agricultural systems. It is rooted in the understanding that agricultural practices should work in harmony with natural ecosystems, promoting biodiversity, soil health, and overall ecosystem resilience. Agroecology places a strong emphasis on ecological processes and a holistic approach to farming, considering social, economic, and environmental factors.

Principles of Agroecology

Agroecology is guided by several key principles that form the foundation of sustainable agriculture. These principles include:

1. **Diversity:** Agroecology promotes the use of diverse crops and livestock to maximize ecological interactions and reduce reliance on external inputs. Diverse agricultural systems are more resilient to pests, diseases, and climate change, ensuring long-term productivity.

2. **Sustainability:** Agroecology seeks to optimize resource use efficiency and minimize negative environmental impacts. It promotes the use of renewable inputs, such as organic fertilizers and biological pest control methods, to minimize reliance on synthetic chemicals.

3. **Soil Health:** Agroecology recognizes the critical role of soil in providing nutrients, water, and a supportive environment for plant growth. It emphasizes the use of practices that enhance soil organic matter, such as cover cropping and crop rotation, to improve soil fertility, structure, and water-holding capacity.

4. **Ecosystem Services:** Agroecology seeks to enhance and sustain ecosystem services, such as pollination, natural pest control, and nutrient cycling. By mimicking natural processes and promoting biodiversity, agroecological systems can better harness these services, reducing the need for external inputs.

5. **Resilience:** Agroecology aims to create farming systems that are resilient to climate variability, pests, and other disturbances. By diversifying crops, managing landscapes, and enhancing soil health, agroecological systems can better absorb and recover from shocks, ensuring the long-term viability of farming.

Agroecological Practices

Agroecology encompasses a wide range of practices that align with its principles and goals. Some of the key practices include:

- **Crop Diversity:** Agroecology promotes the use of diverse crop rotations, polycultures, and agroforestry systems. These practices not only enhance biodiversity but also reduce pest and disease pressure, improve soil fertility, and increase nutrient cycling.

- **Soil Management:** Agroecological approaches prioritize soil health through practices such as cover cropping, mulching, and composting. These practices increase soil organic matter, improve water retention, and enhance nutrient availability, reducing the need for synthetic fertilizers.

- **Water Management:** Agroecology emphasizes efficient water use through practices like drip irrigation, rainwater harvesting, and contour farming. By minimizing water waste and enhancing water infiltration, farmers can conserve water resources and improve drought resilience.

- **Biological Pest Control:** Agroecology encourages the use of natural enemies, such as predatory insects and beneficial microbes, to manage pests. This reduces reliance on synthetic pesticides, minimizes harm to non-target organisms, and promotes ecological balance.

+ **Agroforestry:** Agroecological systems incorporate trees and woody perennials within agricultural landscapes, providing multiple benefits. Trees can stabilize soils, conserve water, enhance biodiversity, and provide additional income through the production of fruits, timber, or non-timber forest products.

Agroecology and Sustainable Development

Agroecology has gained significant attention as a promising approach to achieving sustainable development goals. By promoting sustainable agricultural practices, it addresses multiple dimensions of sustainability, including environmental, social, and economic aspects. Some of the key contributions of agroecology to sustainable development are:

+ **Food Security:** Agroecology emphasizes diversified farming systems that enhance food availability and access. By promoting local food production, it reduces dependence on external sources, improves dietary diversity, and strengthens local food systems.

+ **Climate Change Mitigation:** Agroecology contributes to climate change mitigation efforts by sequestering carbon in soils, reducing greenhouse gas emissions, and enhancing the resilience of agricultural systems to climate impacts. By optimizing resource use and promoting natural carbon sinks, it helps combat climate change.

+ **Rural Livelihoods:** Agroecology prioritizes the well-being of farmers and rural communities by promoting sustainable livelihoods. By diversifying income sources, reducing input costs, and fostering local value chains, it enhances the economic viability of farming and improves rural quality of life.

+ **Biodiversity Conservation:** Agroecological practices contribute to the conservation of biodiversity by providing habitats, promoting ecological connectivity, and reducing the use of agrochemicals that harm non-target organisms. By preserving and enhancing biodiversity, it supports the long-term sustainability of ecosystems.

+ **Resilience to Shocks:** Agroecological systems are more resilient to shocks like extreme weather events, pests, and market fluctuations. By diversifying crops, increasing ecological interactions, and improving resource use efficiency, agroecology enhances the capacity of farmers to withstand and recover from disturbances.

Case Study: Organic Farming

Organic farming is an example of agroecological practice that has gained significant popularity. Organic farmers prioritize the use of natural inputs, such as compost, crop rotations, and biological pest control, while avoiding synthetic pesticides and genetically modified organisms. Organic farming promotes soil health, biodiversity, and sustainable resource use.

Organic farming practices have been shown to have several benefits. For example, studies have demonstrated that organic farms typically have higher levels of soil organic matter, improved water infiltration, and reduced soil erosion compared to conventional farms. Organic farmers also rely on natural pest control methods, such as insect predators and traps, reducing the environmental impact of chemical pesticides.

Additionally, organic farming supports biodiversity by providing habitats, food sources, and nesting sites for beneficial insects, birds, and mammals. This contributes to enhanced pollination, natural pest control, and overall ecosystem resilience.

From a consumer perspective, organic farming provides assurance of food produced without synthetic pesticides and genetically modified organisms. This appeals to individuals seeking healthier and more environmentally conscious food choices.

However, it is important to note that organic farming also faces some challenges. It typically requires more land and labor compared to conventional farming, which can limit its scalability and increase production costs. Additionally, organic farmers need to carefully manage nutrient inputs to ensure optimal crop productivity, as synthetic fertilizers are not used. This requires a deep understanding of ecological processes and nutrient cycling.

Despite these challenges, organic farming serves as a prime example of how agroecological principles and practices can contribute to sustainable agriculture, improved environmental outcomes, and healthier food systems.

Conclusion

Agroecology offers a comprehensive and sustainable approach to agriculture that aligns with the goals of sustainable development. By applying ecological principles and adopting sustainable practices, agroecology promotes resilience, biodiversity, and resource efficiency in farming systems. From diversified crop rotations to soil management techniques, agroecology provides a toolkit of practices that can be customized to local contexts.

The integration of agroecology into agricultural policies and practices is crucial to transform our food systems and ensure long-term sustainability. By investing in research, knowledge transfer, and capacity building, we can support farmers in adopting agroecological approaches and harnessing their benefits. Agroecology has the potential to not only improve environmental outcomes but also address the social and economic challenges faced by farming communities. It is a pathway towards a more sustainable and resilient future for agriculture.

Organic Farming

Organic farming is a method of agricultural production that aims to promote ecosystem health, biodiversity, and sustainability by avoiding the use of synthetic fertilizers, pesticides, genetically modified organisms (GMOs), and other chemical inputs. It is an alternative approach to conventional farming, which heavily relies on the use of synthetic inputs to increase crop yields. Organic farming emphasizes the use of natural inputs, such as organic fertilizers, crop rotation, composting, and biological control methods, to maintain soil fertility and support plant growth.

Principles of Organic Farming

Organic farming is guided by several key principles, including:

1. **Soil Health:** Organic farmers prioritize the health of the soil, recognizing that soil is a living ecosystem that supports plant growth. They focus on enhancing soil fertility and structure through practices such as crop rotation, composting, and the use of organic amendments. This helps maintain nutrient balance, water-holding capacity, and beneficial soil microorganisms.

2. **Biodiversity:** Organic farming promotes biodiversity by providing habitats for native plants and animals. Farmers may create hedgerows, plant cover crops, and maintain wildlife corridors to support beneficial insects, birds, and other organisms that contribute to pest control and pollination.

3. **Ecological Balance:** Organic farmers strive to maintain a balance between the various components of the farm ecosystem. They aim to minimize the use of external inputs and rely on natural ecological processes to regulate pests and diseases, promote plant health, and maintain overall ecosystem resilience.

4. **Prohibition of Synthetic Inputs:** Organic farming prohibits the use of synthetic fertilizers, pesticides, herbicides, and other chemical inputs.

Instead, farmers rely on natural methods such as crop rotation, companion planting, and biological pest control to manage pests and diseases.

5. **Animal Welfare:** Organic farming places a strong emphasis on the ethical treatment and welfare of animals. Livestock raised on organic farms are provided with ample space to move, access to the outdoors, and are fed a diet that meets their nutritional needs without the use of antibiotics or growth hormones.

Benefits of Organic Farming

Organic farming offers several benefits over conventional farming practices:

+ **Environmental Sustainability:** By avoiding the use of synthetic inputs, organic farming reduces potential pollution of water bodies and soil degradation. It promotes the conservation of natural resources, such as water and biodiversity, and helps mitigate climate change by sequestering carbon in the soil.

+ **Improved Soil Health:** Organic farming practices, such as the use of organic matter in the form of compost and cover crops, help improve soil structure, increase water-holding capacity, and enhance nutrient cycling. This leads to higher soil fertility and productivity in the long run.

+ **Reduced Health Risks:** Organic farming avoids the use of synthetic pesticides and genetically modified organisms, which may have adverse effects on human health. By minimizing pesticide residues in food and exposure to harmful chemicals, organic farming reduces the risk of certain health issues.

+ **Enhanced Flavor and Nutritional Value:** Organic produce is often considered to have superior taste and flavor compared to conventionally grown crops. Some studies have also suggested that organic fruits and vegetables may have higher levels of certain nutrients, such as vitamin C, iron, and antioxidants.

+ **Support for Local Economies:** Organic farming often occurs on a smaller scale and is frequently associated with local food systems. By supporting organic farmers, consumers contribute to the development of local economies and help foster community connections.

Challenges and Solutions

While organic farming offers numerous benefits, it also faces several challenges that need to be addressed:

1. **Certification and Standards:** Organic farming requires adherence to specific standards and certification procedures to ensure compliance. The certification process involves costs and administrative efforts that may pose challenges, especially for small-scale and resource-constrained farmers. Streamlining certification processes and providing support for certification costs can help overcome these challenges.

2. **Weed and Pest Management:** Organic farmers face difficulties in managing weeds and pests without the use of synthetic herbicides and pesticides. Implementing integrated pest management strategies, using crop rotation, employing natural predators, and developing resilient crop varieties can help address these challenges effectively.

3. **Yield and Productivity:** Organic farming often yields lower crop yields compared to conventional methods. However, with proper management practices and localized adaptation, organic farming can achieve competitive yields. Research and innovation in organic farming techniques, such as precision farming and agroecological principles, can help improve productivity.

4. **Market Access and Consumer Demand:** Organic produce generally commands higher prices in the market due to increased production costs. However, market access and consumer demand for organic products vary geographically. Strengthening market linkages, promoting consumer awareness, and providing incentives for organic farming can help bridge the gap between supply and demand.

Real-world Example: Organic Farming in India

India has witnessed a significant growth in organic farming in recent years. The country has diverse agro-climatic conditions and a rich tradition of organic farming practices. Sikkim, a state in northeast India, has established itself as the world's first fully organic state, where all agricultural land has transitioned to organic farming.

The success of organic farming in India can be attributed to various factors, including government support, active participation of farmers, and increased consumer demand. The government has implemented various schemes and subsidies to promote organic farming, such as the Paramparagat Krishi Vikas Yojana (PKVY) and the Organic Value Chain Development for the Northeast Region (OVCDNER) program.

Organic farming in India has not only improved soil health and crop quality but has also provided economic opportunities to farmers. It has enabled small-scale farmers to access premium markets and obtain better prices for their produce. Additionally, organic farming has revived traditional farming practices, preserved indigenous seed varieties, and promoted agroecological knowledge among farmers.

Conclusion

Organic farming is an environmentally sustainable approach to agriculture that promotes soil health, biodiversity, and animal welfare. It offers various benefits, including improved soil fertility, reduced health risks, and enhanced flavor and nutritional value of produce. Although it faces challenges, such as certification processes and yield limitations, organic farming is gaining momentum worldwide.

By adopting organic farming practices, farmers can contribute to a more sustainable and resilient food system while protecting the environment and human health. Consumer support for organic products and government interventions play critical roles in fostering the growth of organic farming. As organic farming continues to evolve, research and innovation will help address challenges and further enhance its productivity and economic viability.

Food Security

Food security is a critical issue that affects billions of people around the world. It refers to the availability, access, and utilization of sufficient, safe, and nutritious food to meet the dietary needs and preferences of individuals, ensuring an active and healthy life. Achieving food security is not only crucial for human well-being but also for sustainable development and social stability.

In this section, we will explore the principles and challenges of food security, as well as innovative solutions and strategies to address this global concern.

Principles of Food Security

To understand food security, we need to consider several key principles:

1. **Availability:** Sufficient food should be produced or imported to meet the population's nutritional requirements. This involves sustainable agricultural

practices, including efficient resource management, crop diversification, and technology adoption.

2. **Access**: Individuals and communities should have physical and economic access to food. This means that food must be affordable, and people should have the means to purchase or produce it. Access is influenced by factors such as income, employment, social protection, and functioning markets.

3. **Utilization**: Food must be consumed safely and effectively by the body to provide the necessary nutrients for health and well-being. Enhancing utilization requires education on nutrition, proper food handling, and good hygiene practices.

4. **Stability**: Food availability, access, and utilization should be consistent over time. Stability is vital in the face of environmental shocks, economic fluctuations, and social disruptions. Building resilience in agricultural systems, diversifying food sources, and establishing effective social safety nets can contribute to stability.

Challenges in Achieving Food Security

Several challenges hinder efforts to achieve food security on a global scale. These challenges include:

- **Population Growth**: As the global population continues to grow, the demand for food increases. Feeding a larger population requires sustainable intensification of agricultural practices to produce more food without compromising natural resources.

- **Climate Change**: Rising temperatures, erratic rainfall patterns, and extreme weather events pose significant challenges to agricultural productivity. Adapting agriculture to climate change, promoting climate-resilient crops, and implementing sustainable irrigation systems are crucial for ensuring food security.

- **Poverty**: Poverty and food insecurity are interlinked issues. Many people living in poverty cannot afford an adequate diet, leading to malnutrition and hunger. Poverty eradication efforts, targeted social support programs, and equitable access to resources are essential for addressing this issue.

• **Food Waste and Loss**: A significant amount of food is wasted or lost throughout the supply chain, from production to consumption. Minimizing food waste and loss through improved storage, distribution, and consumer behavior can contribute to increasing food availability.

• **Conflict and Instability**: Conflict and political instability disrupt agricultural production and disrupt food systems. Building peace, strengthening governance, and promoting sustainable agriculture in conflict-affected areas are essential for ensuring food security.

Innovative Solutions for Food Security

Addressing the challenges of food security requires innovative solutions and holistic approaches. Here are some examples of innovative strategies:

1. **Climate-Smart Agriculture**: By combining sustainable agricultural practices, climate adaptation, and mitigation strategies, climate-smart agriculture aims to increase productivity and resilience in the face of climate change. This includes techniques such as conservation agriculture, agroforestry, and precision farming.

2. **Crop Diversification**: Promoting diverse cropping systems can enhance food security by reducing vulnerability to pests, diseases, and market fluctuations. Crop diversification also helps in preserving biodiversity, soil fertility, and cultural heritage.

3. **Digital Technologies**: Information and communication technologies have the potential to revolutionize agriculture and enhance productivity. Sensors, drones, satellite imagery, and mobile applications provide farmers with real-time data on soil moisture, pest outbreaks, and market prices, enabling better decision-making.

4. **Sustainable Intensification**: This approach seeks to produce more food on existing agricultural lands while minimizing adverse environmental impacts. It involves using integrated pest management, precision agriculture, and agroecological principles to increase yields sustainably.

5. **Urban Agriculture**: The cultivation of crops and raising of livestock in urban areas can contribute to food security, especially in densely populated regions. Rooftop gardens, vertical farming, and community gardens provide fresh produce, reduce food miles, and create green spaces.

Case Study: The Green Revolution

The Green Revolution is a notable example of a successful initiative to improve food security. In the mid-20th century, high-yielding crop varieties, along with improved agricultural practices, significantly increased crop yields in many developing countries. This initiative, led by scientists such as Norman Borlaug, played a crucial role in alleviating hunger and poverty.

However, the Green Revolution also faced criticism for its heavy reliance on chemical inputs and monocultures, leading to environmental degradation and loss of biodiversity. Lessons learned from the Green Revolution emphasize the importance of sustainability, inclusivity, and balanced agroecosystems in achieving long-term food security.

Key Takeaways

Food security is a complex issue that requires a multi-dimensional approach. Some key takeaways from this section include:

- Food security involves ensuring availability, access, utilization, and stability of food for all individuals.

- Challenges to food security include population growth, climate change, poverty, food waste, and conflict.

- Innovative solutions for food security include climate-smart agriculture, crop diversification, digital technologies, sustainable intensification, and urban agriculture.

- The Green Revolution serves as a valuable case study, highlighting both successes and limitations in addressing food security.

Addressing the issue of food security will require collaborative efforts and a holistic approach that integrates scientific knowledge, policy frameworks, technological advancements, and local community engagement. By prioritizing sustainable and inclusive practices, we can work towards a future where everyone has access to nutritious and safe food, ensuring the well-being of present and future generations.

Sustainable Cities

In this section, we will explore the concept of sustainable cities and their importance in creating a better future for humanity. As the world population continues to grow, urban areas face numerous challenges, including increasing demand for resources and infrastructure, environmental degradation, and social inequality.   Sustainable cities offer a holistic approach to addressing these challenges by integrating environmental, social, and economic considerations into urban planning and development.

Defining Sustainable Cities

A sustainable city, also known as an eco-city or green city, is designed to minimize its environmental impact while maximizing its social and economic benefits. It is built on the principles of sustainability, emphasizing the efficient use of resources, the reduction of greenhouse gas emissions, the protection of biodiversity, and the promotion of social equity.

At its core, sustainable urban planning aims to create livable spaces that meet the needs of current and future generations. It focuses on creating a harmonious balance between environmental conservation, economic prosperity, and social well-being.

Key Elements of Sustainable Cities

Sustainable cities incorporate various strategies and practices to achieve their goals. Some of the key elements include:

1. Compact Development: Encouraging compact development helps minimize urban sprawl and promotes efficient land use. By concentrating population and activities in central areas, cities can reduce the need for extensive transportation networks and infrastructure.

2. Mixed-Use Zoning: Creating mixed-use zones allows for a diverse range of activities within a given area. By integrating residential, commercial, and recreational spaces, sustainable cities promote walkability, reduce commuting distances, and foster vibrant communities.

3. Sustainable Transportation: Prioritizing sustainable transportation options is crucial in reducing carbon emissions and improving air quality. This includes investing in public transportation systems, promoting cycling and walking, and incorporating electric vehicles into the transport network.

4. Green Infrastructure: Implementing green infrastructure, such as parks, urban forests, and green roofs, improves air and water quality, reduces the urban heat island effect, and provides recreational spaces for residents. These green spaces also support biodiversity and enhance the overall livability of the city.

5. Energy Efficiency: A sustainable city focuses on energy-efficient buildings and infrastructure. This involves using renewable energy sources, improving

building insulation, adopting smart grid systems, and promoting energy-efficient technologies.

6. Waste Management: Implementing effective waste management systems is crucial for sustainable cities. This includes waste reduction, recycling programs, composting facilities, and promoting a circular economy approach.

7. Water Management: Sustainable cities prioritize efficient and responsible water management practices. This includes rainwater harvesting, wastewater treatment, and creating green infrastructure to manage stormwater runoff.

Challenges and Solutions

Building sustainable cities comes with its fair share of challenges. Rapid urbanization, limited resources, and social inequality pose significant hurdles. However, there are various solutions available to address these challenges:

1. Urban Planning and Design: Effective urban planning and design are critical for sustainable cities. This involves engaging communities in the decision-making process, promoting mixed-income housing, providing accessible public services, and creating well-connected neighborhoods.

2. Sustainable Financing: Developing sustainable cities requires significant financial investments. Governments, businesses, and international organizations must collaborate to provide financial resources and incentives for sustainable initiatives.

3. Technology and Innovation: Embracing technology and innovation can provide sustainable solutions for urban challenges. Smart city technologies, renewable energy systems, and digital platforms can improve urban efficiency and enhance residents' quality of life.

4. Education and Awareness: Educating communities about sustainable practices and their benefits is essential. This includes promoting awareness about energy conservation, waste reduction, and sustainable transport options.

5. Collaboration and Partnerships: Collaboration among stakeholders, including governments, businesses, academia, and civil society, is crucial for achieving sustainable urban development. By working together, we can leverage expertise and resources to overcome challenges and create sustainable cities for the future.

Case Study: Curitiba, Brazil

One successful example of a sustainable city is Curitiba, Brazil. Known as the "Green City," Curitiba has implemented various innovative strategies to address urban challenges. Some key initiatives include:

1. Integrated Public Transportation: Curitiba's Bus Rapid Transit (BRT) system is renowned worldwide. It provides efficient and affordable transportation, reducing the need for private vehicles and minimizing traffic congestion.

2. Urban Green Spaces: The city has prioritized the creation of urban green spaces, including parks and tree-lined boulevards. These spaces not only provide recreational areas but also contribute to improved air quality and the preservation of biodiversity.

3. Recycling and Waste Management: Curitiba has implemented a successful recycling program, achieving high rates of waste diversion from landfills. The city has also encouraged composting and waste-to-energy initiatives to manage its waste effectively.

4. Planning and Zoning: Curitiba's urban planning focuses on mixed-use zoning and preserving historical buildings. This creates vibrant and inclusive neighborhoods while preserving the city's cultural heritage.

Curitiba's success story highlights the importance of sustainable urban planning, engagement with the community, and innovative solutions in creating livable and resilient cities.

Conclusion

Sustainable cities are crucial for addressing the numerous challenges associated with rapid urbanization and climate change. By integrating environmental, social, and economic considerations into urban planning and development, sustainable cities provide a blueprint for creating a better future. However, building sustainable cities requires collaboration, innovation, and a long-term commitment from governments, businesses, and communities. With the right strategies and practices in place, we can create cities that are environmentally friendly, socially inclusive, and economically prosperous. Let us strive towards building a sustainable future for all.

Urban Planning

Urban planning is a multidisciplinary field that focuses on the development and management of cities and urban areas. It involves creating plans and strategies to ensure the efficient use of land, infrastructure, resources, and public spaces within urban environments. Urban planning plays a critical role in promoting sustainable development, improving the quality of life for residents, and addressing various social, economic, and environmental challenges.

Principles of Urban Planning

Urban planning is guided by several principles that aim to create livable, inclusive, and sustainable cities. Some of the key principles include:

1. **Compact Cities:** Compact cities promote the efficient use of land by encouraging higher population densities and mixed land-use patterns. This reduces urban sprawl, minimizes the need for private vehicle usage, and increases accessibility to essential services and amenities.

2. **Sustainable Transportation:** Urban planning emphasizes the development of sustainable transportation systems, such as public transit networks, walking and cycling infrastructure, and smart mobility solutions. These initiatives reduce traffic congestion, air pollution, and carbon emissions, while improving mobility and connectivity.

3. **Green Spaces:** Integrating green spaces, parks, and urban forests into urban planning helps improve the quality of life for residents. Green spaces provide recreational opportunities, enhance biodiversity, mitigate the heat island effect, and contribute to overall mental and physical well-being.

4. **Mixed-Use Development:** Encouraging mixed-use development integrates residential, commercial, cultural, and recreational activities within close proximity. This minimizes the need for long commutes, supports local economies, and fosters vibrant and diverse communities.

5. **Public Participation:** Inclusive public participation processes allow residents, stakeholders, and communities to engage in decision-making processes. It ensures that the urban planning process addresses diverse needs, concerns, and aspirations of the people who live and work in the city.

6. **Resilience:** Urban planning considers the resilience of cities to various challenges, including climate change, natural disasters, and economic shocks. It involves designing resilient infrastructure, implementing adaptive strategies, and integrating disaster risk reduction measures.

Challenges in Urban Planning

Urban planning faces numerous challenges that need to be addressed to create sustainable and resilient cities. Some of the key challenges include:

- **Rapid Urbanization:** The unprecedented rate of urbanization puts significant pressure on existing infrastructure, housing, and services. Urban planning must accommodate the needs of the growing population while ensuring equitable access to essential amenities.

- **Income Inequality:** Cities often face income inequality, where certain sections of society have limited access to quality housing, education, healthcare, and employment opportunities. Urban planning needs to address these disparities and promote inclusive growth and social justice.

- **Environmental Degradation:** Urban development can lead to environmental degradation, such as pollution, habitat loss, and depletion of natural resources. Sustainable urban planning aims to minimize these impacts through eco-friendly practices, green infrastructure, and protection of natural areas.

- **Slums and Informal Settlements:** Many cities have slums and informal settlements characterized by substandard living conditions and inadequate access to basic services. Urban planning must prioritize the upgrading and integration of these areas into the formal urban fabric.

- **Climate Change:** Climate change poses significant challenges to urban planning, including increased frequency and intensity of extreme weather events, rising sea levels, and heatwaves. Urban planning must incorporate climate adaptation and mitigation strategies to ensure the long-term sustainability of cities.

Tools and Strategies in Urban Planning

To address the challenges and achieve sustainable urban development, urban planners employ various tools and strategies. Some of the key tools and strategies include:

1. **Land Use Planning:** Land use planning involves allocating land for different purposes, such as residential, commercial, industrial, and recreational. It helps optimize land utilization, control urban sprawl, and ensure the efficient provision of services and infrastructure.

2. **Transit-Oriented Development:** Transit-oriented development (TOD) focuses on creating compact, mixed-use communities around transit nodes. By promoting walking, cycling, and public transportation, TOD reduces the dependency on private vehicles and supports sustainable mobility.

3. **Zoning and Building Codes:** Zoning regulations define the allowable land use and building standards within specific zones. Building codes set construction standards to ensure safety, accessibility, and energy efficiency. These tools guide urban development and promote sustainable built environments.

4. **Urban Design:** Urban design principles shape the physical form, layout, and aesthetics of urban areas. By integrating design considerations such as walkability, active public spaces, and human-scale architecture, urban planners create visually appealing, functional, and livable cities.

5. **Smart Growth:** Smart growth strategies promote development that is environmentally sustainable, economically viable, and socially equitable. It emphasizes compact development, mixed land use, preservation of green spaces, and the efficient use of resources.

6. **Participatory Planning:** Engaging stakeholders, communities, and residents in the planning process fosters a sense of ownership and inclusivity. Participatory planning ensures that the needs and aspirations of various groups are considered, leading to more responsive and people-centered urban development.

Case Study: Curitiba, Brazil

Curitiba, the capital city of the Paraná state in southern Brazil, is a noteworthy example of successful urban planning. The city has implemented innovative strategies to address urban challenges and promote sustainable development.

Transportation System: Curitiba's bus rapid transit (BRT) system is internationally recognized as a model of efficient and sustainable public transportation. It features dedicated bus lanes, integrated fare systems, and well-designed stations, providing a fast, reliable, and affordable mode of transportation.

Green Spaces and Parks: Curitiba boasts an extensive network of parks and green spaces, including the famous Barigui Park and Tanguá Park. These green areas serve as recreational spaces, promote biodiversity, and help mitigate the heat island effect.

Mixed-Use Zoning: The city implemented zoning regulations that encourage mixed-use development along major transportation corridors. This ensures that residents have easy access to commercial and social amenities within walking distance.

Waste Management: Curitiba has implemented an innovative waste management system that emphasizes recycling and citizen participation. The city provides incentives for residents to exchange recyclable material for bus tickets, contributing to both environmental and social sustainability.

Pedestrian-Friendly Streets: The city has prioritized pedestrian-friendly design by creating pedestrian-only streets and implementing traffic calming measures. This encourages walking and cycling, reduces vehicle emissions, and enhances public safety.

Social Housing: Curitiba has implemented successful social housing programs, such as the "Minha Casa, Minha Vida" (My House, My Life), which aim to provide affordable housing to low-income families. These initiatives promote social inclusion, reduce informal settlements, and improve living conditions.

Curitiba's success in urban planning can be attributed to visionary leadership, innovative thinking, citizen engagement, and long-term strategic planning. The city continues to serve as a global inspiration for sustainable urban development.

Conclusion

Urban planning plays a vital role in shaping sustainable and resilient cities. By incorporating principles of compactness, sustainable transportation, green spaces, and public participation, urban planners aim to create cities that are socially inclusive, environmentally conscious, and economically vibrant. Through the application of various tools and strategies, such as land use planning, transit-oriented development, and participatory planning, urban planners can address the challenges posed by rapid urbanization, income inequality, and environmental degradation. The case study of Curitiba, Brazil, highlights the successful implementation of innovative urban planning approaches. As cities continue to grow and face new challenges, urban planning will remain crucial for creating sustainable and livable urban environments.

Green infrastructure

Green infrastructure refers to a network of natural and semi-natural areas, as well as green spaces within urban areas, that are designed and managed to provide multiple ecosystem services. It serves as an interconnected system to mitigate the negative impacts of urbanization and promote sustainability in cities. Green infrastructure includes parks, gardens, street trees, green roofs, permeable pavements, urban wetlands, and other green spaces that help to restore and enhance the natural environment in urban settings. This section explores the principles, benefits, and examples of green infrastructure, highlighting its role in creating resilient and sustainable cities.

Principles of Green Infrastructure

Green infrastructure is guided by several key principles that integrate ecological and landscape planning with urban design and development. These principles include:

1. **Multifunctionality**: Green infrastructure should provide multiple benefits, such as improving air and water quality, reducing urban heat island effects, enhancing biodiversity, and promoting physical and mental well-being. By delivering a range of ecosystem services, it contributes to the overall livability and sustainability of cities.

2. **Connectivity**: Green infrastructure should be designed and managed to create a connected network of green spaces, allowing the movement of organisms and ecological processes. This interconnectedness supports biodiversity conservation, ecological resilience, and the provision of ecosystem services across urban areas.

3. **Integration**: Green infrastructure planning should be integrated into urban planning and development processes, ensuring that it is incorporated from the early stages of design. This integration improves the effectiveness and efficiency of green infrastructure implementation, preventing conflicts with other urban functions.

4. **Collaboration**: Green infrastructure requires collaboration and partnership among various stakeholders, including policymakers, urban planners, landscape architects, community organizations, and residents. Collaboration fosters a shared vision, knowledge exchange, and joint decision-making processes, leading to successful green infrastructure projects.

Benefits of Green Infrastructure

Green infrastructure offers numerous benefits for both the environment and society. Some of the key benefits include:

1. **Climate regulation**: Green infrastructure helps mitigate the impacts of climate change by reducing the urban heat island effect, improving air quality, and regulating surface water runoff. Trees and vegetation provide shade, evaporative cooling, and carbon sequestration, contributing to the overall resilience of cities in the face of climate variability.

2. **Biodiversity conservation**: Green infrastructure provides habitat for native plants, animals, and microorganisms, enhancing biodiversity in urban areas. It

promotes ecological connectivity, enabling the movement of species and the establishment of wildlife corridors. This, in turn, supports pollination, pest control, and other ecological processes that are crucial for ecosystem health.

3. **Water management:** Green infrastructure plays a vital role in managing stormwater runoff, reducing the risk of flooding and improving water quality. Green roofs, rain gardens, and permeable pavements help to absorb and filter rainwater, reducing the strain on traditional drainage systems. Urban wetlands and green spaces also act as natural reservoirs, storing excess water and recharging groundwater.

4. **Human health and well-being:** Access to green spaces has been linked to improved physical and mental health outcomes. Green infrastructure provides opportunities for physical activities, relaxation, and social interaction, enhancing the overall well-being of urban residents. It also contributes to the reduction of air pollution, noise pollution, and stress levels.

Examples of Green Infrastructure

Green infrastructure can take various forms depending on the urban context and specific goals. Some examples of green infrastructure elements and projects include:

1. **Urban parks:** Parks and recreational areas serve as essential components of green infrastructure, providing open spaces for public enjoyment and nature conservation. Well-designed urban parks can also include features such as biofiltration ponds, green roofs, and permeable pavements to manage stormwater runoff.

2. **Green roofs and walls:** Green roofs are vegetated surfaces on the tops of buildings that provide insulation, reduce energy consumption, and manage stormwater runoff. Green walls, also known as vertical gardens, are installations of plants on the exterior walls of buildings, offering similar benefits. These features enhance the visual appeal of urban areas while improving environmental performance.

3. **Urban forests and street trees:** Trees play a crucial role in green infrastructure, providing shade, improving air quality, and reducing the impact of urban heat islands. Urban forests and street trees enhance biodiversity, mitigate climate change, and provide recreational opportunities for urban residents.

4. **Green corridors and wildlife habitats:** Green corridors are linear elements, such as tree-lined streets, that connect different green spaces within urban areas. These corridors facilitate the movement of animals, enable pollination, and create more favorable conditions for urban wildlife. Wildlife habitats, such as urban wetlands and bird sanctuaries, also contribute to biodiversity conservation in cities.

5. **Permeable pavements and rain gardens**: Permeable pavements allow rainwater to infiltrate the ground, reducing surface runoff and replenishing groundwater. Rain gardens, on the other hand, are landscaped areas designed to capture and temporarily hold rainwater, filtering it through the soil. These features improve stormwater management and water quality while beautifying urban spaces.

Challenges and Considerations

Implementing green infrastructure in urban areas can pose certain challenges and considerations. These include:

1. **Limited space**: Urban areas often have limited space for green infrastructure implementation. It requires careful planning and innovative design to integrate green spaces into the urban fabric, especially in densely populated cities.

2. **Maintenance and management**: Green infrastructure requires ongoing maintenance and management to ensure its long-term effectiveness. Regular pruning, watering, and weeding are necessary to keep green spaces healthy and attractive.

3. **Equity and accessibility**: It is crucial to ensure that green infrastructure is accessible to all members of the community, regardless of socio-economic status or location. This requires considering the distribution of green spaces and addressing any potential disparities or barriers to access.

4. **Funding and financing**: Financing green infrastructure projects can be a challenge, as they often require upfront investments. Identifying and securing funding sources, such as public-private partnerships or green bonds, is essential for successful implementation.

5. **Integration with other urban functions**: Green infrastructure should be integrated with other urban functions, such as transportation, housing, and infrastructure planning. This integration requires collaboration among different sectors and professionals to create synergies and maximize the benefits of green infrastructure.

Case Study: The High Line, New York City

The High Line in New York City is a prime example of successful green infrastructure integration in an urban environment. The High Line is a 1.45-mile-long elevated linear park built on a historic freight rail line. It features an innovative design that incorporates native plantings, seating areas, art installations, and public spaces.

The High Line has transformed an abandoned railway into a vibrant green space that attracts millions of visitors each year. It has improved air quality, provided habitat for wildlife, and created a unique recreational area for residents and visitors. The project has also led to economic revitalization in the surrounding neighborhoods, increasing property values and attracting businesses.

The success of the High Line highlights the potential of green infrastructure to transform underutilized urban spaces into valuable community assets. It serves as an inspiration for cities worldwide to embrace green infrastructure as a means to create sustainable and resilient urban environments.

Further Resources

1. Book: "Green Infrastructure: A Landscape Approach" by David C. Rouse (2013).

2. Website: The Nature Conservancy's Green Infrastructure Toolkit (https://www.nature.org/en-us/about-us/who-we-are/ how-we-work/conservation-methods/conservation-strategies/ green-infrastructure-toolkit/)

3. Video: TED Talk by Mitchell Joachim - "Don't Build Your Home, Grow It!" (https://www.ted.com/talks/mitchell_joachim_don_t_build_ your_home_grow_it)

Exercises

1. Research and identify a green infrastructure project in your local area. Describe its main features and the benefits it provides to the community.

2. Design a hypothetical green corridor that connects three existing parks in your city. Consider the type of vegetation, pedestrian and cycling paths, and any additional features that would enhance connectivity and biodiversity.

3. Investigate the potential funding sources for green infrastructure projects in your country or region. Create a list of possible financing options and discuss their advantages and disadvantages.

4. Visit a local urban park or green space and conduct a mini-environmental assessment. Observe and document the biodiversity, recreational activities, and infrastructure elements present. Assess the park's overall sustainability and suggest potential improvements.

5. Watch the TED Talk by Mitchell Joachim titled "Don't Build Your Home, Grow It!" Reflect on the concept of biodesign and discuss its potential applications

in your local area. How could growing homes with green infrastructure contribute to sustainability and resilience?

Sustainable Transportation Systems

Transportation is an integral part of our daily lives, but the conventional transportation systems heavily rely on fossil fuels, leading to various environmental and social challenges. In this section, we will explore the concept of sustainable transportation systems and discuss the principles, strategies, and technologies that can help mitigate the negative impacts of transportation on the environment and society.

Principles of Sustainable Transportation

Sustainable transportation aims to provide accessible, efficient, and environmentally friendly mobility options while minimizing negative impacts on the planet. The following principles are essential for designing sustainable transportation systems:

1. **Reducing greenhouse gas emissions:** Transportation is a major contributor to greenhouse gas emissions, primarily through the combustion of fossil fuels. Sustainable transportation systems prioritize the use of low-carbon or carbon-neutral energy sources to reduce emissions and combat climate change.

2. **Promoting energy efficiency:** Sustainable transportation systems prioritize energy-efficient technologies and practices to minimize energy consumption and waste. This includes promoting fuel-efficient vehicles, optimizing transport routes, and encouraging eco-driving techniques.

3. **Encouraging mode shift:** Sustainable transportation systems aim to shift individuals' travel behavior from private car use to more sustainable modes such as public transportation, cycling, and walking. This helps reduce congestion, air pollution, and energy consumption.

4. **Building compact and mixed-use communities:** Designing communities with a mix of residential, commercial, and recreational areas promotes shorter travel distances and facilitates walking, cycling, and the use of public transportation. Compact communities reduce the need for long and energy-intensive commutes.

5. **Prioritizing active transportation:** Sustainable transportation systems prioritize walking and cycling as primary modes of transportation for short trips. This not only reduces emissions but also promotes physical activity, improves public health, and enhances livability.

6. **Integrating land use and transportation planning:** Efficient land use planning, with a focus on promoting mixed land uses and reducing sprawl, can support sustainable transportation systems. By locating residences, businesses, and amenities in close proximity, the need for long-distance travel is reduced.

Strategies for Sustainable Transportation

To achieve sustainable transportation systems, various strategies can be implemented:

1. **Public Transportation:** Expanding and improving public transportation infrastructure is a key strategy for sustainable transportation. This includes increasing the coverage and frequency of bus and rail networks, implementing bus rapid transit systems, and integrating different modes of public transportation to provide seamless user experiences.

2. **Active Transportation:** Promoting walking and cycling infrastructure is crucial for sustainable transportation. This involves creating dedicated bike lanes, pedestrian-friendly sidewalks, bike-sharing programs, and secure bike parking facilities. Public education campaigns can also encourage more people to choose active modes of transportation.

3. **Electric Vehicles (EVs):** Transitioning to electric vehicles is essential for reducing emissions from transportation. Governments and private sector stakeholders can incentivize the adoption of EVs by providing subsidies, building charging infrastructure, and implementing supportive policies. Battery technology advancements and renewable energy integration are key areas for further research and development.

4. **Intelligent Transportation Systems (ITS):** ITS technologies integrate information and communication technologies into transportation to improve efficiency, safety, and sustainability. This includes intelligent traffic management systems, real-time public transportation information, and smart parking solutions.

5. **Carpooling and Ridesharing:** Encouraging carpooling and ridesharing services can significantly reduce the number of vehicles on the road. This can be achieved through the development of carpooling platforms, dedicated high-occupancy vehicle (HOV) lanes, and employer-based incentive programs.

6. Freight Transportation Optimization: Optimizing freight transportation is vital for sustainable supply chains. This involves promoting more efficient logistics, incorporating alternative fuel vehicles, implementing smart freight management systems, and incentivizing shift towards rail and waterway transportation for long-distance cargo.

Technologies for Sustainable Transportation

In addition to the strategies mentioned above, several emerging technologies can contribute to the development of sustainable transportation systems. Here are some notable examples:

1. Vehicle-to-Grid (V2G) Technology: V2G technology allows bidirectional energy flow between electric vehicles and the grid. It enables EVs to store excess electricity and supply it back to the grid during peak demand periods. V2G can help stabilize the grid, increase renewable energy penetration, and support energy storage.

2. Autonomous Vehicles: Autonomous or self-driving vehicles have the potential to revolutionize transportation systems. With advanced sensors, artificial intelligence, and connectivity, autonomous vehicles can optimize traffic flow, reduce congestion, improve safety, and enhance fuel efficiency through platooning and smart routing.

3. Hyperloop: Hyperloop is a high-speed transportation system that uses low-pressure tubes to propel passenger or cargo pods. With speeds exceeding 600 mph, hyperloop technology has the potential to reduce travel time, energy consumption, and greenhouse gas emissions for long-distance travel.

4. Sustainable Aviation: Aviation is a significant contributor to carbon emissions. Sustainable aviation technologies aim to reduce fuel consumption and emissions through more fuel-efficient aircraft, alternative fuels such as biofuels, and improved air traffic management systems to optimize flight routes.

Case Study: Curitiba's Bus Rapid Transit System

One successful example of a sustainable transportation system is Curitiba's Bus Rapid Transit (BRT) system in Brazil. The city's BRT, known as the Rede Integrada de Transporte (RIT), has transformed public transportation and urban planning in Curitiba. Key features of the system include:

- Exclusive bus lanes with pre-boarding fare collection and rapid boarding systems, reducing waiting times and improving efficiency.

- Integrated fare system allowing passengers to switch between buses and connect different transit lines without additional costs.

- Land use planning ensured that BRT stations were integrated with commercial and residential areas, reducing the need for long commutes.

- Extensive coverage and high frequency of buses, providing convenient and reliable transportation options to all parts of the city.

- Continuous improvements such as dedicated bicycle lanes, green spaces, and pedestrian-friendly streets around BRT stations, promoting active transportation and enhancing quality of life.

The success of Curitiba's BRT system has inspired similar projects worldwide and demonstrated the potential of well-designed sustainable transportation systems in improving urban mobility and reducing reliance on private vehicles.

In conclusion, sustainable transportation systems are essential for mitigating the negative environmental and social impacts of transportation. By incorporating principles such as reducing greenhouse gas emissions, promoting energy efficiency, encouraging mode shift, and integrating land use and transportation planning, we can build more sustainable and livable cities. Implementing strategies such as public transportation expansion, active transportation promotion, electric vehicle adoption, and intelligent transportation systems, along with leveraging emerging technologies, will pave the way towards a greener and more accessible future of transportation.

Sustainable waste management

Waste management is a critical aspect of sustainable development. As the world population continues to grow and urbanization increases, the amount of waste generated is also escalating. In order to protect the environment and human health, it is crucial to adopt sustainable waste management practices. This section will explore various strategies and technologies that can be implemented to achieve sustainable waste management.

The Problem of Waste

The improper disposal of waste poses significant environmental and health risks. Landfills, where most waste ends up, emit greenhouse gases such as methane, a potent contributor to climate change. Improper waste disposal also leads to air, water, and soil pollution, which can harm ecosystems, biodiversity, and human health.

Reduce, Reuse, Recycle

The three Rs - reduce, reuse, and recycle - form the foundation of sustainable waste management.

Reduce: The most effective way to minimize waste is to reduce its generation. This can be achieved by practicing responsible consumption and production. Reducing packaging waste, promoting durable goods, and using digital formats instead of paper are some examples of waste reduction strategies.

Reuse: Reusing products or materials reduces the need for new production and minimizes waste generation. Encouraging the reuse of items, such as shopping bags, containers, and electronic devices, can significantly reduce waste.

Recycle: Recycling involves converting waste materials into new products or raw materials. Materials such as paper, plastic, metal, and glass can be recycled, reducing the demand for virgin resources and minimizing waste going to landfills.

Waste Separation and Segregation

Waste separation and segregation are key steps in sustainable waste management. By separating different types of waste at the source, such as at homes or businesses, materials that can be recycled or reused can be diverted from landfills.

Different waste streams commonly segregated include organic waste, recyclables (paper, plastic, metal, and glass), hazardous waste (chemicals, batteries), and electronic waste (e-waste). Implementation of effective waste separation systems, including collection bins and public awareness campaigns, is crucial for efficient waste management.

Waste-to-Energy Technologies

Waste-to-energy (WTE) technologies can play a significant role in sustainable waste management by converting waste into energy. These technologies provide a way to address two critical issues simultaneously - waste reduction and energy generation.

Incineration: Incineration is a WTE process in which solid waste is burned at high temperatures, producing heat energy. This energy can be harnessed to generate electricity or heat for various applications. Appropriate air emission controls are necessary to minimize the release of pollutants during incineration.

Anaerobic Digestion: Anaerobic digestion is a biological process that decomposes organic waste in the absence of oxygen. This process produces biogas, primarily composed of methane and carbon dioxide. Biogas can be used as a source of renewable energy for electricity generation and heating purposes.

Gasification: Gasification involves the conversion of organic waste into a mixture of carbon monoxide, hydrogen, and methane known as synthesis gas or syngas. Syngas can be further processed to produce electricity, heat, or biofuels.

Composting

Composting is a natural process that decomposes organic waste into nutrient-rich soil, known as compost. It is an environmentally friendly waste management technique and a valuable tool in sustainable agriculture.

During composting, microorganisms break down organic waste, including food scraps, yard waste, and agricultural residues, into stable organic matter. Compost can be applied to soil to improve its structure, fertility, and water retention capacity, reducing the need for chemical fertilizers.

Extended Producer Responsibility (EPR)

Extended Producer Responsibility (EPR) is a policy approach that holds manufacturers responsible for the environmental impact of their products throughout their entire lifecycle, including after they become waste.

EPR encourages manufacturers to design products that are easier to recycle, reuse, or dispose of sustainably. It also promotes the establishment of collection and recycling systems for specific products, such as electronics, batteries, and packaging materials. EPR incentivizes producers to take responsibility for product end-of-life management and reduces the burden on local waste management systems.

Public Awareness and Education

Public awareness and education are crucial for the success of sustainable waste management practices. It is essential to inform and engage individuals, communities, and businesses about the importance of waste reduction, proper waste separation, and the benefits of recycling and composting.

Educational campaigns, workshops, and outreach programs can help raise awareness and encourage behavioral changes. Involving schools, community groups, and local organizations can facilitate the dissemination of information and foster a sense of collective responsibility.

Case Study: Zero Waste Cities

Zero waste cities are emerging as a promising approach to achieve sustainable waste management. These cities aim to minimize waste generation, maximize recycling and reuse, and eliminate the need for landfills and incinerators.

One example is Kamikatsu, a town in Japan that has implemented a zero waste policy. Residents are required to separate waste into 45 different categories for recycling and composting. The town has achieved an impressive 80% waste diversion rate, with the remaining waste sent to a nearby facility that uses anaerobic digestion to convert it into biogas.

By adopting a zero waste approach, cities can significantly reduce their environmental footprint, conserve resources, and create a healthier and more sustainable living environment.

Conclusion

Sustainable waste management is vital for preserving the environment and ensuring a healthy future for generations to come. By following the principles of reduce, reuse, and recycle, implementing waste separation systems, embracing waste-to-energy technologies, promoting composting, and adopting extended producer responsibility, we can move towards a more sustainable and circular waste management system.

Public awareness and education play a pivotal role in driving behavioral changes and fostering a culture of waste reduction and responsible consumption. By working together at individual, community, and global levels, we can successfully address the challenges of waste management and pave the way for a more sustainable future.

Eco-friendly architecture

Eco-friendly architecture, also known as sustainable architecture or green building, aims to minimize the negative impact of buildings on the environment while promoting the health and well-being of occupants. It is an approach to design and construction that incorporates principles of energy efficiency, resource conservation, and environmental responsibility.

Principles of Eco-friendly Architecture

Eco-friendly architecture is guided by several key principles:

1. Energy efficiency: The design and construction of buildings should focus on reducing energy consumption by utilizing passive design strategies. These

strategies may include optimizing building orientation, incorporating proper insulation, and using energy-efficient lighting and appliances.

2. Use of renewable energy sources: Incorporating renewable energy technologies, such as solar panels and wind turbines, can help buildings generate their own clean energy and reduce reliance on fossil fuels.

3. Efficient water management: Implementing water-efficient fixtures, rainwater harvesting systems, and greywater recycling can significantly reduce water consumption in buildings.

4. Sustainable materials: The use of eco-friendly and locally sourced materials helps reduce the carbon footprint associated with the construction process. Additionally, selecting materials with low embodied energy, such as recycled or reclaimed materials, helps conserve resources and minimize waste.

5. Indoor environmental quality: Prioritizing the health and comfort of occupants by improving indoor air quality, optimizing natural lighting, and using non-toxic building materials and finishes.

6. Waste reduction and recycling: Designing buildings with a focus on waste reduction, recycling, and responsible waste management practices, such as designing for deconstruction and implementing on-site composting.

7. Biodiversity and ecosystem integration: Incorporating green spaces, living roofs, and vertical gardens can help increase biodiversity, reduce the urban heat island effect, and provide opportunities for urban agriculture.

Examples of Eco-friendly Architecture

Here are a few examples of eco-friendly architecture in practice:

1. The Edge, Amsterdam: This office building is often referred to as the world's greenest building. It incorporates various sustainable features, including solar panels, smart lighting systems, rainwater harvesting, and a high-performance insulation system.

2. One Angel Square, Manchester: This building utilizes natural ventilation, intelligent lighting systems, and combined heat and power (CHP) generation to minimize energy consumption. It also incorporates a biodiverse roof garden and a rainwater harvesting system.

3. The Bullitt Center, Seattle: This commercial office building is designed to be energy-positive, meaning it generates more energy than it consumes. It features photovoltaic panels, geothermal heating and cooling systems, and extensive use of natural daylighting.

4. Khoo Teck Puat Hospital, Singapore: This hospital incorporates numerous sustainable design strategies, such as efficient lighting systems, natural ventilation, rainwater harvesting, and extensive green spaces that contribute to the healing environment.

Challenges and Benefits

While eco-friendly architecture offers numerous benefits, there are also challenges to overcome:

Challenges:

- Initial costs: The upfront costs of constructing eco-friendly buildings can be higher compared to conventional buildings. However, these costs are often offset by long-term energy savings and operational efficiency.

- Limited awareness and expertise: Many architects and builders may lack the knowledge and expertise in designing and constructing eco-friendly buildings. Increasing awareness and providing training and certifications can address this challenge.

- Building codes and regulations: Current building codes and regulations may not fully support or encourage eco-friendly design and construction. Advocacy for policy changes and adoption of sustainable building practices can help overcome this challenge.

Benefits:

- Energy savings: Eco-friendly buildings typically have lower energy consumption, resulting in reduced utility costs over the lifetime of the building.

- Improved indoor environment: Features such as better air quality, natural lighting, and temperature control contribute to the health and well-being of occupants, leading to increased productivity and overall satisfaction.

+ Reduced environmental impact: By conserving energy and resources, eco-friendly buildings contribute to the reduction of greenhouse gas emissions and the overall environmental footprint.

+ Enhanced market value: Eco-friendly buildings are often sought after by tenants and buyers, as they provide a more sustainable and healthier living or working environment.

Unconventional Approach: Nature-inspired Design

An unconventional yet relevant approach to eco-friendly architecture is nature-inspired design, also known as biomimicry. Biomimicry involves learning from and replicating nature's forms, processes, and patterns to create more sustainable and efficient designs.

For example, the Eastgate Centre in Harare, Zimbabwe, is designed to mimic the passive cooling system of termite mounds. The building utilizes natural ventilation, with cool air entering at the bottom and warm air exiting at the top, reducing the need for mechanical cooling.

By studying nature's solutions to complex problems, such as energy efficient structures, self-cleaning surfaces, and optimized resource utilization, biomimicry offers innovative ways to design eco-friendly buildings that are in harmony with the natural environment.

Conclusion

Eco-friendly architecture plays a crucial role in creating sustainable and resilient buildings that minimize their impact on the environment while providing healthy and comfortable spaces for occupants. By incorporating principles of energy efficiency, renewable energy, water management, sustainable materials, and indoor environmental quality, eco-friendly architecture offers numerous benefits for both the environment and the well-being of individuals and communities. With continued advancements and adoption of eco-friendly practices, we can pave the way for a more sustainable and resilient future.

Advances in Eco Technology

Clean Energy

Renewable Energy Sources

In the pursuit of sustainable development and mitigating the impacts of climate change, the transition to renewable energy sources is crucial. Renewable energy sources are derived from naturally occurring processes that are replenished continuously, making them a sustainable and clean alternative to fossil fuels. This section will explore the various types of renewable energy sources, their principles of operation, advantages, and challenges.

Solar Energy

Solar energy is the most abundant renewable energy source available on Earth. It harnesses the power of the sun to generate electricity or heat. Solar photovoltaic (PV) systems convert sunlight directly into electricity using semiconducting materials like silicon. When sunlight strikes the solar panels, the photons release electrons from the atoms within the material, creating an electric current. This electricity can then be used to power homes, businesses, and other applications.

Solar thermal systems, on the other hand, use sunlight to generate heat. These systems concentrate sunlight onto a receiver to produce high-temperature thermal energy, which can be used for heating water or generating steam to drive turbines for electricity production. Solar thermal systems are commonly used in solar water heaters and concentrated solar power (CSP) plants.

Advantages of solar energy include its infinite availability, minimal environmental impact, and the potential for decentralized energy production. It is a clean source of energy that does not release greenhouse gases or other harmful pollutants during operation. Moreover, solar energy can be harnessed in both large-scale utility projects and small-scale residential installations.

However, solar energy also presents challenges. The availability of solar energy varies with geographical location, seasons, and weather conditions. Additionally, the cost of solar PV technology has been a barrier to widespread adoption, although the prices have significantly decreased in recent years. Efficient energy storage solutions are needed to address the intermittent nature of solar energy and ensure a consistent power supply.

Wind Energy

Wind energy utilizes the kinetic energy of the wind to generate electricity. Wind turbines are the primary technology used for harnessing wind energy. These turbines consist of rotor blades that capture the wind's energy and spin a generator to produce electricity.

The principle of wind energy generation is based on the Bernoulli principle and the transfer of momentum. As the wind blows, it flows over the curved surface of the turbine blades, creating a difference in air pressure. This pressure difference generates a lifting force, causing the blades to rotate. The rotational motion is then converted into electrical energy through the generator.

Wind energy has several advantages. It is an abundant and widely distributed resource, making it a valuable asset for many regions. It produces no greenhouse gas emissions during operation, contributing to reduced carbon footprints and improved air quality. Moreover, wind power offers price stability as the cost of wind turbines and installation continues to decline.

However, there are challenges associated with wind energy. The intermittent nature of wind requires the integration of energy storage technologies or complementary energy sources to ensure a constant power supply. Wind turbines can also cause noise pollution and visual impact, which may raise concerns among local communities. Additionally, proper siting of wind farms is essential to minimize any adverse effects on birds, bats, and their habitats.

Hydroelectric Power

Hydroelectric power harnesses the energy of flowing or falling water to generate electricity. It is currently the most widely used form of renewable energy globally. Hydroelectric power plants typically involve the construction of dams across rivers to create reservoirs. The water stored in the reservoir is released through turbines, spinning them to generate electricity.

The principle of hydroelectric power relies on the conversion of the potential energy of water at a certain elevation into kinetic energy. As water falls or flows through turbines, the kinetic energy of the water is converted into mechanical energy, which is further transformed into electrical energy through generators.

Hydropower offers numerous advantages. It is a reliable and consistent source of energy as water flow can be controlled and adjusted according to demand. It is also a highly efficient technology, with conversion rates of up to 90%. Hydroelectric power plants have long lifespans and low operation and maintenance costs.

However, there are environmental considerations associated with hydroelectric power. The construction of dams can result in ecosystem disruptions, including altered downstream flows, changes in water quality, and impacts on fish migration. Dam failure can also pose a significant risk to downstream communities. Hence, careful evaluation of the environmental, social, and economic impacts is crucial when considering hydroelectric projects.

Geothermal Energy

Geothermal energy harnesses the heat from within the Earth to produce electricity and provide heating or cooling for residential and commercial buildings. This energy originates from the radioactive decay of minerals deep within the Earth's core.

Geothermal power plants utilize natural geothermal reservoirs or drill deep wells to access hot water or steam. The produced geothermal fluid is used directly for heating applications or converted into steam to power turbines, generating electricity.

Geothermal energy has several advantages. It is a clean and reliable energy source, as the Earth's heat is virtually inexhaustible. Geothermal power plants produce minimal greenhouse gas emissions and have significantly lower life-cycle carbon footprints compared to conventional fossil fuel power plants. In addition, geothermal energy can provide local economic benefits by creating jobs and reducing dependence on imported fuels.

However, geothermal energy is not universally available, as it requires specific geological conditions for its effective utilization, limiting its potential in certain regions. The exploration and development costs of geothermal resources can also be high. Additionally, there are potential environmental impacts such as the release of gases and trace amounts of toxic elements during geothermal operations.

Biomass Energy

Biomass energy is derived from organic matter, such as plants, agricultural residues, forestry waste, and dedicated energy crops. It can be used for heating, electricity generation, and transportation fuels.

Biomass power plants typically burn biomass materials to produce steam, which is used to power turbines and generate electricity. The combustion process relies on the principle of converting the chemical energy stored in biomass into thermal energy, mechanical energy, and electrical energy.

Advantages of biomass energy include its potential for utilizing existing waste streams, reducing landfill waste, and displacing fossil fuels. Biomass is considered a carbon-neutral energy source, as the carbon dioxide released during combustion is offset by the carbon absorbed by plants during their growth. Additionally, biomass energy can contribute to rural development by creating jobs and supporting local economies.

However, biomass energy also poses challenges. The supply of biomass resources must be carefully managed to prevent the depletion of natural resources and to ensure sustainable practices. The combustion of biomass can release pollutants, such as particulate matter and greenhouse gases, which need to be controlled through proper emission control technologies. The competition between biomass energy and food production should also be carefully evaluated to avoid negative impacts on food security and land use.

Ocean Energy

Ocean energy encompasses a variety of technologies that harness the power of the ocean's waves, tides, and thermal gradients to generate electricity. While still in the early stages of development, ocean energy has the potential to provide a significant renewable energy resource.

Wave energy converters capture the kinetic energy of ocean waves using offshore devices. As waves pass over or through these devices, their motion is converted into mechanical or hydraulic power, which is then used to generate electricity.

Tidal energy exploits the regular rise and fall of ocean tides to generate electricity. Tidal barrages, similar to hydroelectric dams, use the potential energy of the tides to spin turbines and produce electricity.

Ocean thermal energy conversion (OTEC) utilizes the temperature difference between warm surface water and cold deep water to generate electricity. OTEC systems employ heat exchangers and working fluids to transfer the thermal energy and drive a turbine.

Ocean energy has the advantage of being highly predictable, as the motion of waves and tides follows established patterns. It is a constant and abundant source of energy that has the potential to provide baseload power. Moreover, ocean energy can be harnessed without consuming freshwater resources or occupying valuable land.

However, ocean energy technologies face challenges associated with their deployment in harsh marine environments. The construction and maintenance of devices in the ocean pose logistical and technical difficulties. The impact on marine ecosystems and migratory patterns of marine species should also be carefully assessed to minimize environmental risks.

In conclusion, renewable energy sources such as solar, wind, hydroelectric, geothermal, biomass, and ocean energy play a crucial role in the transition towards a sustainable energy future. Each source has its own unique set of advantages and challenges, which must be carefully evaluated and addressed to maximize their potential. The adoption of renewable energy sources is essential for achieving energy security, reducing greenhouse gas emissions, and mitigating the impacts of climate change. Governments, industries, and individuals must work together to promote and invest in the development and utilization of renewable energy technologies. By embracing renewable energy, we can foster a cleaner, healthier, and more sustainable planet for future generations.

Solar Energy

Solar energy is a crucial and promising source of renewable energy that harnesses the power of the sun to generate electricity or heat. It is one of the most abundant and sustainable resources available to us on Earth. In this section, we will explore the principles of solar energy, its applications, and the advances in technologies that are driving its future growth.

Principles of Solar Energy

Solar energy is derived from the conversion of sunlight into usable energy. The key principle behind this conversion is the photovoltaic effect, which occurs when certain materials, such as silicon, are exposed to sunlight. When photons, particles of light, strike the surface of these materials, they transfer their energy to the electrons in the material, causing them to flow and generate an electric current.

The efficiency of solar panels, which convert sunlight into electricity, is a critical factor in maximizing the amount of energy harnessed from the sun. Advances in solar cell technologies have significantly improved the efficiency of solar panels over the years. Currently, the most common type of solar cell used in panels is the silicon-based photovoltaic cell.

Types of Solar Energy Systems

There are two main types of solar energy systems: solar photovoltaic (PV) systems and solar thermal systems.

Solar PV systems convert sunlight directly into electricity using solar panels made up of interconnected solar cells. These systems are commonly used in residential, commercial, and utility-scale applications. The electricity generated by solar PV systems can be used to power appliances, lights, and other electrical devices. It can also be stored in batteries or fed back into the grid.

Solar thermal systems, on the other hand, capture sunlight to generate heat. This heat can be used directly for heating water or spaces in residential and commercial buildings or converted into electricity through a steam turbine. Solar thermal systems are particularly useful in applications such as water heating, space heating, and industrial processes that require high-temperature heat.

Advances in Solar Energy Technologies

The rapid advancement in solar energy technologies has made solar power more efficient, affordable, and accessible. Here are some notable advances that have contributed to the growth of solar energy:

- **Thin-Film Solar Cells:** Thin-film solar cells use a very thin layer of semiconductor material, which can be deposited on a variety of substrates such as glass or metal. These cells are less expensive to produce than traditional silicon-based cells and have the potential for flexible applications, such as solar panels integrated into building materials and portable solar chargers.

- **Perovskite Solar Cells:** Perovskite solar cells are a type of emerging solar technology that has gained significant attention due to their high efficiency and low manufacturing costs. These cells use a unique class of materials called perovskites, which can be easily processed using low-cost deposition techniques. While still in the early stages of development, perovskite solar cells have the potential to revolutionize the solar industry by providing inexpensive and highly efficient solar power.

- **Solar Tracking Systems:** Solar tracking systems are designed to optimize the angle and orientation of solar panels to maximize sunlight exposure throughout the day. By following the sun's path, solar panels equipped with tracking systems can increase their energy generation by up to 25%

compared to fixed-mount systems. This technology is particularly useful in utility-scale solar power plants where maximizing the energy output is crucial.

+ **Solar-Powered Desalination:** Desalination is the process of removing salt and other impurities from seawater to produce freshwater. Solar desalination systems utilize the sun's energy to power the desalination process, making it a sustainable and environmentally friendly solution to address water scarcity in coastal areas. This technology has the potential to provide clean drinking water to millions of people around the world.

+ **Solar-Powered Vehicles:** Solar energy is also being utilized in the transportation sector to power electric vehicles (EVs). Solar panels integrated into the roof of EVs can help recharge the vehicle's battery and increase its range. Additionally, solar-powered charging stations are being deployed, enabling drivers to replenish their EVs using clean, renewable energy.

Challenges and Future Directions

While solar energy holds great potential, there are several challenges that need to be addressed for its widespread adoption:

+ **Intermittency:** Solar power generation is dependent on sunlight availability, making it intermittent in nature. This intermittency poses challenges for grid stability and reliability. Advances in energy storage systems, such as advanced batteries, are essential to store excess energy during peak sunlight hours and provide electricity during periods of low sunlight.

+ **Cost:** Although the cost of solar energy has significantly declined over the past decade, the upfront cost of installing solar panels is still relatively high. Continued research and innovation are needed to develop cost-effective materials, manufacturing processes, and installation techniques to make solar energy more affordable for a wider range of users.

+ **Integration with Existing Infrastructure:** Integrating solar energy into existing infrastructure, such as power grids, can pose technical challenges. The development of smart grid technologies and grid-scale energy management systems is crucial to ensure seamless integration and optimal utilization of solar energy.

‣ **Environmental Considerations:** While solar energy itself is clean and renewable, the manufacturing and disposal of solar panels can have environmental impacts. Recycling and proper end-of-life management of solar panels are important aspects to ensure the sustainability of solar energy.

In conclusion, solar energy is a key player in the transition to a sustainable energy future. Advancements in solar cell technologies, such as thin-film and perovskite cells, along with the integration of solar power into various sectors like transportation and desalination, are driving the growth of solar energy. Overcoming challenges related to intermittency, cost, integration, and environmental considerations will be crucial for the widespread adoption of solar energy as a clean and reliable source of power.

Summary

In this section, we explored the principles of solar energy and the various types of solar energy systems. We discussed the advancements in solar energy technologies, including thin-film and perovskite solar cells, solar tracking systems, solar-powered desalination, and solar-powered vehicles. We also highlighted the challenges that need to be addressed for the widespread adoption of solar energy, such as intermittency, cost, integration, and environmental considerations. Solar energy holds immense potential as a clean and sustainable source of power, and continued research and innovation will pave the way for a brighter future powered by the sun.

Wind Energy

Wind energy is a renewable source of power that harnesses the kinetic energy of the wind to generate electricity. It is a clean and sustainable form of energy that has gained significant attention in recent years due to its potential to mitigate climate change and reduce dependence on fossil fuels. In this section, we will explore the principles behind wind energy, its benefits and challenges, as well as the technologies and initiatives driving its advancement.

Principles of Wind Energy

Wind energy is based on the principle that moving air possesses kinetic energy. When the wind blows, it causes the rotation of wind turbine blades, which are attached to a generator. As the blades rotate, the kinetic energy of the wind is converted into mechanical energy, which is then transformed into electrical energy by the generator. The electricity generated can be used for various applications, such as powering homes, businesses, and even entire communities.

The amount of power that can be generated from wind depends on several factors, including the speed of the wind and the size and efficiency of the wind turbine. Wind speed is a critical factor, as the power output of a wind turbine increases exponentially with an increase in wind speed. Therefore, wind turbines are typically installed in areas with high and consistent wind speeds, such as coastal regions and open plains.

Advantages of Wind Energy

There are several advantages associated with wind energy:

- **Renewable and Sustainable:** Wind energy is a clean and abundant source of power that does not deplete natural resources or contribute to climate change through greenhouse gas emissions.

- **Reduced Greenhouse Gas Emissions:** Wind energy systems emit little to no greenhouse gases during operation, making them a viable solution for reducing carbon emissions and addressing climate change.

- **Job Creation and Economic Benefits:** The wind energy sector has the potential to create numerous job opportunities, from manufacturing and installation to operations and maintenance. Additionally, wind energy projects can stimulate local economies through investments and taxes.

- **Energy Independence:** Wind energy reduces reliance on fossil fuel imports, enhancing energy security and independence for countries.

- **Low Operational Costs:** Once installed, wind turbines have relatively low operational costs compared to conventional power plants, as wind is a free resource. This can lead to stable and affordable electricity prices for consumers.

Challenges of Wind Energy

While wind energy offers numerous advantages, it also faces certain challenges that must be addressed:

- **Intermittency and Variability:** Wind is an intermittent and variable energy source, as it is dependent on weather conditions. The power output from wind turbines fluctuates, requiring a backup power source or energy storage system to ensure a reliable electricity supply.

- **Land and Resource Requirements:** Large-scale wind farms require vast land areas, which can raise concerns related to land use and environmental impact. Additionally, wind turbines require access to high wind speeds, limiting their deployment to specific locations.

- **Visual and Noise Impacts:** The presence of wind turbines can be visually and acoustically noticeable, leading to potential aesthetic and noise concerns for nearby residents. Proper site selection and community engagement are important factors in addressing these issues.

- **Avian and Bat Interactions:** Wind turbines can pose risks to birds and bats, particularly during migration routes. Understanding and mitigating these interactions are crucial for minimizing ecological impacts.

Technologies and Innovations

The wind energy sector has witnessed significant technological advancements and innovations to overcome the challenges associated with wind power generation.

- **Wind Turbine Design:** Modern wind turbines are designed to operate efficiently at various wind speeds. This is achieved through innovations in blade design, materials, and aerodynamics. Additionally, the use of variable-pitch blades allows turbines to adapt to changing wind conditions.

- **Offshore Wind Farms:** Offshore wind farms have emerged as a promising option to harness wind energy. By installing wind turbines at sea, offshore wind farms can access stronger and more consistent winds, reducing intermittency and addressing land use constraints. However, offshore wind farms face unique technological and logistical challenges, such as foundation design, transmission systems, and installation and maintenance in harsh marine environments.

- **Energy Storage:** Energy storage technologies, such as batteries and pumped hydro storage, play a vital role in enhancing the reliability and grid integration of wind energy. They enable the capture and storage of excess electricity generated during periods of high wind speeds, which can be utilized during low wind periods or high demand periods. Ongoing research and development are focused on improving the efficiency and scalability of energy storage systems.

+ **Smart Grid Integration:** The integration of wind energy into the existing electricity grid requires sophisticated control systems and infrastructure. Smart grid technologies enable the efficient management and distribution of wind-generated electricity, ensuring reliability and stability of the grid. These technologies incorporate real-time data, advanced analytics, and communication systems to optimize the utilization of wind energy and enable demand-response mechanisms.

Case Study: Largest Wind Farm

As an example of the scale and potential of wind energy, let's explore the Gansu Wind Farm, the largest wind farm in the world.

Located in Gansu Province, China, the Gansu Wind Farm spans an area of over 70,000 square kilometers and consists of numerous interconnected wind turbines. With a total installed capacity of 20,000 megawatts, it is capable of producing over 38 terawatt-hours of electricity annually. This wind farm plays a significant role in China's efforts to transition to renewable energy and reduce greenhouse gas emissions.

The Gansu Wind Farm's success can be attributed to China's commitment to renewable energy development, favorable wind resources in the region, and advancements in wind turbine technology. The project not only contributes to the country's renewable energy targets but also provides economic benefits, such as job creation and local investment.

Conclusion

Wind energy is a promising renewable energy source that offers numerous environmental, economic, and social benefits. Although it faces challenges related to intermittency, land requirements, and visual impacts, advancements in technology and innovation are continuously improving its efficiency and reliability.

As countries strive to transition to a more sustainable energy future, wind energy will play a crucial role in reducing greenhouse gas emissions, enhancing energy security, and creating a cleaner and more resilient power system. With ongoing research and development, wind energy has the potential to make a significant contribution to global sustainability efforts.

Hydroelectric power

Hydroelectric power is a renewable energy source that harnesses the energy of flowing or falling water to generate electricity. It is one of the most widely used forms of renewable energy, with hydroelectric power plants being found all over the world. This section will explore the principles, benefits, challenges, and future advancements in hydroelectric power.

Principles of Hydroelectric Power

The principle behind hydroelectric power is relatively simple: the kinetic energy of flowing or falling water is converted into mechanical energy by a turbine, which in turn drives a generator to produce electricity. The basic components of a hydroelectric power plant include a dam, reservoir, turbine, and generator.

When a dam is built across a river, it creates a reservoir or a large artificial lake. The dam controls the flow of water and increases the water level to create potential energy. The stored water in the reservoir is released, usually through a penstock or a large pipe, and the force of the flowing water causes the turbine blades to spin.

The turbine is connected to a generator, which converts the mechanical energy of the turbine into electrical energy. The electricity generated is then transmitted through power lines to homes, businesses, and industries.

Benefits of Hydroelectric Power

Hydroelectric power offers several advantages, making it a popular choice for generating electricity:

1. Renewable and clean energy: As water is constantly replenished by the water cycle, hydroelectric power is considered a renewable energy source. It does not produce greenhouse gas emissions or air pollutants, making it a clean source of energy.

2. Large-scale electricity generation: Hydroelectric power plants can generate a large amount of electricity, making them suitable for supplying power to cities, industries, and even entire countries.

3. Long lifespan and low operating costs: Hydroelectric power plants have a lifespan of 50-100 years, and once constructed, they have relatively low operating costs compared to fossil fuel power plants.

4. Water management: Hydroelectric power plants can serve as multipurpose facilities, providing water for irrigation, flood control, and recreational activities in addition to generating electricity.

5. Energy storage: Hydroelectric power plants with reservoirs can store water and adjust electricity generation according to demand, making it a reliable source of power during peak periods.

Challenges and Future Advancements

Although hydroelectric power has numerous benefits, it also faces certain challenges and limitations:

1. Environmental impact: Constructing large dams can lead to the displacement of communities, loss of ecosystems, and alteration of natural water flow. It is essential to carefully assess and mitigate these impacts through proper environmental planning and management.

2. Limited feasible locations: Not all regions have suitable topography or access to flowing water, limiting the potential for hydroelectric power generation.

3. High initial costs: Building a hydroelectric power plant, especially large-scale ones, requires substantial upfront investment. However, the long-term economic benefits often outweigh these costs.

4. Climate change effects: Climate change can significantly affect water availability, potentially impacting the reliability of hydroelectric power generation in certain regions.

Future advancements in hydroelectric power aim to address these challenges and further enhance the efficiency and sustainability of this energy source. Some of the advancements being explored include:

1. Run-of-river hydroelectric systems: Unlike traditional dams, run-of-river systems do not require large reservoirs, reducing their environmental impact and cost while still harnessing the power of flowing water.

2. Small-scale hydroelectric systems: The development of smaller hydroelectric systems allows for power generation in remote or off-grid locations, providing electricity to communities that are not connected to the central power grid.

3. Fish-friendly turbine designs: Traditional hydroelectric turbines can be harmful to fish populations. New turbine designs, such as fish-friendly turbines, aim to reduce fish mortality and enhance ecosystem sustainability.

4. Pumped-storage hydroelectricity: Pumped-storage hydroelectricity involves using surplus electricity during off-peak periods to pump water into an upper reservoir. When electricity demand is high, the stored water is released to generate electricity. This technology helps balance the grid and store excess renewable energy.

In conclusion, hydroelectric power is a valuable and widely used renewable energy source with numerous benefits. Despite its challenges, ongoing developments and advancements in hydroelectric technology hold the promise of further improving its efficiency, minimizing environmental impacts, and maximizing its contribution to a sustainable energy future.

Energy Storage

In the quest for sustainable energy systems, one of the critical challenges is finding efficient ways to store and manage energy. Energy storage technologies play a crucial role in balancing energy supply and demand, integrating renewable energy sources, and enhancing grid stability. This section will explore various energy storage options, their principles of operation, advantages, drawbacks, and real-world applications.

Battery Technologies

Batteries are widely recognized as one of the most promising energy storage solutions. They convert chemical energy into electrical energy through electrochemical reactions. Several types of batteries are currently in use, each with its unique characteristics and applications.

Lead-Acid Batteries are the oldest and most common rechargeable batteries. They have a simple design, consisting of lead plates immersed in a sulfuric acid electrolyte. Lead-acid batteries are widely used in automotive starting, backup power supply systems, and renewable energy storage. However, their low energy density, short life span, and environmental concerns due to lead and acid leakage have limited their application in certain sectors.

Lithium-Ion Batteries have gained significant attention in recent years due to their high energy density, long cycle life, and suitability for various applications. They utilize an intercalation mechanism to shuttle lithium ions between two electrodes, typically a lithium metal oxide cathode and a graphite anode. Lithium-ion batteries power electric vehicles, mobile devices, and are increasingly deployed for energy storage in homes, commercial buildings, and power grids. Despite their advantages, safety concerns, limited availability of lithium resources, and high cost remain challenges.

Flow Batteries operate by storing energy in liquid electrolytes contained in external tanks. The energy is stored in the form of chemical potential, and electrochemical reactions occur when the electrolytes flow through the battery cell. Flow batteries offer several advantages, including scalability, long cycle life, and the ability to decouple energy capacity from power capacity. Vanadium redox flow batteries and zinc-bromine flow batteries are among the most extensively studied and commercially deployed flow battery systems.

Sodium-Ion Batteries are a promising alternative to lithium-ion batteries due to the abundance and low cost of sodium resources. They employ a similar working principle to lithium-ion batteries, but with sodium ions shuttling between the electrodes. Sodium-ion batteries have shown potential in large-scale energy storage applications such as renewable integration and grid stabilization. However, issues such as lower energy density and limited commercial availability of sodium-ion cells still need to be addressed.

Solid-State Batteries offer improved safety, higher energy density, and faster charging capabilities compared to traditional liquid electrolyte-based batteries. These batteries use solid electrolytes, such as ceramics or polymers, to enable ion transport. Solid-state batteries are still in the early stages of development, and a substantial amount of research is being conducted to overcome manufacturing challenges and improve performance.

Supercapacitors

Supercapacitors, also known as ultracapacitors or electric double-layer capacitors (EDLCs), store energy through the physical mechanism of charge accumulation at the electrode-electrolyte interface. They differ from batteries in terms of their mechanism and performance characteristics. Supercapacitors offer high power density, fast charge and discharge capabilities, and excellent cycling stability. They are particularly suitable for applications that require short bursts of high power, such as electric vehicles, regenerative braking, and grid stabilization. However, their low energy density limits their use for long-term energy storage.

Pumped Hydro Storage

Pumped hydro storage is a mature and widely deployed energy storage technology that utilizes gravitational potential energy. It involves two water reservoirs situated at different elevations. During periods of excess energy production, water is pumped from the lower reservoir to the upper reservoir, storing the energy. When electricity demand surpasses supply, water is released from the upper reservoir and passes through turbines to generate electricity.

Pumped hydro storage offers large-scale storage capacity, high round-trip efficiency, and long duration discharge capability. It provides grid flexibility and is essential for managing the intermittency of renewable energy sources. However, the geographical constraints, high capital costs, and environmental impacts associated with constructing reservoirs limit the widespread deployment of pumped hydro storage.

Flywheels

Flywheel energy storage systems store energy in the rotational motion of a spinning flywheel. During periods of excess energy, electrical energy is used to accelerate the flywheel to a high rotational speed. When energy is needed, the flywheel releases the stored energy by regenerating electricity. Flywheels offer high power density, rapid response times, and long cycle life.

Flywheel systems are employed in applications that require frequent and rapid charge-discharge cycles, such as grid frequency regulation, uninterruptible power supply systems, and hybrid electric vehicles. Nevertheless, flywheels have limited energy storage capacity and tend to lose energy over time due to friction and bearing losses.

Thermal Energy Storage

Thermal energy storage (TES) technologies store energy in the form of heat. They find applications in both large-scale and small-scale systems, including industrial processes, building heating and cooling, and concentrated solar power plants. TES systems operate by capturing and storing excess thermal energy for later use, providing a flexible means of managing energy demand.

Sensible heat storage involves storing thermal energy by increasing the temperature of a storage medium, such as water, rock, or molten salts. Latent heat storage utilizes phase change materials (PCMs) that absorb or release energy during solid-liquid or liquid-gas phase transitions, respectively. TES systems can enhance the efficiency of energy conversion processes, reduce the reliance on fossil fuels, and contribute to load shifting and grid stability.

Real-World Applications

Energy storage technologies are being deployed worldwide and are revolutionizing the energy landscape. Here are a few real-world examples that highlight the significance of energy storage in various domains:

Tesla Powerpacks in South Australia: In response to severe storms and blackouts in South Australia, Tesla installed the world's largest lithium-ion battery energy storage project. The system, comprising thousands of Powerpacks, provides fast response time, grid stabilization, and reliable power supply during peak demand periods and unexpected outages.

Hawaiian Island Grid Stabilization: The Hawaiian island of Kauai relies heavily on renewable energy generation from solar and wind sources. To maintain grid stability and overcome intermittent energy supply, AES Corporation installed a lithium-ion battery storage system with a capacity of 20 MW and 100 MWh. The project has reduced dependence on fossil fuels and enabled a greater share of renewable energy in the power mix.

Hornsdale Power Reserve in South Australia: Neoen's Hornsdale Power Reserve, featuring Tesla's lithium-ion battery technology, has been instrumental in balancing electricity supply and demand in South Australia's power grid. The project's fast response time and capacity to deliver electricity to the grid within milliseconds have improved grid stability and reduced the frequency of blackouts.

Solar + Storage for Rural Electrification: In regions with limited access to centralized electricity grids, solar energy combined with battery storage systems provides a sustainable and cost-effective solution for rural electrification. These off-grid or mini-grid projects allow communities to access reliable electricity for lighting, communication, and productive use activities, improving their quality of life and economic opportunities.

Overall, energy storage technologies are indispensable for realizing a sustainable and resilient energy future. Continued advancements in energy storage research, development, and deployment are essential to address the challenges associated with integrating a higher share of renewables, enhancing grid resilience, and mitigating climate change.

Battery Technologies

Battery technologies play a crucial role in our pursuit of clean and sustainable energy sources. They enable us to store and efficiently harness energy from renewable sources such as solar and wind, allowing for a more reliable and consistent power supply.   In this section, we will explore the principles, advancements, challenges, and future prospects of battery technologies.

Principles of Battery Operation

To understand the principles of battery operation, we need to delve into the basic electrochemical processes that occur within a battery. Batteries consist of two or more electrochemical cells connected in series or parallel. Each cell consists of an anode (negative electrode) and a cathode (positive electrode) immersed in an electrolyte solution.

The electrochemical reactions that take place in a battery involve the transfer of electrons and ions between the electrodes through the electrolyte. During charging, the cathode undergoes oxidation, releasing electrons into the external circuit, while the anode undergoes reduction, accepting these electrons. This process stores energy in the battery. During discharge, the opposite reactions occur, converting stored chemical energy into electrical energy.

The voltage and capacity of a battery depend on the types of materials used for the electrodes and electrolyte, as well as the design and configuration of the cell. The performance characteristics of batteries, such as energy density, power density, cycle life, and safety, vary based on these parameters.

Advancements in Battery Technologies

Over the years, significant advancements have been made in battery technologies, driving their increased efficiency, capacity, and safety. Some of the notable advancements include:

Lithium-ion Batteries: Lithium-ion (Li-ion) batteries revolutionized portable electronics and electric vehicle (EV) industries due to their high energy density and long cycle life. Li-ion batteries use lithium compounds as the active material in the electrodes, allowing for efficient ion movement and high charge carrying capacity. Moreover, advancements in electrode materials, such as lithium nickel manganese cobalt oxide (NMC) and lithium iron phosphate (LiFePO4), have improved the overall performance and safety of Li-ion batteries.

Solid-State Batteries: Solid-state batteries are a promising advancement in battery technology. Unlike traditional Li-ion batteries, solid-state batteries replace the liquid electrolyte with a solid electrolyte, enhancing safety, energy density, and lifespan. Solid-state batteries can use lithium or other metals as the active material, enabling higher energy storage capacity. Additionally, solid-state batteries exhibit better thermal stability, reducing the risk of thermal runaway and fire hazards.

Flow Batteries: Flow batteries offer a unique approach to energy storage, especially for large-scale applications. They rely on the circulation of electrolyte solutions through separate tanks, making it possible to decouple energy capacity from power output. By adjusting the size of the electrolyte tanks, flow batteries can provide a scalable solution for renewable energy integration and grid-level storage. Vanadium redox flow batteries (VRFB) are a commonly used type of flow battery.

Advancements in Anode and Cathode Materials: Researchers continue to explore new materials for anodes and cathodes to improve battery performance. For example, silicon has shown promise as an anode material due to its high energy storage capacity. However, it faces challenges related to volume expansion during charge-discharge cycles. Similarly, cathode materials like lithium-sulfur and lithium-air are being investigated for their high theoretical energy densities, but they still need to overcome practical challenges such as limited cycle life and poor stability.

Challenges and Future Prospects

While battery technologies have significantly advanced, several challenges remain to be addressed for their widespread adoption and further improvement:

Cost and Scalability: The cost of battery production needs to be reduced to make energy storage more economically viable. This includes optimizing manufacturing processes, sourcing raw materials sustainably, and developing recycling systems to minimize waste and resource depletion. Scalability is also crucial to meet the increasing demand for energy storage in various sectors.

Energy Density and Power Density: Improving energy and power density is essential to increase the range and performance of electric vehicles, support renewable energy integration, and enhance the overall efficiency of energy storage systems. Advancements in materials and cell designs are necessary to achieve these goals.

Safety and Environmental Impact: Ensuring the safety of batteries throughout their life cycle, from manufacturing to disposal, is paramount. This includes addressing issues such as thermal runaway, fire hazards, and the use of environmentally friendly materials. Developing more sustainable and efficient recycling processes is also crucial to minimize the environmental impact of battery waste.

Exploration of New Technologies: Continued exploration of emerging technologies, such as solid-state batteries, lithium-sulfur batteries, and beyond-lithium-ion chemistries, holds great promise for further improving energy storage. These technologies offer the potential to enhance energy density, cycle life, and safety while overcoming the limitations of existing battery systems.

Real-World Example: Tesla Powerwall

An excellent example of the application of battery technologies in the real world is the Tesla Powerwall. The Powerwall is a rechargeable lithium-ion battery system designed for residential energy storage. It stores excess solar energy generated during the day and makes it available during the night or during power outages.

The Powerwall utilizes advanced battery management systems to optimize energy storage and discharge, enabling homeowners to reduce their reliance on the electrical grid and lower their energy costs. It also supports grid services by participating in demand response programs and providing backup power during peak demand periods.

By integrating battery technologies into residential energy systems, the Tesla Powerwall exemplifies how batteries can contribute to a more sustainable and resilient energy future.

Exercises

1. Calculate the specific energy and specific power of a battery given its capacity and voltage.

2. Discuss the environmental and social implications of battery production and disposal.

3. Compare the pros and cons of lithium-ion batteries and solid-state batteries.

4. Research and present a case study on the use of flow batteries for grid-level energy storage.

5. Analyze the impact of electric vehicles on the demand for battery technologies.

Further Reading

1. Armand, M., & Tarascon, J.M. (2008). Building better batteries. Nature, 451(7179), 652-657.

2. Dudney, N.J., & Liang, C. (2012). Batteries—Challenges and prospects. In M.L. Trudeau & M. Scrosati (Eds.), Solid State Electrochemistry and its Applications to Sensors and Electronic Devices (pp. 143-178). Springer.

3. Manthiram, A. (2020). An Outlook on Lithium-Ion Battery Technology. ACS Energy Letters, 5(8), 2468-2485.

4. Yang, Y., Zheng, G., & Cui, Y. (2020). Breaking down solid-state batteries. Joule, 4(10), 2058-2065.

5. Zhang, H., & Liao, X. (2022). Progress and challenges of flow batteries for large-scale energy storage. Chemical Reviews, 122(1), 1090-1133.

Remember, battery technologies are constantly evolving, and staying updated with the latest research and developments is crucial for a comprehensive understanding of this field.

Pumped Hydro Storage

Pumped hydro storage is a widely used method for storing energy in large-scale electricity systems. It is a form of hydroelectric energy storage that uses gravitational potential energy to store excess electricity generated during low-demand periods and release it during high-demand periods. Pumped hydro storage plays a critical role in balancing the supply and demand of electricity in the grid and ensuring grid stability.

Principle of Pumped Hydro Storage

The principle behind pumped hydro storage is relatively simple. During periods of low electricity demand or when there is excess electricity generation from renewable sources such as wind or solar power, the excess electricity is used to pump water from a lower reservoir to a higher reservoir, thereby storing the energy as gravitational potential energy. The process essentially requires two reservoirs at different elevations, connected by a waterway and containing turbines and pumps.

Operation of Pumped Hydro Storage

When there is a high demand for electricity, the stored energy is released by allowing the water from the upper reservoir to flow back down through the waterway, driving turbines to generate electricity. The turbines are connected to generators, which convert the mechanical energy from the flowing water into electrical energy.

Efficiency and Capacity of Pumped Hydro Storage

Pumped hydro storage systems are known for their high efficiency, typically reaching 70-85%. The efficiency of the system is determined by various factors such as the size of the storage reservoirs, the elevation difference between the upper and lower reservoirs, and the hydraulic efficiency of the turbines and pumps.

The capacity of a pumped hydro storage system depends on the size of the reservoirs and the elevation difference. The larger the reservoirs and the greater the elevation difference, the higher the potential energy storage capacity. Pumped hydro storage systems can have capacities ranging from tens to thousands of megawatts, making them suitable for large-scale energy storage applications.

Advantages of Pumped Hydro Storage

Pumped hydro storage offers several advantages over other forms of energy storage. First, it has a long lifespan and can operate for several decades without significant degradation. It also has a fast response time, allowing for quick adjustments to changes in electricity demand or supply.

Furthermore, pumped hydro storage has a high energy density, meaning it can store large amounts of energy in a relatively small space. This makes it particularly suitable for grid-scale energy storage where space may be limited.

Challenges and Considerations

While pumped hydro storage has many advantages, there are also challenges and considerations to keep in mind. One major challenge is site availability. Pumped hydro storage requires specific geographic conditions with suitable elevations and available water sources. Identifying suitable sites for new pumped hydro storage facilities can be challenging, especially in densely populated areas.

Additionally, pumped hydro storage systems can have high upfront costs and require significant capital investment. The construction of reservoirs, tunnels, and other infrastructure can be expensive and time-consuming. However, once built, the operational costs are relatively low, making it cost-effective in the long run.

Real-World Example: Dinorwig Power Station

An exemplary real-world example of a pumped hydro storage facility is the Dinorwig Power Station in Wales, UK. It is one of the largest pumped hydro storage facilities in Europe and has a capacity of 1.7 GW. The facility has been operational since 1984 and plays a crucial role in the UK's electricity grid by supplying additional power during peak demand periods.

The Dinorwig Power Station features a lower reservoir located inside a mountain and an upper reservoir located on the surface. It utilizes surplus electricity from the grid to pump water from the lower reservoir to the upper reservoir during off-peak periods. During periods of high demand, the stored water is released back to the lower reservoir, generating electricity by passing through turbines.

Exercise

Consider a hypothetical pumped hydro storage system with a total elevation difference between the upper and lower reservoirs of 500 meters. The upper reservoir has a storage capacity of 5 million cubic meters of water. Determine the potential energy storage capacity of this system.

Solution to Exercise

The potential energy storage capacity of the pumped hydro storage system can be calculated using the formula:

$$\text{Energy storage capacity} = \text{Storage capacity} \times g \times \text{elevation difference},$$

where g is the acceleration due to gravity.
Plugging in the given values, we have:

$$\text{Energy storage capacity} = 5,000,000 \, \text{m}^3 \times 9.8 \, \text{m/s}^2 \times 500 \, \text{m} = 2.45 \times 10^9 \, \text{Joules}.$$

Therefore, the potential energy storage capacity of the system is approximately 2.45 gigajoules (GJ).

Fuel cells

Fuel cells are devices that convert the chemical energy stored in fuels directly into electrical energy through an electrochemical reaction. They offer a promising solution to the growing demands for clean and sustainable energy sources. Fuel cells have the potential to revolutionize energy production by providing efficient and low-emission power generation.

Principles of fuel cells

Fuel cells operate based on the principles of electrochemistry. They consist of an anode, a cathode, and an electrolyte. The anode is the negative electrode where the fuel is oxidized, and the cathode is the positive electrode where the oxidant is reduced. The electrolyte acts as a medium for the ions to travel between the anode and cathode.

The overall chemical reaction in a fuel cell can be represented as follows:

$$\text{Fuel} + \text{Oxidant} \rightarrow \text{Products} + \text{Electricity}$$

Different types of fuel cells exist, depending on the type of electrolyte used. Some common types include proton exchange membrane fuel cells (PEMFCs), alkaline fuel cells (AFCs), solid oxide fuel cells (SOFCs), and molten carbonate fuel cells (MCFCs).

Advantages of fuel cells

Fuel cells offer several advantages over traditional energy conversion methods:

+ **High energy efficiency:** Fuel cells can achieve energy conversion efficiencies of up to 60%, compared to around 30-40% for conventional power generation.

+ **Low environmental impact:** Fuel cells produce minimal greenhouse gas emissions since they do not burn fossil fuels. The only byproducts of the chemical reaction are water and heat.

+ **Versatile fuel options:** Fuel cells can use a wide range of fuels such as hydrogen, natural gas, methane, ethanol, and even biogas. This flexibility allows for the utilization of renewable and sustainable energy sources.

+ **Quiet and vibration-free operation:** Fuel cells operate silently and with minimal vibration, making them suitable for various applications, including portable devices and transportation.

+ **Modular and scalable design:** Fuel cells can be easily interconnected to meet different power demands. They can range in size from small, portable units to large-scale power plants.

Challenges and ongoing research

Despite their numerous advantages, fuel cells still face challenges that hinder their widespread implementation. Some of these challenges include:

+ **Cost:** The cost of fuel cells, especially those based on expensive materials such as platinum, remains a significant barrier to their commercialization. Research is focused on developing alternative, low-cost catalysts and improving manufacturing processes.

- **Infrastructural requirements**: Fuel cells require an infrastructure for fuel storage, distribution, and refueling. Creating a reliable and widespread infrastructure for fuels such as hydrogen is a challenge that needs to be addressed.

- **Durability and lifetime**: Improving the durability and long-term performance of fuel cells is a critical area of research. Extending their lifespan and ensuring their reliability under various operating conditions is crucial for their viability.

To overcome these challenges, ongoing research is being conducted in various areas related to fuel cells. These include:

- **Catalyst development**: Researchers are exploring new catalyst materials, such as non-precious metals or metal-free catalysts, to reduce the reliance on expensive and rare resources.

- **Fuel flexibility**: Efforts are being made to develop fuel cells that can utilize a wider range of fuels, including renewable and carbon-neutral sources.

- **System integration**: Integrating fuel cells with other energy conversion and storage technologies, such as batteries and supercapacitors, can enhance their overall performance and reliability.

- **Infrastructure development**: Research is focused on developing cost-effective and efficient infrastructure for fuel storage, distribution, and refueling, with a particular emphasis on hydrogen.

By addressing these challenges and further advancing the research in fuel cells, they can play a significant role in the transition towards a sustainable energy future.

Case study: Fuel cell vehicles

One exciting application of fuel cells is in the transportation sector, particularly in fuel cell vehicles (FCVs). FCVs use fuel cells to convert hydrogen into electricity, powering an electric motor that drives the vehicle. They offer several advantages over conventional internal combustion engine vehicles:

- **Zero-emission operation**: FCVs produce zero tailpipe emissions. The only byproduct is water vapor, making them environmentally friendly.

- **Longer driving range**: FCVs have a longer driving range compared to battery electric vehicles. They can be refueled quickly, similar to traditional gasoline-powered vehicles.

- **Fast refueling**: Refueling a FCV with hydrogen takes only a few minutes, providing a convenient alternative to recharging batteries for long-distance travel.

However, FCVs also face challenges in terms of hydrogen infrastructure, cost, and public acceptance. Efforts are underway to expand the hydrogen refueling infrastructure and reduce the cost of fuel cells to make FCVs more accessible to consumers.

Conclusion

Fuel cells hold great promise in providing clean and efficient energy solutions. Their ability to convert various fuels into electricity with high efficiency and low emissions makes them an attractive alternative to conventional power generation methods. Ongoing research and development efforts are focused on addressing the challenges associated with fuel cells, such as cost and infrastructural requirements, to accelerate their deployment in various sectors, including transportation, residential, and industrial applications. The advancement of fuel cell technology will contribute significantly to achieving a sustainable and carbon-neutral future.

Energy Efficiency

Energy efficiency is a key concept in achieving sustainability and reducing the impact of human activities on the environment. It refers to the ability to accomplish a desired task or fulfill a particular function with the least amount of energy input. In other words, energy efficiency aims to maximize output while minimizing energy consumption, waste, and associated costs.

The importance of energy efficiency is undeniable. It reduces greenhouse gas emissions, helps in the conservation of natural resources, improves energy security, and enhances economic competitiveness. Moreover, energy efficiency plays a crucial role in mitigating climate change and achieving the sustainable development goals set by the United Nations.

Theoretical Background

The principle of energy efficiency is rooted in the laws of thermodynamics, particularly the first and second laws. The first law, also known as the law of energy conservation, states that energy cannot be created or destroyed; it can only be converted from one form to another. Therefore, any energy input into a system must equal the output plus the energy lost as waste or heat.

The second law of thermodynamics states that in any energy conversion process, the total entropy of the system and its surroundings always increases. It implies that some energy is always dissipated, making it impossible to achieve a perfect or 100% energy conversion efficiency.

Based on these principles, energy efficiency can be quantified using different metrics. Some common metrics include energy conversion efficiency, energy use intensity, and energy efficiency ratio. These metrics are used to evaluate the efficiency of various energy-consuming systems, such as appliances, buildings, transportation, and industrial processes.

Benefits of Energy Efficiency

Energy efficiency offers numerous benefits at both individual and societal levels. Let's explore some of these benefits:

+ **Reduced Energy Consumption:** Improving energy efficiency helps to reduce overall energy consumption. This leads to a decrease in energy dependence, promoting energy security and reducing the need for new energy infrastructure.

+ **Cost Savings:** Energy-efficient technologies and practices can significantly reduce energy bills for households, businesses, and industries. By using less energy to achieve the same or even better results, substantial cost savings can be realized in the long run.

+ **Environmental Protection:** Energy efficiency directly contributes to the reduction of greenhouse gas emissions, air pollution, and other environmental impacts associated with energy generation and consumption. It helps to combat climate change and preserve valuable natural resources.

+ **Job Creation:** The implementation of energy-efficient measures and technologies creates job opportunities in various sectors, including manufacturing, construction, engineering, and research. These jobs often require specialized skills and contribute to economic growth.

+ **Improved Comfort and Productivity:** Energy-efficient buildings and workplaces offer enhanced indoor environmental quality and occupant comfort.   Comfortable and healthy environments have been shown to improve productivity, well-being, and overall satisfaction.

Energy Efficiency Challenges

Despite the benefits, there are several challenges to achieving energy efficiency at a large scale. These challenges include:

+ **Cost and Financing:**   Upgrading existing infrastructure or adopting energy-efficient technologies can be costly, especially for low-income households and developing countries. Access to financing and incentives is crucial to overcome these barriers.

+ **Behavioral and Cultural Factors:** Human behavior and cultural norms often resist changes in energy consumption patterns.   People may be reluctant to adopt energy-efficient practices due to lack of awareness, perceived inconvenience, or perceived high upfront costs.

+ **Lack of Information and Education:** Many individuals and organizations are unaware of the potential energy-saving opportunities or lack the necessary information to make informed decisions.   Increasing public awareness and providing education and training can help overcome this challenge.

+ **Technical Barriers:**   Energy efficiency improvements may require the deployment of advanced technologies, which could face technical barriers, such as the limited availability of certain technologies, inadequate infrastructure, or compatibility issues with existing systems.

Energy Efficiency Strategies

Various strategies and measures are being implemented to promote energy efficiency across different sectors. Here are some examples:

+ **Energy-Efficient Appliances:** The development and use of energy-efficient appliances, such as refrigerators, air conditioners, and lighting systems, can significantly reduce the energy consumption in households and commercial buildings.

+ **Building Design and Retrofits:** Energy-efficient building design, including proper insulation, efficient heating, ventilation, and air conditioning (HVAC) systems, and the use of smart technologies, can reduce energy demand in both residential and commercial buildings. Retrofitting existing buildings with energy-efficient features can also yield significant energy savings.

+ **Transportation Efficiency:** Promoting energy-efficient transportation systems, such as electric vehicles, hybrid cars, and efficient public transportation networks, can help reduce fuel consumption and emissions in the transportation sector.

+ **Industrial Process Optimization:** Industries can improve energy efficiency by optimizing production processes, such as using energy-efficient equipment, implementing waste-heat recovery systems, and adopting advanced process controls.

+ **Awareness and Education:** Raising awareness and educating individuals, communities, and businesses about the importance of energy efficiency and providing information on available energy-saving measures can drive behavioral changes and foster energy-efficient practices.

+ **Policy and Incentives:** Governments and regulatory bodies play a crucial role in promoting energy efficiency through policies, regulations, and financial incentives. These could include energy efficiency standards, tax incentives, rebates, and grants for energy-efficient upgrades.

Case Study: Energy-Efficient Lighting

As an example, let's consider the transition from traditional incandescent light bulbs to energy-efficient alternatives, such as light-emitting diode (LED) bulbs. Incandescent bulbs convert only a small fraction of the electrical energy they consume into visible light, with the rest being wasted as heat. On the other hand, LED bulbs are much more efficient, converting a higher proportion of energy into light.

By replacing incandescent bulbs with LED bulbs, significant energy savings can be achieved. For instance, a typical LED bulb may consume up to 80% less energy compared to an incandescent bulb and last much longer. This not only reduces electricity bills but also decreases the demand for electricity generation and reduces greenhouse gas emissions.

Furthermore, LED bulbs provide other benefits, such as better light quality, lower maintenance costs, and reduced fire hazards. Although the upfront cost of LED bulbs is higher, the long-term cost savings and environmental benefits outweigh the initial investment.

Conclusion

Energy efficiency is a vital tool for achieving sustainability and mitigating climate change. The concept is based on the laws of thermodynamics and aims to maximize output while minimizing energy consumption and waste. Energy efficiency offers various benefits, including cost savings, environmental protection, and improved comfort and productivity. However, challenges such as cost barriers, behavioral factors, and lack of information need to be addressed. By implementing energy-efficient strategies and measures, we can collectively reduce energy consumption, promote economic growth, and pave the way for a sustainable future.

Building design

Building design plays a crucial role in promoting sustainability and reducing the environmental impact of the built environment. In this section, we will discuss the key principles, strategies, and technologies involved in designing green and energy-efficient buildings.

Principles of sustainable building design

Sustainable building design aims to achieve a balance between environmental responsibility, social equity, and economic viability. Here are the key principles guiding sustainable building design:

1. **Energy efficiency**: Designing buildings to minimize energy consumption is essential for reducing greenhouse gas emissions and dependence on fossil fuels. This can be achieved through proper insulation, efficient HVAC systems, and the use of energy-efficient lighting and appliances.

2. **Passive design**: Passive design focuses on maximizing the use of natural resources, such as sunlight, ventilation, and shading, to reduce the need for artificial heating, cooling, and lighting. It involves strategic orientation, window placement, and the use of shading devices to optimize natural light and thermal comfort.

3. **Water conservation**: Efficient use of water is another vital aspect of sustainable building design. This can be achieved through the use of low-flow fixtures, rainwater harvesting systems, and the implementation of water recycling and reuse techniques.

4. **Material selection**: Choosing environmentally-friendly and locally-sourced materials can significantly reduce the carbon footprint of a building. The use of recycled or reclaimed materials, low-emission products, and sustainable wood products can minimize the environmental impact of construction.

5. **Waste reduction**: Designing for waste reduction involves implementing strategies for reducing construction and operational waste. This includes the use of prefabricated construction methods, recycling and sorting systems, and designing for deconstruction and material reuse.

6. **Indoor environmental quality**: Providing a healthy and comfortable indoor environment is crucial for the occupants' well-being. This can be achieved through adequate ventilation, optimal indoor air quality, natural lighting, and the use of low-toxicity building materials.

7. **Lifecycle assessment**: Assessing the environmental impact of a building throughout its entire lifecycle, from construction to operation and eventual demolition, helps identify areas for improvement and optimization. Lifecycle assessment considers factors such as energy consumption, water usage, material extraction, and waste generation.

Innovative technologies for sustainable building design

Several innovative technologies can be integrated into building design to enhance sustainability and energy efficiency. Here are some examples:

1. **Solar panels**: The integration of solar photovoltaic (PV) panels in building design allows for the generation of clean and renewable electricity. These panels can be incorporated into the building's facade, roof, or shading devices to maximize energy production.

2. **Green roofs**: Green roofs are vegetated surfaces installed on top of buildings. They provide insulation, absorb rainwater, reduce the urban heat island effect, and enhance biodiversity. Green roofs also help improve air quality and provide additional recreational and green space.

3. **Energy-efficient glazing**: High-performance windows with advanced glazing technologies help mitigate heat gain and loss, reducing the reliance on artificial heating and cooling systems. These windows can have low-emissivity coatings, multiple panes, and gas fills to enhance thermal insulation and maximize natural light.

4. **Smart building controls:** The integration of smart building controls, including occupancy sensors, automated lighting, and advanced HVAC systems, allows for optimal energy management. These systems can adjust lighting, heating, and cooling based on occupancy and usage patterns, resulting in significant energy savings.

5. **Building-integrated vegetation:** Incorporating vegetation systems within the building envelope can provide a range of benefits, including improved air quality, thermal insulation, noise reduction, and aesthetic appeal. Examples include living walls, vertical gardens, and interior green spaces.

6. **Advanced insulation materials:** The use of high-performance insulation materials, such as aerogels or vacuum-insulated panels, can enhance the thermal efficiency of buildings and reduce energy losses.

7. **Greywater systems:** Greywater systems allow for the collection and treatment of non-potable water from sources like sinks, showers, and washing machines. Treated greywater can be reused for irrigation, toilet flushing, or other non-potable purposes, reducing water demand.

Case study: The Edge, Amsterdam

To illustrate the principles and technologies discussed, let's explore a real-world example: The Edge, a sustainable office building located in Amsterdam, Netherlands.

Passive design and energy efficiency: The Edge incorporates passive design strategies such as optimal orientation, triple-glazed facades with external sunshades, and efficient lighting systems. It also utilizes smart building controls, including occupancy sensors and smartphone apps, to optimize energy consumption and enable personalized workspace control.

Renewable energy generation: The building features one of the largest building-integrated solar panel arrays in Europe, covering the roof and south-facing facades. This solar energy system generates more electricity than the building requires, allowing for the export of surplus energy.

Energy storage and management: The Edge utilizes an advanced energy storage system based on lithium-ion batteries. These batteries store excess solar energy and provide it during times of high demand. The building's energy management system optimizes energy usage based on real-time data, resulting in efficient energy distribution.

Indoor environmental quality: The Edge provides a comfortable indoor environment through features such as efficient ventilation systems, advanced air purification systems, and individualized climate control for occupants.

Smart technologies and data analytics: The building employs various sensors and meters to collect data on energy usage, occupancy patterns, and indoor environmental quality. This data is analyzed to identify areas for improvement and optimize building performance further.

Through the integration of sustainable building design principles and innovative technologies, The Edge has achieved exceptional energy efficiency, reduced environmental impact, and improved occupant well-being.

Conclusion

Building design plays a crucial role in promoting sustainability and energy efficiency. By adopting principles such as energy efficiency, passive design, water conservation, and waste reduction, along with the incorporation of innovative technologies like solar panels, green roofs, and smart building controls, we can create buildings that minimize their environmental impact and provide healthy indoor environments.

Sustainable building design not only reduces energy consumption and greenhouse gas emissions but also improves occupant comfort and well-being. It is essential for creating a more sustainable and resilient built environment for the future.

Key Takeaways:

- Sustainable building design aims to balance environmental responsibility, social equity, and economic viability.

- Principles include energy efficiency, passive design, water conservation, material selection, waste reduction, indoor environmental quality, and lifecycle assessment.

- Innovative technologies for building design include solar panels, green roofs, energy-efficient glazing, smart building controls, and building-integrated vegetation.

- The Edge in Amsterdam is an exemplary sustainable building that incorporates passive design, renewable energy generation, and advanced energy management systems.

- Sustainable building design promotes energy efficiency, reduces environmental impact, and improves occupant well-being.

Resources:

+ U.S. Green Building Council (USGBC) - www.usgbc.org

+ World Green Building Council (WorldGBC) - www.worldgbc.org

+ International Living Future Institute (ILFI) - www.living-future.org

+ Building Research Establishment Environmental Assessment Method (BREEAM) - www.breeam.com

Exercises:

1. Choose a local building in your community and assess its sustainability features based on the principles discussed in this section. Identify areas for improvement and propose sustainable design strategies.

2. Research and compare different types of insulation materials used in buildings. Assess their thermal performance, environmental impact, and cost-effectiveness.

3. Investigate the impact of urban heat islands and the role of building design in mitigating their effects. Propose design strategies to reduce urban heat island effects in cities.

Building design is a constantly evolving domain, driven by technological advancements, environmental considerations, and societal needs. By incorporating sustainable design principles and innovative technologies, we can create buildings that are not only energy-efficient and environmentally friendly but also provide healthy and comfortable spaces for occupants. The future of building design lies in our ability to embrace these principles, integrate emerging technologies, and collaborate towards a more sustainable and resilient built environment.

Industrial Processes

Industrial processes play a significant role in the global economy, providing goods and services that sustain our daily lives. However, these processes often have negative impacts on the environment, including resource depletion, pollution, and greenhouse gas emissions. To achieve sustainability, it is crucial to develop and implement eco-friendly practices in industrial processes. In this section, we will explore various advancements and strategies that aim to minimize the environmental impact of industrial activities.

Resource Efficiency

One of the key challenges in industrial processes is the efficient use of resources. Many industries rely on non-renewable resources, such as fossil fuels and minerals, which are rapidly depleting. To address this issue, resource efficiency focuses on minimizing waste and optimizing resource use.

One approach to improving resource efficiency is through the implementation of cleaner production techniques. This involves the adoption of technologies and practices that reduce the consumption of energy and raw materials, and minimize the generation of waste and emissions. For example, process modifications can be made to improve energy efficiency, such as heat recovery systems, cogeneration, and the use of renewable energy sources.

Another strategy is the adoption of circular economy principles. This approach aims to close the resource loop by promoting recycling, reuse, and remanufacturing. By designing products for longevity and recyclability, and by implementing effective recycling and waste management systems, industrial processes can greatly reduce the demand for raw materials and minimize waste generation.

Emission Reduction

Industrial activities are responsible for a significant portion of global greenhouse gas emissions, contributing to climate change. To mitigate these emissions, industries need to adopt cleaner technologies and practices that reduce their carbon footprint.

One effective method is to transition to cleaner energy sources. Renewable energy technologies, such as solar and wind power, can replace fossil fuel-based electricity generation, reducing both emissions and dependence on finite resources. In addition to on-site renewable energy generation, companies can also purchase renewable energy credits or enter into power purchase agreements with renewable energy providers.

Industrial processes can also reduce emissions through energy efficiency improvements. Energy audits and process optimization can identify areas of high energy consumption and implement energy-saving measures. Energy-efficient equipment and technologies, such as high-efficiency motors and lighting systems, can be used to reduce energy demand and associated emissions.

Furthermore, carbon capture and storage (CCS) technologies can help mitigate emissions from industrial processes that are difficult to decarbonize. CCS involves capturing carbon dioxide emissions from industrial flue gases and storing them underground, preventing their release into the atmosphere. While CCS is still a developing technology, it holds significant potential for reducing emissions from industries such as cement production and steelmaking.

Waste Management

Effective waste management is essential for sustainable industrial processes. Waste generated by industrial activities can have harmful effects on the environment and public health if not properly managed. Therefore, industries need to implement strategies to minimize waste generation and responsibly handle and dispose of waste.

One approach is to adopt the principles of waste reduction, which include waste prevention, recycling, and proper disposal. Waste prevention aims to minimize the generation of waste at its source through process optimization and product design. This can be achieved through techniques such as lean manufacturing, where processes are streamlined to reduce material waste and improve efficiency.

Recycling plays a critical role in waste management, as it reduces the demand for raw materials and decreases energy consumption and emissions associated with extraction and production. Industries can implement recycling programs for materials such as paper, plastics, and metals, as well as explore innovative technologies for recycling complex materials.

For waste that cannot be recycled or prevented, proper disposal is crucial. Industries must adhere to regulations and standards for the safe handling and disposal of hazardous waste. Technologies such as incineration, anaerobic digestion, and landfill gas capture can be employed to minimize the environmental impact of waste disposal.

Green Chemistry

Green chemistry aims to develop and utilize chemical processes that are environmentally friendly, efficient, and sustainable. It focuses on the design and use of chemicals that minimize the use and generation of hazardous substances, and promote the use of safer alternatives.

One aspect of green chemistry is the development of alternative solvents that are less toxic and more sustainable than traditional organic solvents. These solvents can be used in various industrial processes, such as pharmaceutical manufacturing and cleaning products production. Bio-based solvents, for example, are derived from renewable resources and offer lower toxicity and improved biodegradability.

Another area of focus is the development of catalysts that enhance the efficiency of chemical reactions, reducing energy consumption and waste generation. Catalysts can enable processes to occur at lower temperatures, reducing the need for high-energy heating. They can also increase the selectivity of reactions, minimizing the formation of unwanted by-products.

In addition, green chemistry emphasizes the use of renewable feedstocks in chemical production. By replacing fossil fuel-based feedstocks with biomass-derived or waste-derived materials, industries can reduce their carbon footprint and reliance on non-renewable resources.

Overall, integrating green chemistry principles into industrial processes can lead to significant environmental benefits, including reduced waste, emissions, and human and ecological toxicity.

Case Study: Sustainable Manufacturing in the Auto Industry

The automotive industry is one of the largest industrial sectors globally, and it has a significant environmental impact. However, many automobile manufacturers have made efforts to adopt sustainable practices in their industrial processes.

One such example is the implementation of lightweight materials in vehicle production. By using lightweight materials such as aluminum and carbon fiber composites, manufacturers can reduce the weight of vehicles, leading to improved fuel efficiency and reduced emissions. Additionally, the use of recycled and bio-based materials in vehicle interiors can further decrease the environmental impact of manufacturing.

Moreover, the auto industry has been investing in cleaner and more energy-efficient technologies. Electric vehicle (EV) production has gained momentum, and several automakers have shifted their focus towards the development and manufacturing of EVs. The production of electric vehicles significantly reduces greenhouse gas emissions compared to conventional internal combustion engine vehicles.

Furthermore, sustainable manufacturing practices have been implemented in the automotive industry through the adoption of circular economy principles. Manufacturers are increasingly implementing recycling programs for end-of-life vehicles, ensuring that valuable materials are recovered and reused. This not only reduces the demand for virgin materials but also minimizes waste generation and disposal.

In conclusion, sustainable industrial processes are crucial for achieving environmental sustainability. Resource efficiency, emission reduction, waste management, and green chemistry are key elements in creating eco-friendly industries. By adopting these principles and strategies, industries can minimize their environmental impact and contribute to a more sustainable future.

Summary

In this section, we explored the importance of eco-friendly industrial processes and discussed various advancements and strategies for achieving sustainability. Resource efficiency, emission reduction, waste management, and green chemistry were highlighted as key aspects of sustainable industrial practices.

Implementing cleaner production techniques and circular economy principles can enhance resource efficiency and minimize waste generation. Transitioning to cleaner energy sources, improving energy efficiency, and adopting carbon capture and storage technologies can reduce emissions from industrial processes. Effective waste management practices, including waste prevention, recycling, and proper disposal, are essential for minimizing environmental impacts.

Green chemistry principles, such as the use of alternative solvents, catalysts, and renewable feedstocks, promote the development of environmentally friendly chemical processes. We also examined a case study on sustainable manufacturing in the auto industry, which showcased the application of sustainable practices in a large-scale industrial sector.

By adopting these eco-friendly practices and embracing technological innovations, industrial processes can contribute to a more sustainable future while continuing to meet the needs of society. It is essential for industries to prioritize sustainability and work towards minimizing their environmental footprint for the well-being of the planet and future generations.

Transportation systems

Transportation systems play a crucial role in modern society, enabling the movement of people and goods over various distances. However, conventional transportation methods heavily rely on fossil fuels, resulting in significant environmental impacts such as air pollution and greenhouse gas emissions. To address these challenges and promote sustainability, advancements in eco-friendly transportation technologies and infrastructure are necessary. This section explores various sustainable transportation options and their potential to contribute to a greener future.

Electric vehicles

Electric vehicles (EVs) have gained significant attention as a cleaner alternative to traditional gasoline-powered cars. Unlike internal combustion engines, EVs rely on electric motors powered by rechargeable batteries. This eliminates tailpipe emissions, reducing air pollution and improving local air quality. Additionally, the increasing use of renewable energy sources for electricity generation contributes to a significant reduction in greenhouse gas emissions associated with EVs.

One of the main challenges with EVs is the limited range and the need for frequent recharging. However, advancements in battery technology are extending the range of EVs and reducing charging time. Moreover, the development of fast-charging stations enables quick recharging, making EVs more practical for everyday use.

Charging infrastructure

To support the widespread adoption of EVs, a robust and accessible charging infrastructure is essential. Charging stations can be categorized into three levels based on the power output and charging time:

1. Level 1: Also known as trickle charging, this level provides the lowest power output and is suitable for home charging. It uses a standard household electrical outlet (120 volts) and charges the vehicle slowly over a few hours.

2. Level 2: This level offers a higher power output (240 volts) and reduces the charging time significantly. Level 2 chargers are typically installed in public spaces, workplaces, and residential complexes.

3. Level 3: Commonly known as DC fast charging, this level provides the highest power output and can charge an EV to 80% capacity in a relatively short period, usually less than an hour. Level 3 chargers are installed along major highways and in commercial areas.

The expansion of charging infrastructure is essential for encouraging EV adoption and alleviating range anxiety among potential EV owners. Governments, private companies, and organizations are investing in the installation of charging stations, aiming to create a comprehensive network that ensures easy access to charging facilities.

Battery technology

The performance and range of EVs heavily depend on the type and capacity of the batteries used. Currently, lithium-ion batteries dominate the EV market due to their high energy density, longevity, and relatively low self-discharge rate. However, research and development efforts are ongoing to improve battery technology further.

One promising technology is solid-state batteries, which offer higher energy density, faster charging times, and improved safety compared to conventional lithium-ion batteries. Solid-state batteries utilize solid electrolytes instead of liquid electrolytes, eliminating the risk of leakage and enhancing thermal stability. However, commercialization and cost reduction remain significant challenges before solid-state batteries become widely available.

Benefits and challenges

The adoption of sustainable transportation systems, such as electric vehicles, presents numerous benefits:

+ Reduced greenhouse gas emissions: By shifting from fossil fuel-powered vehicles to EVs, the transportation sector can significantly contribute to mitigating climate change.

+ Improved air quality: EVs produce zero tailpipe emissions, reducing air pollution and improving public health, especially in urban areas.

+ Energy efficiency: EVs have higher energy efficiency compared to internal combustion engine vehicles, as electric motors convert a higher percentage of energy into motion.

+ Lower operating costs: EVs have fewer moving parts and require less maintenance, resulting in lower operating and maintenance costs over their lifetime.

However, several challenges need to be addressed for the widespread adoption of sustainable transportation systems:

+ Infrastructure development: The establishment of a comprehensive charging infrastructure network is crucial for eliminating range anxiety and providing convenient charging options for EV owners.

- Affordability: The upfront cost of EVs is often higher than conventional vehicles due to the cost of batteries. However, technological advancements and government incentives are gradually reducing the price gap.

- Limited battery range: Although battery technology has made significant progress, the range of EVs is still limited compared to traditional vehicles. Continued research and development efforts are essential to extend the range and improve battery performance.

- Sustainable energy sources: While EVs themselves produce zero emissions, the environmental impact depends on the source of electricity generation. Shifting to renewable energy sources for electricity production is vital for maximizing the environmental benefits of EVs.

Public transportation

In addition to personal vehicles, sustainable transportation systems also encompass public transportation options. Efficient and eco-friendly public transportation can greatly reduce the number of private vehicles on the road, alleviate traffic congestion, and lower carbon emissions.

Bus rapid transit (BRT) is a sustainable transport system that combines the advantages of buses and light rail systems. BRT systems provide dedicated lanes for buses, allowing them to bypass traffic congestion and maintain a regular schedule. Furthermore, BRT vehicles are often powered by clean energy sources, such as electricity or biofuels, resulting in reduced greenhouse gas emissions.

Light rail systems are another sustainable alternative to traditional transportation methods. Light rail vehicles run on electrified rails, eliminating direct emissions. They provide a higher passenger capacity compared to buses and offer a reliable and efficient mode of transportation, particularly in densely populated urban areas.

Sustainable aviation

The aviation industry is a significant contributor to greenhouse gas emissions and air pollution. To address these environmental challenges, sustainable aviation initiatives are focused on developing alternative fuels, improving aircraft design, and implementing efficient air traffic management systems.

Biofuels are considered a promising option for reducing the carbon footprint of aviation. Biofuels produced from renewable sources, such as plant biomass or algae, have the potential to offer substantial greenhouse gas emissions reductions compared to traditional jet fuels.

Moreover, advancements in aircraft design, including more aerodynamic shapes, lighter materials, and improved engine efficiency, can contribute to reducing fuel consumption and emissions. Additionally, the implementation of sophisticated air traffic management systems can optimize flight routes, reducing unnecessary fuel consumption and emissions.

Resource optimization

Sustainable transportation also involves optimizing resource use and reducing waste. One approach is the implementation of *car-sharing* or *ride-sharing* programs, reducing the number of private vehicles on the road and promoting a more efficient use of vehicles. These programs encourage individuals to share rides, maximizing the passenger capacity of vehicles and minimizing empty seats on the road.

Furthermore, *intelligent transportation systems* (ITS) utilize advanced technologies to optimize traffic management, reduce congestion, and improve overall transportation efficiency. ITS applications include real-time traffic monitoring, adaptive traffic signal control, and dynamic route planning for public transportation.

Conclusion

Transportation systems play a pivotal role in shaping a sustainable future. Advancements in electric vehicles, improved charging infrastructure, and the use of renewable energy sources are transforming the way we travel and reducing the environmental impact of transportation. Additionally, sustainable public transportation, initiatives in aviation, and resource optimization strategies contribute to creating more eco-friendly transportation systems. As technology continues to evolve and awareness of sustainability grows, the transportation sector holds great potential for advancing towards a greener and more efficient future.

Box: *Case Study: The Role of Sustainable Transportation in Curbing Urban Air Pollution*

Urban air pollution is a significant health concern in many cities around the world. One of the primary contributors to air pollution is the transportation sector, particularly emissions from vehicles. Implementing sustainable transportation options can play a crucial role in curbing urban air pollution and improving public health.

For instance, Bogotá, the capital city of Colombia, has been recognized for its successful implementation of sustainable transportation initiatives. The city established a bus rapid transit (BRT) system, called TransMilenio, which transformed the transportation landscape by providing an efficient, affordable, and low-emission mode of transportation. TransMilenio offers dedicated bus lanes, reducing travel times and minimizing air pollution.

In addition to BRT, Bogotá also introduced an extensive network of cycling paths and bike-sharing programs, encouraging residents to opt for non-motorized modes of transportation. These efforts have not only reduced air pollution and greenhouse gas emissions but also improved public health by promoting physical activity.

The case of Bogotá highlights the potential of sustainable transportation systems in combating urban air pollution and creating healthier and more livable cities. Implementing a comprehensive and well-integrated sustainable transportation strategy can significantly contribute to reducing air pollution, mitigating climate change, and improving the quality of life in urban areas.

Green Transportation

Electric vehicles

Electric vehicles (EVs) are vehicles powered by electric motors that run on electricity stored in rechargeable batteries. They are an important component of sustainable transportation and have gained significant attention and popularity in recent years. This section will provide an overview of electric vehicles, including their benefits, challenges, and their role in achieving a sustainable transportation system.

Background

The development of electric vehicles dates back to the early 19th century, with the first electric car invented in 1837 by Robert Anderson. However, it was not until the late 20th century that EVs started to gain momentum. With advancements in battery technology and concerns about environmental pollution caused by conventional vehicles, the need for clean and efficient transportation became apparent.

Principles of Electric Vehicles

Electric vehicles operate on the principles of electromagnetism and energy conversion. They use an electric motor to convert electrical energy stored in batteries into mechanical energy, which propels the vehicle. The key components of an electric vehicle include:

- **Electric motor:** The electric motor is the heart of an electric vehicle. It converts electrical energy into mechanical energy to drive the wheels of the vehicle.

- **Battery pack:** The battery pack is responsible for storing and supplying electrical energy to the electric motor. It is typically made up of a series of rechargeable batteries.

- **Power electronics:** The power electronics system controls the flow of electrical energy between the battery pack and the electric motor. It includes components such as inverters, converters, and controllers.

- **Charging system:** Electric vehicles need to be charged from an external power source. The charging system comprises a charging port, on-board charger, and associated electronics to facilitate the charging process.

Benefits of Electric Vehicles

Electric vehicles offer numerous benefits compared to traditional internal combustion engine vehicles. Some of the key benefits include:

- **Reduced greenhouse gas emissions:** Electric vehicles produce zero tailpipe emissions, helping to reduce greenhouse gas emissions and combat climate change.

- **Improved air quality:** By eliminating exhaust emissions, electric vehicles contribute to improved air quality, especially in densely populated urban areas.

- **Energy efficiency:** Electric vehicles are generally more energy-efficient than internal combustion engine vehicles. They convert a higher percentage of the energy from the grid to power at the wheels.

- **Reduced dependency on fossil fuels:** Electric vehicles reduce reliance on fossil fuels, as they can be powered by renewable energy sources like solar or wind power.

- **Lower operating costs:** Electric vehicles have lower operating costs compared to conventional vehicles, primarily due to lower fuel and maintenance costs.

- **Quiet operation:** Electric vehicles produce less noise pollution, leading to quieter urban environments.

Challenges of Electric Vehicles

Although electric vehicles offer numerous benefits, they also face several challenges that hinder their widespread adoption. Some of the key challenges include:

+ **Limited driving range:** The driving range of electric vehicles is typically shorter compared to conventional vehicles. However, advancements in battery technology are continuously improving the range.

+ **Charging infrastructure:** The availability of charging infrastructure is a crucial factor for electric vehicle adoption. The limited number of charging stations and the time required for charging can be perceived as barriers by potential EV owners.

+ **High vehicle cost:** Electric vehicles tend to have a higher upfront cost compared to traditional vehicles. However, as battery costs decrease and economies of scale improve, the price of EVs is expected to become more affordable.

+ **Battery technology limitations:** Battery technology still faces limitations such as energy density, charging time, and lifespan. Ongoing research and development efforts aim to overcome these limitations.

+ **Grid infrastructure:** Widespread adoption of electric vehicles could pose challenges to the existing electric grid infrastructure. The increased demand for electricity could require upgrades to the grid and smart charging solutions to manage peak loads.

Examples of Electric Vehicles

There are various types of electric vehicles available on the market today. Some of the most common types include:

+ **Battery Electric Vehicles (BEVs):** These vehicles are powered solely by an electric motor and rely on the energy stored in the battery pack.

+ **Plug-in Hybrid Electric Vehicles (PHEVs):** PHEVs have both an electric motor and an internal combustion engine. They can be charged from an external power source and also use gasoline for extended range.

+ **Hybrid Electric Vehicles (HEVs):** HEVs combine an internal combustion engine with an electric motor. The electric motor assists the engine during acceleration and deceleration, improving fuel efficiency.

Integration of Electric Vehicles in Sustainable Transportation

Electric vehicles play a crucial role in achieving a sustainable transportation system. Their integration requires a multi-dimensional approach that addresses key aspects such as:

- **Charging infrastructure development:** The expansion of charging infrastructure is vital for enabling widespread adoption of electric vehicles. This includes establishing public charging stations, workplace charging, and residential charging solutions.

- **Policy support:** Governments and policymakers can provide incentives and regulatory frameworks that promote the adoption of electric vehicles. These may include tax credits, subsidies, and initiatives to improve charging infrastructure.

- **Renewable energy integration:** To maximize the environmental benefits of electric vehicles, the integration and expansion of renewable energy sources for electricity generation are necessary.

- **Smart grid solutions:** Advanced technologies and systems that enable smart charging, vehicle-to-grid integration, and demand response can help optimize the integration of electric vehicles into the grid.

- **Education and awareness:** Raising public awareness about the benefits of electric vehicles and addressing misconceptions is important for driving consumer adoption and acceptance.

Conclusion

Electric vehicles have the potential to revolutionize the transportation sector by significantly reducing greenhouse gas emissions and promoting sustainable mobility. Despite the challenges they face, ongoing advancements in technology and supportive policies are driving their adoption. As the world transitions to a cleaner and more sustainable future, electric vehicles are poised to play a crucial role in creating a greener transportation system.

Key Takeaways:

+ Electric vehicles are powered by electric motors and use rechargeable batteries to store energy.

+ EVs offer several benefits such as reduced emissions, improved air quality, and lower operating costs.

+ Challenges include limited driving range, a lack of charging infrastructure, high vehicle costs, and battery technology limitations.

+ Integration of electric vehicles requires the development of charging infrastructure, policy support, renewable energy integration, smart grid solutions, and education and awareness.

Discussion Questions:

1. How can governments encourage the widespread adoption of electric vehicles?

2. What are the main barriers to the proliferation of charging infrastructure for electric vehicles?

3. How does the lifecycle analysis of electric vehicles compare to conventional vehicles in terms of environmental impact?

4. What role do electric vehicles play in achieving the United Nations Sustainable Development Goals?

Further Reading:

1. International Energy Agency (IEA) - "Global EV Outlook 2021"

2. U.S. Department of Energy - "Electric Vehicles: Benefits, Challenges, and Diverse Market Growth"

3. Union of Concerned Scientists - "Cleaner Cars from Cradle to Grave: How Electric Cars Beat Gasoline Cars on Lifetime Global Warming Emissions"

Fun Fact:

The world's fastest electric car, the Tesla Roadster, can accelerate from 0 to 60 miles per hour in under 2 seconds, making it faster than many high-performance gasoline-powered supercars.

Charging infrastructure

In the transition to a sustainable future, the development of efficient and accessible charging infrastructure is a key component of enabling widespread adoption of electric vehicles (EVs). Charging infrastructure refers to the network of charging stations and related technologies that support the charging and availability of electric vehicles.

Importance of charging infrastructure

The success of electric vehicles depends heavily on a robust charging infrastructure. A widespread and reliable charging network is crucial for several reasons:

+ **Range anxiety reduction:** A well-developed charging infrastructure alleviates concerns about the limited range of electric vehicles. It provides the confidence that electric vehicle owners can find accessible charging stations, allowing them to travel longer distances without worrying about running out of power.

+ **Market confidence:** The availability of an accessible and convenient charging network gives consumers the assurance that electric vehicles are a viable alternative to traditional petrol or diesel vehicles. This confidence can drive increased demand for electric vehicles and encourage manufacturers to invest in producing more electric vehicle models.

+ **Emissions reduction:** Electric vehicles offer significant environmental benefits, primarily by reducing greenhouse gas emissions. However, the emissions reduction potential can only be fully realized if vehicles are charged using renewable energy sources. An extensive charging infrastructure can integrate with clean energy sources, thereby maximizing the environmental benefits of electric vehicles.

+ **Urban air quality improvement:** Electric vehicles produce zero tailpipe emissions, which contributes to improving urban air quality. By providing convenient charging options, charging infrastructure promotes the use of electric vehicles, reducing harmful pollutants such as nitrogen oxides and particulate matter in densely populated areas.

Types of charging infrastructure

The charging infrastructure for electric vehicles comprises various types of charging stations, classified based on the charging power and time required to charge the vehicle. The three primary types of charging infrastructure are:

1. **Level 1 (L1) Charging**: L1 charging refers to the use of a standard household electrical outlet (120 volts AC) to charge an electric vehicle. This type of charging is the slowest, typically delivering a charging rate of around 3-5 miles of range per hour. L1 charging is most suitable for overnight charging at home or when the vehicle is parked for an extended period.

2. **Level 2 (L2) Charging**: L2 charging uses a higher-powered dedicated charging unit, typically operating at 240 volts AC. This type of charging offers faster charging rates, delivering around 10-30 miles of range per hour. L2 charging stations are commonly installed at workplaces, public parking lots, and residential areas to provide convenient charging options throughout the day.

3. **Level 3 (L3) Charging** or **Direct Current (DC) Fast Charging**: L3 charging, also known as DC fast charging, provides high-power charging by directly converting AC power to DC power. DC fast chargers can deliver significantly higher charging rates, allowing for an average of 60-80 miles of range in a 20-minute charging session. These charging stations are typically located along highways or in commercial areas to facilitate long-distance travel and reduce charging time.

Charging station technologies

Charging stations incorporate various technologies to facilitate safe and efficient charging processes. These technologies ensure optimal power delivery, user safety, and integration with the electricity grid. The key technologies utilized in charging stations are:

1. **EV supply equipment (EVSE)**: EVSE acts as an interface between the electric vehicle and the charging station. It ensures that power is delivered at the correct voltage and current levels. EVSE incorporates safety features such as ground fault detection and prevents unauthorized access to charging equipment.

2. **Smart charging:** Smart charging technology enables intelligent management of charging stations, optimizing power usage and facilitating load management. It allows for features such as scheduling charging at off-peak hours, integrating with renewable energy sources, and balancing power demand to prevent grid overloading.

3. **Vehicle-to-Grid (V2G) integration:** V2G technology enables bidirectional power flow between the electric vehicle and the electricity grid. It allows electric vehicles to not only take power from the grid but also feed excess energy back into the grid when needed. V2G integration can provide grid stabilization, demand response, and the potential for additional revenue streams for electric vehicle owners.

4. **Wireless charging:** Wireless charging eliminates the need for physical connectors by using induction or magnetic resonance technologies to transfer power from the charging station to the electric vehicle. It provides convenience and ease of use, as vehicles can simply park over a charging pad without the need for physical plugging.

Challenges and solutions

Despite the growing importance of charging infrastructure, several challenges need to be addressed to ensure its efficient deployment and operation. These challenges include:

+ **Availability and accessibility:** The availability and accessibility of charging stations need to be increased, especially in urban areas, to cater to the growing number of electric vehicles. This can be achieved through public-private partnerships, government incentives, and regulations that encourage the installation of charging stations at strategic locations.

+ **Interoperability:** Interoperability refers to the ability of electric vehicles to charge at any charging station, regardless of the charging infrastructure provider. Standardization of charging connectors, communication protocols, and payment systems is essential to ensure seamless interoperability and enhance the user experience.

+ **Grid integration:** The integration of charging infrastructure with the electricity grid is vital to manage power demand and prevent grid congestion. Advanced metering infrastructure, grid management systems, and load management techniques should be employed to optimize the charging process and balance the electricity load.

+ **Power demand and load management:** The widespread adoption of electric vehicles can significantly increase power demand, particularly during peak charging periods. Load management techniques, time-of-use pricing, and demand response programs can help distribute the charging load more evenly and prevent strain on the grid.

Addressing these challenges requires collaboration between stakeholders, including governments, electric utilities, charging infrastructure providers, and electric vehicle manufacturers. By implementing innovative solutions and investing in charging infrastructure, we can accelerate the transition toward sustainable transportation and reduce our dependence on fossil fuels.

Example: Tesla Supercharger network

An exemplary charging infrastructure network is the Tesla Supercharger network. Tesla has established a vast network of high-power Supercharger stations globally, strategically located along major travel routes and in densely populated areas. The Tesla Supercharger network primarily utilizes DC fast charging technology, allowing Tesla vehicles to quickly recharge their batteries during long-distance journeys.

The integration of advanced technologies enables Tesla Superchargers to provide an exceptional charging experience. The Supercharger stations utilize smart charging capabilities, taking advantage of renewable energy sources and optimizing charging rates based on individual vehicle needs. Furthermore, Tesla's proprietary charging connectors offer convenient and reliable charging for Tesla owners, ensuring interoperability within the Supercharger network.

The success of the Tesla Supercharger network demonstrates the importance of investing in charging infrastructure and the potential for private companies to lead the way in accelerating the adoption of electric vehicles. It serves as a model for other charging infrastructure providers and highlights the benefits of fast and accessible charging networks for electric vehicle users.

Battery Technology

Battery technology plays a crucial role in the field of clean energy and sustainable development. It has become increasingly important as the world seeks to transition from fossil fuels to renewable energy sources. In this section, we will explore the principles, advancements, challenges, and applications of battery technology.

Principles of Battery Technology

A battery is an electrochemical device that stores and releases energy by converting chemical energy into electrical energy. It consists of one or more electrochemical cells connected in series or parallel. Each cell contains two electrodes: an anode (negative electrode) and a cathode (positive electrode), separated by an electrolyte.

The principle behind battery operation is based on redox reactions (reduction-oxidation reactions) that involve the transfer of electrons between the electrodes. During discharge, the anode releases electrons, which flow through an external circuit to the cathode, generating an electric current. At the same time, the chemical reactions within the battery drive the ions in the electrolyte to move from the cathode to the anode. This flow of ions maintains the electrical neutrality of the overall system. During charging, the process is reversed, with the flow of electrons and ions being reversed.

Types of Batteries

There are various types of batteries with different chemistries, each offering different performance characteristics, energy storage capacities, and applications. Let's explore some of the commonly used battery technologies.

+ **Lithium-ion batteries:** Lithium-ion batteries are widely used in portable electronics, electric vehicles, and renewable energy systems. They offer high energy density, long cycle life, and low self-discharge rates. The positive electrode is typically made of lithium cobalt oxide, lithium iron phosphate, or lithium manganese oxide, while the negative electrode is composed of graphite. The electrolyte is a lithium salt dissolved in an organic solvent.

+ **Lead-acid batteries:** Lead-acid batteries are the oldest type of rechargeable batteries and are commonly used in automotive applications and backup power systems. They consist of a lead dioxide positive electrode, a lead negative electrode, and a sulfuric acid electrolyte. They are relatively inexpensive but have lower energy density and shorter cycle life compared to lithium-ion batteries.

+ **Nickel-metal hydride batteries:** Nickel-metal hydride (NiMH) batteries are commonly used in hybrid vehicles, portable electronics, and power tools.

They offer higher energy density and longer cycle life compared to lead-acid batteries but have lower energy density and shorter cycle life compared to lithium-ion batteries. The positive electrode is made of a nickel oxyhydroxide electrode, while the negative electrode contains a hydrogen-absorbing alloy. The electrolyte is usually potassium hydroxide.

+ **Flow batteries:** Flow batteries are a type of rechargeable battery that uses two electrolyte solutions stored in separate tanks. The electrolytes flow through separate electrodes and are circulated by pumps. Flow batteries offer the advantage of decoupling energy and power capacities, making them suitable for large-scale energy storage applications. Vanadium redox flow batteries are the most commonly used type of flow battery.

Advancements in Battery Technology

In recent years, there have been significant advancements in battery technology, driven by the increasing demand for energy storage and the need for more efficient and sustainable solutions. Some key advancements are:

+ **Increased energy density:** Researchers are continuously working on improving the energy density of batteries, allowing for longer-lasting and more powerful energy storage. This has led to the development of high-capacity lithium-ion batteries and the exploration of new battery chemistries.

+ **Fast-charging capabilities:** One of the major challenges with battery technology has been the time required for charging. Recent advancements have focused on developing batteries that can be charged at faster rates without compromising their performance or cycle life. This has opened up opportunities for electric vehicles and renewable energy systems to become more practical and efficient.

+ **Longer cycle life:** Battery degradation over time, resulting in a reduced capacity and performance, has been a significant concern. Researchers are working on improving battery materials, electrode designs, and electrolyte chemistries to increase the cycle life of batteries, allowing for more sustainable and cost-effective solutions.

+ **Safety improvements:** Safety is a critical aspect of battery technology, especially considering the incidents of battery fires and explosions. Advances in battery management systems, thermal management, and the use of new materials help enhance the safety of batteries, making them more reliable and secure.

Challenges and Future Directions

While battery technology has made significant progress, there are still challenges that need to be addressed. Some of the key challenges include:

+ **Cost:** The cost of advanced battery technologies, such as lithium-ion batteries, remains relatively high, making them less accessible for widespread adoption. Researchers are exploring scalable production methods, alternative materials, and recycling techniques to reduce the cost of batteries.

+ **Resource limitations:** Some battery technologies, such as lithium-ion batteries, rely on limited resources like lithium and cobalt. The increased demand for batteries may put a strain on these resources. Research efforts are focused on developing alternative battery chemistries that use abundant and sustainable materials.

+ **Environmental impact:** The production, use, and disposal of batteries can have adverse environmental impacts. This includes the extraction and processing of raw materials, energy-intensive manufacturing processes, and the potential for toxic waste. Battery technologies need to be developed with a focus on minimizing their environmental footprint.

In the future, battery technology is expected to continue evolving and playing a crucial role in various sectors, including energy storage, transportation, and portable electronics. Further advancements in materials science, electrochemistry, and manufacturing techniques will enable the development of next-generation batteries with higher energy densities, longer cycle lives, faster charging capabilities, and improved sustainability.

Real-world Example: Tesla Powerwall

One of the most notable applications of battery technology in recent years is the Tesla Powerwall. The Powerwall is a lithium-ion battery system designed for residential energy storage. It allows homeowners to store excess energy generated by solar panels during the day and use it when the demand is higher or when the grid is down.

The Powerwall aims to reduce reliance on the conventional electrical grid, increase energy independence, and enable a more sustainable and cost-effective energy solution. It also plays a significant role in the integration of renewable energy sources into the grid, as it can store excess energy during times of low demand and supply it back during peak hours.

The success of the Powerwall has prompted other companies to develop similar residential battery systems, further driving the adoption of clean energy and enhancing the overall resilience of energy systems.

Conclusion

Battery technology is a critical component of the clean energy revolution and sustainable development. Advancements in battery technology have enabled the growth of renewable energy systems, electric vehicles, and energy storage solutions. However, challenges related to cost, resources, and environmental impact need to be addressed to achieve widespread adoption and ensure a sustainable future.

By continuously pushing the boundaries of materials science, electrochemistry, and manufacturing techniques, researchers and engineers can unlock the full potential of battery technology and pave the way for a greener and more sustainable world. With further advancements, batteries will play a crucial role in achieving a carbon-neutral future and addressing the pressing global challenges of climate change and energy transition.

Benefits and challenges

In the realm of green transportation, electric vehicles (EVs) have gained significant attention as a sustainable alternative to traditional gasoline-powered cars. This section explores the benefits and challenges associated with the adoption of EVs.

Benefits of Electric Vehicles

1. **Environmental Benefits:** One of the key advantages of EVs is their lower carbon footprint compared to internal combustion engine vehicles. By eliminating tailpipe emissions, EVs contribute to improved air quality and reduced greenhouse gas emissions. This is particularly important in urban areas where air pollution is a major concern.

2. **Energy Efficiency:** EVs are more energy-efficient than conventional vehicles since they convert a higher percentage of electrical energy from the grid to power the wheels. Internal combustion engines, on the other hand, waste a significant amount of energy as heat. This efficiency can help reduce overall energy consumption and dependence on fossil fuels.

3. **Reduced Noise Pollution:** EVs operate quietly because they do not have an engine with lots of moving parts. This can lead to a reduction in noise pollution, especially in densely populated areas or during nighttime driving.

4. **Lower Operating Costs:** EVs have lower operating costs compared to traditional vehicles. The cost of electricity is typically lower than gasoline or diesel, resulting in lower fuel expenses. Additionally, EVs have fewer moving parts, reducing the need for maintenance and the associated costs.

5. **Energy Independence:** By transitioning to electric vehicles, countries can reduce their dependence on imported oil, promoting energy independence and increasing national security.

Challenges of Electric Vehicles

1. **Limited Driving Range:** One of the main challenges of EVs is their limited driving range per charge. Although the range has improved over the years, it is still a concern for long-distance travel. Range anxiety, the fear of running out of charge, is a significant barrier to the widespread adoption of EVs.

2. **Charging Infrastructure:** The availability and accessibility of charging stations is crucial for the successful integration of EVs into society. Currently, charging infrastructure is not as widespread as conventional gas stations, which hinders the convenience and adoption of EVs, particularly for those without access to home charging.

3. **Charging Time:** Compared to refueling a conventional vehicle, charging an EV takes significantly longer. While overnight charging at home is feasible for many, longer charging times can be a challenge when traveling long distances or during on-the-go charging.

4. **Upfront Cost:** EVs often have a higher upfront cost compared to their gasoline-powered counterparts. Although the total cost of ownership can be lower due to savings on fuel and maintenance, the higher initial investment can be a deterrent for potential buyers.

5. **Battery Technology:** The performance, efficiency, and lifespan of EVs heavily rely on battery technology. While significant advancements have been made, challenges remain in terms of energy density, cost, and environmental impact of battery production and disposal.

6. **Supply Chain and Raw Materials:** EV production requires a secure and sustainable supply chain for raw materials such as lithium, cobalt, and rare earth elements. Ensuring responsible sourcing and minimizing the environmental impact of mining and processing these materials is essential.

Despite these challenges, the benefits and potential of electric vehicles make them a promising solution for sustainable transportation. As technology continues to advance and infrastructural improvements take place, the integration of EVs into society will become more seamless, contributing to a greener and cleaner future.

Example: One example of the benefits of electric vehicles is the city of Oslo, Norway. The city has made a concerted effort to promote EV adoption by offering free parking, access to bus lanes, and exemption from toll charges for EV owners. As a result, EVs now make up a significant portion of vehicles in the city, leading to improved air quality and reduced traffic congestion.

Key Takeaways: - Electric vehicles offer environmental benefits, energy efficiency, reduced noise pollution, lower operating costs, and energy independence. - Challenges include limited driving range, charging infrastructure, charging time, upfront cost, battery technology, and raw materials supply chains. - The benefits and challenges of EVs should be addressed through technological advancements, supportive policies, and investments in charging infrastructure.

Further Resources: 1. *Electric Vehicles: Benefits, Challenges, and their Role in the Future of Transportation* - A report by the International Energy Agency (IEA). 2. *The Future of Electric Vehicles: Economic, Environmental, and Social Impacts* - A research paper by the World Economic Forum. 3. *Electric Vehicle Adoption: Costs, Benefits, and Best Practices* - A guidebook by the International Council on Clean Transportation (ICCT).

Exercises: 1. Research and analyze the current charging infrastructure in your local area. What improvements are needed to support the wider adoption of EVs? 2. Conduct a cost comparison between owning an electric vehicle and a gasoline-powered vehicle over a 5-year period, taking into account the upfront cost, fuel/maintenance costs, and any available subsidies or incentives. Present your findings in a tabular format. 3. Investigate the environmental impact of battery production and disposal. What strategies or technologies can be implemented to mitigate these impacts?

Public Transportation

Public transportation plays a crucial role in achieving sustainable urban development and reducing the environmental impact of transportation systems. It provides an alternative to individual car usage, which can help reduce traffic congestion, air pollution, and greenhouse gas emissions. In this section, we will explore the various aspects of public transportation, including its benefits, challenges, and strategies for improving its efficiency and accessibility.

Benefits of Public Transportation

Public transportation has numerous benefits for individuals, communities, and the environment. Here are some key advantages:

+ **Reduced congestion:** By encouraging people to use public transportation instead of private vehicles, congestion on roads and highways can be alleviated. This leads to smoother traffic flow, shorter travel times, and less frustration for commuters.

+ **Environmental benefits:** Public transportation produces significantly lower greenhouse gas emissions per passenger compared to private vehicles. It helps reduce air pollution, noise pollution, and overall carbon footprint, contributing to improved air quality and public health.

+ **Cost savings:** Using public transportation can be more cost-effective than owning and maintaining a private vehicle. It eliminates expenses such as fuel, insurance, and parking fees, making transportation more affordable for individuals and families.

+ **Improved accessibility:** Public transportation provides mobility options to people who do not have access to private vehicles, including low-income individuals, senior citizens, and people with disabilities. It ensures equitable access to essential services, education, employment, and social activities.

+ **Promotes community interaction:** Public transportation facilitates social connections by bringing people from different backgrounds and neighborhoods together. It creates opportunities for interaction and fosters a sense of community cohesion.

Challenges in Public Transportation

While public transportation offers various benefits, it also faces several challenges that require attention and innovative solutions. Let's discuss some of these challenges:

+ **Funding and investment:** Public transportation systems require substantial investment in infrastructure, vehicles, and maintenance. Securing funding can be challenging, especially for infrastructure improvements and expansion projects. Adequate financial resources need to be allocated to ensure the efficient functioning and development of public transportation systems.

+ **Lack of integration and connectivity:** Public transportation networks often face issues of limited integration and connectivity. Inefficient connections between different modes of transportation, such as buses, trains, and trams, can discourage people from using public transportation. Seamless integration between different modes and improved connectivity to various destinations is crucial for enhancing the attractiveness of public transportation.

+ **Limited coverage and frequency:** Public transportation services need to be accessible and convenient for users. However, in some areas, the coverage of public transportation may be limited, resulting in longer travel times and inconvenience for commuters. Additionally, infrequent service frequencies may discourage people from relying on public transportation as their primary mode of travel.

+ **Perception and image:** Public transportation often faces negative perceptions regarding comfort, safety, and reliability. Improving the perception and image of public transportation through enhancements in service quality, cleanliness, and security can help attract more people to use it.

+ **Inefficient operations and maintenance:** Public transportation systems must be operated and maintained efficiently to ensure seamless and reliable service. Challenges such as inadequate staffing, lack of regular maintenance, and outdated technologies can affect the overall performance and quality of public transportation services.

Strategies for Improving Public Transportation

To address the challenges mentioned earlier and improve public transportation systems, various strategies can be implemented. Let's explore some key strategies:

+ **Integrated and multimodal systems:** Developing integrated and multimodal transportation systems ensures seamless connectivity between various modes of public transportation. This includes integrating buses, trams, trains, and other modes with efficient transfer points and convenient schedules. Implementing smart technologies for real-time information about arrival times and transfers can further enhance the usability of these systems.

+ **Infrastructure development:** Investing in the development and improvement of public transportation infrastructure is essential. This includes building new rail and bus lines, expanding existing networks, and optimizing infrastructure for pedestrian and bicycle connectivity. The integration of transportation hubs, such as bus and rail stations, with other amenities like shopping centers or residential areas, can create vibrant transit-oriented developments.

* **Service quality and reliability:** Ensuring high service quality and reliability is crucial for attracting and retaining public transportation users. This includes implementing schedules that cater to peak-hour demands, maintaining clean and comfortable vehicles, and providing reliable real-time information to passengers. Continuous monitoring and improvement of service standards can help enhance the overall user experience.

* **Accessibility and inclusivity:** Public transportation should be accessible to all individuals, regardless of their physical abilities or socio-economic status. This involves providing barrier-free access to vehicles and stations, offering specialized services for people with disabilities, and ensuring affordability through fare subsidy programs. Active engagement with diverse communities can help identify and address specific accessibility needs.

* **Technology adoption:** Embracing emerging technologies can revolutionize public transportation systems. Intelligent transportation systems, including smart fare collection, automated vehicle location, and predictive analytics, can enhance operational efficiency, improve planning, and optimize resource allocation. The integration of electric buses and hybrid vehicles can also contribute to reducing emissions and environmental impact.

* **Collaboration and partnerships:** Public transportation agencies can collaborate with other stakeholders, such as employers, educational institutions, and community organizations, to develop tailored transportation solutions. Partnerships for innovative funding mechanisms, data sharing, and demand management strategies can lead to more sustainable and effective public transportation systems.

Case Study: Curitiba's Bus Rapid Transit (BRT)

An excellent example of a successful public transportation system is Curitiba's Bus Rapid Transit (BRT) in Brazil. Curitiba faced significant urbanization challenges, including high car ownership rates and traffic congestion. The BRT system was introduced as an efficient and cost-effective alternative to address these issues. Here are some key features of Curitiba's BRT system:

- **Dedicated bus lanes:** Curitiba's BRT system includes dedicated lanes exclusively for buses, allowing them to bypass traffic congestion and operate with greater speed and reliability.

- **Prepaid boarding:** Passengers pay the fare before boarding the bus at stations, reducing boarding times and minimizing delays. The fare collection process is efficient and helps maintain the system's punctuality.

- **Level boarding platforms:** BRT stations have level boarding platforms that align with the bus doors, enabling quick and easy boarding for passengers, including those with mobility impairments.

- **Integrated land use planning:** Curitiba's BRT system is integrated with land use planning, ensuring that transit-oriented developments are strategically located along the bus routes. This promotes compact and mixed-use development, reducing the need for long-distance travel.

- **Green spaces and pedestrian-friendly design:** The BRT corridors in Curitiba feature green spaces and pedestrian-friendly designs, creating a pleasant urban environment. This approach enhances the overall public transportation experience and encourages pedestrian and bicycle use.

- **Continuous improvement and expansion:** Curitiba's BRT system has undergone continuous improvement and expansion over the years. Additional routes and stations have been added, and the technology and operations have been upgraded to ensure better service quality and efficiency.

Curitiba's BRT system has not only transformed public transportation in the city but also served as a model for other cities worldwide. It highlights the importance of innovative planning, efficient operations, and a strong commitment to sustainable urban development.

Conclusion

Public transportation plays a vital role in achieving sustainable and inclusive urban transport systems. By reducing congestion, air pollution, and reliance on private vehicles, public transportation offers significant environmental and societal benefits. However, it also faces various challenges, including funding, integration, and perception issues. Implementing strategies such as integrated and multimodal systems, infrastructure development, and technology adoption can help overcome these challenges and improve public transportation efficiency. As exemplified by Curitiba's BRT system, innovative and well-planned public transportation solutions can transform cities and contribute to a more sustainable future.

Bus Rapid Transit

Bus rapid transit (BRT) is a high-capacity public transportation system that aims to provide efficient, reliable, and sustainable urban mobility. It combines the benefits of both buses and rail systems, offering a cost-effective alternative to traditional transit systems. BRT typically operates in dedicated lanes, with fast boarding and alighting processes, and prioritized signal control. This section will explore the key features, benefits, challenges, and examples of BRT implementation.

Key Features of BRT

1. **Dedicated Bus Lanes:** BRT systems have dedicated lanes separated from general traffic, allowing buses to bypass congestion and provide faster travel times.

2. **High-Quality Stations:** BRT stations are designed to provide a comfortable and convenient experience for passengers, with features such as sheltered waiting areas, seating, real-time information displays, and easy accessibility.

3. **Pre-Paid Fare Collection:** BRT systems often adopt a pre-paid fare collection system, where passengers pay their fare before boarding the bus. This reduces boarding time, ensuring faster service and minimizing delays.

4. **Bus Priority at Intersections:** BRT systems utilize signal priority techniques to give buses preferential treatment at intersections. This includes extended green lights or dedicated lanes, enabling buses to maintain their schedules.

5. **Enhanced Bus Vehicles:** BRT vehicles are typically longer and have higher passenger capacities compared to regular buses. They are designed to provide a comfortable and efficient travel experience, often equipped with features such as low floors for easy accessibility, air conditioning, and priority seating for seniors and disabled passengers.

6. **Integrated Network:** BRT systems are integrated into the overall public transportation network, connecting with other modes of transportation such as trains, subways, and bike-sharing programs. This integration encourages multi-modal travel and provides seamless connectivity for passengers.

Benefits of BRT

1. **Improved Travel Time and Reliability:** BRT systems reduce travel time by utilizing dedicated lanes and signal priority, avoiding traffic congestion. This

results in faster and more reliable service, attracting more passengers to choose public transportation over private vehicles.

2. **Reduced Congestion and Emissions:** With increased efficiency and capacity, BRT systems help reduce traffic congestion by effectively moving more people in fewer vehicles. This reduction in traffic leads to decreased greenhouse gas emissions and air pollution, contributing to a cleaner and healthier urban environment.

3. **Affordability and Flexibility:** BRT is a cost-effective alternative to building expensive rail infrastructure. The flexibility of BRT allows for easier adaptation and expansion in response to changing needs, making it a more financially viable option for cities.

4. **Accessibility and Equity:** BRT improves accessibility for all citizens, including those with limited mobility, by providing low-floor buses and well-designed stations. It ensures equitable access to public transportation and promotes social inclusion.

5. **Economic Development and Livability:** BRT systems can stimulate economic development by improving access to employment centers, educational institutions, and other key destinations. By reducing the dependence on private vehicles, BRT also helps create more livable and pedestrian-friendly urban environments.

Challenges and Solutions

1. **Limited Right-of-Way:** One of the primary challenges in implementing BRT is acquiring dedicated lanes within limited urban space. This can be mitigated by reallocating road space, prioritizing public transportation over private vehicles, and integrating BRT into existing road infrastructure.

2. **Integration with Existing Transit Systems:** BRT implementation requires seamless integration with existing transit systems to ensure efficient transfers and connectivity. This can be achieved through coordinated planning, schedules, and fare systems.

3. **Public Perception and Acceptance:** Convincing the public to switch from private vehicles to public transportation can be a challenge. Education campaigns, effective communication, and showcasing the benefits of BRT are essential to gain public support and increase ridership.

4. **Operational and Maintenance Costs:** BRT systems require continuous investment for operation and maintenance. Public-private partnerships, farebox revenue, and government subsidies can help ensure the financial sustainability of BRT projects.

Examples of Successful BRT Implementation

1. **TransMilenio, Bogotá, Colombia:** The TransMilenio BRT in Bogotá is one of the most successful BRT systems globally. It has dedicated bus lanes, fast boarding systems, and an integrated fare collection system. TransMilenio has significantly reduced travel times, improved air quality, and transformed the transportation landscape of Bogotá.

2. **Curitiba, Brazil:** The Curitiba BRT system is the pioneer of BRT and has become a model for many cities worldwide. It features dedicated bus lanes, well-designed stations with off-board fare collection, and an integrated transport system. Curitiba's BRT has been instrumental in reducing traffic congestion and promoting sustainable urban development.

3. **Istanbul, Turkey:** Istanbul's BRT system, known as Metrobus, has been successful in addressing the city's transportation challenges. It has its own dedicated lanes, modern buses, and an efficient signal priority system. The Metrobus has improved travel times, reduced congestion, and provided a reliable alternative for millions of commuters in Istanbul.

4. **TransJakarta, Jakarta, Indonesia:** The TransJakarta BRT system has transformed public transportation in Jakarta. With dedicated bus lanes, comfortable stations, and a well-integrated ticketing system, it has improved mobility, reduced traffic congestion, and cut emissions in the city.

5. **BRT TransOeste, Rio de Janeiro, Brazil:** The BRT TransOeste in Rio de Janeiro has helped revolutionize public transportation in the city. It connects various neighborhoods and major mobility hubs, providing faster, safer, and more reliable travel options. The BRT TransOeste has played a key role in the city's efforts to host major sporting events and improve urban mobility.

Conclusion

Bus rapid transit offers an efficient, sustainable, and cost-effective solution for urban transportation challenges. By incorporating dedicated lanes, high-quality stations, and integrated networks, BRT systems can significantly reduce travel times, congestion, and emissions while improving accessibility and connectivity. Successful implementation of BRT requires careful planning, stakeholder engagement, and continuous investment. With the examples of successful BRT systems around the world, cities can learn from best practices and embrace this innovative approach to enhance their public transportation systems.

Light Rail Systems

Light rail systems are an important part of sustainable transportation infrastructure in urban areas. They provide an efficient and environmentally friendly alternative to private vehicles and help reduce traffic congestion and carbon emissions. In this section, we will explore the principles, benefits, challenges, and future developments of light rail systems.

Principles of Light Rail Systems

Light rail systems, also known as tram or streetcar systems, operate on dedicated tracks, typically in urban areas. They are designed to provide reliable and frequent service, connecting key destinations within a city or region. Light rail vehicles are powered by electricity and can carry a significant number of passengers compared to buses or cars.

The principles that guide the design and operation of light rail systems include:

- **Efficiency:** Light rail systems are designed to provide a smooth and efficient mode of transportation. They use dedicated tracks and have priority at intersections, allowing them to avoid traffic congestion and provide faster travel times.

- **Accessibility:** Light rail systems are designed to be accessible to all passengers, including individuals with disabilities or limited mobility. They feature low-floor boarding and level platforms, making it easy for passengers to get on and off the trains.

- **Integration:** Light rail systems are integrated into the existing transportation network, with connections to other modes of public transit such as buses, trains, and bike-sharing programs. This integration allows passengers to have seamless journeys across different modes of transportation.

- **Environmental Sustainability:** Light rail systems contribute to reducing greenhouse gas emissions and improving air quality. As they are powered by electricity, they produce zero emissions at the point of use. Additionally, their capacity to carry a large number of passengers reduces the number of private vehicles on the road, further reducing emissions.

- **Urban Development:** Light rail systems have the potential to drive urban development and revitalization. The presence of a light rail line can attract commercial and residential investments along its route, leading to increased economic activity and improved livability.

Benefits of Light Rail Systems

Light rail systems offer numerous benefits for both passengers and the communities they serve. Some of the key benefits include:

+ **Reliability:** Light rail systems operate on fixed schedules and dedicated tracks, providing reliable and predictable service. Passengers can plan their journeys with confidence, knowing when the next train will arrive.

+ **Efficiency:** Light rail systems can carry a large number of passengers, especially during peak hours, reducing the need for multiple individual vehicles and alleviating traffic congestion. This leads to more efficient use of road space and improved travel times for all road users.

+ **Environmental Impact:** Light rail systems have a significantly lower environmental impact compared to private vehicles. By reducing the number of cars on the road, they help reduce traffic-related carbon emissions and air pollution. Additionally, as they operate on electricity, they can be powered by renewable energy sources, further reducing their carbon footprint.

+ **Affordability:** Light rail systems offer an affordable mode of transportation for passengers. They often have lower fares compared to other modes of public transit, making them accessible to a wide range of individuals.

+ **Safety:** Light rail systems are designed with safety in mind. They have dedicated tracks and signal systems to prevent collisions with other vehicles or pedestrians. Additionally, the presence of multiple staff members on trains ensures the safety and security of passengers.

Challenges and Future Developments

While light rail systems offer various benefits, they also face challenges and require continuous development to improve their efficiency and effectiveness. Some of the challenges include:

+ **Initial Investment:** The construction of light rail systems requires significant upfront investment. This includes the cost of building new tracks, purchasing vehicles, and developing supporting infrastructure. Securing funding for these projects can be a challenge, especially for cash-strapped municipalities.

- **Right-of-Way Constraints:** Light rail systems often require dedicated tracks, which can be challenging to accommodate in dense urban areas with limited available space. Finding the right-of-way for new lines and negotiating with property owners can be complex and time-consuming.

- **Maintenance and Operations:** Light rail systems require regular maintenance to ensure smooth and safe operations. This includes track maintenance, vehicle maintenance, and ensuring the electrical infrastructure is in good working condition. Maintaining a reliable and efficient service can be costly and time-intensive.

- **Public Perception:** The public perception of light rail systems can vary, with some communities expressing concerns over the potential impacts on neighborhood character, parking availability, and disruptions during construction. Effective public engagement and education campaigns are crucial to address these concerns and build support for light rail projects.

Despite these challenges, ongoing developments in technology and planning are shaping the future of light rail systems. Some of the key developments include:

- **Smart Technologies and Automation:** Light rail systems are incorporating smart technologies to improve efficiency and safety. This includes real-time passenger information systems, automatic fare collection, and integrated traffic management systems. Automation technologies, such as driverless trains, are also being explored to enhance reliability and capacity.

- **Integration with Sustainable Infrastructure:** Light rail systems are being designed and integrated with other sustainable infrastructure elements. This includes the integration of solar panels on station roofs to generate renewable energy, the provision of bicycle parking facilities at stations, and the development of green spaces and pedestrian-friendly environments around stations.

- **Expansion and Interconnectivity:** Light rail systems are expanding to serve larger geographic areas and improve connectivity between neighborhoods, cities, and regions. The integration of light rail networks with regional rail and high-speed rail systems is being explored to enhance intercity and interregional travel options.

- **Community-led Design and Planning:** The involvement of communities in the design and planning of light rail systems is gaining importance.

Participatory design processes and the inclusion of community perspectives help ensure that light rail systems meet the specific needs and expectations of the communities they serve.

Case Study: Portland Streetcar, USA

The Portland Streetcar, a light rail system in Portland, Oregon, is an excellent example of a successful and sustainable light rail system. The streetcar line was first introduced in 2001 and has since expanded to multiple lines serving different neighborhoods and destinations within the city.

The Portland Streetcar system offers numerous benefits to the community, including:

+ **Improved Mobility:** The streetcar provides convenient and frequent service, connecting residents, workers, and tourists to key destinations within the city center and surrounding neighborhoods. It has become an integral part of Portland's transportation network.

+ **Revitalization of Neighborhoods:** The streetcar line has played a crucial role in revitalizing neighborhoods along its route. The presence of the streetcar has attracted commercial and residential investments, leading to increased economic activity and improved livability.

+ **Reduced Traffic Congestion:** By providing an alternative to private vehicles, the streetcar has helped reduce traffic congestion in downtown Portland. This has contributed to improved air quality and a more sustainable transportation system.

+ **Accessibility and Inclusivity:** The streetcar system features low-floor vehicles and level boarding platforms, making it accessible to individuals with disabilities or limited mobility. The system has been designed with inclusivity in mind, ensuring that everyone can benefit from its services.

The success of the Portland Streetcar can be attributed to effective planning, community engagement, and ongoing support from the city government. It demonstrates the potential of light rail systems to transform urban transportation and contribute to sustainable urban development.

Conclusion

Light rail systems are an essential component of sustainable transportation infrastructure. They offer an efficient, reliable, and environmentally friendly mode of travel, reducing traffic congestion and carbon emissions. While light rail systems face challenges, ongoing developments and advancements are shaping their future. With continued investment and community support, light rail systems have the potential to play a significant role in building sustainable and livable cities.

Bike Sharing Programs

Bike sharing programs have become increasingly popular in many cities around the world as a sustainable and efficient mode of transportation. These programs provide access to bicycles for short-term use, allowing people to easily travel short distances without relying on private vehicles or public transportation. In this section, we will explore the principles, benefits, challenges, and examples of bike sharing programs.

Principles of Bike Sharing Programs

Bike sharing programs are based on the principle of providing shared access to bicycles for urban transportation. The key principles underlying bike sharing programs include:

1. **Accessibility:** Bike sharing programs aim to provide easy access to bicycles for all members of the community. This includes ensuring a sufficient number of bike docking stations, strategically located throughout the city, and a user-friendly registration and rental process.

2. **Sustainability:** Bike sharing programs promote sustainable transportation by reducing reliance on private vehicles, thereby decreasing traffic congestion and carbon emissions. By encouraging cycling as a mode of transportation, these programs contribute to a cleaner and greener urban environment.

3. **Affordability:** Bike sharing programs are designed to be affordable for the general public. Users typically pay a small fee for a specified rental period, which can range from a few minutes to several hours. Some programs also offer discounted rates or subscriptions for frequent users.

4. **Integration with Public Transportation:** Bike sharing programs often integrate with existing public transportation systems, such as buses and trains. This allows users to easily combine biking with other modes of

transportation, providing a more flexible and convenient commuting experience.

5. **Infrastructure and Safety**: An essential aspect of bike sharing programs is the provision of safe cycling infrastructure, including bike lanes and dedicated paths. Programs also prioritize bike maintenance and safety measures, such as regular inspections and helmet availability.

Benefits of Bike Sharing Programs

Bike sharing programs offer numerous benefits for both individuals and communities. Some of the major benefits include:

1. **Improved Health and Well-being**: Cycling is a physical activity that promotes cardiovascular fitness, muscle strength, and mental well-being. Bike sharing programs encourage people to incorporate exercise into their daily routines, leading to better health outcomes.

2. **Reduced Traffic Congestion**: By providing an alternative to private vehicles, bike sharing programs help alleviate traffic congestion in urban areas. This leads to faster and more efficient transportation for both cyclists and motorists.

3. **Environmental Sustainability**: Cycling produces zero emissions, making it an environmentally-friendly transportation option. Bike sharing programs contribute to reducing air and noise pollution, as well as carbon footprint, thereby promoting sustainability and combating climate change.

4. **Cost Savings**: Using a bike sharing program can be more cost-effective than owning a private vehicle or relying on public transportation for short trips. Users only pay for the time they use the bike, eliminating the need for parking fees or monthly passes.

5. **Last-mile Connectivity**: Bike sharing programs offer a convenient solution for the "last-mile" problem, where traditional public transportation options may not provide direct access to a final destination. Users can pick up a bike from a docking station near a transit stop and easily reach their destination.

6. **Promotion of Tourism**: Bike sharing programs can attract tourists and enhance their experience by providing a sustainable and enjoyable way to explore cities. Visitors can easily rent a bike and explore local attractions at their own pace, promoting tourism and economic growth.

Challenges and Solutions

While bike sharing programs have numerous benefits, they also face several challenges that need to be addressed for successful implementation. Some of the common challenges include:

1. **Bike Theft and Vandalism:** Bicycles are susceptible to theft and vandalism, which can impact the availability and usability of bike sharing programs. Implementing security measures such as GPS tracking, secure locking mechanisms, and public awareness campaigns can help mitigate these issues.

2. **Inadequate Infrastructure:** Insufficient cycling infrastructure, such as bike lanes and parking spaces, can hinder the growth and adoption of bike sharing programs. Collaboration between city planners, transportation authorities, and bike sharing operators is essential to develop and expand the necessary infrastructure.

3. **Unbalanced Bike Distribution:** Uneven distribution of bikes across docking stations can result in a lack of availability in some areas and overcrowding in others. Real-time monitoring systems and rebalancing strategies, such as incentivizing users to return bikes to empty stations, can help maintain a balanced bike fleet.

4. **Safety Concerns:** Safety is a significant concern for both cyclists and pedestrians. Providing comprehensive safety education and awareness campaigns, enforcing traffic regulations, and improving infrastructure for cyclists can contribute to safer cycling experiences.

Examples of Bike Sharing Programs

Bike sharing programs have been successfully implemented in numerous cities worldwide. Some notable examples include:

1. **Citi Bike:** Citi Bike is one of the largest bike sharing programs in the United States, operating in New York City, Jersey City, and the Miami metropolitan area. It provides thousands of bikes across hundreds of docking stations, offering a convenient and sustainable transportation option for urban residents and tourists.

2. **Vélib':** Launched in Paris, Vélib' is one of the earliest and most successful bike sharing programs in the world. It features a vast network of docking stations

and bicycles, making it an integral part of the city's transportation system. Vélib' has inspired similar programs in many other cities globally.

3. **Santander Cycles:** Formerly known as Barclays Cycle Hire, Santander Cycles is a bike sharing program operating in London, United Kingdom. With thousands of bikes available, it allows users to rent a bike from one station and return it to another, providing a flexible transportation option for Londoners and visitors.

4. **oBike:** oBike is a dockless bike sharing program based in Singapore. Unlike traditional programs with fixed docking stations, oBike allows users to locate and unlock a bike using a smartphone app, making it highly convenient and flexible. Users can pick up and drop off bikes anywhere within designated zones.

Conclusion

Bike sharing programs offer a sustainable, affordable, and flexible transportation solution for urban areas. These programs not only reduce traffic congestion, air pollution, and carbon emissions but also promote health and well-being. However, successful implementation requires addressing challenges such as bike theft, infrastructure development, and safety concerns. By learning from successful examples and adopting innovative strategies, cities can maximize the benefits of bike sharing programs and accelerate the transition towards a more sustainable future.

Sustainable aviation

Aviation plays a significant role in global transportation and economic development, but it also contributes to environmental degradation and climate change. To ensure a sustainable future, it is crucial to explore innovative technologies and practices that minimize the environmental impact of aviation. This section will discuss various strategies for achieving sustainable aviation, including both technological advancements and operational improvements.

Environmental impact of aviation

Aviation has a substantial environmental footprint due to its carbon dioxide (CO_2) emissions, air and noise pollution, and the consumption of non-renewable resources. The burning of jet fuel, which is primarily derived from fossil fuels, releases greenhouse gases such as CO_2, methane (CH_4), and nitrous oxide (N_2O) into the atmosphere. These emissions contribute to climate change, and the aviation industry is responsible for a significant portion of global CO_2 emissions. In addition to CO_2, aviation also emits other pollutants such as nitrogen oxides (NO_X), sulfur dioxide (SO_2), and particulate matter (PM).

Technological advancements

To mitigate the environmental impact of aviation, significant research and development efforts are being made to develop sustainable technologies. One of the most promising areas of research is the development of alternative fuels for aircraft. Sustainable aviation fuels (SAFs) are derived from renewable sources such as bioenergy crops, algae, or waste materials. These fuels can significantly reduce CO_2 emissions compared to conventional jet fuels. SAFs can be drop-in replacements for traditional jet fuels, requiring no modifications to existing aircraft or infrastructure.

In addition to alternative fuels, technological advancements in aircraft design are crucial for sustainable aviation. Improving aerodynamics, reducing weight, and using lighter materials can enhance fuel efficiency and reduce emissions. Advances in engine design, such as the development of more efficient turbofan engines, can also contribute to lower fuel consumption and emissions.

Operational improvements

Besides technological advancements, operational improvements are essential for achieving sustainable aviation. One key aspect is optimizing flight routes and air traffic management. By using advanced navigation systems and sophisticated routing algorithms, airlines can reduce flight distances, minimize fuel consumption, and decrease emissions. Collaborative decision-making between airlines, air traffic control, and airports can also help optimize aircraft movements on the ground and in the air.

Another operational aspect is reducing aircraft idle time. By minimizing the time spent idling on the ground and in the air, airlines can decrease fuel consumption and emissions. Additionally, implementing more efficient aircraft turnaround processes and reducing taxiing time can contribute to fuel savings and environmental benefits.

Government policies and industry initiatives

Governments and the aviation industry are taking steps to promote sustainable aviation through policy measures and industry initiatives. Several countries have introduced carbon pricing mechanisms or emissions trading systems to incentivize the reduction of CO_2 emissions from aviation. International organizations such as the International Civil Aviation Organization (ICAO) are working towards implementing global market-based measures to address aviation emissions.

Industry initiatives such as the Sustainable Aviation Fuel Users Group (SAFUG) bring together airlines, fuel producers, and other stakeholders to promote the deployment of SAFs. These collaborations aim to increase the availability and affordability of sustainable aviation fuels, accelerating their adoption in the industry.

Challenges and considerations

While sustainable aviation holds great promise, there are challenges and considerations that need to be addressed. First, the production of sustainable aviation fuels at scale is still a challenge due to high costs and limited availability. Economic and logistical barriers must be overcome to enable widespread adoption.

Second, the implementation of technological advancements, such as new engine designs or aircraft configurations, requires significant investments. Industry-wide collaboration, research funding, and supportive policies are essential for the successful deployment of these technologies.

Third, there is a need for international cooperation and standardization to ensure the seamless adoption of sustainable aviation practices. Harmonizing regulations, sharing best practices, and establishing common sustainability metrics are vital for global progress.

Case study: Electric aviation

One emerging technology in sustainable aviation is electric propulsion. Electric aircraft have the potential to significantly reduce emissions and noise compared to traditional combustion engines. Several small electric aircraft are already in operation, demonstrating the feasibility of this technology.

However, electric aviation still faces challenges such as limited battery capacity and weight, which restrict the range and payload of electric aircraft. The development of advanced battery technologies, such as high-energy-density lithium-ion batteries or solid-state batteries, is crucial for overcoming these limitations.

Electric vertical takeoff and landing (eVTOL) aircraft, also known as flying taxis, are another application of electric aviation. These aircraft, which are being developed by several companies, aim to revolutionize urban transportation by providing on-demand, environmentally friendly aerial mobility.

In conclusion, sustainable aviation is crucial for mitigating the environmental impact of the aviation industry. Through technological advancements, operational improvements, government policies, and industry initiatives, we can work towards a future where aviation is more sustainable. While challenges exist, continued research and collaborative efforts can pave the way for a greener and more efficient aviation sector.

Biofuels

Biofuels are alternative fuels produced from organic matter, such as plants, algae, and animal waste. They have gained considerable attention in recent years due to their potential to reduce greenhouse gas emissions and dependence on fossil fuels. This section will explore the different types of biofuels, their production processes, benefits and challenges, and their role in achieving sustainable energy solutions.

Types of Biofuels

There are three main types of biofuels: bioethanol, biodiesel, and biogas.

1. **Bioethanol:** Bioethanol is a renewable fuel produced from the fermentation of carbohydrates found in crops such as sugarcane, corn, and wheat. It can be blended with gasoline or used as a standalone fuel for vehicles. The production of bioethanol involves four main steps: feedstock preparation, fermentation, distillation, and dehydration. The feedstock, rich in sugars, is first processed to extract the sugars, which are then fermented by microorganisms to produce ethanol. The ethanol is then purified through distillation and further dehydrated to increase its energy content.

2. **Biodiesel:** Biodiesel is a renewable fuel derived from vegetable oils or animal fats. It can be used in diesel engines either in its pure form or blended with petroleum diesel. Biodiesel production involves a process called transesterification, where the triglycerides in the feedstock are chemically reacted with an alcohol, usually methanol, in the presence of a catalyst. This reaction produces biodiesel (fatty acid methyl esters) and glycerin as a byproduct. The biodiesel is then purified and can be used as a fuel.

3. **Biogas:** Biogas is a mixture of methane and carbon dioxide produced through the anaerobic digestion of organic material, such as agricultural residues,

food waste, and sewage sludge. The bioconversion of organic matter into biogas occurs in a controlled environment without the presence of oxygen. Biogas can be used directly as a fuel for heating, cooking, or electricity generation. It can also be upgraded to biomethane, a purified form of biogas that has similar properties to natural gas and can be injected into the natural gas grid or used as a transportation fuel.

Benefits and Challenges of Biofuels

Biofuels offer several benefits that make them an attractive option for achieving sustainable energy solutions:

1. **Renewable and Low Carbon:** Biofuels are derived from organic matter, which can be replenished through agricultural practices. Unlike fossil fuels, biofuels have a significantly lower carbon footprint, as the carbon dioxide released during combustion is balanced by the carbon dioxide absorbed by the crops during their growth.

2. **Energy Security:** By diversifying energy sources and reducing dependence on imported fossil fuels, biofuels can enhance the energy security of a country. Since they can be produced domestically, biofuels contribute to a more self-sufficient and resilient energy system.

3. **Rural Development:** Biofuel production can revitalize rural economies by creating jobs and income opportunities in agricultural communities. Farmers can grow feedstock crops, such as corn or sugarcane, and sell them to biofuel producers, providing an additional source of income.

Despite these benefits, biofuels also face several challenges that need to be addressed:

1. **Land Use and Food Security:** The cultivation of crops for biofuel production can compete with food production, leading to concerns about food security. It is crucial to strike a balance between dedicated energy crop cultivation and food production to avoid adverse impacts on global food supplies.

2. **Sustainability:** The production of biofuels should adhere to sustainable practices to avoid negative environmental impacts. This includes responsible land use, minimal use of water resources, and avoidance of harmful inputs such as pesticides and fertilizers.

3. **Indirect Land Use Change (ILUC):** The expansion of biofuel crops can lead to indirect land use change, where agricultural activities shift to previously untouched lands, such as forests or grasslands. This can result in deforestation or loss of biodiversity, offsetting the environmental benefits of biofuel production.

Examples and Real-World Applications

Biofuels have gained traction in various sectors and countries, demonstrating their potential as a sustainable energy source. Some notable examples include:

1. Brazil's Ethanol Program: Brazil has embraced bioethanol as a primary fuel source for transportation, particularly in the form of sugarcane ethanol. The country's ethanol program significantly reduced its dependence on imported oil and helped to lower greenhouse gas emissions from the transportation sector.

2. European Union's Biodiesel Industry: The European Union has been at the forefront of biodiesel production and consumption. It has implemented policies and incentives to promote the use of biodiesel made from vegetable oils, resulting in a thriving industry that contributes to greenhouse gas emissions reduction and energy diversification.

3. Biogas in Sweden: Sweden has been successful in utilizing biogas for various applications, including as a transportation fuel and for heating purposes. The country's biogas production primarily relies on the anaerobic digestion of organic waste, providing an environmentally friendly solution for waste management and renewable energy generation.

Resources and Further Reading

To explore the topic of biofuels further, the following resources are recommended:

+ *Bioenergy and Sustainability: Bridging the Gap*, Edited by Birgit Kamm and Michael Kamm

+ *Biofuels, Solar and Wind as Renewable Energy Systems: Benefits and Risks*, by David Pimentel

+ *Biofuels: Greenhouse Gas Mitigation and Global Warming*, Edited by Rattan Lal and David Hansen

+ *The Biofuel Delusion: The Fallacy of Large-Scale Agro-Biofuels Production*, by Mario Giampietro and Kozo Mayumi

Conclusion

Biofuels have emerged as a promising alternative to fossil fuels, offering renewable and low-carbon energy solutions for various applications. Their production from organic matter provides opportunities for sustainable development, energy security, and reduced greenhouse gas emissions. However, challenges related to land use, food security, sustainability, and indirect land use change need to be addressed. By focusing on responsible production practices and careful resource management, biofuels can contribute to a more sustainable and resilient energy future.

Aircraft Design

Aircraft design plays a crucial role in advancing eco-science and ensuring sustainable transportation systems. In this section, we will explore the principles, challenges, and advancements in aircraft design that contribute to environmental sustainability.

Principles of Aircraft Design

Aircraft design involves the integration of various engineering disciplines to create efficient and environmentally friendly flying machines. The following are some key principles considered in aircraft design:

1. **Aerodynamics:** Aerodynamics is the study of how air flows around an object, in this case, an aircraft. The design of aircraft wings, fuselage, and other components is optimized to minimize drag, maximize lift, and improve fuel efficiency.

2. **Structural Design:** Aircraft structural design focuses on creating lightweight yet robust structures that can withstand the forces experienced during flight. The use of advanced materials, such as carbon fiber composites, allows for weight reduction while maintaining structural integrity.

3. **Propulsion System:** The propulsion system of an aircraft is crucial for determining its fuel efficiency and environmental impact. Modern aircraft employ advanced engines, such as turbofans or turboprops, that offer better fuel economy and lower emissions compared to older engine designs.

4. **Efficient Energy Management:** Aircraft design incorporates efficient energy management systems to reduce energy wastage. This includes optimizing the electrical system, minimizing leakage currents, and utilizing regenerative braking or energy storage technologies.

Challenges in Aircraft Design

Designing environmentally sustainable aircraft faces several challenges, including:

1. **Fuel Efficiency**: Improving the fuel efficiency of aircraft is crucial for reducing greenhouse gas emissions and minimizing the carbon footprint of air travel. Designing aerodynamically efficient airframes, reducing weight, and developing advanced propulsion systems are key strategies to address this challenge.

2. **Emission Reduction**: Aircraft engines emit pollutants, such as nitrogen oxides (NOx) and particulate matter, contributing to air pollution and climate change. Aircraft design must focus on reducing emissions through advanced combustion techniques, improved engine designs, and use of alternative fuels.

3. **Noise Reduction**: Aircraft noise pollution is a significant concern, particularly in densely populated areas around airports. Aircraft design should consider noise reduction measures, including optimizing engine design, improving aerodynamics, and utilizing noise-absorbing materials.

4. **End-of-Life Considerations**: Aircraft have a limited operational lifespan, and their disposal can impact the environment. Designing aircraft with recyclable materials and considering end-of-life recycling or repurposing strategies can help minimize environmental impact.

Advancements in Aircraft Design for Sustainability

To address the challenges mentioned above, significant advancements have been made in aircraft design. Here are some notable examples:

1. **Efficient Engine Designs**: Modern aircraft engines incorporate technologies like high-bypass ratio turbofans, which provide better fuel efficiency and reduced noise. Additionally, continuous improvements in engine combustion processes have led to lower emissions.

2. **Lightweight Materials**: The use of lightweight materials, such as carbon fiber composites and aluminum alloys, has reduced the weight of aircraft structures. This weight reduction translates into fuel savings and lower emissions during flight.

3. **Hybrid Electric Propulsion**: Hybrid electric aircraft designs combine traditional fossil fuels with electric propulsion systems. Electric motors assist during takeoff and landing, reducing fuel consumption and emissions. Advances in battery technology have made hybrid electric aircraft viable options for shorter flights.

4. **Advanced Wing Designs:** Wing designs, such as winglets and laminar flow control, have improved aerodynamic efficiency, reducing drag and improving fuel economy. The development of flexible wings that adapt to different flight conditions further enhances performance.

5. **Alternative Fuels:** The aviation industry is actively exploring and adopting alternative fuels, such as biofuels, to reduce reliance on fossil fuels and lower emissions. These fuels can be derived from organic waste or grown crops and offer the potential for significant emissions reductions.

6. **Noise Reduction Techniques:** Engine and airframe design advancements, including improved insulation, serrated engine nacelles, and acoustic linings, help reduce aircraft noise levels. This allows for quieter takeoffs and landings, minimizing noise pollution in surrounding areas.

Example: Sustainable Aircraft Design Initiative

One notable initiative in sustainable aircraft design is the European Union's Clean Sky program. Clean Sky aims to develop and validate novel technologies for more environmentally friendly aircraft. This program focuses on reducing CO2 emissions, noise levels, and fuel consumption by targeting various aspects of aircraft design and operation.

Clean Sky's research areas include aerodynamics, propulsion systems, aircraft systems, structures, and manufacturing. Through collaboration between major aviation industry stakeholders, research organizations, and universities, Clean Sky fosters innovation and advances the state of the art in aircraft design to achieve enhanced sustainability.

Conclusion

Aircraft design has come a long way in addressing the environmental challenges associated with air travel. By incorporating principles of aerodynamics, structural design, propulsion systems, and energy management, design engineers can improve fuel efficiency, reduce emissions, and minimize noise pollution.

Advancements in engine technology, lightweight materials, hybrid electric propulsion, alternative fuels, and noise reduction techniques have all contributed to more sustainable aircraft designs. Ongoing research and collaborative initiatives, such as the Clean Sky program, continue to push the boundaries of eco-friendly aviation.

By embracing and implementing these advancements, the aviation industry can continue to evolve towards more sustainable and environmentally responsible air transportation systems.

Air traffic management

Air traffic management is a crucial aspect of the aviation industry, ensuring the safe and efficient movement of aircraft in the airspace. It involves various processes and technologies that enable the management of air traffic flow, communication between aircraft and air traffic control, and navigation of aircraft.

Challenges in air traffic management

The increasing number of flights and congested airspace pose significant challenges for air traffic management. Some of the key challenges include:

1. **Airspace capacity management:** With the growing demand for air travel, it is essential to optimize the utilization of available airspace capacity. This involves efficient planning and coordination of flight routes, airspace sectors, and arrival and departure procedures to avoid congestion.

2. **Air traffic flow management:** Managing the flow of air traffic is crucial for maintaining safety and avoiding delays. This includes sequencing and spacing of aircraft, rerouting in case of disruptions or weather conditions, and effective utilization of airport runways and taxiways.

3. **Communication and coordination:** Effective communication and coordination between aircraft and air traffic control are vital for the safe and efficient operation of the airspace. This involves clear and concise instructions, real-time information exchange, and adherence to standardized protocols.

4. **Safety and security:** Ensuring the safety and security of air travel is a top priority in air traffic management. This includes implementing measures to prevent collisions, addressing airspace incursions, and monitoring for potential security threats.

5. **Environmental impact:** Aircraft emissions contribute to environmental pollution and climate change. Therefore, air traffic management also needs to address the environmental impact of aviation by promoting efficient routing, reducing fuel consumption, and considering alternative energy sources.

Principles and solutions in air traffic management

To address the challenges mentioned above, air traffic management incorporates several principles and solutions. These include:

1. **Airspace design and optimization:** Airspace design plays a crucial role in managing air traffic efficiently. It involves segmenting the airspace into sectors and optimizing routes to ensure smooth flow. Advanced technologies such as Performance-Based Navigation (PBN) and Free Route Airspace (FRA) allow for flexible and optimal route planning.

2. **Air traffic control automation:** Automation technologies and systems, such as the Automatic Dependent Surveillance-Broadcast (ADS-B) and Collaborative Decision-Making (CDM), are used to improve air traffic control operations. These technologies enhance situational awareness, reduce workload, and enable better coordination and decision-making.

3. **Enhanced communication and navigation systems:** Modern communication and navigation systems, such as Very High Frequency (VHF) radio, Ground-Based Augmentation System (GBAS), and satellite-based systems like the Global Navigation Satellite System (GNSS), enable more reliable and accurate communication and navigation between aircraft and air traffic control.

4. **Collaborative decision-making:** Collaborative decision-making involves close coordination and cooperation between air traffic control, airlines, and other stakeholders. It aims to improve the efficiency of air traffic management by sharing information, making collective decisions, and optimizing resources.

5. **Integration of unmanned aircraft systems (UAS):** With the increasing use of unmanned aircraft systems (UAS), also known as drones, in civilian airspace, air traffic management needs to incorporate appropriate regulations and technologies to ensure safe integration and separation of manned and unmanned aircraft.

Example: Free Route Airspace (FRA)

Free Route Airspace (FRA) is an example of a solution implemented in air traffic management to optimize airspace utilization and reduce flight distances. FRA allows aircraft operators to plan and fly routes freely between defined entry and exit points, without being constrained by predefined airways or routes.

By introducing FRA, aircraft can take more direct routes, saving time, fuel, and reducing emissions. Air traffic control systems have been developed to support FRA, enabling proper coordination and separation of aircraft.

This concept has successfully been implemented in various regions, such as the Free Route Airspace Scandinavia (FRAS) and the Free Route Airspace Maastricht

(FRASMA).

It is vital to note that the implementation of FRA requires careful planning and coordination among airspace users, air traffic control organizations, and other stakeholders to ensure the safe and efficient operation of the airspace.

Conclusion

Air traffic management plays a critical role in ensuring the safe and efficient operation of the airspace. By addressing challenges such as airspace capacity, air traffic flow, communication, safety, and environmental impact, air traffic management aims to optimize the utilization of airspace and improve the overall efficiency of air travel.

Through the principles and solutions discussed, such as airspace design and optimization, automation, enhanced communication and navigation systems, collaborative decision-making, and the integration of unmanned aircraft systems, air traffic management continues to evolve to meet the demands of the aviation industry.

As air travel continues to grow, it is essential to further develop and implement innovative technologies and strategies to ensure the sustainability and effectiveness of air traffic management systems.

Waste Management

Recycling and Composting

Recycling and composting are two essential methods of waste management that contribute to environmental sustainability. These practices help reduce the amount of waste that ends up in landfills, conserve natural resources, and minimize pollution. In this section, we will explore the principles, processes, benefits, challenges, and future directions of recycling and composting.

Principles of Recycling

Recycling is the process of converting waste materials into new products, thereby reducing the consumption of raw materials and energy. The principles underlying recycling include the following:

- **Source separation:** Waste materials are sorted and separated at their point of generation into different categories such as paper, plastic, glass, and metal. This allows for more efficient recycling processes and reduces contamination.

+ **Material recovery:** The collected waste materials are processed and transformed into raw materials suitable for manufacturing new products. This may involve cleaning, shredding, melting, or other physical or chemical processes, depending on the material type.

+ **Manufacturing of recycled products:** The recovered raw materials are used to produce new products, thus closing the loop and reducing the need for extracting and processing virgin resources.

Processes of Recycling

The recycling process consists of several steps, including collection, sorting, processing, and manufacturing. Let's explore each of these steps in detail:

+ **Collection:** Waste materials are collected from households, businesses, and public spaces through various methods, such as curbside collection, drop-off centers, or recycling bins. Effective collection systems are crucial for maximizing recycling rates.

+ **Sorting:** Collected materials are sorted into different categories based on their material type. This can be done manually or using automated sorting technologies like conveyor belts, magnets, optical sensors, and air classification systems.

+ **Processing:** The sorted materials undergo different processing techniques to prepare them for reuse. For instance, paper products are pulped, metals are melted, plastics are shredded or melted and re-pelletized, and glass is crushed and cleaned.

+ **Manufacturing:** The processed materials are then used as feedstock for manufacturing new products. For example, recycled paper can be used to produce newspapers, recycled plastic can be transformed into new plastic bottles, and recycled glass can be used to make new glass containers.

Benefits of Recycling

Recycling offers numerous environmental, economic, and social benefits:

+ **Conservation of resources:** Recycling reduces the demand for virgin materials, such as trees for paper, ores for metals, and crude oil for plastics. This helps conserve natural resources and protects fragile ecosystems.

+ **Energy savings:** Recycling typically requires less energy than manufacturing products from raw materials. For example, producing recycled aluminum requires just 5% of the energy needed for primary aluminum production.

+ **Reduction of landfill waste:** By diverting waste materials from landfills, recycling helps reduce the volume of waste that needs to be disposed of. This extends the lifespan of landfills and minimizes the release of harmful substances into the environment.

+ **Greenhouse gas emissions reduction:** Recycling reduces greenhouse gas emissions associated with the extraction, transportation, and processing of raw materials. For example, recycling one ton of paper can save 17 mature trees and reduce greenhouse gas emissions by 1.6 metric tons of carbon dioxide equivalent.

+ **Job creation and economic growth:** The recycling industry creates jobs in collection, sorting, processing, and manufacturing. It also contributes to the local economy by generating revenue through the sale of recycled materials and the production of new products.

Challenges and Considerations

While recycling offers significant benefits, it also faces various challenges and considerations:

+ **Contamination:** Contamination of recyclable materials with non-recyclables is a major concern. Proper education and awareness programs are essential to educate the public about proper sorting and the types of materials accepted for recycling.

+ **Lack of infrastructure:** Inadequate recycling infrastructure, particularly in developing countries, poses a challenge. Investment in collection systems, processing facilities, and market development for recycled products is crucial to promote recycling.

+ **Market demand and economics:** The demand for recycled products is influenced by market fluctuations and consumer preferences. Stable markets and competitive pricing are essential for the long-term viability of recycling programs.

+ **Technology limitations:** Some materials are more challenging to recycle due to technological limitations or high costs. For example, the recycling of certain types of plastics or composite materials poses technical challenges that require further research and innovation.

Case Study: Plastic Recycling

Plastic recycling is a critical aspect of waste management due to the extensive use of plastics and their environmental impact. Although recycling plastic poses several challenges, innovative solutions are being developed to address them.

One example is the recycling of polyethylene terephthalate (PET) bottles, commonly used for beverage containers. The recycling process involves the following steps:

+ **Collection:** Empty PET bottles are collected from households, public spaces, and recycling centers.

+ **Sorting:** The collected bottles are sorted based on their color and grade. Contamination from other plastics or non-recyclables is minimized through manual or automated sorting processes.

+ **Processing:** The sorted bottles are cleaned, crushed into flakes, and then washed to remove labels, adhesives, and contaminants.

+ **Manufacturing:** The clean PET flakes are melted and extruded into thin fibers, which are then used to produce new products such as polyester fibers for clothing, carpets, and other applications.

Innovative technologies are being developed to improve plastic recycling, including chemical recycling methods that can convert plastic waste into valuable chemicals or feedstock for new plastics.

Future Directions

The future of recycling lies in embracing advanced technologies and adopting sustainable practices. Some key areas for future development include:

+ **Improving recycling rates:** Efforts should be made to increase recycling rates by implementing effective collection systems, promoting public awareness and participation, and exploring innovative recycling technologies.

+ **Advancing recycling technologies:** Research and development should focus on improving the efficiency and effectiveness of recycling processes, particularly for challenging materials like plastics or electronic waste.

+ **Closing the loop:** Encouraging the use of recycled materials in the manufacturing sector can create a circular economy, where products are designed for easy disassembly and recycling.

+ **Promoting sustainable consumption:** A shift towards sustainable consumption patterns, such as reducing waste generation, reusing products, and embracing eco-friendly alternatives, can complement recycling efforts.

Conclusion

Recycling and composting play a crucial role in waste management and environmental sustainability. These practices conserve resources, reduce pollution, and have economic benefits. Although there are challenges to overcome, the development and adoption of advanced technologies and sustainable practices offer promising solutions. By embracing recycling and composting, we can create a more sustainable future for generations to come.

Materials Recovery Facilities

Materials recovery facilities (MRFs) play a crucial role in waste management by facilitating the separation and recovery of valuable materials from mixed waste. These facilities are designed to process the incoming waste stream and separate recyclable materials from non-recyclable waste. MRFs are an essential component of the circular economy, as they promote resource conservation and reduce the amount of waste sent to landfills.

Working Principles of MRFs

The primary goal of a MRF is to recover as many recyclable materials as possible, while minimizing contamination and maximizing the value of the recovered materials. MRFs employ a combination of mechanical, manual, and advanced sorting technologies to achieve this objective.

The process at a MRF typically involves the following steps:

1. Collection and Sorting: Mixed waste is collected from households or commercial establishments and transported to the MRF. Upon arrival, the waste is unloaded and sorted to remove large items and non-processable materials.

2. Shredding and Size Reduction: The remaining waste is then shredded and reduced in size to facilitate further processing.

3. Mechanical Screening: The shredded waste passes through a series of screens and separators. These mechanical devices help separate different types of materials based on size and density. For example, screens can separate larger items such as bottles and cans from smaller particles.

4. Magnetic Separation: Magnetic separators are used to remove ferrous metals, such as iron and steel, from the waste stream. These metals can be easily separated due to their magnetic properties.

5. Eddy Current Separation: Eddy current separators generate a magnetic field that repels non-ferrous metals, such as aluminum and copper. This process helps separate these valuable metals from the waste stream.

6. Optical Sorting: Advanced optical sorting technologies, such as near-infrared (NIR) sensors, are used to detect and sort different types of plastics based on their chemical composition. These sensors can differentiate between various types of plastics, enabling efficient separation and recycling.

7. Manual Sorting: Trained workers manually sort through the waste stream to remove any remaining contaminants and recover materials that may have been missed by the automated sorting systems. Manual sorting is crucial to ensure the quality of the recovered materials and minimize contamination.

8. Commodity Baling: Once the materials have been sorted, they are compacted and baled for easier transportation and storage. Baling also helps maximize the value of the recovered materials.

9. Market Sale: The baled materials are then sold to recycling companies or manufacturers who will process them further into new products.

Challenges and Solutions

Operating a MRF comes with its own set of challenges. Some of the common challenges faced by MRFs include:

1. Contamination: Contamination of the waste stream can compromise the quality of the recovered materials. It is essential to educate the public about proper waste sorting and recycling practices to minimize contamination.

2. Technology Limitations: MRFs heavily rely on sorting technologies, and advancements in these technologies are critical for improving efficiency and increasing the recovery rate. Ongoing research and development efforts are necessary to overcome existing limitations and improve the performance of MRFs.

3. Market Demand: The demand for recycled materials fluctuates based on market conditions. MRFs need to constantly adapt to changes in market demand to ensure the economic viability of the recycling process.

To address these challenges, MRFs can implement the following solutions:

1. Public Education: Educating the public about proper waste sorting and recycling practices is crucial to reduce contamination. This can be done through awareness campaigns, educational programs, and clear instructions on waste disposal bins.

2. Collaboration and Partnerships: MRFs can collaborate with local governments, recycling companies, and other stakeholders to improve waste management practices and create a sustainable recycling infrastructure.

3. Technology Upgrades: Investing in advanced sorting technologies can improve the efficiency and accuracy of the sorting process. This includes incorporating new sensor technologies like hyperspectral imaging and machine learning algorithms for better material identification and separation.

4. Market Development: MRFs can work closely with manufacturers and industry stakeholders to develop new markets for recycled materials. This can involve exploring innovative uses for recycled materials or creating demand for specific types of recycled products.

Case Study: The Recycle Central Facility

One notable example of a MRF is the Recycle Central Facility in San Francisco, California. It is one of the largest and most advanced MRFs in North America. The facility processes mixed waste from residential, commercial, and industrial sources and recovers a wide range of recyclable materials.

The Recycle Central Facility employs a combination of mechanical, manual, and advanced sorting technologies to sort and recover materials. The facility has multiple sorting lines equipped with optical sorters, screens, magnets, and manual sorting stations. It can process up to 750 tons of materials per day.

To minimize contamination, the facility conducts regular audits and provides feedback to residents and businesses on their recycling practices. It also collaborates with local recyclers and manufacturers to create a closed-loop recycling system.

The Recycle Central Facility serves as a model for other MRFs worldwide, showcasing the importance of advanced sorting technologies, effective public education, and collaboration to achieve high recycling rates and reduce the environmental impact of waste.

Conclusion

Materials recovery facilities play a pivotal role in waste management and the transition to a circular economy. By separating and recovering valuable materials from mixed waste, MRFs contribute to resource conservation and reduce the burden on landfills. However, the successful operation of MRFs relies on efficient sorting technologies, public education, and collaboration with stakeholders. Continued advancements in MRF technology and practices are essential to maximize the recovery of recyclable materials and achieve sustainable waste management.

Composting methods

Composting is a natural process that converts organic waste into a nutrient-rich material called compost. It is an essential component of sustainable waste management and plays a crucial role in reducing the environmental impact of waste disposal. Composting methods vary depending on factors such as the type of organic waste, available space, and desired end product. In this section, we will explore different composting methods commonly used in eco science and their suitability for different situations.

Traditional Composting

Traditional composting is a simple and widely practiced method that involves the decomposition of organic waste in a pile or bin. It requires a mix of organic materials such as yard trimmings, vegetable scraps, and leaves, as well as air, water, and microbes. The organic waste is arranged in layers, ensuring a balance between carbon-rich (browns) and nitrogen-rich (greens) materials. Browns include dried leaves and woody materials, while greens consist of fresh grass clippings and vegetable scraps. Water is added periodically to maintain the moisture level, and the pile is turned occasionally to provide aeration.

The decomposition process in traditional composting is aerobic, meaning it occurs in the presence of oxygen. The microbes naturally present in the organic waste break down the materials, producing heat as a byproduct. This heat helps to speed up the decomposition process and kill potential pathogens and weed seeds. Over time, the organic waste transforms into mature compost, which is dark, crumbly, and earthy in texture.

Traditional composting is suitable for backyard or small-scale composting where space is limited. It is a cost-effective method that requires minimal equipment and can be easily incorporated into household waste management practices. However, it may not be suitable for large-scale operations or urban areas with space constraints.

Vermicomposting

Vermicomposting is a specialized form of composting that utilizes earthworms to accelerate the decomposition process. Earthworms, particularly red wigglers (Eisenia fetida), feed on organic waste, breaking it down into smaller particles. Their digestive system enhances the microbial activity and nutrient cycling, resulting in nutrient-rich vermicompost.

To start vermicomposting, a suitable container such as a bin or a worm tower is prepared. The container is filled with a bedding material, which can be a mixture of shredded newspaper, cardboard, and coconut coir. Moisture is added to the bedding to create a suitable environment for the worms. Organic waste, such as fruit and vegetable scraps, coffee grounds, and tea bags, is then added to the container. The worms are introduced into the container, and the process begins.

Vermicomposting requires specific conditions to ensure the well-being of the worms. The temperature should be maintained between 18°C and 27°C, and the moisture level should be kept moist but not waterlogged. Overfeeding should be avoided to prevent odors and pest problems. Maintaining a proper balance between the bedding and the amount of organic waste is crucial for the success of vermicomposting.

Vermicomposting is particularly suitable for households and small-scale operations, as it can be done indoors and requires minimal space. The resulting vermicompost is high in nutrients, making it an excellent soil amendment for gardening and agricultural purposes.

Aerated Static Pile Composting

Aerated static pile (ASP) composting is a method that combines the principles of traditional composting with forced aeration. In this method, organic waste is piled in a specific area and mixed thoroughly to ensure a homogeneous mixture. A perforated pipe or a network of pipes is inserted into the pile to provide a continuous supply of air.

The aeration system in ASP composting promotes the growth of aerobic microorganisms, which thrive in the presence of oxygen. The continuous airflow helps to maintain an optimal temperature and moisture level, enhancing the decomposition process. The pile is periodically turned to further enhance oxygen diffusion and mix the decomposing materials.

ASP composting is suitable for large-scale composting operations, such as municipal composting facilities or commercial composting operations. It offers several advantages, including higher composting rates, reduced odors, and better control over the composting process. However, it requires proper monitoring and management to ensure adequate aeration and prevent the accumulation of moisture or the development of anaerobic conditions.

In-vessel Composting

In-vessel composting refers to composting methods that take place within a closed container or vessel. The container provides a controlled environment for the composting process, allowing for better regulation of temperature, moisture, and aeration. There are various types of in-vessel composting systems, including rotating drums, agitated beds, and forced aeration systems.

In-vessel composting offers several advantages, including faster composting rates, reduced odors, and the ability to process a wide range of organic waste, including meat, dairy, and food scraps. The controlled environment allows for better management of the composting process and enables composting throughout the year, regardless of external weather conditions.

However, in-vessel composting systems can be costly to implement and require specialized equipment and facilities. They are typically used in large-scale operations, such as industrial composting facilities or centralized waste management systems.

Bokashi Composting

Bokashi composting is a unique composting method that relies on fermentation rather than decomposition. It originated in Japan and involves the use of a specific type of microbial inoculant called bokashi bran. Bokashi bran consists of a mixture of beneficial microorganisms, such as lactic acid bacteria and yeast, which ferment the organic waste.

To start bokashi composting, organic waste, including food scraps and paper, is collected in a sealed container. Each layer of waste is sprinkled with a small amount of bokashi bran, which initiates the fermentation process. The container is then sealed tightly to create an anaerobic environment.

The fermentation process in bokashi composting produces organic acids and enzymes, which break down the organic waste. The resulting fermented waste, known as bokashi pre-compost, is not fully decomposed but can be further composted or buried in soil. It enriches the soil with beneficial microorganisms and nutrients.

Bokashi composting is suitable for urban areas or situations where space is limited, as it can be done indoors and does not produce odors. It is a convenient method for composting food waste since it can handle a wide range of materials, including cooked food, meat, and dairy products. However, the bokashi pre-compost should be further processed through traditional composting or buried in soil to complete the composting process.

Conclusion

Composting methods play a crucial role in eco science by providing sustainable solutions for organic waste management. Traditional composting, vermicomposting, aerated static pile composting, in-vessel composting, and bokashi composting are some of the commonly used methods. Each method has its advantages and suitability depending on the scale of operation, available space, and type of organic waste. By adopting appropriate composting methods, we can reduce waste, conserve resources, and contribute to the creation of a more sustainable future.

Circular Economy

In this section, we will explore the concept of a circular economy, which is a key aspect of eco science and sustainable development. We will discuss the principles and benefits of a circular economy, as well as examples and strategies for its implementation.

Introduction to Circular Economy

A circular economy is an economic system that aims to minimize waste and maximize the efficient use of resources. Unlike the traditional linear economy, which follows a "take-make-dispose" model, a circular economy promotes a closed-loop system where materials and resources are kept in use for as long as possible, through recycling, reuse, and regeneration.

The concept of a circular economy is based on the principles of environmental sustainability, resource efficiency, and waste reduction. It recognizes that the Earth's resources are finite and that traditional linear economic models contribute to depletion and environmental degradation. By transitioning to a circular economy, we can create a more sustainable and resilient future.

Principles of Circular Economy

The circular economy is guided by several key principles:

1. **Design for longevity and durability:** Products should be designed to last longer, be repairable, and have easily replaceable components. This reduces the need for constant consumption and disposal of goods.

2. **Design for resource efficiency:** The design of products should consider the entire lifecycle, from production to end-of-life, with a focus on reducing resource consumption, waste, and pollution.

3. **Reduce, reuse, and recycle:** Promoting the reduction of waste generation is essential. Reusable and recyclable materials and products should be prioritized, and waste should be minimized through proper recycling and waste management systems.

4. **Resource recovery and regeneration:** Materials and resources should be recovered and regenerated through recycling or other processes, enabling them to be used as raw materials for new products. This reduces the reliance on virgin resources and minimizes the environmental impact of extraction.

5. **Collaborative and sharing economy:** Encouraging collaborative and sharing business models, such as sharing platforms or product-as-a-service, can optimize resource utilization and reduce overall consumption.

6. **Renewable energy and clean technologies:** Promoting the use of renewable energy sources and clean technologies helps reduce the carbon footprint and reliance on fossil fuels in production processes.

Implementation of Circular Economy

Implementing a circular economy requires the involvement of various stakeholders, including governments, businesses, and individuals. Here are some strategies and examples of how a circular economy can be implemented:

1. Extended producer responsibility (EPR): EPR is a policy approach that holds producers responsible for the entire life cycle of their products, including collection and treatment of waste. This incentivizes manufacturers to design products that are more easily reusable and recyclable.

2. Waste management and recycling systems: Proper waste management and recycling systems are crucial for the effective implementation of a circular economy. Separation of recyclable materials, collection systems, and recycling facilities need to be in place to ensure that materials are recovered and reintroduced into the economy.

3. Product and material labeling: Clear labeling of products and materials can inform consumers about their environmental impact and guide their choices towards more sustainable options. This can include information about recyclability, biodegradability, and resource efficiency.

4. Sharing platforms and collaborative consumption: Sharing platforms, such as peer-to-peer renting or car-sharing services, enable the optimal use of resources by allowing multiple individuals to use a product instead of each person owning one.

5. Industrial symbiosis: Industrial symbiosis involves the exchange of by-products, energy, and resources between different industries located in close proximity. This creates a synergy where the waste of one industry becomes a resource for another, reducing the overall environmental impact and maximizing resource efficiency.

6. Sustainable supply chains: Implementing sustainable supply chains involves considering the entire lifecycle of products, from sourcing raw materials to disposal. This includes choosing suppliers that adhere to sustainable practices and ensuring the ethical and responsible management of resources throughout the supply chain.

Benefits of Circular Economy

The transition to a circular economy brings numerous benefits, including:

1. Resource conservation and environmental protection: By maximizing the usage of resources and minimizing waste generation, a circular economy reduces the depletion of natural resources and helps protect the environment from pollution and degradation.

2. Economic growth and job creation: Implementing circular economy practices can generate economic growth and create new job opportunities in sectors such as recycling, renewable energy, and sustainable manufacturing.

3. Cost savings and efficiency: A circular economy reduces the costs associated with raw material extraction and production, as well as waste management. It encourages the efficient use of resources, leading to cost savings for businesses and individuals.

4. Innovation and technological development: The transition to a circular economy requires innovation and the development of new technologies and processes for recycling, regeneration, and resource optimization. This drives technological advancements and promotes sustainable innovation.

Challenges and Considerations

While the concept of a circular economy holds great promise, there are challenges and considerations that need to be addressed:

1.   Complex value chains: Implementing a circular economy requires collaboration and coordination across complex value chains. Multiple stakeholders need to work together, including manufacturers, retailers, waste management systems, and consumers.

2. Behavioral change: Transitioning to a circular economy requires a shift in consumer behavior and attitudes towards consumption and ownership. Education and awareness programs play a crucial role in promoting sustainable choices.

3. Infrastructure and technology: Investments in infrastructure and technology are necessary to support the transition to a circular economy. This includes recycling facilities, waste management systems, and advanced technologies for resource recovery.

4. Policy and regulation: Clear policies and regulations need to be in place to incentivize and regulate the implementation of circular economy practices. This includes promoting extended producer responsibility, establishing recycling targets, and providing economic incentives for sustainable practices.

Conclusion

The concept of a circular economy offers a promising pathway towards a more sustainable and resilient future. By rethinking the current linear economic model and embracing the principles of a circular economy, we can minimize waste, conserve resources, and protect the environment. The successful implementation of a circular economy requires collaboration, innovation, and a cultural shift towards sustainable consumption and production. Through collective efforts, we can create a future where economic prosperity is achieved hand in hand with environmental stewardship.

Waste-to-energy technologies

Waste-to-energy technologies are innovative solutions that allow us to convert waste materials into usable forms of energy. These technologies not only help to reduce the amount of waste that ends up in landfills but also provide a sustainable and renewable source of energy. In this section, we will explore the different waste-to-energy technologies and their potential benefits.

Incineration

Incineration is one of the most well-known waste-to-energy technologies. It involves the combustion of waste materials at high temperatures, generating heat that can then be used to produce electricity or heat buildings. Incineration is a highly efficient process that can handle a wide range of waste types, including municipal solid waste, hazardous waste, and medical waste. However, there are concerns about air pollution and the emission of greenhouse gases during the incineration process.

Anaerobic Digestion

Anaerobic digestion is another waste-to-energy technology that involves the breakdown of organic waste in the absence of oxygen. This process produces biogas, a mixture of methane and carbon dioxide, which can be used as a renewable source of energy. Anaerobic digestion can be applied to various organic waste streams, such as food waste, agricultural waste, and sewage sludge. Additionally, the byproducts of anaerobic digestion, such as digestate, can be used as a fertilizer. This technology not only reduces the volume of waste but also mitigates the release of methane, a potent greenhouse gas.

Gasification

Gasification is a thermochemical process that converts solid waste into a synthetic gas, or syngas, comprising hydrogen, carbon monoxide, and other gases. The syngas can then be used to produce heat, electricity, or biofuels. Gasification is versatile and can handle a wide range of waste materials, including biomass, coal, and plastic waste. Furthermore, the high-temperature process allows for the capture and removal of harmful pollutants, reducing environmental impacts.

Pyrolysis

Pyrolysis is a thermal decomposition process that converts organic waste into biochar, oil, and gas in the absence of oxygen. The biochar can be used as a soil amendment to improve fertility and sequester carbon. The oil and gas produced can be used as fuels for heat and power generation or as feedstock for the chemical industry. Pyrolysis has the advantage of producing biochar, which has potential benefits in carbon sequestration and soil health improvement. However, the process requires careful control of temperature and residence time to optimize the yield and quality of the products.

Plasma Gasification

Plasma gasification is an advanced waste treatment technology that uses high-temperature plasma arcs to convert waste materials into a synthesis gas. The synthesis gas can be further processed and used for electricity generation or as a raw material for the production of chemicals and fuels. Plasma gasification can handle a wide range of waste materials, including municipal solid waste, hazardous waste, and medical waste. This technology has the advantage of being able to break down hazardous components and convert them into inert materials, reducing the environmental impact of the waste.

Benefits and Challenges

Waste-to-energy technologies offer several benefits in terms of waste management, energy production, and environmental sustainability. Firstly, they reduce the volume of waste that needs to be disposed of in landfills, thus conserving valuable land resources. Secondly, waste-to-energy technologies provide a renewable source of energy, which reduces dependence on fossil fuels and reduces greenhouse gas emissions. Additionally, these technologies can help in the transition towards a circular economy by recovering valuable materials from waste streams.

However, there are also challenges associated with waste-to-energy technologies. One challenge is the potential release of pollutants during the conversion process, which requires the implementation of appropriate emission control technologies. Another challenge is the need for a constant supply of waste materials to ensure the continuous operation of the facilities. Additionally, the economic viability of waste-to-energy projects depends on factors such as waste composition, energy prices, and government policies and regulations.

Example: Waste-to-Energy Plant

To illustrate the application of waste-to-energy technologies, let's consider an example of a waste-to-energy plant. The plant is designed to process municipal solid waste and generate electricity.

Firstly, the waste materials are collected and sorted to remove recyclable materials. The remaining waste is fed into the incineration chamber, where it is burned at high temperatures. The heat generated from the incineration process is used to produce steam, which then drives a turbine connected to a generator, producing electricity. The exhaust gases from the incineration process are treated to remove pollutants before being released into the atmosphere.

Additionally, the plant incorporates an anaerobic digestion system to process organic waste. The organic waste is mixed with water and undergoes a controlled anaerobic digestion process, producing biogas. The biogas is then used to generate heat, which can be utilized in the incineration process or converted into electricity.

This waste-to-energy plant demonstrates the integration of different technologies to maximize energy recovery from waste materials while minimizing environmental impacts.

Conclusion

Waste-to-energy technologies provide a sustainable and efficient solution for waste management and energy production. They help to reduce the volume of waste that goes into landfills, mitigate greenhouse gas emissions, and provide a renewable source of energy. However, their successful implementation requires careful consideration of environmental impacts, emission control, and economic viability. By harnessing the potential of waste-to-energy technologies, we can move towards a more sustainable and circular future.

Incineration

Incineration is a waste management method that involves the combustion of solid waste at high temperatures. It is commonly used to reduce the volume of waste and to generate energy. In this section, we will explore the process of incineration, its environmental impacts, and its role in waste management.

Process of Incineration

The process of incineration involves several steps:

1. Waste Reception: Solid waste is collected and transported to the incineration facility. It is important to ensure that only suitable waste is accepted for incineration, as certain materials can release harmful pollutants when burned.

2. Waste Pre-treatment: Before incineration, the waste may undergo pre-treatment to remove any recyclable materials such as metals and plastics. This helps to maximize the recovery of valuable resources.

3. Combustion: The waste is then fed into the incinerator, where it is burned at high temperatures of around 800 to 1000 degrees Celsius. The combustion process is carefully controlled to ensure complete combustion and minimize emissions.

4. Energy Generation: The heat generated from the combustion process is used to produce steam, which drives a turbine and generates electricity. This is known as waste-to-energy (WtE) or energy-from-waste (EfW) conversion.

5. Air Pollution Control: To reduce air emissions, incineration facilities are equipped with sophisticated air pollution control systems. These systems remove pollutants such as particulate matter, acid gases, dioxins, and heavy metals from the flue gases before they are released into the atmosphere.

6. Residue Treatment: After combustion, the remaining ash is treated to remove any hazardous components and then disposed of in a controlled manner, such as in a landfill or used for construction materials.

Environmental Impacts

While incineration offers benefits such as waste volume reduction and energy generation, it also has environmental impacts that need to be considered:

1. Air Emissions: The combustion of waste releases air pollutants, including harmful gases and particulate matter. Effective air pollution control measures are necessary to minimize these emissions and protect air quality.

2. Ash Residue: Incineration produces ash residue, which must be managed properly to prevent the release of hazardous materials into the environment. Ash can contain heavy metals and other pollutants, requiring careful disposal or treatment.

3. Greenhouse Gas Emissions: Although incineration can contribute to reducing greenhouse gas emissions compared to landfilling, it still releases carbon dioxide (CO_2) and other greenhouse gases. The extent of these emissions depends on the composition of the waste and the efficiency of the incineration process.

4. Energy Efficiency: The energy efficiency of incineration is influenced by factors such as the waste composition, the type of incinerator technology used, and the heat recovery systems in place. Maximizing energy recovery is crucial to improving the overall environmental performance of incineration.

Challenges and Considerations

Incineration is a complex waste management method that requires careful planning and consideration of various factors. Some challenges and considerations include:

1. Waste Composition: The composition of waste can vary significantly, and certain materials may not be suitable for incineration due to their potential to release harmful pollutants. Proper waste separation and sorting are essential to ensure that only appropriate waste is incinerated.

2. Public Perception: Incineration facilities can face opposition from local communities due to concerns about air pollution and the potential health impacts of emissions. Public engagement and clear communication about the environmental controls in place are important in addressing these concerns.

3. Technological Advancements: Ongoing research and development are essential to improve incineration technologies, enhance energy recovery, and minimize emissions. Advancements in air pollution control systems and waste treatment methods can contribute to making incineration a more sustainable waste management option.

4. Integrated Waste Management: Incineration should be considered as part of an integrated waste management approach that includes waste reduction, recycling, and composting. By combining different waste management strategies, the overall environmental impact can be minimized.

Case Study: Waste-to-Energy Plant

To illustrate the application of incineration in waste management, let's consider the example of a waste-to-energy plant located in a city. This facility accepts mixed municipal solid waste and uses mass burn incineration technology to generate electricity.

The plant follows a strict waste reception protocol and conducts thorough waste sorting and pre-treatment to remove recyclable materials. The waste is then fed into the incinerator, where it is combusted at high temperatures.

To control air emissions, the plant has installed state-of-the-art air pollution control systems, including electrostatic precipitators and selective catalytic reduction units. These systems effectively remove particulate matter, acid gases, and nitrogen oxides from the flue gases.

The heat generated during combustion is used to produce steam, which drives a turbine and generates electricity. The plant supplies this electricity to the local grid, contributing to the city's energy needs.

The remaining ash from the incineration process is treated to remove any hazardous components. Some of the ash is used in construction materials, while the rest is disposed of in a specialized ash landfill.

The waste-to-energy plant has implemented a comprehensive monitoring and reporting system to track emissions and ensure compliance with environmental regulations. Ongoing research and technological advancements are undertaken to improve energy efficiency and reduce the plant's environmental footprint.

Summary

Incineration is a waste management method that involves the combustion of solid waste at high temperatures. It offers benefits such as waste volume reduction and energy generation but also has environmental impacts that need to be addressed.

The process of incineration includes waste reception, pre-treatment, combustion, energy generation, and air pollution control. Proper waste management and air pollution control measures are necessary to minimize environmental impacts.

Challenges and considerations in incineration include waste composition, public perception, technological advancements, and integrated waste management. By combining different waste management strategies, the overall environmental impact can be minimized.

A case study of a waste-to-energy plant demonstrates the application of incineration in waste management and highlights the importance of proper waste sorting, advanced air pollution control systems, and efficient energy recovery.

In conclusion, incineration, when properly regulated and integrated into a comprehensive waste management system, can be an effective approach to reduce waste, generate energy, and minimize environmental impacts. However, continuous research and innovation are essential to improve the efficiency and sustainability of incineration technologies.

Anaerobic Digestion

Anaerobic digestion is a biological process that breaks down organic waste materials in the absence of oxygen. It involves the decomposition of complex organic compounds into simpler molecules, such as carbon dioxide, methane, and organic acids. This process occurs naturally in oxygen-depleted environments, such as wetlands and the digestive systems of animals.

Anaerobic digestion is an important technology in the field of eco science because it offers a sustainable solution for both waste management and energy production. It not only reduces the volume and environmental impact of organic waste but also generates biogas, a renewable energy source that can be used for heat and electricity generation. In addition, anaerobic digestion produces nutrient-rich digestate, which can be used as a fertilizer.

The principle behind anaerobic digestion is carried out by a complex microbial community called anaerobes. These microorganisms work together in a series of anaerobic stages to convert complex organic compounds into simpler compounds. The process can be divided into four distinct stages, namely hydrolysis, acidogenesis, acetogenesis, and methanogenesis.

During the hydrolysis stage, complex organic matter is broken down into smaller molecules by the action of hydrolytic bacteria. These bacteria secrete enzymes that break down proteins, carbohydrates, and fats into soluble compounds. The breakdown products, such as amino acids, simple sugars, and fatty acids, are then available for the next stage of the process.

In the acidogenesis stage, fermentative bacteria convert the soluble compounds into volatile fatty acids, alcohols, and organic acids. This stage is characterized by the production of intermediates like acetic acid, propionic acid, and butyric acid. These organic acids serve as substrates for the subsequent stages of anaerobic digestion.

The acetogenesis stage involves the conversion of organic acids into acetic acid, hydrogen, and carbon dioxide by acetogenic bacteria. Acetic acid is the main end product of this stage, and it serves as a precursor for methane production in the final stage.

The final stage of anaerobic digestion is methanogenesis. Methanogenic archaea utilize the acetic acid, hydrogen, and carbon dioxide produced in the previous stages to produce methane gas (CH_4) and carbon dioxide (CO_2). Methane, also known as biogas, is a valuable source of renewable energy that can be used for heating, electricity generation, and even as a vehicle fuel.

The efficiency of anaerobic digestion depends on various factors, including the composition of the feedstock, temperature, pH, and retention time. Different types of organic waste can be used as feedstock for anaerobic digestion, including agricultural residues, food waste, sewage sludge, and energy crops. The process can be carried out in anaerobic digesters, which are sealed containers designed to optimize the conditions for microbial activity.

One of the key advantages of anaerobic digestion is its ability to reduce greenhouse gas emissions. By capturing and utilizing methane, which is a potent greenhouse gas, the process helps mitigate climate change. Furthermore, anaerobic digestion reduces the need for landfilling and incineration of organic waste, which can release harmful pollutants into the environment.

However, anaerobic digestion also faces certain challenges and limitations. The process requires careful monitoring and control to maintain optimal conditions for microbial activity. The composition and characteristics of the feedstock can affect the efficiency and stability of the process. High ammonia concentrations, for example, can inhibit the activity of methanogenic bacteria.

To overcome these challenges, various strategies and technologies have been developed. Co-digestion, for instance, involves mixing different types of organic waste to optimize the biogas production. Pre-treatment methods, such as thermal, mechanical, or chemical treatments, can improve the digestibility of certain feedstocks.

In addition, the integration of anaerobic digestion with other waste management technologies, such as composting, can enhance the overall efficiency and sustainability of the process. This allows for the recovery of valuable resources from organic waste, such as compost and biogas, while minimizing environmental impacts.

Overall, anaerobic digestion plays a crucial role in the transition towards a more sustainable and circular economy. It offers a viable solution for the management of organic waste, energy production, and nutrient recovery. By harnessing the power of microorganisms, we can turn waste into a valuable resource and contribute to a more eco-friendly future.

Example Problem: Understanding Biogas Potential

A farmer is considering implementing an anaerobic digestion system on their farm to manage the manure generated by their livestock. The farmer wants to calculate the potential biogas production from the available manure. The manure has a total solids content of 5% and a volatile solids content of 80% on a dry weight basis.

Solution:

To calculate the potential biogas production, we need to consider the volatile solids content of the manure. The volatile solids are the organic components that are converted into biogas during anaerobic digestion.

First, we need to calculate the volatile solids (VS) content of the wet manure. We do this by multiplying the total solids (TS) content by the volatile solids content:

$$VS = TS \times \frac{\text{Volatile solids content}}{100}$$

Assuming the manure has a moisture content of 95% (or a dry matter content of 5%), we can calculate the volatile solids content as follows:

$$VS = 5\% \times \frac{80\%}{100\% - 5\%} = 4.21\%$$

Next, we need to calculate the biogas potential based on the volatile solids content. The approximate biogas potential is around 0.35 m³ of biogas per kg of volatile solids. However, this value can vary depending on the composition of the organic matter.

Assuming a biogas potential of 0.35 m³/kg VS, we can calculate the potential biogas production as follows:

$$\text{Biogas production} = VS \times \text{biogas potential}$$

$$\text{Biogas production} = 4.21\% \times 0.35 \, \text{m}^3/\text{kg} = 0.0147 \, \text{m}^3/\text{kg}$$

Therefore, for every kilogram of volatile solids in the manure, approximately 0.0147 m³ of biogas can be produced. The farmer can use this information to estimate the potential biogas production and evaluate the feasibility of implementing an anaerobic digestion system on their farm.

Note: The actual biogas production can vary depending on several factors, including the specific characteristics of the manure, the operating conditions of the anaerobic digester, and the efficiency of the biogas capture and utilization system. It is important to conduct site-specific assessments and consider the local conditions when estimating biogas potential.

Additional Resources: - Book: "Anaerobic Digestion: Principles and Applications" by G. Lettinga, et al. - Scientific paper: "Anaerobic digestion for bioenergy production: Global status, environmental and techno-economic implications, and government policies" by A. Sheikhzadeh, et al. - Website: International Water Association (IWA) Anaerobic Digestion Specialist Group (https://www.iwa-adsorptive.net/)

Gasification

Gasification is a thermochemical process that converts solid or liquid carbonaceous materials into a gaseous fuel called synthesis gas or syngas. This process involves the partial oxidation of the feedstock at high temperatures in a controlled environment. The resulting syngas can be used as a clean and efficient fuel for various applications, including power generation, heating, and industrial processes.

Principles of Gasification

The principles of gasification are based on the understanding of chemistry, thermodynamics, and fluid dynamics. In gasification, the feedstock is usually a carbon-rich material, such as coal, biomass, or municipal solid waste. The process occurs in a gasifier, which is typically a high-pressure vessel with an oxygen-limited or oxygen-starved environment.

The gasification process involves several key reactions, including pyrolysis, combustion, and reduction. Pyrolysis is the thermal decomposition of the feedstock in the absence of oxygen, resulting in the production of solid char, volatile gases, and tar. Combustion occurs when oxygen is introduced, leading to the partial oxidation of the volatile gases and the release of heat. Reduction is the reaction between carbon monoxide and steam or carbon dioxide, resulting in the production of hydrogen and carbon dioxide.

Gasification Technologies

There are various gasification technologies, including fixed-bed, fluidized bed, entrained flow, and ablative gasifiers. Each technology has its own advantages and disadvantages in terms of efficiency, fuel flexibility, scalability, and environmental impact.

Fixed-bed gasifiers consist of a stationary bed of solid fuel through which gasification agents, such as air, oxygen, or steam, flow. Fluidized bed gasifiers suspend the feedstock particles in a fluidized bed, allowing for better mixing and heat transfer. Entrained flow gasifiers inject the feedstock as fine particles into a high-velocity stream of gas, resulting in rapid mixing and high conversion rates. Ablative gasifiers use a rotating cone or disk to remove the char as it forms, allowing for continuous conversion.

Applications of Gasification

Gasification has a wide range of applications, and its versatility makes it a promising technology for sustainable energy production and waste management. Some of the key applications of gasification include:

1. Power Generation: Syngas produced through gasification can be used to generate electricity in gas turbines or combined cycle power plants. The high energy content and low emissions of syngas make it an attractive fuel option.

2. Chemical Production: Syngas can be further processed to produce various chemicals, such as ammonia, methanol, and synthetic natural gas. These chemicals serve as important building blocks for the production of fertilizers, plastics, and other industrial products.

3. Heating and Industrial Processes: Syngas can be used for heating applications in industries or residential areas. It can also replace fossil fuels in various industrial processes, such as steel production or cement kilns, reducing greenhouse gas emissions.

4. Waste Management: Gasification provides an environmentally friendly solution for waste management by converting municipal solid waste or biomass waste into energy. This can help reduce landfill waste, lower greenhouse gas emissions, and promote a circular economy.

Challenges and Future Directions

While gasification offers many benefits, there are also challenges that need to be addressed for its widespread implementation. Some of these challenges include:

1. Feedstock Quality and Availability: The quality and availability of feedstock greatly impact the efficiency and economics of gasification. Ensuring a consistent supply of suitable feedstock is essential for the success of gasification projects.

2. Tar Formation and Cleanup: Tar is a byproduct of gasification and can cause operational issues by clogging equipment and reducing efficiency. Developing effective tar cleanup technologies is crucial to improve gasification processes.

3. Environmental Impact: Gasification produces various emissions, including particulate matter, sulfur compounds, and nitrogen oxides. Appropriate emission control technologies are necessary to reduce the environmental impact of gasification.

4. Economics and Scale: Gasification plants require significant capital investment, and the economics of gasification are highly dependent on factors such

as feedstock costs, energy prices, and policy incentives. Scaling up gasification technologies to commercial levels is also a challenge that needs to be addressed.

Future directions in gasification include research and development in advanced gasification technologies, such as plasma gasification and supercritical water gasification. These technologies aim to improve efficiency, increase fuel flexibility, and reduce environmental impact. Additionally, integration of gasification with carbon capture and storage (CCS) technologies can help mitigate greenhouse gas emissions and contribute to the transition to a low-carbon economy.

In conclusion, gasification is a promising technology for sustainable energy production and waste management. It offers opportunities for clean energy generation, chemical production, and waste utilization. However, there are challenges that need to be overcome for the widespread adoption of gasification. Continued research, innovation, and policy support are essential to unlock the full potential of gasification and contribute to a more sustainable future.

Landfill Management

Landfills play a crucial role in waste management, particularly for non-hazardous solid waste disposal. Effective landfill management is essential to minimize the negative environmental and health impacts associated with landfills. In this section, we will explore the principles, techniques, and challenges of landfill management and discuss innovative approaches to expand their sustainability.

Principles of Landfill Management

The management of landfills is guided by several principles aimed at minimizing the environmental impact and maximizing resource recovery. These principles include:

1. **Waste Minimization:** The first step in landfill management is to reduce waste at its source through waste minimization strategies. This can include promoting recycling, composting, and waste-to-energy conversion technologies.

2. **Waste Segregation:** Proper waste segregation ensures that the landfill receives waste streams separated by type, such as municipal solid waste, construction and demolition waste, and hazardous waste. This segregation minimizes the potential for contamination and allows for more effective resource recovery.

3. **Engineering Controls:** The design and construction of landfills incorporate engineering controls to prevent the migration of contaminants into the environment. These controls include liners, leachate collection systems, and methane gas capture systems.

4. **Monitoring and Control**: Landfills require continuous monitoring and control to assess the environmental impact and the effectiveness of various management strategies. This includes monitoring of leachate quality, groundwater contamination, and gas emissions.

5. **Landfill Design**: Proper landfill design is critical to ensure the containment and management of waste. Key design aspects include liner selection, landfill cell construction, landfill cover systems, and landfill cap closure.

6. **Landfill Operation**: Effective landfill operation involves activities such as waste compaction, cover placement, and timely closure of landfill cells. Proper operation ensures maximum space utilization and minimizes environmental risks.

7. **Landfill Closure and Post-Closure Care**: After a landfill reaches its capacity, it undergoes closure and post-closure care. Closure involves the installation of a final cap and cover system, while post-closure care ensures ongoing monitoring and maintenance to prevent any environmental impacts.

Landfill Management Techniques

To ensure effective landfill management, various techniques and methods are employed. These techniques address different aspects of landfill operations and their environmental impact. Some important techniques include:

1. **Waste Compaction**: Waste compaction reduces the volume of waste and increases the capacity of landfills. This technique involves the use of heavy machinery to compact waste and limit settlement.

2. **Liner Systems**: Liner systems are an integral part of landfill design, preventing the migration of contaminants into the underlying soil and groundwater. Common liner systems include composite liners composed of geomembranes and geosynthetic clay liners.

3. **Leachate Management**: Leachate is the liquid that drains from landfilled waste. Proper leachate management is crucial to prevent groundwater contamination. Techniques for leachate management include collection systems, leachate treatment plants, and recirculation to enhance degradation.

4. **Landfill Gas Management**: Landfills produce significant amounts of methane gas, a potent greenhouse gas. Landfill gas management involves the installation of methane gas collection systems to capture and utilize the gas. Methane can be used for generating electricity, heat, or converted into compressed natural gas.

5. **Cover Systems**: Cover systems are placed on top of disposed waste to minimize water infiltration and control odors. These systems include a

geomembrane or clay layer covered with soil or other materials. Advanced cover systems, such as evapotranspiration covers, promote natural evaporation processes.

Innovative Approaches to Landfill Management

While traditional landfill management techniques have been effective, there is a growing need for more sustainable and resource-efficient approaches. Several innovative practices are being explored to address the challenges associated with landfills. Here are a few examples:

1. **Landfill Mining:** Landfill mining involves extracting valuable resources, such as metals, plastics, and organic matter, from existing landfills. This approach promotes resource recovery, reduces the need for raw materials, and minimizes the environmental impact of landfilling.

2. **Bioreactor Landfills:** Bioreactor landfills aim to enhance the degradation of organic waste by providing optimal conditions for microbial activity. This includes the controlled addition of liquids or leachate recirculation to accelerate waste decomposition, reduce landfill lifespan, and reduce greenhouse gas emissions.

3. **Phytoremediation:** Phytoremediation involves the use of plants to naturally degrade or stabilize contaminants present in landfilled waste. Certain plant species are well-suited for extracting heavy metals and other pollutants from the soil, contributing to the restoration and remediation of landfill sites.

4. **Improved Landfill Cover Design:** Advanced landfill cover designs, such as capillary break layers and bioactive covers, can enhance landfill performance and reduce environmental impacts. These cover systems minimize the infiltration of water, promote natural degradation processes, and reduce the generation of leachate.

Tackling Challenges in Landfill Management

Landfill management faces several challenges that need to be addressed to ensure long-term sustainability. Some of the key challenges include:

1. **Limited Land Availability:** Finding suitable land for new landfills is becoming increasingly challenging, especially in densely populated areas. Developing innovative techniques for maximizing the capacity of existing landfills and exploring alternative waste management options are essential.

2. **Environmental Impacts:** Landfills have environmental consequences such as groundwater contamination, soil pollution, and emissions of greenhouse gases. Minimizing these impacts through effective management practices and adopting sustainable alternatives is crucial.

3. **Landfill Closure and Aftercare**: Proper closure and post-closure care of landfills are often overlooked, leading to potential environmental risks in the long term. Adequate financial provisions and monitoring systems are needed to ensure the effective closure and management of landfilled areas.

4. **Public Perception and Community Engagement**: Landfills face public resistance due to aesthetic, health, and environmental concerns. Engaging communities in the decision-making process, promoting transparency, and educating the public about proper waste management is vital to mitigate these concerns.

Case Study: The Puente Hills Landfill

The Puente Hills Landfill, located in Los Angeles County, California, is a notable example of landfill management. It was the largest landfill in the United States and operated for over half a century. The landfill, covering approximately 700 acres, received millions of tons of waste annually.

To address environmental concerns and manage the landfill effectively, several strategies were implemented:

1. **Methane Gas Recovery**: A comprehensive gas collection and flaring system were installed to capture and combust methane gas emitted from the landfill. The gas was then utilized to generate electricity for thousands of homes.

2. **Landfill Diversion**: The landfill management prioritized waste diversion practices, such as recycling and composting, to minimize the volume of waste sent to the landfill. This was achieved through public awareness campaigns, improved recycling infrastructure, and the implementation of waste diversion programs.

3. **Closure and Post-Closure Plan**: A detailed plan was developed to ensure the proper closure and ongoing post-closure management of the Puente Hills Landfill. The plan included procedures for final cover installation, leachate and gas collection system shutdown, and long-term monitoring and maintenance.

The successful management of the Puente Hills Landfill demonstrates the importance of implementing comprehensive and sustainable practices to mitigate the environmental impact of landfills.

Exercises

1. Research and discuss one innovative technique for landfill management not covered in this section. Highlight its advantages and potential challenges in implementation.

2. Conduct a case study on a landfill in your region/country. Analyze the existing landfill management practices and suggest improvements for enhanced sustainability.

3. Explore the concept of a circular economy in the context of landfill management. Discuss how the principles of a circular economy can be applied to minimize waste generation and promote resource recovery.

Conclusion

Effective landfill management is critical to minimize the environmental impact of waste disposal. By implementing the principles of waste minimization, waste segregation, engineering controls, and monitoring, landfills can be operated in a more sustainable manner. Innovative approaches such as landfill mining, bioreactor landfills, and phytoremediation are further enhancing sustainability efforts. However, challenges persist, including land scarcity, potential environmental impacts, and public perception. By addressing these challenges and incorporating sustainable practices, we can strive towards a more sustainable and resource-efficient future in landfill management.

Leachate Control

Leachate is a potentially hazardous liquid that forms when water passes through waste materials, such as landfills or compost piles. It can contain various pollutants, including heavy metals, organic compounds, and pathogens, posing a significant risk to the environment and human health if not properly managed. Leachate control refers to the strategies and technologies used to prevent and treat leachate to minimize its impact on the surrounding ecosystems.

Causes and Characteristics of Leachate

Leachate is formed due to the decomposition and breakdown of waste materials. Rainwater or other water sources infiltrate the waste, picking up dissolved and suspended substances along the way. The characteristics of leachate depend on the composition of the waste and environmental factors such as temperature, pH, and contact time.

The main sources and components of leachate include:

+ Organic matter: Decomposing organic waste produces organic acids, which contribute to the high biochemical oxygen demand (BOD) and chemical oxygen demand (COD) of leachate.

+ Inorganic ions: Leachate contains high concentrations of various ions, including ammonia, nitrate, chloride, sulfate, and heavy metals. These ions can contaminate groundwater and surface water bodies.

+ Pathogens: Microorganisms present in the waste can contaminate the leachate, posing a risk to human health and ecosystems.

The management of leachate is critical to prevent its migration to the surrounding environment, where it can contaminate soil, groundwater, and surface water bodies.

Preventive Measures

Effective leachate control starts with preventive measures that aim to minimize the formation and migration of leachate. The following strategies are commonly employed:

1. Liner systems: Landfills are often equipped with liner systems, consisting of impermeable layers such as clay or synthetic geomembranes. These liners prevent the downward movement of leachate, protecting underlying soils and groundwater.

2. Surface water diversion: Proper stormwater management and diversion systems can prevent rainwater from infiltrating waste containment areas, reducing the volume of leachate generated.

3. Waste compaction: Compaction of waste materials improves their density and reduces the void spaces, minimizing the infiltration of water and the potential for leachate generation.

4. Cover systems: Effective cover systems, such as clay caps or synthetic covers, prevent precipitation from infiltrating waste and minimize leachate generation.

Implementing these preventive measures is crucial in reducing the risks associated with leachate generation.

Treatment Technologies

Even with preventive measures in place, leachate management often requires treatment to reduce its pollution potential. Various treatment technologies can be employed, depending on the characteristics and volume of the leachate. The following are some commonly used treatment methods:

- Physical treatment: Physical processes such as sedimentation, filtration, and adsorption can be used to remove suspended solids, large particles, and certain heavy metals from leachate.

- Chemical treatment: Chemical processes including coagulation, flocculation, precipitation, and oxidation can be utilized to remove contaminants such as heavy metals, ammonia, and organic compounds.

- Biological treatment: Biological treatment methods, such as activated sludge process and constructed wetlands, utilize microorganisms to break down organic matter and remove nutrients from leachate.

- Advanced treatment: Technologies like reverse osmosis, membrane filtration, and advanced oxidation processes can be employed for advanced treatment of leachate to meet stringent water quality standards.

It is essential to select appropriate treatment technologies based on the specific characteristics of the leachate and the desired treated effluent quality.

Case Study: Leachate Treatment in Landfills

One example of leachate control is the treatment of leachate generated in landfills. Landfill leachate can be highly contaminated, containing various pollutants that require careful treatment.

A common approach for leachate treatment in landfills is the combination of physical, chemical, and biological methods. The treatment process involves several stages, including:

1. Preliminary treatment: In this stage, large solids and debris are removed through screening or sedimentation.

2. Physical treatment: Physical processes like sedimentation, filtration, and activated carbon adsorption are used to remove suspended solids, organic compounds, and certain heavy metals.

3. Chemical treatment: Coagulants, like ferric chloride or aluminum sulfate, can be added to facilitate the removal of colloidal particles and dissolved substances.

4. Biological treatment: The treated leachate is then subjected to biological treatment, where microorganisms, such as bacteria or fungi, decompose organic matter, ammonia, and other nutrients.

5. Post-treatment: Additional processes like disinfection with chlorine or ultraviolet light may be applied to ensure the removal of pathogens.

The effectiveness of the leachate treatment process should be monitored continuously to ensure compliance with regulatory standards and protect the environment.

Challenges and Future Directions

Leachate control faces several challenges that need to be addressed for better environmental protection and sustainability. Some of these challenges include:

+ Volume management: The treatment and disposal of large volumes of leachate can be challenging in terms of logistics, energy requirements, and costs. Developing efficient and cost-effective treatment systems is essential.

+ Emerging contaminants: New contaminants, such as pharmaceuticals and microplastics, are being detected in leachate. Developing treatment technologies that can effectively remove these emerging contaminants is crucial.

+ Sustainable solutions: Finding sustainable solutions for leachate treatment and management, such as resource recovery from leachate, can help minimize environmental impacts and enhance overall sustainability.

The future of leachate control lies in innovative technologies, improved waste management practices, and a holistic approach to environmental protection. Collaboration between researchers, policymakers, and waste management professionals is essential to develop and implement effective strategies for leachate control.

Summary

Leachate control is a critical aspect of waste management and environmental protection. Preventive measures, such as liner systems and surface water diversion, aim to minimize leachate generation. Treatment technologies, including physical, chemical, and biological processes, are employed to treat the generated leachate and meet water quality standards. Addressing the challenges associated with leachate control and exploring sustainable solutions will contribute to a cleaner and healthier environment for future generations.

Key Terms

+ Leachate: Potentially hazardous liquid formed when water passes through waste materials, containing various pollutants.

+ Liner systems: Impermeable layers in landfills to prevent leachate migration.

+ Physical treatment: Processes like sedimentation, filtration, and adsorption to remove suspended solids and certain contaminants.

+ Chemical treatment: Coagulation, flocculation, oxidation, and precipitation to remove heavy metals, organic compounds, and other contaminants.

+ Biological treatment: Utilization of microorganisms to decompose organic matter and remove nutrients from leachate.

+ Emerging contaminants: Newly identified contaminants, such as pharmaceuticals and microplastics, in leachate.

Further Reading

1. Christensen, T. H., Cossu, R., & Stegmann, R. (Eds.). (2017). *Landfilling of waste: Biogas.*

2. Tchobanoglous, G., Theisen, H. R., & Vigil, S. A. (2014). *Integrated solid waste management: Engineering principles and management issues.*

3. Ghosh, S. K. (Ed.). (2015). *Biological approaches to sustainable soil systems.*

Exercises

1. Explain the main causes and characteristics of leachate.

2. Discuss the preventive measures employed to control leachate generation.

3. Describe three different treatment technologies used for leachate treatment.

4. Conduct research on a case study of leachate treatment in your region. Describe the treatment methods employed and the challenges faced.

5. Discuss the future challenges and directions for leachate control.

Remember to refer to the textbook chapters on principles of eco science and waste management concepts for a more comprehensive understanding of leachate control.

Methane Capture

Methane capture is a crucial process in mitigating greenhouse gas emissions and addressing climate change. Methane (CH_4) is a potent greenhouse gas, with a significantly higher global warming potential than carbon dioxide (CO_2). It is released into the atmosphere from various sources such as landfills, livestock manure, and natural gas systems. Therefore, capturing methane before it enters the atmosphere is of utmost importance.

Importance of Methane Capture

Methane capture plays a crucial role in reducing greenhouse gas emissions for several reasons:

1. **Mitigating Climate Change:** By capturing methane, we can prevent its release into the atmosphere where it contributes to global warming. Methane has a warming potential 28 times higher than CO_2 over a 100-year period. By reducing methane emissions, we can significantly reduce the overall greenhouse gas emissions and slow down climate change.

2. **Improving Air Quality:** Methane is not only a potent greenhouse gas but also a significant contributor to air pollution. It reacts with other pollutants and forms ground-level ozone, which has adverse effects on human health and the environment. By capturing methane, we can reduce the formation of ground-level ozone and improve air quality.

3. **Waste Management:** Many sources of methane emissions, such as landfills and wastewater treatment plants, are byproducts of waste management systems. By capturing methane from these sources, we can not only reduce greenhouse gas emissions but also utilize the captured methane as a valuable energy resource. This process turns waste into a renewable energy source, promoting sustainable practices.

4. **Renewable Energy Generation:** Captured methane can be used as a valuable energy source. Methane can be converted into electricity and heat through

various processes such as combustion, anaerobic digestion, or upgrading to biomethane. These energy sources are cleaner and more sustainable compared to fossil fuel-based energy generation.

Methods of Methane Capture

There are several methods and technologies used for methane capture:

1. **Landfill Gas Capture:** Landfills are one of the largest human-made sources of methane emissions. Landfill gas capture involves extracting methane-rich gas from landfills and utilizing it as an energy source. The gas can be collected using a system of wells and pipes, and then either combusted or used to generate electricity and heat.

2. **Livestock Manure Management:** Livestock, such as cattle and swine, produce significant amounts of methane through their digestive processes, as well as from manure storage systems. Capturing methane from livestock manure management involves using anaerobic digestion systems to break down manure in the absence of oxygen, producing methane-rich biogas. This biogas can be used as an energy source or upgraded to biomethane.

3. **Biogas Capture from Wastewater Treatment Plants:** Wastewater treatment plants produce methane as a byproduct of the decomposition of organic matter. Capturing methane from wastewater treatment plants involves anaerobic digestion of the sludge generated during the treatment process. The resulting biogas can be used for energy generation or upgraded to biomethane.

4. **Coal Mine Methane Capture:** Methane is released during coal mining operations due to the geological properties of coal seams. Coal mine methane capture involves capturing methane released during mining activities to prevent its release into the atmosphere. The captured methane can be used as an energy source or sold commercially.

5. **Natural Gas Systems:** Methane leaks from natural gas extraction, processing, and distribution systems contribute to overall methane emissions. Methane can be captured by using advanced technology and better infrastructure to minimize leaks and capture the released methane. It is crucial to ensure that the entire natural gas supply chain is well-maintained to prevent methane emissions.

Challenges and Considerations

While methane capture presents an effective solution for reducing greenhouse gas emissions, there are challenges and considerations that need to be addressed:

1. **Technical Feasibility:** Each methane capture method requires specific infrastructure and technology. Implementing these systems may require significant investments and expertise. It is crucial to assess the technical feasibility and scalability of the various options to ensure their effectiveness.

2. **Economic Viability:** The economics of methane capture projects depend on factors such as the cost of capturing and utilizing methane, energy prices, and policy incentives. It is essential to consider the economic viability of these projects to encourage their implementation.

3. **Monitoring and Measurement:** Accurate monitoring and measurement of methane emissions are essential to identify significant sources and assess the effectiveness of methane capture systems. Developing robust monitoring techniques and standards is crucial for successful implementation.

4. **Policy and Regulation:** Supportive policies and regulations play a significant role in incentivizing methane capture projects. Governments need to develop and enforce regulations that promote the adoption of methane capture systems and provide financial incentives for cleaner energy generation.

5. **Adoption in Developing Countries:** Many developing countries face challenges in implementing methane capture systems due to limited resources and technical expertise. It is crucial to support capacity building efforts and provide financial and technical assistance to ensure global adoption of methane capture technologies.

Case Study: Landfill Gas Capture

Landfill gas capture is a widely adopted method for methane capture. Let's take a closer look at the process:

1. Landfills receive solid waste, including organic matter, which decomposes under anaerobic conditions, producing methane.

2. Wells and a network of pipes are installed in the landfill to collect the landfill gas, which is a mixture of methane, carbon dioxide, and other trace gases.

3. The collected landfill gas is then transported to a treatment facility, where it is typically compressed, dried, and purified.

4. The purified landfill gas can be used for electricity and heat generation, either through direct combustion or by using it as fuel in internal combustion engines or turbines.

5. In some cases, the collected landfill gas is upgraded to biomethane by removing impurities such as CO_2 and trace gases. This upgraded biomethane can be injected into the natural gas grid or used as a vehicle fuel.

Landfill gas capture not only reduces methane emissions but also provides a renewable energy source. The electricity and heat generated from captured landfill gas can offset the need for energy from fossil fuel sources, further reducing greenhouse gas emissions.

Conclusion

Methane capture is a critical strategy for mitigating climate change, reducing air pollution, and promoting sustainable practices. By capturing methane emissions from various sources, such as landfills, livestock manure, and natural gas systems, we can minimize the release of this potent greenhouse gas into the atmosphere. Methane capture methods, such as landfill gas capture, offer opportunities for renewable energy generation and waste management. However, challenges related to technical feasibility, economic viability, monitoring, and policy support need to be addressed for successful implementation. Methane capture is an essential part of building a sustainable and low-carbon future.

Synthetic Soil Cover

Synthetic soil cover is a technique used in waste management and environmental remediation to prevent or minimize the spread of contaminants from contaminated soil or waste material. It involves the use of synthetic materials to create a protective layer or barrier that separates the contaminants from the surrounding environment. This section will discuss the principles, methods, and benefits of using synthetic soil cover in various applications.

Principles of Synthetic Soil Cover

The use of synthetic soil cover is based on the principle of physical containment. By placing a barrier between the contaminated soil or waste material and the surrounding environment, the movement of contaminants can be controlled and limited. The synthetic material used for the soil cover should be impermeable or have low permeability to effectively prevent the migration of contaminants.

Methods of Implementation

There are different methods of implementing synthetic soil cover, depending on the specific application and site conditions. Some commonly used methods include:

1. **Geomembranes:** Geomembranes are synthetic sheets made of materials such as high-density polyethylene (HDPE) or polyvinyl chloride (PVC). These sheets are placed directly on the contaminated soil, forming a barrier to prevent the movement of contaminants. Geomembranes are often used in landfill capping to isolate the waste materials from the environment.

2. **Geosynthetic Clay Liners:** Geosynthetic clay liners (GCLs) are composite materials consisting of layers of geotextile and bentonite clay. GCLs have high hydraulic conductivity and are used in applications where both containment and drainage are required. They are commonly used in landfill caps and wastewater treatment facilities.

3. **Geomix Systems:** Geomix systems utilize a mixture of synthetic materials, such as geomembranes, geotextiles, and geocomposites, to create a customized soil cover. This method allows for flexibility in design and can be tailored to specific site requirements.

4. **Spray-on Coatings:** Spray-on coatings involve the application of a synthetic material as a liquid or semi-liquid layer on the contaminated soil. The material solidifies upon drying, forming a protective barrier. This method is suitable for irregular surfaces and can be used in soil remediation and containment applications.

5. **Hybrid Systems:** Hybrid systems combine different synthetic materials and techniques to optimize their effectiveness. For example, a combination of geomembranes and clay liners can be used to enhance the barrier properties and drainage capacity of the soil cover.

Benefits and Applications

Synthetic soil cover offers several benefits in waste management and environmental remediation. Some of the key benefits include:

1. **Containment of Contaminants:** The primary benefit of synthetic soil cover is the containment of contaminants, preventing their migration and spread to the surrounding environment. This helps to protect groundwater resources and minimize the risk to human health.

2. **Land and Water Protection:** By isolating contaminated soil or waste material, synthetic soil cover helps to protect land and water resources from pollution. This is particularly important in areas with sensitive ecosystems or where the risk of contamination is high.

3. **Versatility and Adaptability:** Synthetic soil cover can be customized and tailored to different applications and site conditions. The choice of materials and methods allows for flexibility in design and adaptation to specific requirements.

4. **Long-Term Stability:** Synthetic materials used in soil cover systems are designed to be durable and resistant to environmental conditions. They provide long-term stability and performance, ensuring the integrity of the containment barrier.

Synthetic soil cover has applications in various sectors, including:

1. **Landfills:** Synthetic soil cover is commonly used in landfill caps to prevent the leaching of contaminants and control gas emissions. It helps to ensure the long-term stability and safety of landfills.

2. **Brownfield Remediation:** Brownfield sites, which are abandoned or underutilized properties with potential contamination, can be remediated using synthetic soil cover. It helps to contain and isolate the contaminants, allowing for the redevelopment and reuse of these sites.

3. **Industrial Facilities:** Synthetic soil cover is employed in industrial facilities, such as factories and chemical plants, to prevent the release of hazardous materials into the environment. It acts as a protective barrier, minimizing the risk of contamination.

4. **Wastewater Treatment:** Synthetic soil cover, especially geosynthetic clay liners, is used in wastewater treatment facilities to contain and manage sludge or other waste materials. It prevents the release of pollutants and protects soil and water resources.

Example: Landfill Capping

A common application of synthetic soil cover is in landfill capping. Landfills receive a large amount of waste materials, which can contaminate the surrounding soil and water if not properly managed. Synthetic soil cover is used in landfill capping to provide a barrier that isolates the waste materials from the environment.

In this example, a landfill has reached its capacity and needs to be closed and capped. The process involves the following steps:

1. **Excavation and Preparation:** The top layer of the landfill is excavated to remove any remaining waste materials. The surface is then graded and compacted to create a stable base for the soil cover system.

2. **Installation of Synthetic Materials:** Geomembranes, such as HDPE sheets, are laid over the compacted base. The sheets are overlapped and welded together to ensure a continuous barrier. Geotextiles may be used as an additional layer to protect the geomembrane from punctures.

3. **Drainage System:** A layer of drainage material, such as gravel or geocomposite drainage mats, is installed on top of the geomembrane. This allows for the collection and removal of leachate that may accumulate within the landfill.

4. **Final Soil Cover:** A layer of soil is placed on top of the synthetic materials to provide additional protection and support vegetation growth. This final soil cover helps to integrate the landfill with the surrounding landscape.

5. **Monitoring and Maintenance:** Regular monitoring and maintenance of the synthetic soil cover system is essential to ensure its long-term effectiveness. This may include inspections, leak detection surveys, and repair or replacement of damaged components.

By implementing synthetic soil cover in landfill capping, the risk of groundwater contamination and the release of landfill gases are minimized. This promotes environmental protection and ensures the long-term sustainability of the landfill site.

Conclusion

Synthetic soil cover is a valuable technique in waste management and environmental remediation. By using synthetic materials as a barrier, it helps to contain and isolate contaminants, protecting the environment and human health. The versatility and adaptability of synthetic soil cover make it suitable for various applications, such as landfill capping, brownfield remediation, and industrial facilities. Proper design, installation, and maintenance of the soil cover system are crucial to its long-term effectiveness. Synthetic soil cover plays a vital role in promoting sustainability and ensuring a cleaner and healthier environment for future generations.

Water and Air Pollution Control

Water Pollution

Water pollution refers to the contamination of water bodies, such as rivers, lakes, oceans, and groundwater, with harmful substances. It is a significant environmental issue worldwide, impacting not only the health of ecosystems but also human well-being. In this section, we will explore various sources of water pollution, its effects, and strategies to mitigate and prevent it.

Sources of Water Pollution

Water pollution can originate from both point sources and nonpoint sources. Point sources refer to specific, identifiable locations where pollutants are discharged directly into water bodies. Examples of point sources include industrial wastewater discharges, sewage treatment plants, and oil spills. Nonpoint sources, on the other hand, are diffuse and do not have a single point of discharge. They include pollution from agricultural runoff, urban stormwater runoff, and atmospheric deposition.

Types of Water Pollutants

Water pollutants can be categorized into several types, including organic pollutants, inorganic pollutants, and pathogens.

Organic pollutants: These are substances derived from living organisms or human activities. They can include pesticides, herbicides, industrial chemicals, and pharmaceuticals. Organic pollutants are a major concern as they can persist in water bodies for a long time, leading to bioaccumulation in aquatic organisms and posing a risk to human health.

Inorganic pollutants: Inorganic pollutants are chemical compounds that do not contain carbon atoms. They can include heavy metals (such as lead, mercury, and cadmium), acids, salts, and other toxic compounds. Industrial processes, mining activities, and improper waste disposal are primary sources of inorganic pollutants in water bodies.

Pathogens: Pathogens are microorganisms, including bacteria, viruses, and protozoa, that can cause waterborne diseases. Inadequate sanitation and sewage systems are the main sources of pathogenic contamination in water.

Effects of Water Pollution

Water pollution has severe consequences for both ecosystems and human health. Some of the key effects include:

Ecosystems: Water pollution disrupts aquatic ecosystems by reducing biodiversity, causing habitat degradation, and altering the balance of aquatic organisms. Excessive nutrient inputs, such as nitrogen and phosphorus from agricultural runoff or sewage, can lead to eutrophication, a process that depletes oxygen levels in water bodies and creates dead zones where aquatic life cannot survive.

Human Health: Contaminated water can pose significant health risks to humans. Consumption of polluted water can lead to various waterborne diseases, including diarrhea, cholera, dysentery, and hepatitis. In addition, exposure to certain pollutants, such as heavy metals, can cause long-term health effects, including organ damage, developmental issues, and increased cancer risk.

Water Pollution Control

Efforts to control water pollution involve a combination of regulatory measures, technological solutions, and public awareness. Some key strategies include:

Wastewater Treatment: Effective wastewater treatment is crucial for removing pollutants before they are discharged into water bodies. Treatment processes involve physical, chemical, and biological methods to remove suspended solids, organic matter, and nutrients. Advanced treatment technologies, such as membrane filtration and disinfection, are also used to ensure the safety of treated water.

Stormwater Management: Proper management of stormwater runoff can minimize pollution inputs. Implementation of green infrastructure techniques, such as rain gardens, pervious pavements, and constructed wetlands, helps to capture and treat stormwater, reducing the load of pollutants entering water bodies.

Improved Agricultural Practices: Agricultural activities contribute significantly to water pollution through the runoff of fertilizers and pesticides.

Promoting sustainable agricultural practices, such as precision farming, crop rotation, and integrated pest management, can help reduce nutrient and chemical runoff and protect water quality.

Pollution Prevention: Prevention is often more effective and economical than cleanup. Implementing pollution prevention practices in industries and households can help minimize the release of harmful substances into water bodies. This includes proper waste management, recycling programs, and the use of eco-friendly products.

Case Study: The Clean Water Act

One prominent example of legislation aimed at controlling water pollution is the Clean Water Act (CWA) in the United States. Enacted in 1972, the CWA established a framework for regulating pollutant discharges into U.S. waters. It set water quality standards, created a permitting system for point source discharges, and established water quality monitoring and enforcement programs.

Under the CWA, the Environmental Protection Agency (EPA) has the authority to regulate and enforce various aspects of water pollution control. The legislation has led to significant improvements in water quality across the country, with increased monitoring and stricter standards for point source dischargers. However, challenges remain, particularly in addressing nonpoint source pollution and emerging contaminants.

Conclusion

Water pollution poses a significant threat to ecosystems and human health worldwide. Understanding the sources, types, and effects of water pollution is crucial for developing effective strategies to mitigate and prevent it. By implementing appropriate wastewater treatment, stormwater management, and pollution prevention measures, we can work towards preserving clean and sustainable water resources for future generations.

Point source pollution

Point source pollution refers to the discharge of pollutants from a single identifiable source, such as a factory, power plant, or sewage treatment plant. These sources release pollutants directly into water bodies or the atmosphere. Examples of point source pollutants include industrial chemicals, heavy metals, nutrients, and pathogens.

Understanding point source pollution

To better understand point source pollution, it is important to examine the principles of environmental science and pollution control. Environmental science is an interdisciplinary field that combines elements of biology, chemistry, physics, and geology to study the environment and its interactions with humans and other organisms. Pollution control focuses on mitigating the harmful effects of pollution through various strategies.

Regulation and monitoring

One key aspect of addressing point source pollution is the regulation of discharges. Governments and regulatory agencies set limits on the types and amounts of pollutants that can be released into the environment. These regulations aim to protect human health and the environment, and they often require industries to obtain permits for their discharges.

Monitoring plays a crucial role in enforcing regulations and identifying sources of pollution. Regular monitoring of water bodies and air quality allows authorities to detect and track pollutants, ensuring compliance with regulations. Advanced techniques, such as remote sensing and water quality sensors, provide real-time data on pollution levels.

Impact on water bodies

Point source pollution can have significant effects on water bodies, including rivers, lakes, and oceans. Pollutants discharged from factories and sewage treatment plants can degrade water quality, harm aquatic life, and disrupt ecosystems.

One example of point source pollution is industrial wastewater. Industries generate wastewater during manufacturing processes, which may contain toxic chemicals and heavy metals. When released into water bodies without proper treatment, these pollutants can accumulate in aquatic organisms, leading to ecological imbalances and harming human health if consumed.

Water treatment technologies

To mitigate the impacts of point source pollution on water bodies, various water treatment technologies are employed. These technologies aim to remove or reduce the concentration of pollutants in wastewater before it is discharged back into the environment.

Common water treatment processes include physical, chemical, and biological methods. Physical methods, such as sedimentation and filtration, remove solid particles and suspended matter from wastewater. Chemical methods, such as coagulation and disinfection, use chemicals to precipitate contaminants or kill microorganisms. Biological methods, such as activated sludge processes and constructed wetlands, harness the power of microorganisms to break down organic matter and remove nutrients.

Case study: The Chesapeake Bay

The Chesapeake Bay, located on the East Coast of the United States, has suffered from point source pollution for decades. In the 20th century, excessive nutrient pollution from agricultural runoff, wastewater treatment plants, and industrial discharges caused harmful algal blooms and oxygen depletion in the bay.

To address this issue, the Chesapeake Bay Program, a regional partnership, was established in 1983. The program focuses on reducing nutrient pollution and improving water quality in the bay. It implements a Total Maximum Daily Load (TMDL) approach, which sets limits on the amount of nutrients that can enter the bay from various sources.

The Chesapeake Bay Program also promotes best management practices among farmers to minimize nutrient runoff, works with wastewater treatment plants to upgrade their facilities, and implements stormwater management strategies to reduce runoff from urban areas. These efforts have resulted in measurable improvements in the Chesapeake Bay's water quality and the recovery of its ecosystems.

Challenges and future directions

Despite significant progress in addressing point source pollution, challenges remain. Enforcement of regulations, especially in developing countries, can be inadequate due to limited resources and capacity. Additionally, emerging pollutants, such as pharmaceuticals and microplastics, pose new challenges in pollution control.

To overcome these challenges, future directions in point source pollution management include:

* Advancements in wastewater treatment technologies to remove emerging pollutants effectively.

* Enhanced international cooperation and knowledge sharing to address transboundary point source pollution.

- ✦ Integration of nature-based solutions, such as constructed wetlands and green infrastructure, into point source pollution management strategies.

- ✦ Promotion of sustainable industrial practices, such as green chemistry and circular economy principles, to minimize pollution at the source.

- ✦ Continued monitoring and assessment of pollutant levels to identify emerging threats and assess the effectiveness of pollution control measures.

In conclusion, point source pollution is a significant environmental issue that requires comprehensive regulation, monitoring, and pollution control measures. By implementing effective strategies and technologies, we can mitigate the impacts of point source pollution and protect our valuable water resources.

Nonpoint Source Pollution

Nonpoint source pollution refers to the contamination of water bodies, such as rivers, lakes, and oceans, caused by the runoff of pollutants from multiple and diffuse sources. Unlike point source pollution, which originates from a single identifiable source, nonpoint source pollution comes from various activities and areas, making it challenging to regulate and control. This section will explore the causes and impacts of nonpoint source pollution and discuss some strategies for its prevention and mitigation.

Causes of Nonpoint Source Pollution

Nonpoint source pollution can result from numerous human activities and natural processes. Some common sources include:

1. Agriculture: The use of fertilizers, pesticides, and herbicides in agricultural practices can contribute to nonpoint source pollution. When it rains, these chemicals can be washed away from fields and end up in nearby water bodies.

2. Urban Areas: Stormwater runoff from urban areas often contains pollutants such as oil, grease, heavy metals, and sediments. These contaminants can come from paved surfaces, rooftops, and construction sites.

3. Forestry: Logging operations, especially when poorly managed, can lead to soil erosion and the release of sediments into nearby streams and rivers.

4. Construction: Construction activities generate sediment-laden runoff that can enter water bodies, causing water quality degradation.

5. Mining: Activities related to mining, including excavation, processing, and waste disposal, can introduce pollutants into nearby water sources.

6. Septic Systems: Improperly designed or maintained septic systems can leak wastewater into the groundwater, contaminating nearby wells and streams.

Impacts of Nonpoint Source Pollution

Nonpoint source pollution can have detrimental effects on both the environment and human health. Some of the key impacts include:

1. Water Quality Degradation: Nonpoint source pollution can result in elevated levels of nutrients like nitrogen and phosphorus in water bodies, leading to eutrophication. Excessive nutrient levels can cause algal blooms, oxygen depletion, and harm aquatic organisms.

2. Habitat Destruction: Sediment runoff can smother aquatic habitats, disrupting the natural balance and diversity of aquatic ecosystems. It can also impact the reproductive and feeding habits of various species.

3. Loss of Biodiversity: The introduction of pollutants from nonpoint sources can lead to the decline or loss of certain aquatic species, which can have ripple effects throughout the food chain.

4. Drinking Water Contamination: Nonpoint source pollution can contaminate drinking water sources, making it unsafe for consumption. This poses a risk to human health, as exposure to certain pollutants can have adverse effects on the nervous and digestive systems.

Prevention and Mitigation Strategies

Addressing nonpoint source pollution requires a multi-faceted approach that includes both structural and non-structural measures. Some strategies that can be employed to prevent and mitigate nonpoint source pollution are:

1. Best Management Practices (BMPs): Implementing BMPs in agriculture, construction, and urban areas can help reduce the amount of runoff and pollutants reaching water bodies. Examples of BMPs include erosion control measures, vegetated buffer strips, and stormwater management systems.

2. Soil and Water Conservation: Promoting soil conservation practices, such as contour plowing, terracing, and conservation tillage, can help prevent soil erosion and reduce sediment runoff.

3. Riparian Zone Restoration: Restoring and maintaining riparian zones, the areas along rivers and streams, can act as natural filters, reducing the amount of pollutants entering water bodies.

4. Education and Outreach: Raising awareness among the public, farmers, and other stakeholders about the impacts of nonpoint source pollution is crucial.

Educating people about proper waste disposal, the use of environmentally friendly agricultural practices, and the importance of preserving water quality can lead to behavioral changes that reduce pollution.

5. Monitoring and Assessment: Continuous monitoring and assessment of water quality, sedimentation rates, and land use practices can help identify areas of concern and evaluate the effectiveness of pollution prevention strategies.

Case Study: Nonpoint Source Pollution in the Chesapeake Bay

The Chesapeake Bay, located in the eastern United States, has been significantly impacted by nonpoint source pollution. The bay's watershed, covering parts of six states, experiences runoff from urban areas, agriculture, forestry, and other land uses.

To address the issue, the Chesapeake Bay Program was established, which brings together federal and state agencies, local governments, non-profit organizations, and academic institutions. The program aims to reduce nutrient and sediment pollution in the bay through various measures, including the implementation of nutrient management plans on farms, the improvement of wastewater treatment plants, and the restoration of riparian buffers.

Through collaborative efforts and the involvement of stakeholders, the Chesapeake Bay Program has made significant progress in reducing nonpoint source pollution. However, ongoing monitoring and adaptation of strategies are necessary to ensure the long-term health of the bay and its ecosystem.

Conclusion

Nonpoint source pollution poses significant challenges for water resource management and environmental sustainability. Recognizing the diverse sources of pollution and implementing various prevention and mitigation strategies are essential for protecting water bodies and ensuring their long-term health. By taking collective action and adopting sustainable practices, we can minimize the impacts of nonpoint source pollution and promote a cleaner and healthier environment for future generations.

Water Treatment Technologies

Water is a vital resource for all living beings, and its quality plays a crucial role in ensuring public health and environmental sustainability. Water treatment technologies are essential for purifying and removing contaminants from various water sources, such as rivers, lakes, and groundwater, before it can be used for drinking, industrial processes, agriculture, and other purposes. In this section, we will explore the different types of water treatment technologies and their applications.

Water Pollution and Contaminants

Before delving into water treatment technologies, let's first understand the common sources of water pollution and the types of contaminants that may be present in water sources.

Water pollution can arise from both point source and nonpoint source pollution. Point source pollution refers to the release of pollutants from a single identifiable source, such as industrial discharge pipes, sewage treatment plants, or oil spills. On the other hand, nonpoint source pollution refers to pollutants that enter water bodies from diffuse sources, such as runoff from agricultural fields, urban areas, or construction sites.

Contaminants in water can be broadly classified into physical, chemical, and biological contaminants. Physical contaminants include suspended solids, sediments, and debris that affect water clarity and may require filtration or sedimentation processes for removal. Chemical contaminants encompass a wide range of substances, including heavy metals, organic compounds, pesticides, and chemical byproducts. Biological contaminants include bacteria, viruses, and parasites that can cause waterborne diseases.

Water Treatment Processes

Water treatment technologies employ various processes to remove contaminants and ensure safe water supply. The treatment processes can be categorized into the following stages:

1. **Preliminary Treatment:** The water undergoes physical removal of large debris, such as sticks, leaves, and rocks, through processes such as screening and sedimentation. This step helps protect downstream treatment units from damage and reduces the load on subsequent treatment stages.

2. **Coagulation and Flocculation:** In this stage, chemicals like aluminum sulfate (alum) or ferric chloride are added to the water to destabilize suspended particles. The destabilized particles are then agglomerated into larger floc particles through gentle mixing, known as flocculation. The floc particles can be easily removed by sedimentation or filtration.

3. **Sedimentation:** During sedimentation, the water is held in large tanks to allow the heavier floc particles to settle at the bottom. This process utilizes gravitational force to separate the settled particles from the water, which is known as clarification.

4. **Filtration:** Filtration further removes smaller suspended particles that may have escaped the sedimentation process. It involves passing water through porous media, such as sand, gravel, or activated carbon, to trap remaining impurities. Depending on the level of filtration required, different types of filters, including rapid sand filters, multimedia filters, and membrane filters, can be used.

5. **Disinfection:** Disinfection is a critical step to eliminate or inactivate pathogenic microorganisms present in the water. Common disinfection methods include chlorination, ozonation, ultraviolet (UV) irradiation, and the use of chlorine dioxide. Disinfection helps prevent waterborne diseases and ensures the safety of drinking water.

6. **pH Adjustment:** pH adjustment may be necessary to meet regulatory standards or optimize treatment processes. Depending on the pH of the source water and treatment objectives, chemicals such as lime or sulfuric acid can be added to achieve the desired pH level.

Advanced Water Treatment Technologies

In addition to the conventional water treatment processes mentioned above, several advanced water treatment technologies have emerged to address specific challenges and contaminants. Let's explore a few of these advanced technologies:

Membrane Filtration: Membrane filtration is a high-precision filtration process that uses thin porous membranes to separate particles and solutes from water. This technology includes processes such as reverse osmosis, nanofiltration, and ultrafiltration. Membrane filtration is effective in removing dissolved salts, organic compounds, and microorganisms, making it suitable for desalination, wastewater treatment, and producing high-purity water for industrial processes.

Advanced Oxidation Processes (AOPs): AOPs involve the use of powerful oxidants to degrade and remove persistent organic pollutants present in water. These processes generate hydroxyl radicals ($OH\cdot$) that can break down organic compounds into non-toxic substances. AOPs, such as ozonation, $UV/H2O2$ (ultraviolet light with hydrogen peroxide), and photocatalysis, are effective in treating water contaminated with pharmaceuticals, pesticides, and other organic contaminants.

Activated Carbon Adsorption: Activated carbon has a large surface area and can adsorb a wide range of organic compounds, including volatile organic compounds (VOCs), taste and odor-causing compounds, and certain pesticides. In this process, water passes through a bed of activated carbon, and the contaminants adhere to the carbon surface through physical adsorption. Activated carbon adsorption is commonly used as a polishing step in water treatment to improve taste, odor, and remove residual organic impurities.

Electrocoagulation: Electrocoagulation is an electrochemical process that uses metal ions generated from sacrificial electrodes to coagulate and remove contaminants from water. Electric current promotes the formation of metal hydroxides, which act as coagulants, destabilizing and agglomerating suspended particles and dissolved pollutants. Electrocoagulation is particularly useful for treating industrial wastewaters, oily wastewater, and water contaminated with heavy metals.

Water Treatment Challenges and Future Outlook

While water treatment technologies have made significant advancements, several challenges persist in achieving efficient and sustainable water treatment:

* **Emerging Contaminants:** The detection and removal of emerging contaminants, such as pharmaceuticals, personal care products, and endocrine-disrupting compounds, pose challenges due to their complex nature and low concentration levels. Research and development efforts are focused on improving technologies to detect and effectively remove these contaminants.

* **Energy Consumption and Environmental Impacts:** Some water treatment technologies, such as membrane filtration and advanced oxidation processes, require significant energy inputs. Developing energy-efficient treatment processes and exploring alternative energy sources can help reduce the environmental footprint of water treatment operations.

* **Water Scarcity and Water Reuse:** As water scarcity becomes a global concern, increasing emphasis is being placed on water reuse and recycling. Advanced treatment processes, including reverse osmosis and advanced oxidation, are being employed to treat wastewater and produce high-quality recycled water for various non-potable applications.

+ **Climate Change and Water Availability:** Climate change is impacting water availability, with some regions experiencing prolonged droughts or increased frequency of extreme weather events. Water treatment systems must adapt to changing water availability patterns and develop resilience measures to ensure water supply in the face of climate uncertainties.

In conclusion, water treatment technologies play a vital role in ensuring the availability of safe and clean water for various purposes. From conventional processes like coagulation and sedimentation to advanced techniques such as membrane filtration and advanced oxidation processes, these technologies help remove contaminants, improve water quality, and prevent waterborne diseases. However, ongoing research, innovation, and sustainable practices are essential to address emerging challenges and make water treatment processes more efficient, cost-effective, and environmentally friendly.

Further Reading

For further reading on water treatment technologies, the following resources are recommended:

+ United States Environmental Protection Agency (EPA) - Water Treatment Technology Fact Sheets: An extensive collection of fact sheets providing detailed information on various water treatment technologies.

+ World Health Organization (WHO) - Guidelines for Drinking-Water Quality: A comprehensive guide that sets out the standards for drinking water quality and provides information on water treatment technologies.

+ Water Environment Federation (WEF) - Water Reuse: A compilation of resources and best practices related to water reuse and recycling for sustainable water management.

+ American Water Works Association (AWWA) - Water Science: A peer-reviewed journal publishing research articles on water treatment technologies, water quality, and related topics.

Remember, water treatment is crucial for a sustainable future, and understanding the principles and applications of water treatment technologies is essential for practitioners, policymakers, and anyone concerned with preserving our most precious resource - water.

Air Pollution

Air pollution is a pressing environmental issue that affects the quality of the air we breathe. It is caused by the release of harmful substances, such as gases, particles, and biological materials, into the atmosphere. These pollutants can have detrimental effects on human health, ecosystems, and the overall environment. In this section, we will explore the sources, impacts, and control measures of air pollution.

Sources of Air Pollution

Air pollution can originate from both natural and human activities. Natural sources include volcanic eruptions, forest fires, and dust storms. However, the majority of air pollution is caused by human activities, such as industrial processes, transportation, energy production, and agriculture. These activities release pollutants into the air, contributing to the deterioration of air quality.

Industrial activities are a significant source of air pollution. Factories and power plants emit large amounts of pollutants, including sulfur dioxide (SO_2), nitrogen oxides (NO_x), carbon monoxide (CO), and particulate matter (PM). These pollutants can cause respiratory problems, cardiovascular diseases, and even premature death.

Transportation, especially vehicles running on fossil fuels, is another major contributor to air pollution. Exhaust gases from cars, trucks, and motorcycles release nitrogen oxides, carbon monoxide, and particulate matter into the atmosphere. Additionally, the combustion of gasoline and diesel fuels also produces greenhouse gases, contributing to climate change.

Energy production from fossil fuels, such as coal and natural gas, releases pollutants like sulfur dioxide, nitrogen oxides, and particulate matter. These emissions not only contribute to air pollution but also contribute to the formation of acid rain and smog.

Agricultural activities, such as livestock farming and the use of chemical fertilizers, can also release pollutants into the air. Methane (CH_4) emissions from livestock contribute to greenhouse gas emissions, while the use of fertilizers can release nitrogen oxides into the atmosphere.

Impacts of Air Pollution

Air pollution has a wide range of impacts on human health, ecosystems, and the environment as a whole. The effects can be immediate, such as respiratory problems and allergies, or long-term, such as chronic respiratory diseases and cancer.

Exposure to air pollutants can cause respiratory issues, including coughing, wheezing, and shortness of breath. Fine particulate matter can penetrate deep into the lungs, causing inflammation and exacerbating existing respiratory conditions such as asthma and bronchitis. Long-term exposure to air pollution has been linked to the development of chronic diseases, such as lung cancer, cardiovascular diseases, and even neurological disorders.

Air pollution also has detrimental effects on ecosystems. Acid rain, which is caused by the deposition of sulfur dioxide and nitrogen oxides onto the Earth's surface, can damage forests, soils, and bodies of water. It can lead to the loss of biodiversity and disruption of ecosystems. In addition, air pollutants can harm plant health, inhibit photosynthesis, and reduce crop yields.

Furthermore, air pollution contributes to climate change by contributing to the accumulation of greenhouse gases in the atmosphere. Greenhouse gases, such as carbon dioxide (CO_2) and methane (CH_4), trap heat in the Earth's atmosphere, leading to global warming and climate-related impacts, including rising sea levels, extreme weather events, and habitat loss.

Control Measures for Air Pollution

To address the challenges posed by air pollution, various control measures and regulations have been implemented at local, national, and international levels. These measures aim to reduce the emission of pollutants and improve air quality.

One effective strategy is the use of emission control technologies in industrial processes and power plants. Technologies, such as scrubbers and catalytic converters, help remove or convert harmful pollutants before they are released into the atmosphere. Additionally, the adoption of cleaner and more efficient industrial practices can significantly reduce air pollution.

In the transportation sector, regulatory measures have been implemented to reduce vehicle emissions. These include the enforcement of vehicle emission standards, the promotion of electric and hybrid vehicles, and the improvement of public transportation systems. Investing in sustainable transportation infrastructure, such as bike lanes and pedestrian-friendly cities, can further reduce air pollution.

Transitioning to cleaner and renewable energy sources is another crucial step in combating air pollution. Increasing the use of solar and wind energy, as well as investing in energy-efficient technologies, can significantly reduce the emission of pollutants from energy production.

Furthermore, sustainable agricultural practices can help mitigate air pollution from the agricultural sector. Implementing practices such as organic farming, precision agriculture, and proper waste management can reduce the release of harmful gases and particulate matter.

International cooperation and collaboration are essential in addressing air pollution on a global scale. Agreements and treaties, such as the Paris Agreement, aim to mitigate climate change by reducing greenhouse gas emissions. Similarly, international efforts to phase out ozone-depleting substances have led to significant improvements in air quality.

Education and public awareness also play a crucial role in combating air pollution. By raising awareness about the sources and impacts of air pollution, individuals can make informed choices to reduce their own contribution. Engaging communities in air quality monitoring and reporting can also help identify pollution hotspots and drive policy changes.

Case Study: Beijing's Air Pollution Challenge

A prominent example of air pollution challenges can be seen in Beijing, China. In recent years, the city has faced severe smog episodes, primarily attributed to industrial activities, vehicle emissions, and coal-fired power plants. The high levels of particulate matter, especially $PM_{2.5}$, have posed significant health risks to the population.

To combat air pollution, the Chinese government has implemented various measures. These include the introduction of stricter emission standards for vehicles, the closure of highly polluting factories, and the promotion of renewable energy sources. The city has also taken steps to limit coal consumption and improve energy efficiency.

Additionally, Beijing has invested in a comprehensive air quality monitoring network. Real-time data on air pollution levels are made available to the public, enabling individuals to make informed decisions regarding outdoor activities and protective measures. The government's commitment to transparency and accountability has encouraged public participation and engagement in addressing the issue.

Despite the significant progress made in recent years, Beijing continues to face air pollution challenges. However, the combination of regulatory measures, technological advancements, and public involvement serves as an example of how cities can tackle air pollution effectively.

Conclusion

Air pollution is a significant environmental challenge that requires urgent attention and action. Identifying the sources, understanding the impacts, and implementing effective control measures are crucial steps towards improving air quality. By adopting cleaner and sustainable practices in industry, transportation, energy production, and agriculture, we can mitigate the harmful effects of air pollution and strive towards a healthier and more sustainable future.

Air quality monitoring

Air quality monitoring is an essential component of environmental science and plays a crucial role in assessing and managing air pollution. Monitoring air quality allows us to understand the levels of pollutants present in the atmosphere, identify their sources, and evaluate the effectiveness of pollution control measures. In this section, we will explore the principles of air quality monitoring, different monitoring techniques, and their applications.

Principles of air quality monitoring

Air quality monitoring is based on the principles of sampling and analysis. The process involves the collection of air samples from various locations and the measurement of different pollutants. The collected data is then analyzed to determine the concentration of pollutants in the air. The key principles of air quality monitoring include:

- **Representative sampling**: It is important to collect air samples that accurately represent the air quality of a particular area or region. Proper selection of sampling sites is crucial to ensure that the collected data is reliable and representative of the larger area.

- **Measurement accuracy**: Accurate measurement of pollutant concentrations is essential to obtain reliable data. Calibration of monitoring equipment and adherence to standard operating procedures are necessary to ensure measurement accuracy.

- **Quality control**: Quality control measures, such as regular calibration checks, use of certified reference materials, and participation in inter-laboratory comparison programs, are employed to verify the accuracy and reliability of monitoring data.

+ **Data interpretation**: Proper interpretation of monitoring data is crucial to derive meaningful conclusions. Statistical analysis and comparison with air quality standards or guidelines help assess the severity of pollution and its potential health and environmental impacts.

Monitoring techniques

A variety of techniques and instruments are employed for air quality monitoring. These techniques can be broadly classified into two categories: continuous monitoring and passive monitoring.

Continuous monitoring Continuous monitoring involves the real-time measurement of pollutant concentrations using automated instruments. The data collected through continuous monitoring provides information on the short-term variations in pollutant levels. Some commonly used continuous monitoring techniques include:

+ **Ambient air samplers**: These devices continuously collect air samples for analysis. They are equipped with various types of sensors or analyzers specific to different pollutants, such as particulate matter (PM), nitrogen oxides (NOx), sulfur dioxide (SO2), ozone (O3), carbon monoxide (CO), and volatile organic compounds (VOCs).

+ **Meteorological sensors**: These sensors measure parameters such as wind speed, wind direction, temperature, and relative humidity. Meteorological data is crucial for understanding the dispersion and transport of pollutants in the atmosphere.

+ **Remote sensing instruments**: Remote sensing techniques, such as lidar (light detection and ranging) and passive remote sensing, use lasers or satellite-based sensors to measure pollutant concentrations over a larger area. These techniques provide valuable information on air quality at the regional or global scale.

Passive monitoring Passive monitoring involves the use of passive samplers or collector devices that do not require power or continuous operation. These samplers rely on the diffusion of pollutants into a collecting medium over a specified period. The collected samples are then analyzed in a laboratory to determine the average concentration of pollutants during the sampling period.

Passive monitoring techniques are cost-effective and widely used for long-term monitoring. Examples of passive monitoring techniques include:

+ **Diffusive samplers:** Diffusive samplers consist of a diffusion tube filled with an adsorbent material. Pollutants diffuse into the tube, and their concentrations are determined by laboratory analysis of the collected adsorbent material.

+ **Bioindicators:** Plants, lichens, and certain species of moss can be used as bioindicators of air pollution. These organisms accumulate pollutants from the air and can be analyzed to assess air quality over a specific period in a particular location.

Applications of air quality monitoring

Air quality monitoring has numerous applications in both scientific research and regulatory activities. Some of the key applications include:

+ **Health risk assessment:** Monitoring air quality helps in assessing the potential health risks associated with exposure to various pollutants. By measuring pollutant concentrations and comparing them with air quality standards or guidelines, scientists and health professionals can identify areas with high pollution levels and take appropriate measures to protect public health.

+ **Source identification and apportionment:** Air quality monitoring is crucial for identifying and quantifying the sources of pollution. By analyzing pollutant concentrations and their spatial and temporal variations, researchers can determine the contributions of different emission sources, such as industrial activities, vehicular emissions, and biomass burning. This information helps policymakers develop effective pollution control strategies.

+ **Long-term trend analysis:** Monitoring air quality over an extended period allows for the analysis of long-term trends and the evaluation of the effectiveness of pollution control measures. It helps in understanding the impact of changing emissions, climate, and land use on air quality.

+ **Regulatory compliance:** Air quality monitoring data is used by regulatory agencies to assess compliance with air quality standards and guidelines. Industries and other pollution sources are required to monitor and report their emissions to ensure that they meet the prescribed emission limits.

Challenges and future directions

Air quality monitoring faces several challenges and limitations that need to be addressed for more accurate and comprehensive data collection. Some of the key challenges include:

+ **Monitoring infrastructure:** The establishment and maintenance of a robust and extensive monitoring network require significant investments in resources and infrastructure. There is a need for expanding the coverage of monitoring stations, particularly in developing countries and remote areas.

+ **Sensor calibration and accuracy:** Continuous monitoring instruments need regular calibration and maintenance to ensure accurate measurement of pollutant concentrations. Standardization and quality control measures are crucial for obtaining reliable data from different monitoring devices and techniques.

+ **Data integration and analysis:** With advancements in sensor technology and the availability of large datasets, there is a need for improved data integration and analysis techniques. Integration of air quality monitoring data with other environmental and health datasets can provide a more holistic understanding of the impacts of air pollution.

+ **Emerging pollutants:** Monitoring techniques need to be updated and optimized to detect and measure emerging pollutants, such as microplastics, nanoparticles, and pharmaceutical residues. These pollutants pose new challenges due to their complex nature and potential health and environmental impacts.

In future, advancements in sensor technology, satellite-based monitoring, and data analytics will play a significant role in improving air quality monitoring. Integration of air quality monitoring networks with smart city infrastructure and the Internet of Things (IoT) will enable real-time monitoring and better management of urban air quality. Additionally, citizen science initiatives can contribute to expanding the monitoring coverage and engaging communities in air quality monitoring efforts.

Conclusion

Air quality monitoring is an essential tool for understanding, managing, and mitigating air pollution. It provides valuable data for assessing health risks, identifying pollution sources, and evaluating the effectiveness of pollution control measures. Continuous and passive monitoring techniques, coupled with reliable data analysis and interpretation, help in making informed decisions for better air quality management. As technology advances and challenges are addressed, air quality monitoring will continue to evolve, leading to improved pollution control strategies and a healthier environment for future generations.

Emission Control Technologies

Emission control technologies play a vital role in mitigating air pollution and reducing harmful emissions released into the atmosphere. With the increasing concern over climate change and its impacts, it is crucial to develop and implement effective methods to control and reduce emissions from various sources. This section will explore some of the key technologies used for emission control and their applications in different sectors.

Selective Catalytic Reduction (SCR)

Selective Catalytic Reduction (SCR) is a widely used technology for reducing nitrogen oxides (NOx) emissions from stationary sources, such as power plants and industrial facilities. SCR involves the use of a catalyst, typically made of metal oxide, to convert NOx into nitrogen (N2) and water (H2O) through a chemical reaction with a reducing agent, usually ammonia or urea. This process occurs in a specialized catalyst chamber, known as the SCR reactor.

The chemical reaction can be represented by the following equation:

$$4NO + 4NH_3 + O_2 \rightarrow 4N_2 + 6H_2O$$

SCR technology offers high efficiency in reducing NOx emissions, with removal rates exceeding 90%. It has become a standard approach in many industrial processes and power plants globally. However, the availability of ammonia or urea as the reducing agent and the need for a catalyst make SCR technology more suitable for stationary sources than for mobile applications.

Diesel Particulate Filters (DPF)

Diesel engines are a significant source of particulate matter (PM) emissions, which can pose serious health risks. Diesel Particulate Filters (DPF) are effective technologies used to capture and remove PM from the exhaust gases of diesel engines.

DPFs are typically made of a ceramic or metal honeycomb structure with fine channels coated with a catalyst material. The structure acts as a filter, trapping the solid particles while allowing the exhaust gases to pass through.

To prevent excessive pressure drop and filter clogging, the collected particles need to be periodically burned off. This process is called regeneration and can be achieved through two main methods: passive regeneration and active regeneration.

Passive regeneration occurs when the exhaust gas temperature is high enough to oxidize the collected particles and regenerate the DPF naturally. On the other hand, active regeneration involves raising the temperature inside the DPF to initiate the combustion of the accumulated particles. This is achieved by injecting a small amount of fuel either directly into the exhaust or into the engine cylinders during the exhaust stroke.

DPFs have proven effective in reducing PM emissions from diesel engines by up to 90%. However, they require regular maintenance and can be prone to clogging if not properly managed.

Flue Gas Desulfurization (FGD)

Flue Gas Desulfurization (FGD) is a technology used to remove sulfur dioxide (SO2) emissions from flue gases produced by fossil fuel combustion, particularly in power plants and industrial boilers.

FGD systems use a variety of methods to remove SO2, including wet scrubbing, dry scrubbing, and semi-dry scrubbing. Wet scrubbing, the most commonly used method, involves spraying a slurry of water and a sorbent, such as limestone or lime, into the flue gas stream. The sorbent reacts with the SO2, forming a solid product that can be disposed of safely.

The chemical reaction between the sorbent and SO2 can be represented by the following equation:

$$CaCO3 + SO2 + \frac{1}{2}O2 + H2O \rightarrow CaSO4 \cdot 2H2O + CO2$$

FGD systems can achieve removal efficiencies of up to 95% for SO2 emissions. They are crucial in reducing acid rain and improving air quality in regions where coal-fired power plants are prevalent.

Volatile Organic Compound (VOC) Control

Volatile Organic Compounds (VOCs) are emitted from various sources, including industrial processes, vehicle emissions, and consumer products. These compounds contribute to the formation of ground-level ozone and are known to have harmful effects on human health and the environment.

Several technologies are used to control VOC emissions, including adsorption, absorption, condensation, and thermal oxidation.

Adsorption involves the use of adsorbent materials, such as activated carbon, to capture VOCs from the exhaust gases. The adsorbent material has a large surface area, allowing it to attract and retain the VOC molecules effectively.

Absorption, on the other hand, is the process of dissolving VOCs in a liquid solvent, typically water or organic solvents. This method is often used for controlling VOC emissions from chemical processes and industrial facilities.

Condensation is employed for VOCs with high boiling points. It involves cooling the exhaust gases to a temperature below the boiling point of the VOC, causing it to condense into a liquid form for separation and collection.

Thermal oxidation, also known as incineration, is an effective method for destroying VOCs through high-temperature combustion. The high temperatures break down the VOC molecules, converting them into carbon dioxide (CO_2) and water vapor (H_2O).

Each VOC control technology has its advantages and limitations, and the selection of the appropriate method depends on the specific application and emission characteristics.

Alternative Energy Sources

Apart from controlling emissions at the source, another effective strategy for emission control is the utilization of alternative energy sources that produce fewer or no emissions. Renewable energy sources, such as solar power, wind energy, hydroelectric power, and biomass, offer cleaner and sustainable alternatives to fossil fuels.

Solar power involves the conversion of sunlight into electricity through the use of photovoltaic cells or concentrated solar power systems. This technology produces no emissions during operation, making it an environmentally friendly energy source.

Wind energy harnesses the power of wind to generate electricity through wind turbines. Wind power is a clean and renewable energy resource, with no direct emissions during operation.

Hydroelectric power utilizes the flow of water in rivers and dams to generate electricity. It is a reliable and clean energy source without direct emissions.

Biomass energy involves the use of organic materials, such as wood, crop residues, and organic waste, to produce heat, electricity, or biofuels. While biomass combustion does release carbon dioxide, it can be considered a carbon-neutral energy source if the biomass is sustainably sourced.

The adoption of alternative energy sources not only helps reduce emissions but also contributes to the diversification of the energy mix, increasing energy security and promoting sustainable development.

In conclusion, emission control technologies are crucial in mitigating air pollution and reducing harmful emissions. The technologies discussed in this section, including SCR, DPF, FGD, VOC control, and alternative energy sources, play a significant role in controlling emissions from various sources. As the global focus on environmental sustainability intensifies, the continued development and implementation of effective emission control technologies are essential for a cleaner and healthier future.

Indoor Air Quality

Indoor air quality (IAQ) refers to the quality of air inside buildings and structures, specifically to the health and comfort of the occupants. With people spending a significant amount of time indoors, IAQ has become a major concern due to the potential health risks associated with poor air quality. Various factors can affect IAQ, including building materials, furniture, cleaning products, and ventilation systems. In this section, we will explore the importance of indoor air quality, the factors influencing it, and strategies to improve it.

Importance of Indoor Air Quality

Indoor air quality is crucial because poor air quality can have adverse effects on human health and well-being. Exposure to pollutants and allergens indoors can lead to respiratory problems, allergies, asthma, and other health issues. Inadequate ventilation, which can result in the accumulation of indoor pollutants, can also contribute to the spread of infectious diseases.

Furthermore, poor IAQ can have a significant impact on the productivity and comfort of occupants. Studies have shown that indoor environments with good air quality can enhance cognitive function, concentration, and overall performance. On the other hand, exposure to pollutants can cause discomfort, fatigue, and decreased productivity.

Indoor Air Pollutants

Indoor air pollutants can originate from various sources, including building materials, cleaning products, combustion appliances, furniture, and outdoor air pollution. Some common indoor air pollutants include:

+ **Volatile Organic Compounds (VOCs):** VOCs are emitted as gases from certain solids or liquids, such as paints, varnishes, cleaning products, and adhesives. Exposure to high levels of VOCs can cause eye, nose, and throat irritation, headaches, and even long-term health effects.

+ **Particulate Matter (PM):** PM consists of tiny particles suspended in the air, such as dust, pollen, and smoke. These particles can be inhaled and can aggravate respiratory conditions or cause respiratory symptoms.

+ **Biological Contaminants:** Biological contaminants include mold, bacteria, viruses, and allergens. Mold growth in damp indoor environments can lead to respiratory issues and allergic reactions, while bacteria and viruses can cause illnesses.

+ **Radon:** Radon is a colorless, odorless, and tasteless radioactive gas that can seep into buildings from the ground. Prolonged exposure to high levels of radon can increase the risk of lung cancer.

It is essential to identify and control these indoor pollutants to maintain good indoor air quality.

Improving Indoor Air Quality

There are several strategies and measures that can be implemented to improve indoor air quality:

+ **Source Control:** The most effective approach is to eliminate or minimize the use of pollutants at their source. This can involve using low-emitting materials, choosing environmentally friendly cleaning products, and ensuring proper ventilation during painting or renovation activities.

+ **Ventilation:** Good ventilation is crucial for maintaining good air quality. It helps to remove contaminants and bring in fresh outdoor air. Natural ventilation, such as opening windows, can be effective, but mechanical ventilation systems should be used in buildings where natural ventilation is not feasible. Regular maintenance and cleaning of ventilation systems are essential to prevent the buildup of contaminants.

+ **Air Filtration:** Air filtration systems can remove particulate matter and some gaseous pollutants from indoor air. High-efficiency particulate air (HEPA) filters are commonly used to capture fine particles.

+ **Humidity Control:** Maintaining appropriate humidity levels can help prevent the growth of mold and bacteria. Relative humidity should be kept between 30% and 60% to minimize the risk of moisture-related issues.

+ **Regular Cleaning:** Regular cleaning of indoor spaces, including dusting and vacuuming, helps to remove dust, allergens, and other particles that can contribute to poor air quality.

+ **Educating Occupants:** Raising awareness among building occupants about the importance of IAQ and providing guidelines for maintaining good air quality can help ensure their active participation in maintaining a healthy indoor environment.

Case Study: IAQ in Schools

Indoor air quality is a significant concern in educational institutions, as children spend a considerable amount of time in school buildings. Poor IAQ in schools can have short-term and long-term effects on students' health and academic performance.

Several factors can contribute to poor IAQ in schools, including inadequate ventilation, the presence of allergens and asthma triggers, poor maintenance of HVAC systems, and the use of potentially harmful cleaning products. These issues can result in increased absenteeism, respiratory problems, and decreased cognitive function among students.

To address these concerns, educational institutions should prioritize IAQ management strategies. This can involve regular monitoring of air quality, proper maintenance of ventilation systems, using low-emitting materials for construction and renovation, implementing effective cleaning and maintenance protocols, and educating students and staff on IAQ-related practices.

By proactively managing IAQ in schools, we can provide a healthy learning environment for students, ensure their well-being, and promote better academic performance.

Conclusion

Indoor air quality is a critical aspect of building design, construction, and maintenance. Poor IAQ can have adverse effects on human health, comfort, and productivity. By understanding the sources of indoor air pollutants and implementing effective strategies to improve IAQ, we can create healthier indoor environments. Through proper ventilation, source control, air filtration, and education, we can ensure that buildings provide safe and comfortable spaces for occupants. By prioritizing indoor air quality, we can move towards a more sustainable and healthy future.

Soil Contamination

Soil contamination is a significant environmental issue that arises from the introduction of pollutants into the soil. Pollutants can include heavy metals, pesticides, industrial chemicals, petroleum products, and radioactive materials. Contaminated soil poses serious threats to human health, ecosystems, and agricultural productivity. In this section, we will explore the causes and consequences of soil contamination, as well as various techniques for remediation and prevention.

Causes of Soil Contamination

Soil contamination can occur due to both natural and anthropogenic activities. Natural sources of soil contamination include volcanic eruptions, weathering of rocks, and the accumulation of metals and radionuclides. However, the majority of soil contamination is a result of human activities. Some common causes of anthropogenic soil contamination are:

1. Industrial activities: Industrial processes such as mining, manufacturing, and waste disposal can release toxic chemicals and heavy metals into the soil. These pollutants can persist in the environment for long periods and accumulate over time.

2. Agricultural practices: Intensive agricultural practices, including excessive use of pesticides, herbicides, and fertilizers, can lead to soil contamination. Runoff from agricultural fields can carry these chemicals into nearby soils.

3. Accidental spills and leaks: Accidental spills of petroleum products, chemicals, and hazardous waste can contaminate the soil. Leaking underground storage tanks, pipelines, and improper disposal of hazardous materials contribute to soil contamination.

4. Urbanization and construction: Urban development and construction activities can result in soil contamination from the disposal of construction materials, waste, and improper handling of pollutants such as asbestos and lead-based paints.

5. Atmospheric deposition: Airborne pollutants from industrial emissions, vehicle exhaust, and agricultural practices can be deposited onto the soil surface. These pollutants can infiltrate the soil and cause contamination.

Consequences of Soil Contamination

Soil contamination has wide-ranging consequences for human health, ecosystems, and agricultural productivity. Some of the key consequences are:

1. Human health risks: Contaminated soil can pose serious health risks to humans through direct contact, inhalation of contaminated dust, or consumption of plants and animals grown on contaminated soil. Exposure to heavy metals, pesticides, and other toxic chemicals can cause various health issues, including cancers, neurological disorders, and developmental problems.

2. Environmental impacts: Soil contamination can have detrimental effects on the environment. It can negatively impact soil fertility, reducing crop yields and agricultural productivity. Contaminated soils can also affect the growth and survival of plants by interfering with their nutrient uptake and physiological processes. Furthermore, pollutants in the soil can leach into groundwater, rivers, and other surface water bodies, leading to water contamination and harming aquatic ecosystems.

3. Ecological disturbances: Soil contamination disrupts the balance of ecosystems by affecting soil-dwelling organisms, such as earthworms, microbes, and beneficial fungi. These organisms play crucial roles in nutrient cycling, soil structure formation, and plant growth. The loss of biodiversity in contaminated soils can have cascading effects on higher trophic levels and disrupt ecosystem functioning.

Remediation of Soil Contamination

Remediation of soil contamination is essential to restore soil quality and mitigate the associated environmental and health risks. Several techniques are used for soil remediation, depending on the type and extent of contamination. Some commonly employed techniques are:

1. Soil excavation and disposal: For localized contamination, excavation and removal of contaminated soil followed by proper disposal in designated facilities is an effective remediation method. This technique is commonly used for soil contaminated with petroleum products or heavy metals.

2. Soil washing: Soil washing involves the physical separation of contaminants from soil particles. It is particularly effective for soils contaminated with heavy

metals and organic compounds. Contaminated soil is treated with water or chemical solutions, and the resulting slurry is separated into clean soil and liquid waste fractions.

3. Bioremediation: Bioremediation utilizes microorganisms, plants, or their enzymes to degrade or immobilize contaminants in the soil. Microorganisms can break down organic pollutants, while plants can stabilize contaminants through processes like phytoremediation. Bioremediation is cost-effective and environmentally friendly, making it an attractive option for soil remediation.

4. Soil vapor extraction: This technique involves the extraction of volatile contaminants from the soil by applying a vacuum. It is particularly effective for soils contaminated with volatile organic compounds. As the vacuum is applied, contaminants evaporate and are collected for treatment or disposal.

5. Soil stabilization: Soil stabilization aims to immobilize contaminants in the soil to prevent their migration. This technique involves adding chemical amendments to the soil, such as solidifying agents or adsorbents. These amendments bind with contaminants, reducing their mobility and bioavailability.

Prevention of Soil Contamination

Preventing soil contamination is crucial for maintaining soil health and preventing the associated risks. Some key strategies for preventing soil contamination include:

1. Proper waste management: Proper management and disposal of hazardous waste, industrial byproducts, and municipal solid waste can minimize the risk of soil contamination. Implementing robust waste management practices, such as recycling, reuse, and safe disposal, can prevent the release of pollutants into the environment.

2. Sustainable agricultural practices: Adopting sustainable agricultural practices, such as integrated pest management, organic farming, and precision agriculture, can reduce the use of chemical inputs and minimize soil contamination. Additionally, implementing erosion control measures and proper nutrient management can prevent soil erosion and nutrient runoff into nearby soils.

3. Containment and spill prevention: Implementing appropriate containment systems and leak detection measures for storage tanks, pipelines, and industrial facilities can prevent accidental spills and leaks. Regular inspections and maintenance of these systems are essential to minimize the risk of soil contamination.

4. Remediation of brownfields: Brownfields are abandoned or underutilized properties with potential soil contamination. Redeveloping brownfield sites

through proper assessment and remediation can revitalize urban areas while preventing further soil contamination.

5. Education and awareness: Educating the public, agricultural communities, and industrial sectors about the impacts of soil contamination and the importance of sustainable practices is crucial for prevention. Promoting awareness campaigns, providing technical guidance, and enforcing environmental regulations can significantly contribute to preventing soil contamination.

Given the widespread nature and adverse impacts of soil contamination, addressing this issue requires a holistic and interdisciplinary approach. Collaboration between scientists, policymakers, industries, and communities is essential to develop effective strategies for soil remediation, prevention, and sustainable soil management. By adopting sustainable practices, implementing proper waste management, and fostering environmental stewardship, we can ensure healthier soils for present and future generations.

Summary

Soil contamination is a result of the introduction of pollutants into the soil, often from human activities such as industrial processes, agricultural practices, and accidental spills. Contaminated soil poses significant risks to human health, ecosystems, and agricultural productivity. Remediation techniques include soil excavation, soil washing, bioremediation, soil vapor extraction, and soil stabilization. Prevention strategies involve proper waste management, sustainable agricultural practices, containment and spill prevention, remediation of brownfields, and education and awareness. Effective soil management requires collaboration and interdisciplinary approaches to mitigate the adverse impacts of soil contamination and protect our environment for future generations.

Remediation Techniques

Remediation is the process of addressing and resolving environmental pollution and contamination issues. It involves the use of various techniques to restore ecosystems, clean up polluted sites, and protect human health. In this section, we will explore different remediation techniques that are commonly used to mitigate pollution and restore degraded environments.

Containment

Containment is a common technique used to prevent the spread and migration of pollutants in the environment. It involves the construction of physical barriers to isolate contaminated areas. Containment methods include the installation of impermeable liners, caps, and barriers to prevent the movement of pollutants from soil, water, or air.

For example, in the case of contaminated groundwater, a technique called groundwater containment can be used. This involves the construction of underground barriers such as slurry walls or sheet piles to prevent the spread of contaminants. Containment is often combined with other remediation techniques to ensure the effectiveness of the overall remediation process.

Excavation and Removal

Excavation and removal is a widely used technique for remediating contaminated soil or sediments. It involves the physical removal of contaminated material from a site. Excavated soil or sediment is then transported to a treatment facility or a proper disposal site.

Excavation and removal is particularly effective for addressing localized contamination, such as industrial spills or leaking underground storage tanks. The technique allows for the removal of the source of contamination, minimizing the risk of further spread. However, it may not be suitable for large-scale contamination or areas with sensitive ecosystems.

Bioremediation

Bioremediation is a sustainable and cost-effective technique that utilizes microorganisms to break down or transform contaminants into less harmful substances. It harnesses the natural ability of certain microorganisms, such as bacteria and fungi, to degrade a wide range of pollutants.

There are two main types of bioremediation: in situ and ex situ. In situ bioremediation involves the treatment of contaminants in their original location, while ex situ bioremediation involves the removal of contaminated soil or water for treatment at a separate location.

Bioremediation can be enhanced through various methods, including the addition of nutrients to stimulate microbial activity, adjusting environmental conditions (e.g., temperature, pH) to optimize microbial growth, and introducing genetically engineered microorganisms with enhanced degradation capabilities.

For instance, in the case of oil spills, certain strains of bacteria can be used to degrade hydrocarbons present in the spilled oil. These bacteria break down the contaminants into non-toxic byproducts, such as carbon dioxide and water.

Chemical Oxidation

Chemical oxidation is a technique used to treat contaminated soil and groundwater by introducing chemical agents that react with and degrade contaminants. This process typically involves the addition of oxidizing agents, such as hydrogen peroxide or ozone, to the contaminated area.

Oxidation reactions break down or transform contaminants into less harmful substances. Chemical oxidation can effectively treat a wide range of organic contaminants, including petroleum hydrocarbons, chlorinated solvents, and pesticides.

One example of chemical oxidation is the use of potassium permanganate to treat sites contaminated with chlorinated solvents. When potassium permanganate comes into contact with the contaminants, it undergoes a chemical reaction that breaks down the pollutants into harmless byproducts.

Phytoremediation

Phytoremediation is a technique that utilizes plants to remove, degrade, or immobilize contaminants in soil, groundwater, or sediments. Certain plant species have the ability to absorb, metabolize, or break down various pollutants, making them effective in cleaning up contaminated environments.

There are several different mechanisms by which plants can remediate contaminants. Phytoextraction involves the uptake and accumulation of contaminants in the plant's roots, stems, and leaves. Phytodegradation refers to the breakdown of contaminants by enzymes produced by the plants. Phytostabilization involves the immobilization or containment of contaminants within the plant or in the rhizosphere.

Phytoremediation can be used for a wide range of contaminants, including heavy metals, organic pollutants, and even radioactive materials. It is often used in conjunction with other remediation techniques to enhance the overall effectiveness of the remediation process.

For example, certain plant species, such as sunflowers, have been used to remove heavy metals from soil. These plants accumulate the metals in their tissues, which can then be harvested and properly disposed of, effectively reducing the concentration of contaminants in the soil.

Electrokinetic Remediation

Electrokinetic remediation utilizes an electric field to enhance the removal of contaminants from soil or groundwater. The technique involves the insertion of electrodes into the contaminated material and the application of an electric potential. The electric field created drives the movement of contaminants towards the electrodes, where they can be collected and removed.

This technique is particularly effective for treating soils contaminated with heavy metals, as well as certain organic pollutants. It can also be used to control the migration of contaminants in the subsurface.

Electrokinetic remediation is often used in combination with other remediation techniques, such as bioremediation or chemical oxidation, to enhance their effectiveness. It can be applied in both in situ and ex situ remediation settings.

Hydraulic Control

Hydraulic control techniques are used to manage the flow and movement of contaminated groundwater, preventing further spread and minimizing the impact of contamination. These techniques involve the use of hydraulic barriers, such as permeable reactive barriers (PRBs) or pumping wells, to control the movement of groundwater and capture contaminants.

Permeable reactive barriers are constructed by placing reactive materials, such as zerovalent iron or activated carbon, in the flow path of the groundwater. As the contaminated water passes through the barrier, the reactive materials chemically react or adsorb the contaminants, reducing their concentration in the groundwater.

Pumping wells can be used to extract contaminated groundwater, preventing it from spreading to unaffected areas. The extracted groundwater can then be treated or disposed of properly.

Combination and Integration

In many cases, a combination and integration of multiple remediation techniques may be necessary to address complex contamination problems effectively. The selection of the appropriate combination depends on factors such as the type and extent of contamination, site-specific conditions, and economic feasibility.

For example, a combination of excavation and removal, followed by in situ bioremediation, can be employed to treat contaminated soil. The excavation and removal technique can be used to remove the bulk of contaminated material, reducing the source of contamination. In situ bioremediation can then be implemented to further degrade the remaining contaminants.

Integration of different techniques can also help to enhance the overall efficiency and effectiveness of the remediation process. For instance, using phytoremediation in conjunction with electrokinetic remediation can accelerate the removal of contaminants from soil and groundwater, as the plants help to immobilize or degrade the contaminants while the electric field drives their movement towards the electrodes.

It is crucial to carefully evaluate site-specific conditions, consult with experts, and consider the long-term sustainability of the chosen remediation approach when deciding on the appropriate combination or integration of techniques.

Conclusion

Remediation techniques play a vital role in addressing environmental pollution and restoring degraded ecosystems. These techniques employ a range of approaches, from containment and excavation to bioremediation and phytoremediation, to mitigate the impacts of pollution and ensure the long-term sustainability of our environment.

Understanding and applying the principles and practices of remediation are essential for environmental scientists, engineers, and policy-makers. By utilizing a combination of techniques, tailoring solutions to site-specific conditions, and embracing innovative approaches, we can effectively remediate contaminated sites and contribute to a healthier and more sustainable future.

Phytoremediation

Phytoremediation is an innovative and sustainable approach to environmental remediation, which utilizes plants to clean up contaminated soil, water, and air. It is a cost-effective and environmentally friendly alternative to traditional remediation methods, such as excavation and chemical treatment.

Principles of Phytoremediation

Phytoremediation works on the principle that certain plants have the ability to absorb, metabolize, and/or detoxify contaminants in their tissues. This process can occur through various mechanisms, including phytoextraction, rhizodegradation, phytostabilization, and phytovolatilization.

- **Phytoextraction:** This process involves the uptake and accumulation of contaminants from the soil or water into the plant shoots. The contaminants can then be harvested, removed from the site, and disposed of safely. Phytoextraction is particularly effective for heavy metals, such as lead, cadmium, and arsenic.

+ **Rhizodegradation:** In this process, plants release substances (e.g., enzymes, organic acids) through their roots into the soil, which stimulate the growth of beneficial microorganisms. These microorganisms can then degrade or detoxify contaminants in the vicinity of the plant roots. Rhizodegradation is commonly used for organic pollutants, such as petroleum hydrocarbons and chlorinated solvents.

+ **Phytostabilization:** Phytostabilization involves the immobilization or containment of contaminants in the soil, reducing their availability for uptake by plants or leaching into groundwater. Certain plants have the ability to trap contaminants in their roots or modify the soil conditions to reduce the mobility of contaminants. Phytostabilization is often used for metals and metalloids.

+ **Phytovolatilization:** This process involves the uptake of contaminants by plants, followed by their transformation into volatile compounds, which are released into the air. Contaminants that can undergo phytovolatilization include organic compounds, such as benzene and trichloroethylene. This process is particularly useful for addressing air pollution and indoor air quality issues.

Applications of Phytoremediation

Phytoremediation has been successfully applied in various environmental contexts, ranging from industrial sites to agricultural lands and urban areas. Some notable applications include:

+ **Heavy metal contamination:** Phytoremediation has been used to clean up sites contaminated with heavy metals, such as mining areas and industrial waste sites. For example, Indian mustard (Brassica juncea) has been used to remove lead, zinc, and cadmium from contaminated soils.

+ **Organic pollutant remediation:** Plants like hybrid poplars (Populus spp.) and willows (Salix spp.) have been used to remediate sites contaminated with organic pollutants, such as hydrocarbons and pesticides. These plants promote microbial degradation in the rhizosphere and enhance the breakdown of contaminants.

+ **Wastewater treatment:** Phytoremediation has been employed as an effective and sustainable method for treating wastewater. Certain plants, such as water hyacinth (Eichhornia crassipes) and duckweed (Lemna spp.), can absorb and accumulate pollutants from wastewater, thereby improving its quality.

+ **Landfill leachate treatment:** Phytoremediation has been utilized in landfill sites to treat leachate, which is the liquid generated from the decomposition of waste. Plants with high transpiration rates, such as poplars and willows, are capable of extracting water from the ground, thereby reducing the volume of leachate.

+ **Soil erosion control:** Planting vegetation with deep root systems, such as grasses and legumes, can help stabilize soil and prevent erosion. These plants enhance soil structure, increase organic matter content, and improve water infiltration rates.

+ **Urban air pollution mitigation:** Trees and other green infrastructure can help mitigate air pollution in urban areas by capturing and trapping particulate matter, absorbing carbon dioxide, and releasing oxygen. Planting trees along busy roads and in urban parks can improve air quality and promote a healthier environment.

Challenges and Considerations

While phytoremediation offers great potential for environmental cleanup, there are several challenges and considerations that need to be addressed:

+ **Contaminant specificity:** Different plants have varying capabilities to remediate specific contaminants. Site-specific factors, such as soil types, pH, and nutrient levels, also influence the effectiveness of phytoremediation. Therefore, the selection of appropriate plant species for a particular site is crucial.

+ **Time frame:** Phytoremediation is generally a slower process compared to traditional remediation methods. It may take several years or even decades to achieve significant results. Long-term monitoring and management are therefore necessary to ensure the effectiveness of phytoremediation projects.

+ **Scale-up and applicability:** While phytoremediation has been successfully applied at small to moderate scales, its feasibility at larger scales is still being explored. Additionally, certain contaminants may not be amenable to phytoremediation due to their high toxicity or mobility.

✦ **Regulatory considerations:** The use of phytoremediation as a remediation technique may require regulatory approval and compliance with relevant environmental guidelines. Stakeholder engagement and public perception of phytoremediation projects also need to be considered.

Conclusion

Phytoremediation offers a promising and sustainable approach to environmental remediation by harnessing the natural capabilities of plants. It provides a cost-effective alternative to traditional methods while promoting ecological restoration and biodiversity conservation. However, careful consideration of site-specific factors, plant selection, and long-term management is essential for the success of phytoremediation projects. Continued research and development in this field will contribute to the advancement of eco science and the achievement of a more sustainable future.

Soil Amendments

Soil amendments play a crucial role in improving soil fertility and maintaining optimal soil conditions for plant growth. They can greatly enhance the physical, chemical, and biological properties of the soil, leading to increased crop yields and overall sustainability in agriculture. In this section, we will explore different types of soil amendments, their benefits, and their application methods.

Types of Soil Amendments

There are several types of soil amendments that can be used to improve soil quality:

1. **Organic Amendments:** These include materials such as compost, manure, and plant residues. Organic amendments contribute organic matter to the soil, improving its structure and water-holding capacity. They also supply essential nutrients and enhance microbial activity, promoting a healthy soil ecosystem.

2. **Inorganic Amendments:** This category includes materials like lime, gypsum, and elemental sulfur. Inorganic amendments are used to adjust soil pH and improve nutrient availability. Lime is often added to acidic soils to raise pH, while sulfur is used to lower pH in alkaline soils. Gypsum is employed to improve soil structure and drainage.

3. **Mineral Amendments:** These substances, such as rock phosphate and greensand, provide essential minerals and trace elements to the soil. Mineral amendments can correct nutrient deficiencies in the soil and promote healthy plant growth.

4. **Biochar:** Biochar is a type of charcoal produced from the pyrolysis of organic materials, such as agricultural waste. It is used as a soil amendment to improve fertility, water retention, and nutrient availability. Biochar also has the ability to sequester carbon dioxide, contributing to climate change mitigation.

5. **Microbial Amendments:** Microbial amendments contain beneficial microorganisms, such as mycorrhizal fungi and rhizobacteria. These microorganisms form symbiotic relationships with plants and enhance their nutrient uptake and disease resistance. Microbial amendments are often used in organic farming systems to promote soil health and reduce the need for synthetic fertilizers and pesticides.

Benefits of Soil Amendments

Soil amendments offer a wide range of benefits, both for the soil ecosystem and for crop production:

+ **Improved Soil Structure**: Organic amendments improve soil aggregation, creating a crumbly texture that enhances water infiltration and aeration. This allows plant roots to penetrate the soil more easily and promotes the growth of beneficial soil organisms.

+ **Enhanced Nutrient Availability**: Soil amendments provide essential nutrients to plants, either directly or by improving nutrient availability in the soil. Organic amendments release nutrients slowly over time, while inorganic amendments can quickly correct nutrient imbalances in the soil.

+ **Increased Water-Holding Capacity**: Organic amendments improve the water-holding capacity of sandy soils, reducing water loss through leaching. They also enhance the drainage of clay soils, preventing waterlogging and improving root development.

+ **Promotion of Beneficial Microorganisms**: Organic amendments supply a source of food and habitat for beneficial soil microorganisms. These microorganisms, in turn, enhance nutrient cycling, break down organic matter, and suppress the growth of plant pathogens.

+ **Reduction of Soil Erosion**: Soil amendments, particularly organic mulches, help prevent soil erosion by reducing the impact of rainfall, wind, and water runoff. They protect the soil surface, stabilize slopes, and promote the formation of soil aggregates.

Application Methods

The application of soil amendments depends on the type of amendment and the specific requirements of the soil and crops. Here are some common application methods:

1. **Incorporation**: Organic amendments can be mixed into the soil by tilling or plowing. This method ensures uniform distribution of the amendment throughout the root zone.

2. **Surface Application:** Some amendments, such as compost and mulch, can be applied to the soil surface. Surface application helps conserve soil moisture, prevents weed growth, and improves soil structure as the amendment breaks down over time.

3. **Topdressing:** Inorganic amendments, like lime or gypsum, can be spread on the soil surface without incorporation. They slowly dissolve and move into the soil with rainfall or irrigation.

4. **Injection:** Liquid amendments, such as liquid fertilizers or microbial inoculants, can be injected into the soil through irrigation systems or specialized equipment. This method delivers the amendment directly to the root zone, minimizing losses and increasing efficiency.

5. **Seed Coating:** Microbial amendments, like rhizobacteria or mycorrhizal fungi, can be coated onto seeds before planting. This ensures close contact between the inoculant and the emerging roots, promoting early establishment and symbiotic interactions.

Challenges and Considerations

While soil amendments offer significant benefits, there are also certain challenges and considerations to keep in mind:

+ **Nutrient Imbalances:** Poorly planned or excessive application of amendments can lead to nutrient imbalances in the soil. It is crucial to conduct soil tests and follow recommended application rates to avoid overloading the soil with nutrients.

+ **Environmental Impacts:** Some soil amendments, such as chemical fertilizers, can contribute to water pollution if not managed properly. It is important to adhere to best management practices and consider the environmental impact of amendments.

+ **Cost and Availability:** The cost and availability of soil amendments can vary depending on the region and the specific amendment. It is essential to assess the economic feasibility and availability of amendments before implementing a soil amendment program.

+ **Long-Term Sustainability:** Achieving long-term sustainability in agriculture requires a systems approach that integrates soil amendments with other practices, such as crop rotation, cover cropping, and integrated pest management. Soil amendments should be seen as part of a broader strategy for soil health and ecosystem sustainability.

Real-World Example: Biochar in Amazonian Dark Earths

An intriguing example of soil amendment is found in the Amazon rainforest, where indigenous people created fertile soils known as Amazonian Dark Earths (ADEs) or Terra Preta. These soils, enriched with charcoal-like material (biochar), have sustained agriculture for centuries. The biochar acts as a long-lasting carbon sink, retains water and nutrients, and supports microbial activity. ADEs continue to inspire modern applications of biochar as a sustainable soil amendment worldwide.

Further Reading

+ Lehmann, J., Joseph, S. (eds.). (2009). *Biochar for Environmental Management: Science, Technology and Implementation.* Earthscan.

+ Magdoff, F., van Es, H. (2000). *Building Soils for Better Crops*, 3rd edition. Sustainable Agriculture Network.

+ Brady, N.C., Weil, R.R. (2014). *The Nature and Properties of Soils*, 14th edition. Pearson Education.

Eco Science in Policy and Governance

Environmental Policy

International Agreements and Treaties

International agreements and treaties play a crucial role in addressing global environmental challenges and promoting sustainable development. These agreements are essential for fostering cooperation among nations and establishing a framework for collective action. In this section, we will explore some of the key international agreements and treaties that have shaped the field of eco science and have had a significant impact on environmental policy and governance.

Convention on Biological Diversity

The Convention on Biological Diversity (CBD) is an international treaty adopted in 1992 with the objective of conserving biodiversity, promoting its sustainable use, and ensuring the fair and equitable sharing of the benefits arising from the utilization of genetic resources. The CBD recognizes that biodiversity is fundamental to human well-being and is threatened by human activities.

The CBD has three main objectives:

1. Conservation of biological diversity: This objective is aimed at conserving and protecting ecosystems, species, and genetic resources to maintain the Earth's biodiversity.

2. Sustainable use of biological resources: This objective emphasizes the need to use natural resources in a way that ensures their long-term sustainability and promotes the equitable sharing of benefits derived from their utilization.

3. Fair and equitable sharing of benefits: This objective aims to ensure that the benefits derived from the use of genetic resources are shared fairly and equitably with the countries providing those resources.

The CBD has brought significant attention to biodiversity conservation and has led to the development of national biodiversity strategies and action plans by participating countries. It has also facilitated the establishment of protected areas, the implementation of sustainable land use practices, and the integration of biodiversity considerations into various sectors, including agriculture, forestry, and fisheries.

Paris Agreement

The Paris Agreement is an international treaty adopted in 2015 under the United Nations Framework Convention on Climate Change (UNFCCC). It aims to limit global warming to well below 2 degrees Celsius above pre-industrial levels and to pursue efforts to limit the temperature increase to 1.5 degrees Celsius.

The key elements of the Paris Agreement include:

1. Nationally Determined Contributions (NDCs): Parties to the agreement are required to prepare and communicate their NDCs, which outline their efforts to mitigate greenhouse gas emissions and adapt to the impacts of climate change.

2. Transparency and accountability: The agreement emphasizes the importance of transparency and regular reporting by parties on their implementation efforts, including mitigation and adaptation measures.

3. Long-term goal: The agreement sets a long-term goal of reaching global peaking of greenhouse gas emissions as soon as possible to achieve a balance between sources and sinks of greenhouse gases.

4. Climate finance: The agreement establishes a mechanism to support developing countries in their efforts to mitigate and adapt to climate change through financial and technological assistance.

The Paris Agreement represents a significant step forward in global efforts to combat climate change and transition to a low-carbon and climate-resilient future. It has mobilized governments, businesses, and civil society organizations to take ambitious actions to reduce greenhouse gas emissions and enhance climate resilience.

Kyoto Protocol

The Kyoto Protocol is an international treaty adopted in 1997 under the UNFCCC. It sets binding targets for 37 developed countries and the European Union to reduce their greenhouse gas emissions. The protocol follows the principle of "common but differentiated responsibilities," recognizing that developed countries have historically contributed the most to greenhouse gas emissions and have greater financial and technological capabilities to address climate change.

Under the Kyoto Protocol, participating countries have committed to reducing their overall emissions of six greenhouse gases by an average of 5% below 1990 levels during the period 2008-2012. To achieve these reductions, countries can implement various mechanisms, such as emissions trading, joint implementation, and the Clean Development Mechanism.

The Kyoto Protocol has played a critical role in raising awareness about the importance of reducing greenhouse gas emissions and has created a framework for international cooperation on climate change mitigation. However, its effectiveness has been limited by the withdrawal of some key countries and the lack of binding commitments for major emitting nations, such as the United States and China.

Other International Agreements and Treaties

In addition to the CBD, Paris Agreement, and Kyoto Protocol, several other international agreements and treaties have contributed to the development of environmental policy and governance. Some notable examples include:

1. **Montreal Protocol:** The Montreal Protocol on Substances that Deplete the Ozone Layer is an international agreement aimed at protecting the ozone layer by phasing out the production and consumption of ozone-depleting substances, such as chlorofluorocarbons (CFCs) and hydrochlorofluorocarbons (HCFCs).

2. **Ramser Convention:** The Convention on Wetlands of International Importance, also known as the Ramser Convention, is an international treaty for the conservation and wise use of wetlands. It promotes the sustainable management of wetlands and recognizes their ecological, economic, and cultural value.

3. **United Nations Framework Convention on Climate Change (UNFCCC):** The UNFCCC is an international environmental treaty aimed at preventing dangerous human interference with the climate system. It provides the

framework for international cooperation on climate change and serves as the foundation for the Paris Agreement and the Kyoto Protocol.

4. **United Nations Convention to Combat Desertification (UNCCD):** The UNCCD is a global agreement that addresses the challenges of desertification, land degradation, and drought. It promotes sustainable land management practices and the rehabilitation of degraded ecosystems in dryland areas.

These international agreements and treaties highlight the importance of cooperation and collaboration at the global level to address environmental challenges and achieve sustainability. They provide a framework for countries to work together, share knowledge and resources, and develop common strategies for environmental protection and sustainable development.

Challenges and Limitations

While international agreements and treaties have made significant contributions to environmental governance, they also face challenges and limitations. Some of these challenges include:

1. **Enforcement and compliance:** Ensuring compliance with international agreements can be challenging, as there is often a lack of effective mechanisms for monitoring and enforcing commitments.

2. **Divergent interests and priorities:** Negotiating and reaching consensus among a large number of countries with diverse interests and priorities can be a complex and time-consuming process.

3. **Lack of participation:** Some countries, especially developing nations, may face challenges in fully participating in the decision-making processes and implementing the commitments due to resource constraints and capacity limitations.

4. **Inadequate funding:** Many international agreements and treaties rely on financial support from developed countries to assist developing nations in implementing measures for biodiversity conservation, climate adaptation, and mitigation. However, funding commitments are often inadequate or inconsistent.

Addressing these challenges requires continued efforts to strengthen international cooperation, enhance transparency and accountability, and mobilize resources for implementing the commitments made under these agreements. It also requires fostering dialogue and collaboration between governments, civil society organizations, and the private sector to align interests and work towards common goals.

Conclusion

International agreements and treaties play a crucial role in shaping global environmental policy and governance. They provide a framework for addressing environmental challenges, promoting sustainable development, and fostering cooperation among nations. The Convention on Biological Diversity, Paris Agreement, Kyoto Protocol, and other international agreements have made significant contributions to biodiversity conservation, climate change mitigation and adaptation, and the protection of critical ecosystems.

While these agreements face challenges and limitations, they provide a platform for dialogue, knowledge sharing, and collective action. They serve as a reminder of the importance of international cooperation and the need for global solutions to address global environmental challenges. Moving forward, it is essential to continue strengthening these agreements, mobilizing resources, and engaging all stakeholders to build a more sustainable and resilient future.

Paris Agreement

The Paris Agreement is an international treaty aimed at addressing climate change and its impacts. It was adopted on December 12, 2015, at the United Nations Climate Change Conference (COP21) in Paris, France. The agreement represents a critical milestone in global efforts to combat climate change and limit the rise in global temperatures to well below 2 degrees Celsius above pre-industrial levels.

Background

Climate change is a major global challenge with far-reaching environmental, social, and economic consequences. The Earth's climate is changing primarily due to human activities, such as the burning of fossil fuels, deforestation, and industrial processes, leading to an increase in greenhouse gas (GHG) emissions. These emissions trap heat in the atmosphere and cause the planet to warm, resulting in climate-related impacts such as rising sea levels, extreme weather events, and disruptions to ecosystems and biodiversity.

In response to the urgent need to address climate change, the United Nations Framework Convention on Climate Change (UNFCCC) was established in 1992. The UNFCCC is an international treaty that aims to stabilize greenhouse gas concentrations in the atmosphere at a level that would prevent dangerous anthropogenic interference with the climate system.

Key Principles and Objectives

The Paris Agreement builds upon the principles and objectives of the UNFCCC and sets out a comprehensive framework for global climate action. It aims to:

1. Limit global temperature rise: The agreement aims to hold the increase in the global average temperature to well below 2 degrees Celsius above pre-industrial levels and to pursue efforts to limit the temperature increase to 1.5 degrees Celsius. This ambitious target recognizes the need to avoid catastrophic impacts of climate change.

2. Adaptation to climate change: The agreement recognizes the importance of enhancing adaptive capacity, strengthening resilience, and reducing vulnerability to the impacts of climate change. It emphasizes the need for support to developing countries to undertake adaptation measures.

3. Mitigation of greenhouse gas emissions: The agreement requires all countries to contribute to global efforts to mitigate greenhouse gas emissions. Each country is expected to prepare and communicate its nationally determined contributions (NDCs), which outline their efforts to reduce emissions and adapt to the impacts of climate change.

4. Financing and technology transfer: The agreement acknowledges the need for financial resources, technology transfer, and capacity-building support to enable developing countries to take climate action. Developed countries are expected to provide financial support to developing countries to assist them in fulfilling their climate commitments.

5. Transparency and accountability: The agreement establishes a robust transparency framework to ensure transparency and accountability of climate actions. It promotes the enhanced transparency of NDCs, as well as the measurement, reporting, and verification of greenhouse gas emissions and climate finance.

Implementation and Challenges

Implementing the Paris Agreement requires concerted efforts from all countries, as well as the involvement of non-state actors, including businesses, cities, regions, and civil society organizations. The agreement provides a flexible framework that allows countries to determine their own climate actions based on their national circumstances and capabilities.

One of the main challenges in implementing the agreement is the gap between the current commitments made by countries and the emission reductions required to achieve the goals of the Paris Agreement. The agreement emphasizes the need for regular review and revision of NDCs to ensure that they are in line with the long-term temperature goals.

Another challenge is the mobilization of financial resources to support climate actions in developing countries. The Paris Agreement established the Green Climate Fund (GCF) as a mechanism to assist developing countries in implementing their climate commitments. However, there is still a need to scale up financial resources and enhance access to funding for climate projects.

Furthermore, coordination and cooperation among countries is crucial to address the global nature of climate change. The Paris Agreement encourages international collaboration in areas such as capacity-building, technology transfer, and knowledge sharing.

Examples of Action

Since the adoption of the Paris Agreement, countries have taken various actions to reduce greenhouse gas emissions and adapt to the impacts of climate change. Here are some examples:

1. Renewable energy transition: Many countries have increased their investment in renewable energy sources such as solar, wind, and hydropower. For example, Germany has significantly expanded its use of solar and wind energy, while Costa Rica has achieved nearly 100% renewable electricity generation.

2. Energy efficiency improvements: Countries have implemented measures to improve energy efficiency and reduce emissions in sectors such as buildings, transportation, and industry. For instance, Sweden has implemented policies to promote energy-efficient buildings and has reduced energy consumption in the building sector by more than 30%.

3. Afforestation and reforestation: Several countries have undertaken afforestation and reforestation projects to increase forest cover and sequester carbon dioxide from the atmosphere. China has implemented the largest afforestation program in the world, aiming to increase forest cover by 40 million hectares by 2020.

4. Carbon pricing: Some countries and regions have implemented carbon pricing mechanisms, such as carbon taxes and emissions trading systems, to

incentivize the reduction of greenhouse gas emissions. For example, the European Union has implemented a cap-and-trade system that covers various sectors of the economy.

These examples illustrate the diverse range of actions that countries are taking to address climate change in line with the objectives of the Paris Agreement. However, more ambitious and accelerated efforts are still needed to achieve the long-term goals of the agreement.

Caveats and Criticisms

While the Paris Agreement has been widely hailed as a landmark achievement, it is not without its limitations and criticisms. Some key caveats and criticisms include:

1.   Lack of binding commitments: The Paris Agreement is a voluntary agreement, and the commitments made by countries are not legally binding. This raises concerns about the enforceability of the agreement and the ability to hold countries accountable for their climate actions.

2. Insufficient emission reduction targets: The current commitments made by countries under the Paris Agreement are not sufficient to limit global temperature rise to well below 2 degrees Celsius. There is a need for more ambitious emission reduction targets to bridge the emissions gap.

3.   Financing challenges: Mobilizing the necessary financial resources to support climate actions in developing countries remains a significant challenge. The current level of climate financing falls short of the estimated needs, particularly for adaptation measures.

4.   Equity and fairness concerns: The burden of climate change impacts and mitigation efforts falls disproportionately on vulnerable and developing countries. Critics argue that the Paris Agreement does not adequately address the equity and fairness considerations in climate action.

5.   Lack of inclusion of subnational actors: While the Paris Agreement acknowledges the role of non-state actors, such as cities and businesses, in addressing climate change, it does not provide a formal mechanism for their inclusion in the negotiation process.

Addressing these caveats and criticisms requires ongoing dialogue, engagement, and collaboration among countries and stakeholders. It is crucial to continuously evaluate and enhance the effectiveness of the Paris Agreement in driving global climate action.

Convention on Biological Diversity

The Convention on Biological Diversity (CBD) is an international treaty that aims to conserve biodiversity, ensure sustainable use of its components, and promote fair and equitable access to genetic resources. The CBD was opened for signature at the Earth Summit in Rio de Janeiro, Brazil, in 1992 and entered into force in 1993.

Background

Biodiversity refers to the variety of life on Earth, encompassing all living organisms, their genetic diversity, and the ecological systems they are part of. It is a critical component of our planet's natural capital, providing essential ecosystem services such as regulation of climate, purification of air and water, pollination of crops, and nutrient cycling.

However, biodiversity is facing unprecedented threats due to human activities, such as habitat destruction, pollution, overharvesting, and climate change. Recognizing the urgent need to address these challenges, the CBD was established as a global framework to safeguard biodiversity and promote its sustainable use.

Principles

The CBD is based on several fundamental principles that guide its implementation:

Sovereignty: States have the sovereign right to exploit their resources within the framework of their environmental policies and obligations.

Conservation: The CBD promotes the conservation of biodiversity, taking into account the ecological, genetic, social, economic, scientific, educational, cultural, recreational, and aesthetic values of biological diversity.

Sustainable Use: The CBD recognizes the importance of sustainable use of biological resources, ensuring long-term socio-economic benefits and ecological sustainability.

Fair and Equitable Access: The CBD emphasizes the fair and equitable sharing of benefits arising from the use of genetic resources, based on appropriate access to these resources and the transfer of relevant technologies.

Precautionary Approach: The CBD encourages the application of the precautionary approach when scientific evidence is insufficient, to avoid potential risks to biodiversity.

Ecosystem Approach: The CBD promotes the integration of biodiversity considerations into all sectors of society, adopting a holistic ecosystem approach to decision-making.

Objectives

The CBD has three main objectives:

Conservation of Biodiversity: The CBD seeks to conserve biodiversity and the ecological processes that sustain it. It aims to maintain and restore ecosystems, protect species and their habitats, and safeguard genetic diversity.

Sustainable Use of Biodiversity: The CBD promotes the sustainable use of biological resources, ensuring their long-term availability for the benefit of present and future generations. It supports the development of sustainable practices in sectors such as agriculture, forestry, fisheries, and tourism.

Fair and Equitable Sharing of Benefits: The CBD emphasizes the fair and equitable sharing of benefits arising from the use of genetic resources. It recognizes the role of indigenous and local communities in the conservation and sustainable use of biodiversity and their rights to access and share in the benefits.

Implementation

To achieve its objectives, the CBD establishes a framework for action at national, regional, and global levels. Parties to the CBD are required to develop national biodiversity strategies and action plans, incorporate biodiversity considerations into relevant sectors, and report on their progress.

The CBD also encourages collaboration among countries to address transboundary issues and promote international cooperation in the conservation and sustainable use of biodiversity. It facilitates the exchange of information, technology transfer, and capacity-building initiatives to support developing countries in implementing the convention.

The CBD operates through its Conference of the Parties (COP), which meets regularly to review progress, adopt decisions, and discuss emerging issues. The COP has established subsidiary bodies, such as the Subsidiary Body on Scientific, Technical, and Technological Advice (SBSTTA), to provide scientific and technical guidance.

Challenges and Opportunities

The CBD faces several challenges in its implementation:

Loss of Biodiversity: Despite efforts to conserve biodiversity, species continue to decline at an alarming rate. More comprehensive and effective conservation measures are needed to halt and reverse this trend.

Access and Benefit Sharing: Ensuring fair and equitable sharing of benefits from the use of genetic resources remains a complex issue, requiring strengthened cooperation, capacity-building, and effective legal frameworks.

Integration of Biodiversity: Integrating biodiversity considerations into various sectors, such as agriculture, energy, and infrastructure, poses challenges but also offers opportunities for mainstreaming biodiversity into sustainable development strategies.

Knowledge Gaps: There is still much to learn about the ecological processes supporting biodiversity and the impacts of human activities on ecosystems. Bridging these knowledge gaps is crucial for informed decision-making.

Conclusion

The Convention on Biological Diversity plays a vital role in addressing the global biodiversity crisis. It provides a framework for international cooperation, policy development, and action to conserve biodiversity and promote its sustainable use. By implementing the CBD's principles and objectives, we can strive towards a future where biodiversity is valued, protected, and utilized in a sustainable manner for the well-being of all species, including humans.

Resource:

CBD. (2021). Convention Text. Retrieved from `https://www.cbd.int/convention/convention-text`.

Exercise:

1. Choose a local or regional biodiversity hotspot in your country. Describe the threats it faces and suggest conservation measures that could be implemented to protect its unique species and ecosystems.

Kyoto Protocol

The Kyoto Protocol is a significant international agreement aimed at addressing the issue of climate change. It was adopted in 1997 under the United Nations Framework Convention on Climate Change (UNFCCC) and entered into force in 2005. The protocol sets binding targets for the reduction of greenhouse gas (GHG) emissions by industrialized countries.

Background and Objectives

The Kyoto Protocol was a response to the growing concerns over global warming and its potential impacts on the environment and human societies. It recognized that human activities, particularly the burning of fossil fuels, are the primary drivers of climate change due to the release of GHGs such as carbon dioxide (CO2), methane (CH4), and nitrous oxide (N2O).

The main objective of the protocol is to stabilize GHG concentrations in the atmosphere at a level that would prevent dangerous anthropogenic interference with the climate system. To achieve this, the protocol sets specific emission reduction targets for Annex I countries, which are mainly developed nations.

Key Provisions

The Kyoto Protocol introduced several key provisions to guide the efforts of participating countries in reducing GHG emissions. Some of the notable provisions include:

1. **Emission Reduction Targets**: Annex I countries are assigned individual targets for reducing their emissions of GHGs. These targets are legally binding and are expressed as a percentage reduction from their 1990 levels. The overall goal of the protocol is to collectively achieve a 5.2% reduction in GHG emissions by 2012.

2. **Flexibility Mechanisms**: The protocol includes three flexibility mechanisms to assist countries in meeting their emission reduction targets more efficiently. These mechanisms are:

   + **Emissions Trading**: Countries can buy and sell emission allowances to each other, allowing for flexibility in meeting their targets.

   + **Clean Development Mechanism (CDM)**: Annex I countries can invest in emission reduction projects in developing countries and receive certified emission reduction credits in return.

   + **Joint Implementation (JI)**: Annex I countries can undertake emission reduction projects in other Annex I countries and receive emission reduction units as credits.

3. **Compliance Mechanism**: The protocol establishes a compliance system to ensure that countries fulfill their emission reduction commitments. It includes procedures for reporting and reviewing emissions data, as well as penalties and corrective actions for non-compliance.

4. **Adaptation Fund**: The protocol created the Adaptation Fund to support developing countries in adapting to the adverse effects of climate change. The fund is financed by a share of proceeds from CDM projects.

Critics and Challenges

While the Kyoto Protocol has been a significant step towards addressing climate change, it has faced criticism and challenges. Some of the main criticisms include:

1. **Limited Coverage:** The protocol only imposes binding emission reduction targets on developed nations (Annex I countries), leaving out major emitting countries such as the United States, China, and India. This limited coverage undermines the effectiveness of the protocol in addressing global emissions.

2. **Lack of Commitment:** Some critics argue that the emission reduction targets set by the protocol are not ambitious enough to sufficiently mitigate climate change. They believe that more aggressive and comprehensive actions are needed to adequately address the issue.

3. **Withdrawals and Non-Compliance:** Several countries, including Canada, Japan, and Russia, have withdrawn from the protocol or indicated their unwillingness to participate in the second commitment period. Additionally, some countries have struggled to meet their emission reduction targets, highlighting the challenges of compliance.

Real-World Example: European Union Emissions Trading Scheme (EU ETS)

The European Union Emissions Trading Scheme (EU ETS) is one of the notable implementations of the Kyoto Protocol's emissions trading mechanism. It is the largest cap-and-trade system in the world, covering more than 11,000 installations in the European Union.

Under the EU ETS, emissions allowances are allocated to participating installations, which include power plants, industrial facilities, and airlines. These allowances can be traded among participants, creating a market for emission reductions. The scheme aims to incentivize emission reductions by putting a price on carbon.

The EU ETS has faced challenges, including issues with the initial allocation of allowances and fraud attempts. However, it has also demonstrated the potential for emissions trading to contribute to the reduction of GHG emissions.

Key Takeaways

The Kyoto Protocol is an important international agreement that aims to address climate change by setting binding emission reduction targets for developed countries. It introduced flexibility mechanisms such as emissions trading, CDM, and JI to support countries in meeting their targets. However, the protocol has faced criticism for its limited coverage and perceived lack of ambition. The EU ETS serves as a real-world example of the protocol's emissions trading mechanism in action.

The next section will delve into the concept of national environmental policies and their role in promoting sustainability and addressing climate change.

National Environmental Policies

National environmental policies play a crucial role in addressing environmental challenges and promoting sustainability within a country's borders. These policies are developed and implemented by governments to regulate various aspects of environmental protection, conservation, and sustainable development. They provide a framework for decision-making, set targets and standards, allocate resources, and establish mechanisms for monitoring and enforcement.

Importance of National Environmental Policies

National environmental policies are essential for several reasons. Firstly, they help in managing the use of natural resources and minimizing negative impacts on the environment. They ensure that activities such as industrial production, land use, and resource extraction are carried out sustainably, minimizing pollution, deforestation, habitat destruction, and water scarcity.

Secondly, national environmental policies aim to protect and conserve biodiversity and ecosystems. They help in establishing protected areas, promoting habitat restoration, and conserving endangered species. These policies also focus on ecosystem services like clean air, water, and soil, which are vital for human well-being.

Thirdly, national environmental policies address the growing concern of climate change. They set targets for greenhouse gas reduction, promote renewable energy sources, and encourage energy efficiency. These policies also support adaptation measures to minimize the impacts of climate change and enhance resilience.

Finally, national environmental policies contribute to sustainable development by integrating environmental objectives with social and economic goals. They promote sustainable agriculture, sustainable infrastructure development, sustainable consumption and production patterns, and support the transition to a green economy.

Key Components of National Environmental Policies

National environmental policies typically consist of the following key components:

1. **Legislative Framework:** This component includes the development and enactment of laws and regulations that govern environmental protection. It outlines the rights and responsibilities of stakeholders, sets standards and criteria for environmental management, and provides legal mechanisms for monitoring, enforcement, and penalties.

2. **Institutional Arrangements:** This component involves the establishment of institutions responsible for implementing environmental policies. It includes the creation of government departments or agencies dedicated to environmental protection, as well as the collaboration with other sectors, such as agriculture, industry, and health, to ensure a holistic and integrated approach.

3. **Environmental Impact Assessment (EIA):** National environmental policies often require the conduct of EIAs for proposed projects or activities that may have significant environmental impacts. EIAs assess the potential environmental effects of a project, identify alternatives, and recommend mitigation measures to minimize adverse impacts.

4. **Pollution Control Measures:** This component focuses on the regulation and control of various sources of pollution, including air, water, and soil. It sets emission limits, regulates wastewater discharge, enforces waste management practices, and promotes the use of cleaner technologies to minimize pollution and protect public health.

5. **Biodiversity Conservation:** Biodiversity conservation is a vital aspect of national environmental policies. This component aims to protect and preserve ecosystems, habitats, and species. It includes the establishment of protected areas, the development of species recovery plans, the regulation of wildlife trade, and the integration of biodiversity considerations into land-use planning.

6. **Climate Change Mitigation and Adaptation:** National environmental policies address the challenges of climate change by setting targets for greenhouse gas reduction, promoting renewable energy sources, and encouraging energy efficiency. They also incorporate adaptation measures to enhance resilience, such as the development of climate change action plans and the promotion of climate-smart agriculture.

7. **Public Participation and Stakeholder Engagement:** National environmental policies recognize the importance of engaging the public and stakeholders in decision-making processes. This component ensures transparency, inclusiveness, and accountability, allowing for public input, consultation, and involvement in the development and implementation of environmental policies.

8. **Monitoring, Reporting, and Enforcement:** To ensure the effectiveness of national environmental policies, monitoring, reporting, and enforcement mechanisms are essential. This component involves the establishment of

monitoring networks, the collection and analysis of data on environmental indicators, the reporting of progress towards targets, and the enforcement of regulations through inspections, penalties, and legal actions.

Examples of National Environmental Policies

National environmental policies vary across countries, depending on their specific environmental challenges, political context, and socio-economic conditions. Here are a few examples of national environmental policies from different regions:

1. **Germany:** Germany's environmental policy, known as the Environmental Action Program, focuses on sustainable development, climate protection, and resource efficiency. It includes targets for greenhouse gas reduction, the promotion of renewable energy sources, waste management regulations, and the integration of environmental considerations into urban planning.

2. **Brazil:** Brazil's environmental policy emphasizes the protection and conservation of the Amazon rainforest, one of the world's most biodiverse regions. It includes measures to combat deforestation, promote sustainable land use practices, support indigenous land rights, and regulate activities such as mining and agriculture in the Amazon region.

3. **China:** China's environmental policy has recently undergone significant developments in response to its environmental challenges. The policy includes measures to reduce air pollution, promote renewable energy sources, increase energy efficiency, and address water scarcity. It also incorporates the concept of ecological civilization, which seeks to balance economic development and environmental protection.

4. **Costa Rica:** Costa Rica is known for its progressive environmental policies and commitment to conservation. The country has established a network of national parks and protected areas, promotes sustainable tourism, and has set targets for carbon neutrality by 2050. Costa Rica's environmental policies also involve payment for ecosystem services, which provide incentives for landowners to conserve forests and protect water sources.

Challenges and Future Directions

Despite the progress made through national environmental policies, significant challenges remain. These challenges include limited enforcement capacity, inadequate funding and resources, conflicting priorities, and the need for improved coordination and integration across different sectors and levels of governance.

In the future, national environmental policies need to address emerging environmental issues such as plastic pollution, the depletion of natural resources, and the impacts of urbanization. They also need to be adaptive and responsive to changing environmental conditions and scientific knowledge. Collaboration and cooperation among countries are crucial for addressing transboundary environmental challenges and achieving global sustainability goals.

Conclusion

National environmental policies are essential for promoting environmental protection, conservation, and sustainable development within a country. They provide a framework for decision-making, set targets and standards, allocate resources, and establish mechanisms for monitoring and enforcement. These policies play a crucial role in managing natural resources, protecting biodiversity, addressing climate change, and integrating environmental objectives with social and economic goals. However, challenges remain, and future policies need to be adaptive and collaborative to address emerging environmental issues and achieve global sustainability. National environmental policies are a critical tool in achieving a sustainable future for our planet.

Environmental impact assessment

Environmental impact assessment (EIA) is a critical tool used to identify and evaluate the potential environmental consequences of proposed projects, policies, or programs. It is a systematic process that considers the social, economic, and environmental impacts and benefits of various development activities. EIA plays a crucial role in ensuring sustainable development by providing decision-makers with the necessary information to avoid, minimize, or mitigate adverse environmental effects.

Principles of Environmental Impact Assessment

EIA is guided by several fundamental principles that are essential for its effective implementation. These principles include:

- **Holistic approach:** EIA considers the interrelationships between all components of the environment, including natural, social, and economic aspects. It recognizes that changes in one component can have ripple effects on others.

+ **Participation and transparency:** EIA involves the participation of all stakeholders, including local communities, indigenous groups, and other interested parties. The process should be transparent, allowing for public input and access to information.

+ **Precautionary principle:** EIA adopts a precautionary approach to decision-making. It recognizes that in the absence of complete scientific certainty, preventive measures should be taken to avoid potential harm to the environment.

+ **Best available technology and practices:** EIA assesses the impacts of proposed projects using the best available knowledge, technology, and practices. It considers innovative solutions and alternatives that minimize harm to the environment.

+ **Sustainability:** EIA aims to promote sustainable development by assessing the long-term impacts and benefits of proposed projects. It evaluates the project's compatibility with environmental, social, and economic goals and identifies measures to enhance sustainability.

+ **Adaptive management:** EIA recognizes that decision-making is an ongoing process. It promotes adaptive management approaches that allow for monitoring, evaluation, and adjustment of activities to achieve desired environmental outcomes.

Process of Environmental Impact Assessment

The process of EIA involves a series of steps that systematically evaluate the potential environmental impacts of a proposed project. The steps may vary depending on national or regional regulations, but generally include the following:

1. **Screening:** The screening stage determines whether a project requires a full EIA or can be exempted. It involves assessing the project's characteristics, scale, and location to determine its potential environmental significance.

2. **Scoping:** Scoping defines the scope and boundaries of the EIA study. It identifies the key environmental issues to be considered and establishes the baseline conditions against which the impacts will be assessed. Stakeholder consultation is an integral part of this stage.

3. **Impact assessment:** The impact assessment phase evaluates the potential environmental impacts of the proposed project. It involves identifying and predicting the likely impacts on various environmental components, such as air quality, water resources, biodiversity, and socio-economic aspects.

4. **Mitigation and alternatives:** Based on the identified impacts, mitigation measures are proposed to avoid or minimize adverse effects on the environment. Alternatives to the proposed project are also assessed to evaluate their environmental benefits and feasibility.

5. **Environmental management plan:** An environmental management plan outlines the measures to be implemented during project construction, operation, and decommissioning phases to prevent, mitigate, or compensate for any negative environmental impacts.

6. **Monitoring and evaluation:** Monitoring and evaluation are critical components of EIA to ensure that the predicted impacts and mitigation measures are effectively implemented. Regular monitoring helps identify any deviations from the expected outcomes and enables adaptive management.

7. **Public participation:** Throughout the EIA process, public participation allows interested individuals and communities to provide input, voice concerns, and contribute to decision-making. It promotes transparency, inclusivity, and accountability.

8. **Decision-making and follow-up:** The final decision on the proposed project is made based on the findings of the EIA. The decision-maker considers the environmental impacts, mitigation measures, and public input. Following the decision, implementation, enforcement, and follow-up processes are undertaken to ensure compliance with the approved measures.

Challenges and Limitations

Although EIA is a valuable tool for environmental decision-making, it faces several challenges and limitations that need to be addressed:

+ **Data availability and quality:** EIA relies heavily on data regarding the environment, baseline conditions, and predicted impacts. Limited data availability and poor data quality can hinder the accuracy and reliability of the assessment.

+ **Interdisciplinary coordination:** EIA requires collaboration among experts from different disciplines, such as ecology, economics, sociology, and engineering. Coordinating interdisciplinary inputs and ensuring effective communication can be challenging.

+ **Cumulative impacts:** EIA often focuses on individual projects, neglecting the cumulative impacts of multiple projects in a given area. Cumulative impacts can result in synergistic effects that significantly affect the environment but may not be adequately addressed through the standard EIA process.

+ **Time and cost constraints:** EIA can be time-consuming and expensive, particularly for large-scale projects. Balancing the need for a thorough assessment with the project's timeline and budget constraints can be a significant challenge.

+ **Political and social factors:** EIA can be influenced by political and social factors, leading to biased decision-making. It is crucial to ensure the independence and integrity of the assessment process.

+ **Lack of enforcement and compliance:** The effectiveness of EIA depends on proper enforcement and compliance with the recommended mitigation measures. Inadequate follow-up, monitoring, and enforcement can undermine the achievement of desired environmental outcomes.

Case Study: Environmental Impact Assessment of a Dam Construction Project

To illustrate the process of EIA, let's consider a case study of a proposed dam construction project. The project aims to generate hydropower to meet the growing energy demand of a region. The following steps would be involved in the EIA for this project:

Screening: The project scale and potential environmental impacts are evaluated to determine if a full EIA is necessary.

Scoping: Stakeholder consultation is conducted to identify key environmental issues, such as the potential impacts on aquatic ecosystems, displacement of local communities, and alteration of water flow. Baseline conditions are established, including the biodiversity, water quality, and socio-economic aspects.

Impact assessment: Various environmental components are evaluated, such as the impacts on fish populations, water availability downstream, and socio-economic factors like livelihoods and cultural heritage.

Mitigation and alternatives: Measures to minimize the impacts are proposed, such as fish ladders to facilitate fish migration and compensation programs for affected communities. Alternative energy sources, like solar or wind power, are also considered.

Environmental management plan: An environmental management plan is developed to guide the construction, operation, and decommissioning phases of the dam. It includes measures to monitor water quality, protect biodiversity, and address social impacts.

Monitoring and evaluation: Throughout the project's life cycle, monitoring programs are implemented to assess the actual environmental impacts and the effectiveness of mitigation measures. Adaptive management allows for adjustments to be made if unforeseen impacts arise.

Public participation: Local communities, indigenous groups, and other stakeholders are involved in the decision-making process. Public input is sought, and concerns are addressed to ensure transparency and accountability.

Decision-making and follow-up: The final decision on the dam construction project takes into account the findings of the EIA, including the predicted impacts and proposed mitigation measures. Regulatory authorities enforce compliance with the environmental management plan throughout the project's lifecycle.

Conclusion

Environmental impact assessment is a crucial tool in achieving sustainable development by ensuring that environmental considerations are integrated into decision-making processes. It helps identify and mitigate potential adverse effects on the environment, while also promoting public participation and transparency. The process of EIA follows a systematic approach, considering the holistic assessment of impacts, best available practices, and sustainable alternatives. However, challenges such as data availability, interdisciplinary coordination, and political influences need to be addressed to enhance the effectiveness of EIA.

Environmental legislation

Environmental legislation plays a crucial role in ensuring the protection and preservation of our natural environment. It encompasses the laws, regulations, and policies that are put in place by governments at various levels to regulate and control activities that may have an impact on the environment. In this section, we will explore the principles of environmental legislation, its importance, and its implementation.

Principles of environmental legislation

Environmental legislation is guided by several key principles that form the foundation of its development and implementation. These principles include:

1. **Precautionary principle:** This principle states that if there is a threat of serious or irreversible damage to the environment, lack of full scientific certainty should not be used as a reason to postpone measures to prevent or minimize that damage. In other words, it is better to err on the side of caution and take proactive measures to protect the environment.

2. **Polluter pays principle:** According to this principle, those who cause pollution or environmental damage should bear the costs associated with preventing, controlling, and cleaning up that damage. This principle aims to incentivize individuals, businesses, and industries to adopt cleaner and more sustainable practices by holding them accountable for their actions.

3. **Integration principle:** The integration principle emphasizes the need to consider environmental concerns in all areas of decision-making. It calls for the integration of environmental considerations into policies, plans, and programs related to sectors such as agriculture, industry, energy, and transport. This principle recognizes that environmental protection cannot be achieved in isolation from other societal goals.

4. **Public participation principle:** This principle recognizes the importance of involving the public in decision-making processes concerning the environment. It promotes transparency, accountability, and inclusiveness by giving individuals and communities the opportunity to express their opinions, share information, and contribute to the development and implementation of environmental laws and regulations.

Importance of environmental legislation

Environmental legislation is essential for several reasons. Firstly, it provides a legal framework for the protection and conservation of natural resources, biodiversity, and ecosystems. It sets standards and guidelines for activities such as waste management, pollution control, land use planning, and the exploitation of natural resources, ensuring that these activities are carried out in a sustainable and responsible manner.

Secondly, environmental legislation helps to prevent, mitigate, and address environmental issues and challenges. By setting limits on pollution levels, regulating emissions from industries, and establishing guidelines for the management of hazardous substances, it helps to prevent further degradation of the environment. It also provides mechanisms for addressing environmental emergencies and responding to environmental disasters.

Thirdly, environmental legislation contributes to the achievement of sustainable development goals. It promotes the integration of environmental considerations into economic, social, and development policies, thereby fostering a balance between economic growth, social well-being, and environmental protection. By incentivizing sustainable practices and discouraging environmentally harmful activities, it creates a conducive environment for the transition to a green and sustainable economy.

Implementation of environmental legislation

The implementation of environmental legislation involves various stages and processes. These include:

1. **Legislative drafting**: This stage involves the development and drafting of laws and regulations. It requires the involvement of policymakers, legal experts, scientists, and other stakeholders to ensure that the legislation is comprehensive, effective, and enforceable. During this stage, considerations are given to the objectives of the legislation, its scope, the entities it applies to, and the penalties for non-compliance.

2. **Enforcement and compliance**: Once environmental legislation is enacted, it needs to be effectively enforced to ensure compliance. Enforcement agencies, such as environmental protection agencies or departments, are responsible for monitoring and enforcing compliance with the legislation. They may conduct inspections, issue permits and licenses, and take enforcement actions against violators. Compliance with environmental legislation is crucial to ensuring that its objectives are met and environmental protection is achieved.

3. **Monitoring and evaluation**: Monitoring and evaluation play a vital role in assessing the effectiveness of environmental legislation. This involves the collection of data, analysis of trends, and regular reporting on key environmental indicators. Monitoring and evaluation help identify gaps and shortcomings in the legislation, evaluate its impact on the environment, and inform decision-makers on the need for amendments or adjustments.

4. **Capacity building and public awareness**: The successful implementation of environmental legislation requires capacity building and public awareness

initiatives. These initiatives aim to enhance the understanding of the legislation among relevant stakeholders, including government officials, industry representatives, and the general public. Capacity building programs can include training workshops, seminars, and educational campaigns to ensure that individuals and organizations have the knowledge and skills to comply with the legislation.

5. **International cooperation and coordination**: Given the global nature of many environmental issues, international cooperation and coordination are crucial in implementing environmental legislation effectively. This involves harmonizing environmental standards and regulations across countries, sharing best practices, and collaborating on transboundary issues such as air and water pollution. International agreements and treaties, such as the Paris Agreement and the Convention on Biological Diversity, provide a framework for such cooperation and coordination.

Case study: The Clean Air Act

One example of significant environmental legislation is the Clean Air Act, which was enacted in the United States in 1970 and has been amended several times since then. The Clean Air Act aims to protect and improve air quality by setting standards for the control of pollutants and by regulating emissions from industrial sources, motor vehicles, and other pollution sources.

Under the Clean Air Act, the Environmental Protection Agency (EPA) is responsible for establishing and enforcing national ambient air quality standards, regulating the emissions of hazardous air pollutants, and implementing programs to address acid rain, ozone depletion, and climate change. The Act also includes provisions for public participation, compliance monitoring, and enforcement actions against violators.

The Clean Air Act has had a significant impact on air quality in the United States. It has resulted in the reduction of harmful pollutants such as lead, sulfur dioxide, and particulate matter, leading to improvements in public health and environmental quality. The Act has also stimulated technological innovation in pollution control and has provided economic benefits through the creation of green jobs and the growth of the clean energy sector.

However, challenges remain in the implementation of the Clean Air Act. New pollutants and sources of pollution continue to emerge, requiring ongoing updates and amendments to the legislation. Enforcement and compliance issues also persist, with some industries and regions struggling to meet the required standards. These challenges highlight the need for continuous monitoring, evaluation, and improvement of environmental legislation to address emerging issues and ensure its effectiveness.

Conclusion

Environmental legislation is a critical component of sustainable development and the preservation of our natural environment. It provides a legal framework for the protection and conservation of natural resources, regulates human activities that may impact the environment, and promotes the integration of environmental considerations into decision-making processes.

The principles of environmental legislation, such as the precautionary principle and the polluter pays principle, guide its development and implementation. Through effective enforcement, monitoring, and evaluation, environmental legislation can help prevent environmental degradation, mitigate the impacts of pollution, and promote sustainable practices.

However, the successful implementation of environmental legislation requires the involvement of various stakeholders, including government bodies, industry representatives, and the general public. Capacity building and public awareness initiatives play a vital role in ensuring compliance with the legislation, while international cooperation and coordination are crucial in addressing global environmental challenges.

The Clean Air Act serves as an example of significant environmental legislation and demonstrates the potential impact of well-designed and enforced regulations. Despite challenges, environmental legislation remains an essential tool for achieving environmental sustainability and ensuring a better future for our planet. It is therefore crucial to continuously strive for the improvement and strengthening of environmental legislation to address emerging issues and protect our natural environment.

Regulatory Frameworks

Regulatory frameworks play a crucial role in ensuring environmental protection and sustainability. These frameworks establish rules and standards that govern various aspects of human activities, with the aim of preventing pollution, conserving natural resources, and promoting sustainable development. In this section, we will explore the key elements of regulatory frameworks and their importance in eco science and environmental governance.

Definition and Purpose

Regulatory frameworks refer to a set of laws, policies, regulations, and guidelines that are designed to manage and control human activities with potential environmental impacts. These frameworks are developed and implemented by governments, international organizations, and other stakeholders to achieve specific environmental goals. The main purpose of regulatory frameworks is to ensure compliance with environmental laws, promote responsible behavior, and protect the environment for present and future generations.

Elements of Regulatory Frameworks

Regulatory frameworks consist of various elements that work together to achieve environmental objectives. These elements include:

+ **Legislation**: Environmental legislation forms the foundation of regulatory frameworks. It defines the legal obligations, rights, and responsibilities of individuals, organizations, and governments regarding environmental protection. Legislation may cover a wide range of areas, such as air and water quality, waste management, conservation of biodiversity, and climate change mitigation.

+ **Regulations**: Regulations are specific rules and standards derived from legislation. They provide detailed guidance on how to comply with environmental requirements. Regulations may set limits on pollutant emissions, prescribe waste disposal methods, or outline procedures for environmental impact assessments.

+ **Permits and Licenses**: Regulatory frameworks often require individuals and organizations to obtain permits or licenses before engaging in certain activities with potential environmental impacts. These permits ensure that activities comply with environmental standards and may include conditions or restrictions to minimize negative impacts.

+ **Monitoring and Enforcement**: Effective regulatory frameworks require robust monitoring and enforcement mechanisms. Monitoring involves the collection and analysis of data to assess compliance and measure environmental performance. Enforcement ensures that non-compliance is addressed through penalties, sanctions, or other legal measures.

- **Advisory Bodies**: Regulatory frameworks may include advisory bodies or expert committees that provide scientific and technical advice to policymakers. These bodies help in the development of regulations, evaluation of environmental risks, and decision-making processes.

- **Public Participation**: Public participation is an essential element of regulatory frameworks. It involves engaging the public, affected communities, and other stakeholders in decision-making processes. Public participation ensures transparency, accountability, and better environmental outcomes.

- **International Cooperation**: Given the global nature of many environmental issues, regulatory frameworks often involve international cooperation. This includes bilateral or multilateral agreements, treaties, and conventions that promote collective action and harmonization of environmental standards.

Importance of Regulatory Frameworks

Regulatory frameworks are crucial for several reasons:

- **Environmental Protection**: Regulatory frameworks help prevent pollution, conserve natural resources, and protect ecosystems. They provide a legal basis for addressing environmental challenges and ensuring that human activities are conducted in a sustainable manner.

- **Safeguarding Human Health**: Many environmental issues, such as air and water pollution, have direct implications for human health. Regulatory frameworks aim to minimize exposure to harmful pollutants and ensure that human health is prioritized in decision-making processes.

- **Promoting Sustainability**: Regulatory frameworks are instrumental in promoting sustainable development. They help integrate environmental considerations into economic activities, ensuring that development is balanced and does not compromise the needs of future generations.

- **Creating a Level Playing Field**: Regulatory frameworks establish a level playing field for businesses and industries. By setting uniform standards and requirements, they promote fair competition and discourage harmful practices that may provide short-term gains at the expense of the environment.

+ **Enabling International Cooperation**: Regulatory frameworks facilitate international cooperation by providing a common framework for addressing global environmental challenges. They promote collaboration, information sharing, and the adoption of best practices across nations.

Challenges and Solutions

Despite their importance, regulatory frameworks face several challenges in practice. These challenges include:

+ **Lack of Implementation**: In some cases, regulatory frameworks are not effectively implemented or enforced due to resource constraints, weak institutional capacity, or inadequate monitoring mechanisms. This can undermine their effectiveness in achieving environmental objectives.

+ **Regulatory Capture**: Regulatory capture occurs when regulatory agencies are influenced or controlled by the industries they are supposed to regulate. This can lead to the weakening or manipulation of regulations, resulting in inadequate environmental protection.

+ **Complexity and Fragmentation**: Environmental issues are often complex and interconnected, requiring integrated approaches and coordination across different sectors and levels of governance. Regulatory frameworks may suffer from fragmentation and lack of coherence, making it challenging to address environmental problems comprehensively.

To overcome these challenges, several solutions can be implemented:

+ **Capacity Building**: Enhancing institutional capacity and providing adequate resources are crucial for effective implementation of regulatory frameworks. This includes training regulatory officials, improving monitoring and enforcement capabilities, and fostering collaboration among relevant stakeholders.

+ **Transparency and Accountability**: Transparency and accountability are vital to ensuring the integrity of regulatory frameworks. This can be achieved through public disclosure of information, regular reporting on compliance, and independent oversight or audit mechanisms.

+ **Strengthening International Cooperation**: Given the transboundary nature of many environmental issues, strengthening international cooperation is essential. This involves harmonizing standards, sharing best practices, and providing support to developing countries to implement effective regulatory frameworks.

+ **Integrated Approaches:** Addressing complex environmental challenges requires integrated approaches that consider social, economic, and environmental factors. Regulatory frameworks should incorporate interdisciplinary approaches, promote stakeholder engagement, and encourage the use of innovative solutions.

Case Study: The European Union's Environmental Regulatory Framework

As a real-world example, let's explore the environmental regulatory framework of the European Union (EU). The EU has developed a comprehensive set of regulations and directives to address various environmental issues across its member states. The framework includes legislation on air and water quality, waste management, biodiversity conservation, and climate change mitigation.

One key example is the EU's Water Framework Directive, which aims to achieve good water quality across Europe through integrated river basin management. The directive establishes a framework for the protection and sustainable use of surface waters and groundwater, setting guidelines for monitoring, protection, and restoration of water bodies.

Another significant regulation is the EU Emissions Trading System (ETS), a cap-and-trade scheme aimed at reducing greenhouse gas emissions in the European Union. The ETS sets a carbon dioxide emission cap for certain industries and allows companies to buy and sell emission allowances, creating economic incentives for emission reductions.

These examples demonstrate how the EU's regulatory framework combines legislation, regulations, monitoring mechanisms, and international cooperation to achieve environmental protection and sustainability.

Conclusion

Regulatory frameworks are essential tools for achieving environmental protection, promoting sustainability, and ensuring responsible behavior. They provide a legal basis for managing human activities with environmental impacts and help address complex environmental challenges. Effective regulatory frameworks require robust legislation, regulations, monitoring mechanisms, public participation, and international cooperation. Overcoming challenges such as lack of implementation, regulatory capture, and complexity requires capacity building, transparency, and integrated approaches.   By adopting strong and well-implemented regulatory frameworks, we can strive towards a more sustainable and environmentally conscious world.

Exercises

1. Explain the main elements of regulatory frameworks.

2. Discuss the importance of regulatory frameworks in environmental protection and sustainability.

3. What are some challenges faced by regulatory frameworks? How can these challenges be addressed?

4. Choose a specific country and research its environmental regulatory framework. Discuss its strengths and weaknesses.

5. Can you think of any innovative approaches or technologies that can enhance the effectiveness of regulatory frameworks?

Additional Resources

+ Environmental Regulatory Frameworks: Principles and Practice by Richard Macrory and Richard L. Ottinger.

+ Environmental Law and Policy by Markell and Babcock.

+ The Oxford Handbook of Environmental Political Theory edited by Teena Gabrielson, Cheryl Hall, John M. Meyer, and David Schlosberg.

+ United Nations Environment Programme (UNEP) - Environmental Governance.

+ European Environment Agency (EEA) - Environmental Policy and Governance.

Tricks and Caveats

+ Keep in mind that regulatory frameworks may vary across different countries and regions. It is important to consider the specific context and legal framework when studying or analyzing regulatory frameworks.

+ The effectiveness of regulatory frameworks depends not only on their design but also on implementation and enforcement. It is necessary to evaluate not only the existence of regulations but also their practical implications and impact on the environment.

◆ Stakeholder engagement and public participation are critical for the development and implementation of effective regulatory frameworks. Involving affected communities and other stakeholders in decision-making processes can lead to more inclusive and sustainable outcomes.

◆ Regulatory frameworks often require periodic updates and revisions to keep up with evolving environmental challenges and scientific knowledge. It is important to review and adapt regulatory frameworks to ensure their continued effectiveness.

Summary

In this section, we explored the importance of regulatory frameworks in environmental protection and sustainability. We discussed the main elements of regulatory frameworks, including legislation, regulations, monitoring mechanisms, and public participation. We also highlighted the challenges faced by regulatory frameworks, such as lack of implementation and regulatory capture, and provided solutions to overcome these challenges. Finally, we examined the European Union's environmental regulatory framework as a case study and provided exercises, additional resources, and tricks and caveats to enhance understanding. Regulatory frameworks play a vital role in achieving environmental goals and ensuring a more sustainable future. By continuously improving and implementing effective regulatory frameworks, we can better protect the environment and promote sustainable development.

Environmental governance frameworks

Environmental governance refers to the system of rules, regulations, and institutions that shape and govern environmental decision-making and actions. It encompasses the processes and mechanisms through which environmental policies are developed, implemented, and enforced, as well as the interactions between various stakeholders involved.

In order to effectively address environmental challenges and achieve sustainable development, robust governance frameworks are necessary. These frameworks provide a structure and set of principles for managing natural resources and addressing environmental issues at different scales, from local to global. They also emphasize the participation and collaboration of diverse stakeholders, including governments, businesses, civil society organizations, and local communities.

Principles of environmental governance

Several key principles underpin effective environmental governance frameworks:

1. **Transparency:** Ensuring openness, accessibility, and availability of information related to environmental policies, decision-making processes, and outcomes. This facilitates public awareness, understanding, and involvement in environmental matters.

2. **Accountability:** Holding individuals, organizations, and institutions responsible for their environmental actions or inactions. This contributes to the promotion of sustainable practices and the prevention of environmental harm.

3. **Participation:** Encouraging the active involvement of stakeholders in environmental decision-making at all levels. This includes civil society, local communities, non-governmental organizations (NGOs), and marginalized groups, ensuring their voices are considered in policy formulation and implementation.

4. **Integration:** Promoting interdisciplinary and cross-sectoral approaches to environmental decision-making. This involves incorporating environmental considerations into various sectors, such as energy, agriculture, transport, and urban planning, to achieve more coherent and sustainable outcomes.

5. **Precaution:** Emphasizing the importance of taking preventive action in the face of potential environmental risks or uncertainties. This principle highlights the need to adopt precautionary measures to protect the environment and human health, even in the absence of full scientific certainty.

6. **Sustainability:** Ensuring that environmental governance frameworks are designed and implemented in a manner that supports long-term environmental, social, and economic sustainability. This involves balancing the needs of present and future generations and optimizing the use of natural resources.

7. **Equity and justice:** Addressing environmental issues in a fair and just manner, taking into account the distribution of environmental costs, benefits, and risks among different social groups. This principle seeks to promote social equity and environmental justice, particularly for marginalized and vulnerable populations.

Frameworks for environmental governance

Various frameworks have been established to guide environmental governance efforts. These frameworks provide a structured approach to decision-making, policy formulation, and implementation, helping to address complex environmental issues. Some prominent frameworks include:

1. **Integrated Environmental Management (IEM):** IEM is an approach that seeks to integrate environmental considerations into all stages of decision-making and policy development. It emphasizes the coordination and collaboration of different sectors and stakeholders to ensure sustainable outcomes.

2. **Ecosystem-based Management (EBM):** EBM is a holistic approach that focuses on the management of ecosystems as a whole, rather than individual components. It considers the interdependencies and interactions between different species, habitats, and ecological processes, aiming to maintain the health and resilience of ecosystems.

3. **Adaptive Governance:** Adaptive governance recognizes the dynamic and complex nature of environmental systems and acknowledges the need for flexible and adaptive approaches. It involves continuous learning, experimentation, and adjustment of policies and practices based on new information, feedback, and changing circumstances.

4. **Multi-Level Governance:** Multi-level governance refers to the distribution of decision-making authority and responsibility across different levels of government, from local to national and international. It recognizes that some environmental issues transcend national boundaries and require collective actions and cooperation across jurisdictions.

5. **Public-Private Partnerships (PPPs):** PPPs involve collaboration between public and private entities to achieve mutual environmental goals. These partnerships leverage the resources, expertise, and innovation of both sectors to address environmental challenges more effectively.

6. **Indigenous Governance:** Indigenous governance recognizes the unique perspectives, knowledge, and rights of Indigenous peoples in environmental decision-making. It values traditional ecological knowledge and promotes the active participation of Indigenous communities in managing and conserving natural resources.

Challenges and opportunities

While environmental governance frameworks provide a valuable foundation for addressing environmental challenges, they also face several challenges. These challenges include:

+ **Fragmentation:** Environmental decision-making often involves multiple sectors and levels of government, leading to fragmentation and lack of coordination. This can hinder effective implementation and lead to conflicting policies and actions.

+ **Power imbalances:** Power imbalances among different stakeholders may limit the participation and influence of marginalized groups, hindering inclusive and equitable governance. Addressing these power imbalances is essential for ensuring fair and effective environmental decision-making.

+ **Enforcement and compliance:** Weak enforcement mechanisms and insufficient compliance with environmental regulations can undermine the effectiveness of governance frameworks. Strengthening enforcement and promoting compliance is crucial for achieving environmental goals.

+ **Data and knowledge gaps:** Incomplete or unreliable data and knowledge gaps can impede evidence-based decision-making and hinder the formulation of effective policies. Bridging these gaps through data collection, research, and knowledge sharing is essential for informed and effective governance.

+ **Political will:** Environmental governance requires strong political will and commitment from governments and decision-makers. Lack of political will can hinder the implementation of necessary policies and actions.

Despite these challenges, environmental governance frameworks also present opportunities for transformative change and innovation. By embracing these opportunities, we can enhance the effectiveness and inclusivity of environmental decision-making and pave the way for a more sustainable future.

Case study: The Aarhus Convention

The Aarhus Convention is an international environmental governance framework that focuses on promoting access to information, public participation, and justice in environmental matters. Adopted in 1998, the convention has been ratified by numerous countries in Europe and beyond.

The convention has three pillars: access to information, public participation, and access to justice. It recognizes that transparency, public involvement, and access to justice are crucial for effective environmental governance. The convention empowers individuals and civil society organizations to actively participate in decision-making processes and provides mechanisms for seeking legal remedies in case of environmental harm.

For example, under the Aarhus Convention, public authorities are required to provide access to environmental information upon request. This includes information on the state of the environment, measures taken to protect it, and the potential impact of proposed activities. Such transparency allows the public to be well-informed and participate meaningfully in environmental decision-making processes.

The Aarhus Convention also promotes public participation in environmental decision-making. It encourages governments to involve the public at an early stage, allowing them to contribute to the development, implementation, and review of environmental policies and plans. This inclusive approach ensures that a wide range of perspectives and expertise are considered, leading to more robust and sustainable outcomes.

Furthermore, the convention provides access to justice in environmental matters. It enables individuals and organizations to challenge decisions that violate environmental laws or adversely impact the environment. This access to justice mechanism ensures accountability and helps prevent environmental harm.

The Aarhus Convention serves as an important example of an environmental governance framework that prioritizes transparency, public participation, and access to justice. It demonstrates how such frameworks can empower individuals and civil society organizations to play a meaningful role in environmental decision-making and contribute to sustainable development.

Conclusion

Environmental governance frameworks play a crucial role in addressing environmental challenges and promoting sustainable development. They provide a structure and set of principles for decision-making, policy formulation, and implementation, emphasizing transparency, accountability, participation, and integration.

While these frameworks face challenges such as fragmentation, power imbalances, and weak enforcement, they also present opportunities for transformative change and innovation. By addressing these challenges and seizing these opportunities, we can enhance the effectiveness and inclusivity of environmental decision-making.

The case study of the Aarhus Convention highlights the importance of transparency, public participation, and access to justice in environmental governance. The convention demonstrates how these principles can empower individuals and civil society organizations to contribute to sustainable outcomes and ensure accountability.

As we move forward, it is crucial to continue advancing and refining environmental governance frameworks to address emerging environmental issues and promote equitable and sustainable development. By doing so, we can create a future where environmental concerns are effectively addressed, and the well-being of both present and future generations is safeguarded.

National level governance

National level governance plays a crucial role in shaping environmental policy and ensuring sustainable development within a country. It involves the formulation and implementation of laws, regulations, and strategies that aim to protect the environment, conserve natural resources, and promote sustainable practices. This section explores various aspects of national level governance, including the key components, challenges, and strategies for effective environmental governance.

Key components of national level governance

National level governance for environmental sustainability encompasses a wide range of components, each playing a critical role in achieving the desired outcomes. These components include:

1. **Legislation and regulations:** National governments enact laws and regulations to create a legal framework for environmental protection. These laws address issues such as pollution control, resource management, land use planning, and conservation of biodiversity. Regulations help enforce these laws and set specific standards for industries, businesses, and individuals to adhere to.

2. **Institutional framework:** An effective institutional framework is essential for coordinating and implementing environmental and sustainability policies. This framework includes government agencies responsible for environmental management, such as environmental ministries, regulatory bodies, and research institutions. These agencies collaborate with each

other, civil society organizations, and the private sector to achieve collective goals.

3. **Monitoring and enforcement**: National governments establish monitoring systems to assess environmental conditions, track progress, and identify areas that require intervention. Robust monitoring allows policymakers to make evidence-based decisions and take corrective actions when necessary. Effective enforcement mechanisms, including penalties, incentives, and compliance monitoring, help ensure that environmental regulations are followed.

4. **Public participation**: Public participation is a cornerstone of democratic governance and plays a crucial role in environmental decision-making. Governments engage citizens, communities, and stakeholders in policy development, implementation, and evaluation processes. This involvement improves the legitimacy of environmental governance and leads to more informed and effective outcomes.

5. **Capacity building**: Building institutional and individual capacities is vital for effective environmental governance. National governments invest in training programs, workshops, and educational initiatives to enhance the skills and knowledge of professionals involved in environmental management. Capacity building also extends to raising awareness among the general public about environmental issues and promoting sustainable behaviors.

6. **Science-policy interface**: Effective communication and collaboration between scientists and policymakers are essential for evidence-based decision-making. National governments establish mechanisms to bridge the gap between scientific research and policy formulation. This includes creating platforms for scientists to share their findings, engaging them in advisory roles, and integrating scientific knowledge into policy documents.

7. **Financial mechanisms**: Adequate funding is necessary to support the implementation of environmental policies and programs. National governments allocate budgets for environmental protection initiatives, which may include funding for research, conservation projects, pollution control, and sustainability programs. Governments may also explore innovative financing mechanisms, such as green bonds or carbon pricing, to mobilize additional resources for environmental initiatives.

Challenges in national level governance

National level governance for environmental sustainability faces several challenges that hinder the effective implementation of policies. Understanding these challenges is crucial for developing strategies to overcome them. Some of the key challenges include:

1. **Coordination and integration:** Environmental governance involves multiple government departments, agencies, and stakeholders. Poor coordination and integration among these entities can lead to fragmented decision-making, conflicting policies, and implementation gaps. Overcoming these challenges requires effective mechanisms for interagency coordination and stakeholder engagement.

2. **Lack of resources:** Limited financial and technical resources can impede the implementation of environmental policies. Many developing countries face resource constraints, making it difficult to invest in sustainable development initiatives. Mobilizing resources through international cooperation, public-private partnerships, and innovative financing mechanisms can help overcome this challenge.

3. **Political will:** Environmental sustainability often requires long-term planning and bold actions, which may be challenging in the face of short-term political priorities. Political will and commitment are essential to drive effective environmental governance. Raising awareness among policymakers and demonstrating the socio-economic benefits of sustainability can help foster political support.

4. **Capacity gaps:** Building and maintaining the necessary human and institutional capacities for environmental governance is a continuous challenge. Insufficient technical expertise, limited access to data and information, and outdated governance structures can hinder effective decision-making. Addressing capacity gaps through training programs, knowledge sharing platforms, and institutional reforms is crucial.

5. **Enforcement and compliance:** Environmental regulations are often difficult to enforce, and non-compliance remains a significant challenge. Insufficient enforcement mechanisms, corruption, and lack of public awareness contribute to weak compliance. Strengthening enforcement through regular monitoring, stricter penalties, and public education campaigns can improve compliance rates.

6. **Equity and social justice:** Environmental degradation can disproportionately affect vulnerable communities, exacerbating existing social inequities. Ensuring equitable access to resources, meaningful public participation, and environmental justice in decision-making processes is crucial for effective environmental governance. This requires addressing power imbalances and promoting inclusive policies.

Strategies for effective national level governance

To address the challenges and promote effective environmental governance at the national level, several strategies can be employed:

1. **Integrated policy frameworks:** Developing integrated policy frameworks that incorporate environmental sustainability across sectors is crucial. Policies should be aligned with sustainable development goals and take into account social, economic, and environmental considerations. This can promote coherence, minimize conflicts, and ensure holistic decision-making.

2. **Strengthening institutions:** Building strong and effective institutions is essential for implementing environmental policies. This includes creating dedicated environmental ministries or departments, establishing clear roles and responsibilities, and fostering interagency collaboration. Institutional reforms should also promote transparency, accountability, and the rule of law.

3. **Public participation and stakeholder engagement:** Engaging the public and relevant stakeholders in decision-making processes improves the legitimacy and effectiveness of environmental governance. Governments should create platforms for dialogue, consultative processes, and participatory mechanisms that allow diverse perspectives to be considered. This can foster collective ownership and ensure decisions reflect the priorities and needs of different stakeholders.

4. **Capacity building and education:** Investing in capacity building programs for environmental professionals and raising awareness among the public are critical for effective environmental governance. Training programs, workshops, and knowledge-sharing platforms can enhance technical expertise, build skills, and promote sustainable behaviors. Education systems should also incorporate environmental education at all levels.

5. **Incentives and economic instruments:** Governments can use a mix of incentives and economic instruments to promote sustainable practices. This includes offering tax incentives, subsidies, and grants for businesses and individuals adopting environmentally friendly practices. Economic instruments like carbon pricing or pollution charges can discourage harmful activities and promote sustainable alternatives.

6. **Collaboration and partnership:** Collaboration among governments, civil society organizations, businesses, and academia can strengthen environmental governance efforts. Partnerships can facilitate knowledge sharing, resource mobilization, and innovation. Collaborative initiatives can address complex environmental challenges that require collective action beyond national borders.

7. **Monitoring and evaluation:** Implementing robust monitoring systems and evaluating the effectiveness of environmental policies are crucial for adaptive management. Governments should invest in data collection, monitoring tools, and evaluation frameworks to track progress, identify gaps, and inform policy decisions. Regular evaluation can enable policy learning and improve the efficiency of environmental governance.

Case study: National level governance in Singapore

One example of effective national level governance for environmental sustainability is Singapore. Despite limited land and natural resources, Singapore has achieved remarkable progress in environmental management and sustainable development.

Key initiatives and strategies:

+ **Environmental legislation and regulations:** Singapore has implemented strict environmental regulations and laws to address air and water pollution, waste management, and biodiversity conservation. These regulations are effectively enforced, ensuring compliance and environmental protection.

+ **Institutional framework:** Singapore has established specialized agencies, such as the National Environment Agency and the Public Utilities Board, to manage environmental issues. These agencies work collaboratively and coordinate efforts to achieve environmental objectives.

+ **Public participation:** Singapore actively engages citizens through various platforms, including public consultations, feedback channels, and educational programs. This enables public involvement in decision-making processes and fosters a sense of ownership.

+ **Capacity building:** Singapore invests heavily in training programs and research institutions to develop skills and expertise for environmental management. The emphasis on education and awareness ensures that the population is well-informed about sustainable practices.

+ **Incentives and economic instruments:** Singapore employs economic instruments to promote sustainable behaviors. For example, it imposes a congestion charge to reduce traffic congestion and encourages the use of public transportation. The country also offers grants and incentives for businesses adopting green practices.

+ **Collaboration and partnership:** Singapore actively collaborates with international organizations, academic institutions, and businesses to address global environmental challenges. The country participates in initiatives like the United Nations Environment Programme and engages in joint research projects.

+ **Monitoring and evaluation:** Singapore has a robust monitoring system that tracks environmental indicators and measures progress. Regular evaluation of policies and programs enables evidence-based decision-making and drives continuous improvement.

Outcomes and lessons learned: Singapore's national level governance efforts have yielded several positive outcomes:

+ **Clean and livable environment:** Singapore is known for its clean and green environment, with well-managed parks, efficient waste management, and high air and water quality. The city-state has transformed its once-polluted Singapore River into a clean waterway.

+ **Sustainable urban development:** Despite its rapid urbanization, Singapore has achieved sustainable urban planning and green building design. The integration of nature and green spaces in the urban landscape has enhanced the quality of life for its citizens.

+ **Water and energy security:** Singapore has implemented innovative strategies to achieve water and energy security. The country invests in water recycling, desalination, and rainwater harvesting to meet its water needs. It also promotes energy efficiency and renewable energy to reduce reliance on fossil fuels.

* **Biodiversity conservation:** Singapore's national parks, nature reserves, and wildlife habitats showcase successful biodiversity conservation efforts. The country actively protects its native flora and fauna, mitigating the impact of urbanization on biodiversity.

The case of Singapore demonstrates that effective national level governance, supported by strong leadership, strategic planning, and stakeholder engagement, can result in significant environmental and sustainability outcomes.

Conclusion

National level governance plays a pivotal role in environmental protection, natural resource management, and sustainable development. By establishing and implementing effective environmental policies, national governments can address environmental challenges and promote sustainable practices. Key components of national level governance include legislation and regulations, institutional frameworks, monitoring and enforcement mechanisms, public participation, capacity building, science-policy interface, and financial mechanisms.

However, national level governance faces several challenges, including coordination and integration, resource constraints, political will, capacity gaps, enforcement and compliance issues, and the need for equity and social justice. Overcoming these challenges requires strategies such as integrated policy frameworks, strengthening institutions, promoting public participation, capacity building, using incentives and economic instruments, fostering collaboration and partnership, and implementing robust monitoring and evaluation systems.

The case study of Singapore provides valuable insights into effective national level governance, demonstrating how a small nation can achieve significant environmental and sustainability outcomes through strategic planning, innovative policies, and stakeholder engagement. By learning from such examples and implementing relevant strategies, national governments can work towards a more sustainable future.

Overall, effective national level governance is essential for promoting environmental sustainability, ensuring the well-being of current and future generations, and achieving global sustainability goals.

Regional and global governance

Regional and global governance plays a crucial role in addressing the challenges of sustainability and promoting eco-science advancements. It involves the coordination, cooperation, and regulation of actions at the regional and global levels to achieve common goals and ensure the effective management of global resources. In this section, we will explore the key aspects of regional and global governance, including its importance, principles, structures, and challenges.

Importance of regional and global governance

Regional and global governance is essential for several reasons. Firstly, it enables countries to work together to address transboundary environmental issues that cannot be effectively managed by individual nations alone. Issues such as climate change, pollution, and biodiversity loss require collective action and cooperation among nations to achieve meaningful results. Regional and global governance frameworks provide the necessary platforms for dialogue, negotiation, and the formulation of shared strategies and policies.

Secondly, regional and global governance ensures the equitable distribution of resources and benefits. Many environmental challenges, such as access to clean water, energy, and biodiversity conservation, have uneven impacts on different regions and countries. Effective governance mechanisms can help ensure that these resources are managed and used in a fair and sustainable manner, taking into account the needs and priorities of all stakeholders.

Furthermore, regional and global governance promotes knowledge sharing and capacity building. It allows countries to learn from each other's experiences, best practices, and innovative solutions. By fostering collaboration and sharing of expertise, regional and global governance can accelerate the development and implementation of sustainable technologies, policies, and practices.

Principles of regional and global governance

Several principles underpin regional and global governance in the context of sustainability. These principles guide the actions and decision-making processes of international organizations, governments, and stakeholders involved in regional and global governance initiatives. Here are some key principles:

1. Multilateralism: The principle of multilateralism emphasizes the involvement of multiple actors, including governments, civil society organizations, and the private sector, in decision-making processes. It recognizes that addressing

complex environmental challenges requires the collective efforts of diverse stakeholders.

2. Participation and inclusiveness: Regional and global governance should be inclusive, ensuring the meaningful participation of all relevant stakeholders, including marginalized groups, indigenous communities, and local communities who are directly affected by environmental issues. Inclusiveness helps ensure that decisions reflect a diversity of perspectives and interests.

3. Subsidiarity: The principle of subsidiarity suggests that decisions should be taken at the lowest appropriate level, considering local context and capabilities. Regional and global governance frameworks should support and strengthen the capacities of national and local governments to address environmental challenges effectively.

4. Precautionary principle: The precautionary principle states that in the face of uncertainty, action should be taken to prevent or minimize potential harm to the environment or human health. It emphasizes the need for proactive measures to address emerging environmental threats, even in the absence of scientific certainty.

5. Integration and coherence: Regional and global governance initiatives should promote the integration and coherence of policies and actions across sectors and scales. This requires aligning environmental goals with social and economic objectives, as well as integrating environmental considerations into decision-making processes in other policy areas.

6. Equity and justice: Equity and justice are fundamental principles in regional and global governance. Environmental policies and actions should take into account the needs and interests of present and future generations, ensuring the fair distribution of costs and benefits and addressing social inequities.

Structures and mechanisms of regional and global governance

Regional and global governance structures and mechanisms vary depending on the specific issue being addressed and the organizations involved. Here, we will explore some key structures and mechanisms that contribute to effective regional and global governance:

1. United Nations Environment Programme (UNEP): UNEP plays a central role in coordinating global environmental policies and programs. It provides a platform for international cooperation and supports countries in the implementation of global environmental agreements, such as the Paris Agreement and the Convention on Biological Diversity.

2. Regional environmental agreements: Regional organizations, such as the European Union and the African Union, have established regional environmental

agreements to address specific environmental challenges within their respective regions. These agreements provide frameworks for cooperation, information sharing, and the development of regional policies and strategies.

3. Intergovernmental panels and forums: Intergovernmental panels and forums, such as the Intergovernmental Panel on Climate Change (IPCC) and the United Nations Framework Convention on Climate Change (UNFCCC), bring together scientists, policymakers, and stakeholders to assess scientific knowledge, negotiate agreements, and promote international cooperation.

4. Global environmental funds: Global environmental funds, such as the Global Environment Facility (GEF), provide financial resources to support environmental projects and initiatives in developing countries. These funds play a crucial role in promoting sustainable development and addressing global environmental challenges.

5. Public-private partnerships: Public-private partnerships bring together governments, businesses, and civil society organizations to collaborate on sustainability projects and initiatives. These partnerships leverage the strengths and resources of different stakeholders, driving innovation, and promoting sustainable practices.

Challenges and future directions

Despite the progress made in regional and global governance, several challenges persist. These challenges must be addressed to ensure the effectiveness and sustainability of regional and global governance initiatives:

1. Implementation gaps: One of the significant challenges is the gap between policy formulation and implementation. Many international agreements and frameworks lack the necessary enforcement mechanisms, leading to inadequate implementation at the national level. Bridging this implementation gap requires strengthening monitoring and accountability systems and promoting knowledge sharing among nations.

2. Coordination and coherence: Coordinating diverse actors and policies across different sectors and scales is a complex and ongoing challenge. Lack of coherence and coordination among various initiatives can lead to duplication of efforts, conflicting policies, and limited progress. Enhancing coordination and coherence requires increased collaboration and information sharing among stakeholders.

3. Resource constraints: Limited financial and technical resources pose significant challenges to regional and global governance. Developing countries, in particular, face resource constraints in implementing sustainable practices and

meeting their environmental commitments. Addressing these resource constraints requires increased financial assistance, technology transfer, and capacity building support from developed countries.

4.   Inclusivity and equity: Ensuring the meaningful participation of marginalized groups, indigenous communities, and local communities in regional and global governance processes remains a challenge.   Their perspectives and knowledge are vital for effective decision-making and the development of sustainable solutions.   Efforts should be made to promote inclusivity and mainstream indigenous knowledge and practices in governance frameworks.

In the future, regional and global governance must adapt and evolve to address emerging environmental challenges.   Enhancing the integration of sustainability principles into economic and development policies, promoting sustainable consumption and production patterns, and fostering global collaboration will be key priorities.   Additionally, leveraging emerging technologies, such as artificial intelligence and blockchain, can create innovative solutions to enhance governance processes and achieve sustainability goals.

In conclusion, regional and global governance is crucial for addressing environmental challenges and advancing sustainability.   It plays a key role in facilitating international cooperation, ensuring resource equity and sharing, and promoting knowledge exchange.   By addressing challenges and embracing future directions, regional and global governance can contribute to a more sustainable and resilient future.

Indigenous knowledge in governance

Indigenous knowledge refers to the unique knowledge, practices, and worldviews developed and passed down through generations by indigenous peoples.   It encompasses a deep understanding of the natural environment, cultural traditions, and community organization.   Indigenous knowledge plays a crucial role in governance, offering valuable perspectives and solutions for sustainable development.   This section explores the importance of indigenous knowledge in governance and its integration into environmental policies and decision-making processes.

Recognition and preservation of indigenous knowledge

The recognition and preservation of indigenous knowledge is vital for sustainable governance.   Indigenous peoples possess a wealth of knowledge about their ecosystems, including traditional resource management systems, sustainable agricultural practices, and medicinal plant use. This knowledge is often based on a holistic understanding of nature and interconnectedness with the environment. Governments and policymakers should acknowledge the value of indigenous knowledge and take steps to protect and preserve it.

One challenge in preserving indigenous knowledge is the issue of intellectual property rights. Indigenous knowledge is often held collectively and passed down orally, making it difficult to attribute ownership or protect it from exploitation. Governments should develop mechanisms to recognize and protect the intellectual property rights of indigenous communities, ensuring that their knowledge is not misappropriated or used without their consent.

Integrating indigenous knowledge in policy-making

Integrating indigenous knowledge into policy-making processes can lead to more effective and sustainable governance. Indigenous communities have developed sustainable practices that promote the conservation of natural resources and biodiversity. By incorporating indigenous knowledge into environmental policies, governments can tap into this rich source of expertise and ensure more holistic and culturally appropriate approaches to sustainable development.

One way to integrate indigenous knowledge is through the establishment of mechanisms for meaningful engagement and consultation with indigenous communities. This can involve creating platforms for indigenous representatives to participate in decision-making processes, providing financial and technical support for indigenous-led initiatives, and incorporating traditional ecological knowledge into scientific research and monitoring.

Participatory governance and indigenous self-determination

Indigenous knowledge in governance is closely linked to the concept of participatory governance, which emphasizes the inclusion of diverse voices and the active participation of all stakeholders in decision-making processes. Recognizing the rights of indigenous peoples to self-determination and the governance of their territories is essential for incorporating indigenous knowledge into governance structures.

Participatory governance requires governments to engage in genuine and respectful dialogue with indigenous peoples, ensuring their meaningful participation in the development and implementation of policies and programs that affect their lands and resources. This can include the establishment of co-management agreements, where indigenous communities have active roles in the management of protected areas or the development of land-use plans.

Challenges and opportunities

Despite the recognition of the importance of indigenous knowledge in governance, there are challenges to its effective integration. One challenge is the power asymmetry between indigenous communities and governments, which can hinder indigenous participation and decision-making authority. Addressing this power imbalance requires creating spaces for indigenous voices to be heard and actively addressing historical injustices.

Another challenge is the potential for cultural appropriation or commodification of indigenous knowledge. Governments and researchers must approach indigenous knowledge with respect, seeking consent and engaging in mutually beneficial partnerships that prioritize indigenous control and benefit sharing.

While there are challenges, there are also opportunities for the integration of indigenous knowledge in governance. Indigenous communities can provide valuable insights into sustainable resource management, climate change adaptation, and biodiversity conservation. Building on these strengths, governments can work collaboratively with indigenous communities to develop innovative, culturally appropriate, and sustainable solutions to environmental challenges.

Example: Indigenous fire management

An example of the valuable contribution of indigenous knowledge in governance is indigenous fire management practices. Many indigenous communities have developed sophisticated and sustainable methods of using fire to manage landscapes, reduce the risk of wildfires, and promote ecosystem resilience.

Indigenous fire management involves controlled burning, where fires are intentionally set at specific times and locations to achieve specific ecological objectives. This practice helps prevent the buildup of fuel, supports the regeneration of fire-adapted plant species, and promotes biodiversity.

Integrating indigenous fire management practices into mainstream governance can lead to more effective wildfire prevention and management strategies. By recognizing and supporting indigenous fire practitioners, governments can tap into their knowledge and experience to develop collaborative fire management plans that benefit both ecosystems and communities.

Conclusion

Incorporating indigenous knowledge in governance is essential for sustainable development.	Indigenous communities possess valuable knowledge and perspectives that can contribute to more holistic and culturally appropriate approaches to environmental policies and decision-making processes. Recognizing and preserving indigenous knowledge, integrating it into policy-making, promoting participatory governance, and addressing challenges are vital steps in harnessing the potential of indigenous knowledge for sustainable governance. By embracing indigenous knowledge, governments and societies can work towards a more inclusive, just, and sustainable future.

Corporate Sustainability

Corporate Social Responsibility

Corporate Social Responsibility (CSR) refers to the ethical and responsible behavior of businesses towards society and the environment. It is a concept that emphasizes that companies have a moral obligation to go beyond their financial performance and contribute to the well-being of society. In recent years, CSR has become an important aspect for businesses in order to maintain their social license to operate, attract customers and investors, and build a positive brand reputation.

Principles of Corporate Social Responsibility

There are several key principles that guide corporate social responsibility initiatives:

1. **Ethics and Transparency**: Companies should conduct their business in an ethical and transparent manner.	This includes being honest, fair, and accountable in their operations, as well as disclosing relevant information to stakeholders.

2. **Respect for Stakeholders**: Companies should consider the interests and concerns of all stakeholders, including employees, customers, suppliers, communities, and shareholders. This involves engaging with stakeholders and actively addressing their needs and expectations.

3. **Sustainable Practices**: Companies should adopt sustainable practices that minimize their negative impact on the environment and promote the efficient use of resources. This includes reducing greenhouse gas emissions, conserving energy and water, and managing waste responsibly.

4. **Social Impact:** Companies should actively contribute to the social development and well-being of the communities in which they operate. This can be done through philanthropic activities, community engagement programs, and support for education, healthcare, and other social initiatives.

Benefits of Corporate Social Responsibility

Implementing CSR initiatives can bring numerous benefits to companies:

+ **Enhanced Reputation:** CSR helps build a positive image for companies, attracting customers who prefer to support socially responsible businesses. It also improves brand reputation and can serve as a competitive advantage.

+ **Increased Employee Engagement:** CSR initiatives can boost employee morale and engagement by providing opportunities for them to contribute to meaningful causes. This leads to higher job satisfaction and increased productivity.

+ **Risk Mitigation:** By adopting sustainability practices and engaging with stakeholders, companies can avoid potential negative impacts on their reputation, such as environmental accidents or labor disputes.

+ **Access to Capital:** Many investors are now focusing on environmental, social, and governance (ESG) factors when making investment decisions. Companies with strong CSR programs are more likely to attract sustainable investment capital.

+ **Long-Term Business Viability:** CSR contributes to the long-term success of companies by fostering positive relationships with stakeholders, ensuring regulatory compliance, and addressing societal challenges that could impact business operations.

+ **Innovation and Differentiation:** CSR can drive innovation by encouraging companies to explore sustainable solutions and develop eco-friendly products or services. This can lead to market differentiation and new business opportunities.

Challenges and Limitations of Corporate Social Responsibility

While CSR offers many benefits, it also faces certain challenges and limitations:

+ **Greenwashing:** Some companies engage in greenwashing, which involves making false or exaggerated claims about their environmental or social performance. This undermines the credibility of CSR initiatives and may mislead consumers and investors.

+ **Complex Supply Chains:** For companies with complex supply chains, ensuring responsible practices across all stages can be challenging. It requires close collaboration with suppliers and monitoring compliance with sustainability standards.

+ **Resource Constraints:** Implementing CSR programs can be costly, especially for small and medium-sized enterprises. Limited resources may hinder their ability to invest in sustainability practices or philanthropic activities.

+ **Measuring Impact:** Assessing the impact of CSR initiatives and quantifying their benefits can be challenging. It requires effective monitoring, evaluation, and reporting methods to demonstrate the positive outcomes of such programs.

Case Study: Patagonia

Patagonia, an outdoor apparel company, is often cited as a leading example of corporate social responsibility. The company has made sustainability a core part of its business strategy and has implemented various initiatives to minimize its environmental impact. Patagonia is known for its commitment to transparency, ethical sourcing, and corporate activism.

One of Patagonia's notable initiatives is its Worn Wear program, which encourages customers to repair and reuse their clothing instead of buying new products. This approach promotes circular economy principles and reduces the environmental footprint of the apparel industry. Patagonia also donates a portion of its sales to environmental causes and engages in advocacy for policy changes that support sustainability.

By integrating CSR into its business model, Patagonia has built a loyal customer base and a strong brand reputation. The company's success demonstrates that a commitment to social and environmental responsibility can align with financial success.

Key Takeaways

Corporate social responsibility is a concept that emphasizes the ethical behavior and responsible practices of businesses towards society and the environment. It is guided by principles of ethics, transparency, stakeholder respect, and sustainability. Implementing CSR initiatives can bring benefits such as enhanced reputation, increased employee engagement, risk mitigation, access to capital, and long-term business viability. However, challenges such as greenwashing, complex supply chains, resource constraints, and measuring impact need to be addressed. Companies like Patagonia serve as role models in effectively integrating CSR into their business strategies.

Exercise: Research and discuss another company that is recognized for its corporate social responsibility efforts. What specific initiatives does the company undertake and how does it contribute to sustainable development?

Sustainable Business Practices

Sustainable business practices play a crucial role in the transition towards a more environmentally friendly and socially responsible economy. Companies that embrace sustainability not only contribute to the preservation of our planet and society but also enjoy long-term benefits such as cost savings, reputation enhancement, and increased customer loyalty. In this section, we will explore the principles and strategies behind sustainable business practices, as well as their positive impacts on both business and the environment.

Principles of Sustainable Business

Sustainable business practices are guided by a set of principles that aim to align economic growth with social and environmental well-being. These principles include:

1. **Triple Bottom Line (TBL):** The TBL framework considers three main dimensions of sustainable business: profit, people, and the planet. It emphasizes the need for businesses to measure and report their impact in economic, social, and environmental terms.

2. **Life Cycle Thinking (LCT):** LCT involves assessing the environmental and social impacts of a product or service throughout its entire life cycle, from raw material extraction to disposal. By considering the full life cycle, businesses can identify opportunities for improvement and reduce their overall environmental footprint.

3. **Circular Economy:** The circular economy is an alternative to the traditional linear economy, where resources are used and then discarded. In a circular

economy, materials flow in a continuous loop, minimizing waste generation and maximizing resource efficiency. This principle encourages businesses to design products for durability, repairability, and recyclability.

4. **Stakeholder Engagement:** Businesses should engage with stakeholders, including employees, customers, suppliers, local communities, and NGOs, to understand their concerns and integrate their perspectives into decision-making processes. By involving stakeholders, companies can ensure that their actions align with societal expectations and build mutual trust.

5. **Transparency and Accountability:** Sustainable businesses prioritize transparency by openly sharing their goals, strategies, and performance with stakeholders. They also hold themselves accountable by implementing robust monitoring and reporting systems to track their progress towards sustainability targets.

Strategies for Sustainable Business Practices

Implementing sustainable business practices requires a comprehensive approach that addresses various aspects of operations, including procurement, production, marketing, and supply chain management. Here are some key strategies that businesses can adopt:

1. **Energy Efficiency:** Improving energy efficiency is a fundamental strategy for reducing greenhouse gas emissions and operational costs. Businesses can conduct energy audits, invest in energy-efficient technologies, and promote energy conservation practices among employees.

2. **Renewable Energy:** Transitioning to renewable energy sources, such as solar or wind power, is an effective way to decarbonize operations and reduce reliance on fossil fuels. Businesses can install their renewable energy systems or purchase renewable energy credits to support the development of clean energy infrastructure.

3. **Waste Reduction and Recycling:** Implementing waste management programs that prioritize waste reduction, recycling, and composting can help businesses minimize their environmental impact. Adopting circular economy principles can also encourage the reuse and recycling of materials within the production cycle.

4. **Supply Chain Sustainability**: Businesses can drive sustainability by collaborating with suppliers who share their values. This can involve assessing suppliers' environmental and social performance, supporting fair trade practices, and promoting responsible sourcing of raw materials.

5. **Product Design and Packaging**: Sustainable businesses prioritize product design that minimizes environmental impact, such as using eco-friendly materials, reducing packaging waste, and ensuring products can be easily repaired or recycled. They also explore innovative packaging solutions, such as biodegradable or reusable packaging.

6. **Employee Engagement**: Engaging employees in sustainability initiatives can foster a culture of responsibility and innovation. Businesses can promote sustainable practices among employees through training programs, incentivizing sustainable behavior, and involving employees in decision-making processes.

7. **Ethical Marketing**: Sustainable businesses communicate their commitment to sustainability accurately and honestly. Greenwashing, the practice of misleadingly marketing products or services as environmentally friendly, should be avoided. Transparent and genuine communication builds trust with consumers.

8. **Social Responsibility**: Businesses can actively contribute to social causes by supporting community development projects, promoting diversity and inclusion, investing in employee well-being, and ensuring ethical labor standards throughout their supply chains.

Benefits and Challenges

Embracing sustainable business practices offers numerous benefits for both businesses and the environment. Some key advantages include:

+ **Cost Savings**: Adopting sustainable practices can lead to cost savings through energy efficiency improvements, waste reduction, and process optimization.

+ **Competitive Advantage**: Businesses that demonstrate their commitment to sustainability can differentiate themselves in the market, attract environmentally conscious customers, and gain a competitive edge.

+ **Risk Mitigation:** Sustainable practices can help businesses mitigate risks associated with climate change, resource scarcity, and regulatory changes by building resilience and adapting to new market conditions.

+ **Reputation Enhancement:** Businesses that prioritize sustainability can enhance their reputation among stakeholders, including customers, employees, investors, and communities. This positive reputation can lead to increased brand loyalty and trust.

+ **Innovation and Business Opportunities:** Embracing sustainability often drives innovation, leading to the development of new products, services, and business models that meet evolving market demands.

However, there are also challenges associated with implementing sustainable business practices. Some common challenges include:

+ **Initial Investment:** Adopting sustainable practices may require upfront investments in technologies, infrastructure, and employee training. Overcoming financial barriers can be a challenge, especially for small and medium-sized enterprises.

+ **Complexity of Supply Chains:** Managing sustainability across complex global supply chains can be challenging. Ensuring that suppliers comply with sustainability standards and maintaining transparency throughout the supply chain can present difficulties.

+ **Behavioral Change:** Shifting cultural norms and behaviors within an organization can be a significant challenge. Creating buy-in from employees and stakeholders and fostering a sustainability-focused culture requires effective communication and change management.

+ **Regulatory Uncertainty:** Keeping up with evolving environmental regulations and policies can be challenging for businesses. Adapting to new regulatory frameworks and ensuring compliance can require additional resources and expertise.

Despite these challenges, the benefits of embracing sustainable business practices far outweigh the costs. By integrating sustainability into their core strategies, businesses can contribute to a more sustainable future while simultaneously creating value for their stakeholders and the environment.

Case Study: Patagonia

An excellent example of a company that exemplifies sustainable business practices is Patagonia, the outdoor clothing and gear company. Patagonia has consistently set the bar high by integrating sustainable principles at every level of its operations.

One of Patagonia's notable initiatives is its Common Threads Initiative, which encourages customers to reduce consumption by repairing, reusing, reselling, and recycling garments. Patagonia provides repair services, offers store credit for used products, and promotes the resale of pre-owned items on its website.

Additionally, Patagonia has made significant strides in supply chain transparency and fair labor practices. The company actively works towards ensuring fair wages, safe working conditions, and social welfare for workers in its global supply chain. Patagonia also invests in sustainable materials and technologies to minimize its environmental impact.

Through these sustainable initiatives, Patagonia has not only gained a loyal customer base but has also inspired other companies to follow suit. The company's success demonstrates that sustainable business practices can be both socially responsible and economically viable.

Summary

Sustainable business practices are essential for achieving long-term success while promoting environmental and social well-being. By adopting principles such as the Triple Bottom Line and Circular Economy, businesses can drive positive change and contribute to a more sustainable future.

Strategies like energy efficiency, waste reduction, supply chain sustainability, and employee engagement play a crucial role in implementing sustainable practices. Although businesses may face challenges during the transition, the benefits, including cost savings, competitive advantage, and reputation enhancement, make the effort worthwhile.

As demonstrated by companies like Patagonia, sustainable business practices can create a positive impact on both the environment and society. By embracing sustainability, businesses become key players in fostering a more sustainable and equitable world.

Green Supply Chain Management

Green supply chain management (GSCM) is an approach that integrates environmental concerns into every stage of the supply chain, from raw material extraction to final product disposal. It aims to minimize the environmental impact of the supply chain while creating economic value. GSCM recognizes that the traditional linear supply chain model, which focuses solely on economic efficiency, is no longer sustainable in the face of increasing environmental challenges.

1. Need for Green Supply Chain Management

Traditional supply chains often result in negative environmental impacts such as resource depletion, waste generation, and pollution. These impacts pose risks to the long-term viability of businesses, as well as to the health and well-being of communities and ecosystems. The need for GSCM arises from the urgent need to address these environmental challenges and create a more sustainable future.

2. Principles of Green Supply Chain Management

GSCM is guided by several key principles:

2.1 *Life Cycle Thinking*: GSCM takes into account the entire life cycle of a product, from design and production to use and disposal. This holistic approach helps identify environmental hotspots and drives innovation to reduce environmental impacts.

2.2 *Reverse Logistics*: GSCM emphasizes the efficient management of product returns, recycling, and remanufacturing. By optimizing reverse logistics processes, companies can minimize waste, cut costs, and recover valuable resources.

2.3 *Collaboration and Cooperation*: GSCM requires collaboration among supply chain partners, including suppliers, manufacturers, distributors, and customers. Effective collaboration enables the sharing of information, resources, and best practices, leading to improved environmental performance across the supply chain.

2.4 *Product and Process Innovation*: GSCM encourages the development of environmentally friendly products and processes. This includes the use of renewable materials, eco-design principles, energy-efficient manufacturing, and the adoption of clean technologies.

2.5 *Supplier Engagement*: GSCM involves working closely with suppliers to ensure they adhere to environmental standards and practices. This may involve conducting audits, providing training and support, and incentivizing sustainable practices.

3. Benefits of Green Supply Chain Management

3.1 *Environmental Benefits*: GSCM reduces greenhouse gas emissions, pollution, and waste generation, leading to improved air and water quality, conservation of natural resources, and protection of biodiversity.

3.2 *Cost Savings*: GSCM can result in significant cost savings through improved energy and resource efficiency, waste reduction, and operational optimization. It also helps companies avoid fines and penalties for non-compliance with environmental regulations.

3.3 *Reputation and Market Advantage*: Adopting GSCM practices enhances brand reputation and can attract environmentally conscious customers. It also helps companies differentiate themselves in the market and gain a competitive edge.

3.4 *Stakeholder Engagement*: GSCM promotes transparency and fosters positive relationships with stakeholders such as investors, consumers, communities, and regulatory agencies. This engagement can lead to increased trust and long-term partnerships.

4. Challenges and Solutions

4.1 *Lack of Awareness and Commitment*: One of the primary challenges in implementing GSCM is the lack of awareness and commitment from businesses. To address this, companies should prioritize sustainability as part of their corporate strategy and educate employees about the importance of GSCM.

4.2 *Complexity of Supply Chains*: Global supply chains are often complex, involving numerous stakeholders and spanning multiple countries. This complexity makes it challenging to ensure environmental standards are met throughout the supply chain. Collaboration, transparency, and supplier engagement are key solutions to address this challenge.

4.3 *Measurement and Evaluation*: Measuring and evaluating the environmental performance of the supply chain can be complex. Companies should develop key performance indicators (KPIs) and leverage technology, such as supply chain management software and data analytics, to track and monitor their environmental impact.

4.4 *Regulatory Compliance*: Compliance with environmental regulations can be challenging, especially for companies operating in different jurisdictions. Businesses should stay abreast of relevant regulations and standards, establish robust compliance systems, and seek guidance from experts when needed.

4.5 *Financial Constraints*: Implementing GSCM practices may require upfront investments, which can be a barrier for small and medium-sized enterprises (SMEs). However, various financial incentives, grants, and loans are available to support sustainable initiatives. Collaboration with larger companies can also help SMEs overcome financial constraints.

5. Case Study: Unilever's Sustainable Living Plan

Unilever, a multinational consumer goods company, has implemented a comprehensive GSCM strategy known as the Sustainable Living Plan. This plan integrates sustainability into every aspect of the company's operations and supply chain. Unilever aims to decouple its environmental impact from business growth, reduce its carbon footprint, enhance water stewardship, and improve waste management. Through collaboration with suppliers, innovation, and consumer education, Unilever has demonstrated the feasibility and benefits of GSCM.

Summary

Green supply chain management is an essential component of sustainable business practices. By integrating environmental considerations into the entire supply chain, companies can minimize their environmental impact, enhance operational efficiency, and gain a competitive advantage. Despite the challenges, implementing GSCM brings numerous benefits, including cost savings, improved stakeholder relationships, and a positive brand image. As businesses increasingly recognize the importance of sustainability, GSCM will continue to play a crucial role in shaping a more environmentally friendly future.

Life Cycle Assessment

Life Cycle Assessment (LCA) is a comprehensive method used to assess the environmental impacts associated with a product, process, or activity throughout its entire life cycle. It provides a systematic framework for evaluating and quantifying the environmental inputs and outputs of a system, from raw material extraction to end-of-life disposal.

Principles of Life Cycle Assessment

LCA is based on the following principles:

1. **Goal and scope definition:** Clearly defining the objectives and boundaries of the assessment is critical. This includes determining the purpose of the study, the functional unit, system boundaries, and the life cycle stages to be included.

2. **Inventory analysis:** This stage involves compiling a comprehensive inventory of all inputs (e.g., energy, materials, water) and outputs (e.g., emissions, waste) associated with each life cycle stage. Data can be collected from both primary sources (e.g., actual measurements) and secondary sources (e.g., databases).

3. **Impact assessment:** The inventory data is then evaluated to determine the potential environmental impacts associated with each life cycle stage. This step involves categorizing and characterizing the environmental burdens, such as global warming potential, acidification potential, eutrophication potential, and resource depletion.

4. **Interpretation:** The final step of LCA involves analyzing and interpreting the results to draw conclusions and make informed decisions. This includes

identifying key contributors to environmental impacts, evaluating alternative scenarios, and considering uncertainties and limitations.

Application of Life Cycle Assessment

LCA can be applied to various sectors and contexts, including:

1. **Product design and development:** LCA allows designers to proactively evaluate the environmental implications of alternative design choices. By considering environmental impacts early in the design process, more sustainable products can be developed, reducing resource consumption and minimizing environmental burdens.

2. **Supply chain management:** LCA can assess the environmental performance of different suppliers, helping companies make informed decisions about their procurement practices. It can identify hotspots in the supply chain where improvement efforts can be focused to reduce overall environmental impacts.

3. **Policy-making and regulation:** LCA provides a scientific basis for policy-makers to develop effective regulations and standards. It can help governments prioritize environmental issues, set realistic targets, and evaluate the overall environmental performance of different sectors.

4. **Sustainable procurement:** LCA can be used to assess and compare the environmental performance of different products or services before making purchasing decisions. It allows organizations to consider the entire life cycle, including production, use, and disposal, to choose more sustainable options.

5. **Industrial ecology:** LCA is a valuable tool for analyzing the environmental impacts of industrial processes and identifying opportunities for improvement through process optimization, waste reduction, and material recycling.

Challenges and Limitations

While LCA is a powerful tool for assessing environmental impacts, it does have some challenges and limitations:

1. **Data availability and quality:** LCA relies on accurate and up-to-date data, which may be challenging to obtain, especially for complex supply chains or emerging technologies. Data quality also varies among different sources, which can introduce uncertainties and affect the reliability of the assessment.

2. **System boundaries:** Defining the system boundaries of an LCA study can be subjective and influence the results. It is important to clearly define the scope and functional unit to ensure a meaningful and comparable assessment.

3. **Impact characterization:** The translation of inventory data into environmental impacts involves simplifications and assumptions. The choice of impact assessment method and the assignment of characterization factors can influence the results and the interpretation of the assessment.

4. **Allocation and system expansion:** Allocating environmental burdens when multiple products or co-products are involved in a system is a challenge. System expansion methods may be used to avoid the allocation problem by considering the potential displacement of other products or services.

5. **Lack of social and economic aspects:** LCA primarily focuses on the environmental dimensions of sustainability and often neglects social and economic factors. Integrating social life cycle assessment (SLCA) and life cycle costing (LCC) can provide a more comprehensive assessment of sustainability.

Case Study: Life Cycle Assessment of a Solar Panel

To illustrate the application of LCA, let's consider a case study on the life cycle assessment of a solar panel. The goal of the study is to evaluate the environmental impacts associated with the production, use, and end-of-life management of the solar panel.

The LCA study would involve collecting data on the energy and material inputs during the manufacturing process, the use-phase energy generation, and the eventual disposal or recycling of the panel. Emissions and waste associated with each life cycle stage would also be accounted for.

The inventory analysis would quantify the resource consumption, emissions, and waste generation. Then, an impact assessment would be conducted to characterize the environmental burdens, such as greenhouse gas emissions, water consumption, and resource depletion. The results could be compared with conventional energy sources to assess the environmental benefits of solar energy.

The interpretation phase would involve identifying hotspots in the life cycle and exploring opportunities for improvement. For example, the study may reveal that the manufacturing phase has the highest environmental impact. Potential solutions could include optimizing manufacturing processes, reducing material waste, or increasing the use of renewable energy in production.

By conducting an LCA on the solar panel, manufacturers, policymakers, and consumers can make informed decisions towards more sustainable energy systems.

Conclusion

Life Cycle Assessment is a valuable tool for evaluating the environmental impacts of products, processes, and activities. By providing a holistic perspective, it enables decision-makers to identify opportunities for improvement and make more informed choices towards sustainability. However, it is essential to consider the limitations and challenges associated with LCA to ensure the reliability and relevance of its results. With further advancements in data availability, assessment methodologies, and integration with social and economic aspects, LCA will continue to play a crucial role in promoting sustainable development.

Circular Economy in Business

Circular economy is an innovative and sustainable approach to business that aims to minimize waste and maximize resource efficiency. It is based on the principle of closing the materials loop, where products and materials are reused, recycled, or regenerated, rather than being disposed of as waste. Circular economy strategies can offer numerous benefits to businesses, including cost savings, increased resilience, and reduced environmental impacts.

Principles of Circular Economy

The circular economy is guided by several key principles:

1. **Design for Longevity and Durability:** Products should be designed to last longer, be repairable, and have modular components that can be easily replaced or upgraded. This extends the lifespan of products and reduces the need for new resource extraction.

2. **Optimize Resource Use:** Businesses should strive to use resources efficiently, minimizing waste generation and ensuring that materials are used to their fullest potential. This can be achieved through processes such as remanufacturing, where products are refurbished and repaired to extend their life.

3. **Promote Recycling and Material Recovery:** Businesses should implement strategies to promote the recycling and recovery of materials from products at the end of their life. This can involve setting up take-back systems, establishing partnerships with recycling facilities, and using recycled materials in new product manufacturing.

4. **Utilize Renewable Energy and Resources:** Transitioning to renewable energy sources helps businesses reduce their carbon footprint and dependence on finite resources. Additionally, using renewable resources in product manufacturing promotes sustainability and reduces environmental impacts.

5.   **Foster Collaboration and Innovation**:   Circular economy requires collaboration between businesses, governments, and consumers.   Sharing knowledge, technologies, and best practices is crucial to drive innovation and overcome barriers to circularity.

Business Strategies for Circular Economy

1. **Product Life Extension**: Businesses can implement strategies to extend the lifespan and utility of their products.   For example, offering repair services, providing spare parts, or encouraging customers to return products at the end of their life for refurbishment or recycling.

2. **Closed-Loop Supply Chains**: Creating closed-loop supply chains involves designing products so that their components can be easily disassembled and reused or recycled. By keeping resources within the value chain, businesses can reduce their reliance on virgin materials and minimize waste generation.

3.   **Product-as-a-Service**: Adopting a product-as-a-service model allows businesses to retain ownership of products and offer them as a service to customers.   Customers pay for the use or access to the product, while the responsibility for maintenance, repair, and end-of-life management remains with the business. This encourages product durability and allows for more efficient resource utilization.

4. **Collaborative Consumption**: Collaborative consumption models, such as sharing or rental platforms, enable multiple users to access and utilize products and resources. This reduces the overall demand for new products and encourages the efficient use of existing resources.

5. **Material Recovery and Recycling**: Businesses can implement robust systems for collecting and recycling their own products. This involves designing products for recyclability, establishing collection networks, and partnering with recycling facilities to ensure the proper handling and processing of materials.

Benefits and Challenges

Implementing circular economy strategies in business offers several benefits:

1. **Cost Savings**: By reducing waste generation and optimizing resource use, businesses can achieve significant cost savings. For example, remanufacturing and recycling processes can reduce the need for raw material extraction and minimize production costs.

2. **Resilience and Risk Mitigation:** Circular economy strategies increase a business's resilience by diversifying its resource base and reducing its vulnerability to price fluctuations or disruptions in the supply chain.

3. **Environmental Benefits:** By minimizing waste generation, promoting recycling, and using renewable resources, businesses can significantly reduce their environmental impact, including carbon emissions, water consumption, and pollution.

However, adopting circular economy principles may also pose challenges to businesses:

1. **Technological Barriers:** Implementing circular economy strategies often requires significant investments in new technologies and infrastructure. Businesses may face challenges in identifying suitable technologies and securing investment to support their implementation.

2. **Business Model Innovation:** Circular economy requires a shift in traditional business models, which may be challenging for some organizations. Developing new revenue models, such as product-as-a-service or sharing platforms, requires careful planning and market adaptation.

3. **Regulatory and Policy Support:** To accelerate the transition to a circular economy, businesses need supportive policies and regulations that encourage sustainable practices and provide incentives for innovation. Governments play a crucial role in creating an enabling environment for circular economy adoption.

Case Study: Patagonia

Patagonia, an outdoor clothing and gear company, is a successful example of implementing circular economy principles in business. The company has adopted various strategies to promote sustainability and reduce waste.

1. **Product Durability and Repair:** Patagonia designs its products to withstand the test of time. It encourages customers to repair their products rather than replacing them, offering free repairs for any manufacturing defects.

2. **Worn Wear Program:** Patagonia has established a Worn Wear program, where customers can return their used Patagonia products for repair, resale, or recycling. The program aims to keep products in use for as long as possible and reduce the need for new purchases.

3. **Recycled Materials:** Patagonia prioritizes the use of recycled materials in its products, including recycled polyester made from plastic bottles. This reduces the demand for virgin resources and encourages recycling.

4. **Transparency and Education:** Patagonia is actively engaged in raising awareness about the environmental impact of the fashion industry and the benefits

of a circular economy. The company provides information on its website about its sustainability initiatives and encourages customers to make informed choices.

The success of Patagonia's circular economy strategies demonstrates that businesses can thrive while prioritizing sustainability and environmental responsibility.

Conclusion

The circular economy presents significant opportunities for businesses to reduce waste, increase resource efficiency, and minimize environmental impacts. By embracing circularity, businesses can achieve cost savings, enhance their resilience, and contribute to a more sustainable future. However, the transition to a circular economy requires collaboration, innovation, and supportive policies. Through the adoption of circular economy principles, businesses can play a vital role in driving the transformation towards a more sustainable and regenerative economic model.

Sustainable finance and investment

Sustainable finance and investment play a crucial role in promoting environmental conservation and social responsibility. This section explores the principles, strategies, and challenges associated with sustainable finance and investment. It examines how financial institutions and investors can align their activities with environmental and social goals, thereby contributing to long-term sustainability.

Principles of Sustainable Finance

Sustainable finance operates based on several key principles that integrate environmental, social, and governance (ESG) factors into investment decisions. These principles guide investors and financial institutions towards making responsible and sustainable choices. The following principles are commonly applied in sustainable finance:

1. **Integration of ESG factors:** This principle emphasizes the incorporation of ESG considerations into investment analysis and decision-making processes. Investors evaluate the environmental, social, and governance performance of companies before making investment choices.

2. **Risk management:** Sustainable finance aims to identify and manage financial risks associated with ESG factors. This involves assessing how environmental and social issues can impact the long-term financial performance of an investment.

3. **Stakeholder engagement:** Encouraging active engagement with stakeholders, including communities, customers, and employees, is crucial in sustainable finance. By involving stakeholders in decision-making processes, financial institutions can ensure transparency and accountability.

4. **Impact measurement and reporting:** Measuring and reporting the positive environmental and social impacts of sustainable investments is essential. This principle promotes transparency and accountability, allowing investors to track the progress and effectiveness of their sustainable investment strategies.

5. **Continuous improvement:** Sustainable finance requires a commitment to continuous improvement. Financial institutions and investors should strive to enhance their understanding of ESG issues and adopt more sustainable practices over time.

Investment Strategies in Sustainable Finance

Several investment strategies are employed in sustainable finance to support the achievement of environmental and social objectives. These strategies are designed to deliver financial returns while also generating positive impacts. Here are some commonly used investment strategies:

1. **Socially responsible investing (SRI):** SRI involves selecting investments based on both financial considerations and ESG criteria. Investors screen companies based on their ESG performance, excluding those that engage in activities deemed harmful to society or the environment.

2. **Impact investing:** Impact investing aims to generate measurable positive social and environmental impacts alongside financial returns. Investors actively seek out projects or companies that address specific sustainability challenges, such as renewable energy development or microfinance initiatives.

3. **Green bonds:** Green bonds are fixed-income securities issued to finance projects with environmental benefits. The proceeds from these bonds are dedicated to projects such as renewable energy, energy efficiency, or sustainable infrastructure development.

4. **ESG integration:** ESG integration involves incorporating ESG factors into traditional investment analysis and decision-making processes. Investors

consider ESG risks and opportunities to enhance their understanding of company performance.

5. **Engagement and active ownership:** This strategy involves active engagement with companies through shareholder advocacy to promote more sustainable practices. Shareholders use their voting rights and dialogue with management to push for improved ESG performance.

Challenges in Sustainable Finance and Investment

Despite the growing interest in sustainable finance and investment, several challenges persist in its widespread adoption. These challenges include:

1. **Data availability and quality:** Access to reliable and consistent ESG data is essential for making informed investment decisions. However, data gaps and inconsistencies exist, making it challenging for investors to accurately assess ESG risks and opportunities.

2. **Lack of standardization:** The lack of standardized ESG metrics and reporting frameworks makes it difficult for investors to compare and evaluate performance across companies. Developing globally recognized standards is necessary for meaningful and consistent measurement of sustainability.

3. **Short-termism:** Many financial institutions and investors have a short-term focus, prioritizing immediate financial gains over long-term sustainability. Aligning financial incentives with sustainable outcomes is essential to overcome this challenge.

4. **Complexity of ESG factors:** Assessing and understanding the impact of ESG factors on financial performance can be complex. The integration of ESG factors requires expertise and resources to accurately evaluate and interpret their impact.

5. **Greenwashing:** Greenwashing refers to the practice of falsely presenting a company or investment as environmentally friendly or socially responsible. Robust frameworks and regulations are necessary to prevent greenwashing and ensure the credibility of sustainable finance initiatives.

Case Study: Sustainable Investments in Renewable Energy

An excellent example of sustainable finance and investment is the focus on renewable energy projects. These investments contribute to both environmental sustainability and financial returns. Let's consider a case study on a wind energy project:

Background: A financial institution decides to invest in a wind energy project in a region with significant wind resources. The project aims to generate clean energy and reduce greenhouse gas emissions, contributing to the transition to a low-carbon economy.

Investment Analysis: The financial institution conducts an in-depth analysis of the project, considering various factors such as wind resource assessment, capital costs, operational expenses, and revenue potential. The assessment also includes an evaluation of the environmental and social impacts of the project.

Financial Viability: The financial institution considers the project's financial viability by assessing the expected financial returns from the sale of electricity generated. Factors such as government incentives, power purchase agreements, and long-term revenue stability are evaluated.

Risk Management: The financial institution analyzes the risks associated with the project, including regulatory risks, technological risks, and market risks. This allows them to make informed investment decisions and develop risk mitigation strategies.

Impact Measurement and Reporting: The financial institution sets up a monitoring framework to measure the project's environmental and social impacts. This includes monitoring the amount of clean energy generated and the reduction in greenhouse gas emissions. Regular reporting is done to ensure transparency and accountability.

Long-Term Benefits: The investment in the wind energy project generates both financial returns for the financial institution and significant environmental benefits. It promotes the transition to clean energy, reduces reliance on fossil fuels, and contributes to mitigating climate change.

Resources for Sustainable Finance and Investment

To further explore sustainable finance and investment, here are some resources worth exploring:

- **Principles for Responsible Investment (PRI):** The PRI provides a framework for integrating ESG factors into investment practices. Their website offers a wealth of information and resources on sustainable investment.

+ **Global Sustainable Investment Alliance (GSIA):** The GSIA is a collaboration of sustainable investment organizations worldwide. Their reports provide valuable insights into the current state of sustainable investment globally.

+ **Sustainable Stock Exchanges (SSE) Initiative:** The SSE Initiative promotes sustainable investment and transparency in capital markets. Their website offers resources for investors interested in sustainable finance.

+ **Sustainable Finance Disclosure Regulation (SFDR):** The SFDR is a regulatory framework in Europe aimed at promoting transparency and comparability of sustainable investments. It provides guidelines for investment managers and financial advisors.

Conclusion

Sustainable finance and investment offer a pathway towards aligning financial goals with environmental and social objectives. By integrating ESG factors into investment decision-making and adopting responsible investment strategies, financial institutions and investors can contribute to a more sustainable and resilient future. However, challenges such as data availability, standardization, and short-termism need to be addressed to realize the full potential of sustainable finance. By overcoming these challenges, sustainable finance can drive positive change and help build a more sustainable and inclusive global economy.

Green bonds

Green bonds are a type of financial instrument that are specifically issued to fund environmentally friendly or sustainable projects. They are a form of fixed-income securities, where investors lend money to the issuer in exchange for regular interest payments and the return of the principal amount at maturity. The main difference between green bonds and conventional bonds is that the proceeds from green bonds are exclusively used for projects with positive environmental or climate benefits.

The concept of green bonds emerged as a response to the growing recognition of the need to finance projects that address climate change and promote sustainability. By providing an avenue for investors to support environmentally friendly initiatives, green bonds play a crucial role in mobilizing capital towards sustainable development and the transition to a low-carbon economy.

There are several key features of green bonds that differentiate them from conventional bonds:

1. **Use of proceeds:** Green bonds are specifically issued to finance or refinance projects that have clear environmental benefits. These projects can include renewable energy generation, energy efficiency improvements, sustainable transportation, waste management, and green building construction, among others. The issuer of green bonds is required to provide transparency and accountability regarding the use of proceeds.

2. **Certification and verification:** To ensure the integrity and credibility of green bonds, projects financed by these bonds are typically subject to certification and verification by independent, third-party entities. This process helps to verify that the projects indeed meet defined environmental criteria.

3. **Reporting and transparency:** Issuers of green bonds are expected to provide regular reporting on the environmental impact of the financed projects. This transparency allows investors to assess the environmental benefits associated with their investments and holds issuers accountable for the use of proceeds.

4. **Market development:** The green bond market has experienced significant growth in recent years, driven by increasing investor demand for sustainable investments. The market has witnessed the issuance of green bonds from a diverse range of issuers, including governments, municipalities, corporations, and financial institutions.

Benefits of green bonds

Green bonds offer several benefits for different stakeholders, including investors, issuers, and the environment:

- **Investor demand and diversification:** The demand for green bonds has been rapidly increasing as investors seek to align their investment portfolios with sustainability objectives. Green bonds provide an opportunity for investors to diversify their portfolios and access investments that support environmentally friendly projects.

- **Risk management:** Investing in green bonds can help mitigate climate-related financial risks. By supporting projects that contribute to the transition to a low-carbon economy, investors can reduce exposure to assets that may become stranded or face regulatory and reputational risks due to their environmental impact.

- **Enhanced reputation and brand value:** Issuing green bonds can enhance the reputation and brand value of issuers, demonstrating their commitment to sustainability and responsible business practices. This can help attract socially responsible investors and strengthen relationships with other stakeholders, such as customers and regulators.

- **Environmental impact:** The main benefit of green bonds is their contribution to the financing of environmentally friendly projects. By directing capital towards renewable energy, energy efficiency, and other sustainable initiatives, green bonds play a significant role in addressing climate change, reducing greenhouse gas emissions, and promoting sustainable development.

Challenges and considerations

While green bonds offer numerous benefits, there are also challenges and considerations that need to be addressed:

- **Standardization and harmonization:** The lack of standardized definitions and frameworks for green bonds can lead to inconsistencies in the market. Efforts are being made to develop common principles and guidelines to ensure transparency, credibility, and comparability across green bond offerings.

+ **Greenwashing risks:** Greenwashing refers to the potential for issuers to misrepresent the environmental benefits of projects financed by green bonds. To mitigate greenwashing risks, independent certification and verification processes are essential. Increased transparency and reporting requirements can also help protect investors from misleading claims.

+ **Limited supply of eligible projects:** There is a need for a sufficient supply of eligible projects to meet the growing demand for green bonds. This requires a pipeline of environmentally friendly projects across various sectors and regions. Governments and policymakers play a crucial role in creating an enabling environment for the development of green projects and encouraging private sector investment.

+ **Market liquidity and pricing:** The green bond market is still relatively nascent compared to the broader bond market. As the market matures, enhancing liquidity and price transparency will be important for attracting more investors and achieving competitive pricing.

Case study: Green bonds in the renewable energy sector

An example of the application of green bonds is the financing of renewable energy projects. As the world transitions towards a clean energy future, green bonds have played a significant role in mobilizing capital for renewable energy infrastructure.

For instance, a renewable energy company may issue green bonds to raise funds for the construction of a wind farm. The proceeds from the bond issuance can be used to finance the purchase and installation of wind turbines, transmission infrastructure, and other components of the wind farm. The certification and verification process ensure that the project meets defined environmental criteria, such as the reduction of greenhouse gas emissions and the promotion of sustainable energy generation.

Investors who purchase these green bonds contribute to the development of renewable energy and can expect to receive regular interest payments and the return of the principal amount at maturity. By investing in green bonds, individuals and institutions can support the transition to a low-carbon economy and contribute to mitigating climate change.

Conclusion

Green bonds have emerged as an effective financial instrument for mobilizing capital towards sustainable development and addressing environmental challenges.

By channeling investments into environmentally friendly projects, green bonds facilitate the transition to a low-carbon and sustainable economy. However, to maximize the benefits and ensure the integrity of the green bond market, standardized frameworks, independent verification, and increased transparency are essential. The development of the green bond market presents an opportunity for investors, issuers, and the environment to align their interests and contribute to a more sustainable future.

Socially responsible investing

Socially responsible investing (SRI), also known as sustainable investing or ethical investing, is an investment strategy that considers environmental, social, and governance (ESG) factors in addition to financial returns. SRI aims to generate positive impact and promote sustainable practices by investing in companies that align with certain values and goals.

Background

Traditional investing focuses solely on financial performance and maximizing returns. However, SRI recognizes that investment decisions can have a broader impact on society and the environment. As awareness of environmental issues, social inequality, and corporate governance concerns has grown, so has the demand for investment strategies that align with these values.

SRI incorporates ESG criteria to guide investment decisions, evaluating companies based on their environmental practices, social impact, and commitment to responsible governance. This approach seeks to influence corporate behavior and encourage sustainable practices.

Principles of socially responsible investing

1. Environmental considerations: SRI focuses on investing in companies that demonstrate environmental responsibility and sustainability. This includes evaluating a company's carbon footprint, resource usage, waste management practices, and commitment to renewable energy.

2. Social considerations: SRI takes into account a company's impact on society and human rights. It considers factors such as labor practices, human rights records, employee diversity and equality, community engagement, and product safety.

3. Governance considerations: SRI assesses a company's governance structure, including board independence, executive compensation, transparency, and

accountability. It aims to invest in companies with strong ethical leadership and responsible decision-making processes.

Examples of socially responsible investing strategies

1. Screening: SRI strategies often involve positive and negative screening. Positive screening involves selecting investments based on specific ESG criteria. For example, investors may actively seek companies with strong commitments to renewable energy or community development. Negative screening, on the other hand, excludes investments in companies involved in activities deemed socially or environmentally harmful, such as tobacco, firearms, or fossil fuels.

2. Impact investing: This approach aims to generate positive social and environmental impact alongside financial returns. Impact investors actively seek investment opportunities in companies targeting specific social or environmental objectives, such as clean energy projects, affordable housing, or sustainable agriculture.

3. Shareholder advocacy: SRI investors may engage with companies as shareholders to influence their policies and practices. This can involve voting in favor of resolutions that promote sustainability, engaging in dialogue with company management, and filing shareholder proposals. By actively participating in corporate decision-making, SRI investors can leverage their influence to drive positive change.

Benefits and challenges

Socially responsible investing offers several benefits:

1. Alignment with personal values: SRI allows investors to align their investment decisions with their personal values and beliefs. It provides an opportunity to support companies working towards environmental and social progress.

2. Positive impact: SRI strategies can drive positive change by allocating capital to companies that prioritize sustainability, social responsibility, and good governance. This encourages companies to adopt more responsible practices.

3. Long-term sustainability: Companies focusing on ESG factors are often better positioned for long-term success. Addressing environmental and social risks can help mitigate potential financial and reputational damage, leading to more resilient and sustainable businesses.

However, socially responsible investing also faces certain challenges:

1.   Performance trade-off: Critics argue that SRI strategies may sacrifice financial returns in favor of ethical considerations.   While academic research suggests that SRI does not necessarily result in underperformance, investors need to carefully evaluate potential trade-offs.

2. Lack of standardized metrics: ESG criteria can be subjective, and there is no universally accepted framework for evaluating companies' performance. This poses challenges in comparing investments and measuring the impact of SRI strategies.

3. Greenwashing: Some companies may falsely promote their ESG credentials to appear socially responsible without implementing substantial changes. Investors must exercise due diligence and carefully assess the authenticity and extent of a company's commitment to sustainability.

Resources and tools

Investors interested in socially responsible investing can find a variety of resources and tools to support their decision-making process. These include:

1. ESG ratings and indices: Several organizations provide ESG ratings and indices that assess companies' performance on environmental, social, and governance factors. These ratings can help investors identify companies that align with their values and goals.

2. Impact measurement frameworks: Impact measurement frameworks, such as the Global Reporting Initiative (GRI) and the Sustainability Accounting Standards Board (SASB), provide guidance for assessing the social and environmental impact of investments.

3.   SRI funds and platforms: There are numerous mutual funds and exchange-traded funds (ETFs) that specialize in socially responsible investing. These funds offer a diversified portfolio of companies that meet specific ESG criteria.

4. ESG research providers: Many research firms specialize in evaluating companies' ESG performance. Their reports and analyses can provide valuable insights into a company's sustainability practices and impact.

Case study: The Rise of ESG Integration

ESG integration involves incorporating ESG factors into traditional investment analysis to better assess a company's financial performance and risk profile. This approach recognizes that ESG factors can materially impact a company's long-term prospects and shareholder value.

For instance, consider two companies in the same industry. One company has a robust sustainability strategy, strong corporate governance, and a positive social impact, while the other lags behind in these areas. By integrating ESG analysis, investors can identify the potential risks and opportunities associated with each company's ESG performance. This allows for a more holistic evaluation of investment options.

ESG integration can provide investors with a more complete understanding of a company's risk profile. It helps identify potential risks, such as regulatory non-compliance, reputational damage, or supply chain disruptions. By identifying these risks early on, investors can make more informed investment decisions.

Moreover, ESG integration allows investors to capture potential opportunities that arise from sustainable practices. For example, companies that prioritize energy efficiency or have a strong commitment to renewable energy may be better positioned to benefit from the transition to a low-carbon economy.

Incorporating ESG factors into investment decisions is a growing trend. Increasingly, asset managers and institutional investors are integrating ESG analysis into their investment processes. This shift indicates the recognition that ESG considerations are financially material and can impact long-term investment performance.

Conclusion

Socially responsible investing offers individuals and institutions an opportunity to support sustainable practices, promote positive change, and align their investments with their values. By considering ESG factors alongside traditional financial analysis, investors can have a more comprehensive view of a company's performance, risks, and opportunities.

While socially responsible investing faces challenges such as performance trade-offs and lack of standardized metrics, the growing interest in sustainability and corporate responsibility suggests that SRI will continue to gain prominence. As sustainable investing strategies evolve, investors can contribute to a more resilient, equitable, and sustainable future.

Impact investing

Impact investing is a growing field that combines financial returns with positive social and environmental outcomes. It is a form of investing that seeks to generate measurable and beneficial impacts alongside financial returns. In this section, we will explore the principles and practices of impact investing, its role in driving sustainable development, and the potential challenges and opportunities it presents.

Principles of Impact Investing

Impact investing is guided by several key principles that distinguish it from traditional investing. These principles help investors align their capital with their values and contribute to positive social and environmental change. Here are the four main principles of impact investing:

1. Intentionality: Impact investors have a clear intention to generate positive social and environmental impacts through their investments. They actively seek investment opportunities that address pressing social or environmental challenges.

2. Measurement: Impact investors measure and assess the social and environmental impacts of their investments. They use a variety of metrics and frameworks to quantify and report on the outcomes and effectiveness of their investments.

3. Additionality: Impact investors aim to create additional benefits beyond what would have occurred without their investment. They strive to make a meaningful difference and go beyond business as usual.

4. Accountability: Impact investors are accountable to their stakeholders, including investors, beneficiaries, and society at large. They are transparent about their intentions, objectives, and impact performance.

Examples of Impact Investments

Impact investments can take various forms and span across different sectors, such as sustainable agriculture, renewable energy, affordable housing, education, healthcare, and microfinance. Here are a few examples of impact investments:

1. Impact bonds: Impact bonds are financial instruments that link the success of an investment to the achievement of predetermined social or environmental outcomes. They provide upfront capital to service providers to deliver innovative social programs, with payments tied to specific outcomes.

2. Green bonds: Green bonds are fixed-income financial instruments used to raise capital for climate and environmental projects. The proceeds from green bonds are exclusively allocated to finance projects with positive environmental benefits, such as renewable energy installations or energy efficiency improvements.

3. Microfinance: Microfinance involves providing small loans, savings accounts, and other financial services to low-income individuals and small

businesses in developing countries. By giving access to financial resources, microfinance supports entrepreneurship and helps lift people out of poverty.

4. Social impact funds: Social impact funds pool capital from various investors to finance projects and organizations that generate positive social impacts. These funds focus on specific social objectives, such as reducing homelessness, improving access to education, or supporting sustainable community development.

Challenges and Opportunities

While impact investing offers great potential for driving positive change, it also comes with its own set of challenges and opportunities. Here are a few key considerations:

1. Measurement and reporting: Measuring the social and environmental impacts of investments can be complex and subjective. Developing robust impact measurement frameworks and reporting standards is crucial for ensuring transparency and accountability.

2. Financial returns: Impact investing aims to generate both financial returns and positive social or environmental outcomes. Balancing these dual objectives can sometimes be challenging, as impact investments may require longer time horizons or involve higher risks.

3. Scalability: Scaling up impact investing is essential for addressing the world's most pressing challenges. Finding innovative ways to mobilize larger amounts of capital and attract mainstream investors to the field is essential for achieving significant impact.

4. Collaboration and knowledge sharing: Building partnerships and sharing knowledge and best practices are vital for the growth and effectiveness of impact investing. Collaboration can help address common challenges, foster innovation, and expand the reach and impact of investments.

Case Study: The Rise Fund

One prominent example of impact investing is The Rise Fund, a global impact investment fund managed by TPG Capital. Launched in 2017, The Rise Fund aims to generate both market-rate financial returns and positive social and environmental impact.

The Rise Fund focuses on several key impact sectors, including education, energy, food and agriculture, financial services, healthcare, and technology. It invests in companies that align with the United Nations Sustainable Development

Goals and have the potential to generate measurable social and environmental impact at scale.

The fund employs a rigorous impact measurement and management framework to assess the outcomes and effectiveness of its investments. It tracks key performance indicators related to social, environmental, and governance factors, aiming to provide transparency and accountability to its stakeholders.

Through its investments, The Rise Fund has supported a range of impactful initiatives. For example, it has invested in Solar Home Systems, a company that provides clean energy solutions to off-grid households in Africa, enabling access to affordable and sustainable electricity. This investment not only generates financial returns but also contributes to reducing carbon emissions and improving energy access in underserved communities.

In conclusion, impact investing offers a promising approach to finance projects and businesses that deliver positive social and environmental outcomes. By aligning capital with values, impact investors have the potential to drive meaningful change and contribute to sustainable development. However, realizing the full potential of impact investing requires collaboration, measurement, and a commitment to long-term impact. Together, investors, entrepreneurs, and policymakers can shape a more sustainable and inclusive future through impact investing.

References: 1. Global Impact Investing Network: www.thegiin.org 2. The Rise Fund: www.therisefund.com

Environmental Governance

Stakeholder Engagement

Stakeholder engagement plays a crucial role in environmental governance and decision-making processes. It involves actively involving individuals, groups, and organizations who are affected by or have an interest in environmental issues. Stakeholders can include government agencies, non-profit organizations, local communities, indigenous groups, industry representatives, scientists, and the general public.

The goal of stakeholder engagement is to ensure that different perspectives are considered, and decisions are made in a transparent and inclusive manner. By involving stakeholders, environmental policies and projects can be more effectively implemented and address the diverse needs and concerns of the community.

Importance of Stakeholder Engagement

Stakeholder engagement is important for several reasons. Firstly, it acknowledges that environmental issues often have multiple dimensions and impacts people differently. By involving stakeholders, decision-makers gain a better understanding of the social, economic, and environmental implications of their actions.

Secondly, stakeholder engagement fosters trust and legitimacy in environmental decision-making processes. When stakeholders are included, they feel valued and are more likely to support and cooperate with the outcomes. This leads to the creation of more effective and sustainable policies and projects.

Thirdly, stakeholder engagement enhances the quality of decision-making. By involving a wide range of perspectives and knowledge, stakeholders can contribute valuable insights, expertise, and innovative solutions. This helps to identify potential risks, unintended consequences, and identify the most appropriate course of action.

Challenges in Stakeholder Engagement

While stakeholder engagement has numerous benefits, it also comes with its challenges. These challenges include:

1. **Diverse interests and conflicting priorities:** Stakeholders often have different interests, priorities, and values. Balancing these diverse perspectives can be challenging and require careful negotiation and compromise.

2. **Power imbalances:** Some stakeholders may have more power, resources, or influence than others. This can lead to unequal participation and decision-making processes. Efforts should be made to address power imbalances and ensure marginalized groups have a voice.

3. **Lack of trust and skepticism:** Building trust among stakeholders is crucial but can be difficult, especially in cases where there have been historical conflicts or distrust. Transparency, open communication, and accountability are key to overcoming this challenge.

4. **Time and resource constraints:** Stakeholder engagement requires time, effort, and resources. Limited timeframes and financial constraints can impact the effectiveness and inclusivity of the engagement process. Adequate resources should be allocated to facilitate meaningful participation.

5. **Effective communication:** Communicating complex environmental issues to a diverse group of stakeholders can be challenging. Efforts should be made to ensure information is accessible, clear, and presented in a way that allows meaningful participation.

Strategies for Effective Stakeholder Engagement

To overcome the challenges associated with stakeholder engagement, various strategies can be employed. These strategies include:

1. **Identifying key stakeholders:** Understanding who the key stakeholders are is essential. This involves mapping out relevant individuals, groups, and organizations and determining their interests and potential contributions.

2. **Building relationships:** Developing and nurturing relationships with stakeholders is crucial for effective engagement. This can be achieved through regular communication, inclusive forums, and involvement in decision-making processes from the early stages.

3. **Inclusive and accessible processes:** Ensuring that stakeholder engagement processes are inclusive and accessible is vital. This may involve providing translated materials, using plain language, offering different engagement formats (e.g., public meetings, online platforms), and addressing the needs of marginalized groups.

4. **Meaningful participation:** Stakeholders should be given a meaningful role in the decision-making process by providing opportunities for input, involvement, and feedback. This can include consultation, collaborative decision-making, and the incorporation of stakeholder views into final decisions.

5. **Capacity building:** Supporting stakeholders in understanding and participating effectively in environmental decision-making can be achieved through capacity-building initiatives. This could involve providing training, resources, and technical support to enhance stakeholders' knowledge and skills.

6. **Monitoring and evaluation:** Regularly monitoring and evaluating stakeholder engagement processes is important to assess their effectiveness and make adjustments as needed. This can help identify areas for improvement and inform future engagement initiatives.

Example: Stakeholder Engagement in Urban Planning

To illustrate the importance of stakeholder engagement in a real-world context, let's consider its application in urban planning. Urban planning involves decisions about land use, infrastructure, and transportation systems, which can have significant environmental impacts.

In an urban planning project, engaging stakeholders could involve consulting with residents, community organizations, business owners, environmental groups, and government agencies. By involving these stakeholders, planners can gain insights into community needs, concerns, and ideas for sustainable development.

For example, during the development of a new public transportation system, stakeholder engagement could include public consultations, workshops, and surveys. These activities would allow stakeholders to provide input on route options, accessibility needs, environmental impact considerations, and community integration.

Through stakeholder engagement, potential issues such as displacement of communities, unequal access to transportation, or negative environmental impacts can be identified and addressed. By involving stakeholders in decision-making, the public transportation system can be designed to meet the needs of the community while minimizing its environmental footprint.

In this example, stakeholder engagement not only improves the quality of the urban planning process but also creates a sense of ownership and support among the community. It ensures that the decision-making process is fair, transparent, and inclusive, leading to more sustainable and socially just outcomes.

Summary

Stakeholder engagement is a fundamental aspect of environmental governance and decision-making processes. It helps to ensure that decisions are made in a transparent, inclusive, and effective manner. By involving stakeholders, diverse perspectives and expertise can be incorporated, leading to more sustainable and socially just outcomes.

While stakeholder engagement comes with challenges, strategies such as identifying key stakeholders, building relationships, inclusive processes, meaningful participation, capacity building, and monitoring and evaluation can help overcome these challenges.

In the context of urban planning, stakeholder engagement plays a crucial role in creating sustainable and inclusive cities. By involving community members, businesses, and organizations in decision-making processes, planners can design cities that meet the needs of the community while minimizing environmental impacts.

Overall, stakeholder engagement is an essential tool for fostering collaboration, building trust, and achieving sustainable environmental outcomes. It should be integrated into environmental policies, projects, and governance frameworks to promote effective decision-making and long-term sustainability.

Participatory Decision-Making

Participatory decision-making is a vital aspect of environmental governance, aiming to involve stakeholders in the decision-making process to ensure transparency, inclusivity, and accountability. It recognizes that decisions about environmental issues should not be made solely by a few experts or government officials but should involve a wide range of people affected by the decisions. In this section, we will explore the principles, methods, and benefits of participatory decision-making in the context of eco-science.

Principles of Participatory Decision-Making

Participatory decision-making is based on several key principles:

- **Inclusivity:** It involves diverse stakeholders representing different perspectives, including local communities, indigenous people, NGOs, scientists, policymakers, and industry representatives.

- **Transparency:** The decision-making process should be open and transparent, ensuring that all information and data are accessible to all participants.

- **Equality:** It ensures that all participants have an equal opportunity to voice their opinions, regardless of their social status or power dynamics.

- **Collaboration:** Participants collaborate and work together to find common ground, facilitating the development of collective solutions that consider multiple viewpoints.

- **Empowerment:** It empowers individuals and communities by providing them with the information and skills necessary to actively contribute to the decision-making process.

Methods of Participatory Decision-Making

There are various methods and approaches to facilitate participatory decision-making. Some commonly used methods include:

+ **Public Consultations:** This method involves organizing public meetings or hearings where stakeholders can express their views, concerns, and suggestions on specific environmental issues. It provides a platform for dialogue and informed decision-making.

+ **Stakeholder Workshops:** Workshops bring together stakeholders from different backgrounds to discuss and deliberate on specific issues. These workshops often include facilitated discussions, group exercises, and brainstorming sessions that encourage active participation and collaboration.

+ **Focus Groups:** Focus groups involve gathering a small group of individuals who share similar characteristics or interests. Through guided discussions, participants can provide valuable insights and perspectives on specific topics.

+ **Citizen Science:** Citizen science initiatives engage the public in scientific research and data collection. Citizens contribute their observations and data, which are then used for scientific analysis and decision-making processes.

+ **Deliberative Polling:** Deliberative polling brings together a representative sample of the population to discuss and debate specific issues. It combines opinion surveying with in-depth discussions, allowing participants to gain a deeper understanding of the topic and develop informed opinions.

Benefits of Participatory Decision-Making

Participatory decision-making offers several benefits, both for the decision-making process itself and for the outcomes achieved. These include:

+ **Enhanced Decision Quality:** Including a diverse range of perspectives and knowledge in decision-making leads to more robust and informed decisions. It reduces the likelihood of overlooking important considerations and increases the likelihood of finding innovative solutions.

+ **Legitimacy and Trust:** Involving stakeholders in decision-making processes builds trust and legitimacy. When people are included and their voices are heard, they are more likely to accept and support the decisions made.

+ **Ownership and Commitment:** Participatory decision-making fosters a sense of ownership among participants, as they have actively contributed to the process. This sense of ownership increases commitment and accountability for the implementation of decisions.

+ **Social Learning:** The process of engaging with diverse stakeholders allows for knowledge exchange and social learning. Participants gain insights from each other's experiences and expertise, leading to better-informed decision-making.

+ **Sustainable Solutions:** Participatory decision-making promotes the development of sustainable solutions that take into account the social, economic, and environmental dimensions of a problem. It helps identify and address potential trade-offs and ensures the long-term viability of decisions.

Challenges and Considerations

While participatory decision-making has many potential benefits, it also faces several challenges and considerations. These include:

+ **Power Imbalances:** Power imbalances among stakeholders can hinder effective participation and decision-making. It is crucial to ensure a level playing field and actively involve marginalized groups to avoid reinforcing existing inequalities.

+ **Time and Resource Constraints:** Participatory decision-making processes can be time-consuming and resource-intensive. Sufficient time and resources must be allocated to ensure meaningful engagement and avoid tokenistic participation.

+ **Information Asymmetry:** Participants may have differing levels of knowledge and expertise, which can create information asymmetry. Efforts should be made to provide accessible and understandable information to all participants, bridging the gap in knowledge.

+ **Conflict and Disagreements:** Including multiple stakeholders with conflicting interests can lead to disagreements and conflicts. Facilitation techniques, such as mediation and consensus-building, are needed to navigate these challenges and facilitate constructive dialogue.

+ **Implementation and Accountability:** Participatory decision-making is only effective if decisions are implemented and stakeholders are held accountable. Clear mechanisms for implementation and monitoring must be established to ensure the legitimacy and success of the process.

Case Study: Participatory Forest Management in India

One example of participatory decision-making in action is the Participatory Forest Management (PFM) approach implemented in India. Under this approach, local communities are involved in decision-making processes related to forest management, conservation, and livelihood activities.

The PFM approach recognizes the indigenous knowledge and practices of local communities and involves them in the planning, implementation, and monitoring of forest-related activities. It aims to promote sustainable forest management while addressing the needs and aspirations of local communities.

Through participatory processes such as community meetings, joint forest management committees, and village-level planning, local communities have been able to actively contribute to decision-making processes. This has resulted in improved forest conservation, reduced conflicts, and positive socio-economic outcomes for the communities involved.

The PFM approach in India serves as an example of how participatory decision-making can lead to more sustainable and equitable outcomes, highlighting the importance of including local knowledge and engaging stakeholders in environmental governance.

Summary

Participatory decision-making is a fundamental principle of environmental governance. It involves inclusive, transparent, and collaborative processes that empower stakeholders to actively contribute to decision-making. Participatory decision-making enhances decision quality, legitimacy, and ownership, leading to more sustainable outcomes. However, challenges related to power imbalances, resource constraints, and conflict need to be carefully addressed. Through case studies like the Participatory Forest Management approach in India, we can see how participatory decision-making can lead to positive environmental and social outcomes.

Environmental justice

Environmental justice is a concept that addresses the unequal distribution of environmental benefits and burdens among different communities. It recognizes that marginalized communities, particularly those with a history of systemic discrimination and socioeconomic disadvantage, are often disproportionately exposed to environmental hazards and suffer from the negative impacts of environmental degradation.

Background

The environmental justice movement emerged in the United States in the 1980s, primarily led by grassroots organizations and activists from communities of color and low-income neighborhoods. These communities were found to be disproportionately affected by hazardous waste facilities, polluting industries, and other environmental hazards. The movement sought to address these disparities and advocate for fair treatment and meaningful participation of all people in environmental decision-making processes.

The principles of environmental justice have since been recognized globally, with movements and organizations working towards just and sustainable environmental solutions for marginalized communities around the world.

Principles of Environmental Justice

The principles of environmental justice include:

1. Equity: The fair distribution of environmental benefits and the fair protection from environmental burdens for all individuals, regardless of race, ethnicity, income, or social status.

2. Participation: The right of all individuals to participate in decision-making processes that affect their environment. This includes providing opportunities for meaningful engagement and ensuring that marginalized communities have equal opportunities to voice their concerns, knowledge, and perspectives.

3. Transparency: The free flow of information and public access to information related to environmental issues and decision-making processes. This allows for informed participation and accountability.

4. Accountability: Ensuring that individuals and organizations responsible for environmental harm are held accountable for their actions and that impacted communities have access to remedies and justice.

5. Sustainability: Promoting sustainable practices that protect and enhance the environment while also meeting the needs of present and future generations. This

includes promoting renewable energy, sustainable agriculture, and equitable access to resources.

Environmental Justice Issues

Environmental justice issues can manifest in various forms, including:

1. Proximity to Pollution Sources: Marginalized communities often experience a higher concentration of polluting industries, waste facilities, and other sources of pollution. This results in increased exposure to harmful substances and negative health impacts.

2. Health Disparities: Environmental justice issues are often linked to public health concerns. Marginalized communities may face higher rates of asthma, respiratory diseases, cancer, and other health conditions due to exposure to environmental hazards.

3. Access to Environmental Resources: Some communities lack access to clean air, clean water, green spaces, and other environmental resources that contribute to overall well-being and quality of life.

4. Climate Change Vulnerability: Marginalized communities are often more vulnerable to the impacts of climate change, such as extreme weather events, sea-level rise, and food insecurity. They may have limited resources to adapt and recover from these impacts.

Case Study: Flint Water Crisis

One prominent case study illustrating environmental justice issues is the Flint water crisis. In 2014, the city of Flint, Michigan, switched its water source to the Flint River to cut costs. However, the river water was not properly treated, leading to a series of problems that contaminated the drinking water with lead and other toxins.

The majority of Flint's residents are African American, and many live in poverty. The decision to switch the water source and the inadequate response to the contamination disproportionately harmed these marginalized communities. The crisis highlighted the systemic environmental injustice faced by vulnerable communities and ignited national attention and calls for change.

Addressing Environmental Justice

Addressing environmental justice requires a multi-faceted approach, involving policymakers, communities, organizations, and individuals. Some strategies include:

1. Community Empowerment: Supporting community-led initiatives that promote environmental justice through education, advocacy, and direct engagement. This includes building capacity, providing resources, and amplifying the voices of marginalized communities.

2. Policy and Legislative Changes: Advocating for stronger environmental regulations that prioritize equity and public health. This includes pushing for greater enforcement of environmental laws and regulations to ensure that polluters are held accountable.

3. Environmental Impact Assessment: Incorporating environmental justice considerations into the evaluation of projects and policies. This includes assessing the potential impacts on marginalized communities and ensuring their meaningful participation in the decision-making process.

4. Collaborative Partnerships: Fostering partnerships between government agencies, non-governmental organizations, and community members to address environmental justice issues collectively. This allows for knowledge sharing, resource pooling, and the collective implementation of solutions.

Unconventional Solution: Art and Storytelling

Art and storytelling can play a powerful role in raising awareness about environmental justice issues and inspiring change. Through creative expression, marginalized communities can share their experiences, challenges, and visions for a more just and sustainable world. Art can elicit emotional responses and foster empathy, leading to increased public engagement and support for environmental justice causes.

For example, community-led mural projects can transform public spaces and serve as powerful visual reminders of the need for environmental justice. Storytelling through mediums such as film, theater, and literature can also bring attention to the experiences of marginalized communities and provoke reflection and dialogue.

Conclusion

Environmental justice is an important aspect of sustainable development and eco science. It addresses the inequalities in environmental burdens and benefits that affect marginalized communities. By promoting equity, participation, transparency, accountability, and sustainability, environmental justice aims to ensure that all individuals, regardless of their race, ethnicity, income, or social status, have a voice in environmental decision-making processes and access to a clean and healthy environment.

By understanding and addressing environmental justice issues, we can create a more just and sustainable world for present and future generations. It is crucial to continue working towards equitable and inclusive environmental policies, practices, and initiatives that empower marginalized communities and protect the environment for all.

Ecosystem-based management

Ecosystem-based management (EBM) is an approach to environmental governance that aims to achieve sustainable development by considering the complex interactions between ecological, social, and economic systems. It recognizes that human activities are interconnected with and depend on healthy ecosystems. EBM seeks to preserve the integrity and resilience of ecosystems while meeting the needs and aspirations of present and future generations.

Principles of Ecosystem-based Management

EBM is guided by several key principles:

1. **Holistic approach:** EBM recognizes that ecosystems are interconnected and must be managed as a whole, rather than focusing on individual components or species. It considers the interactions and feedbacks within and between ecological, social, and economic systems.

2. **Adaptive management:** EBM emphasizes the need for flexibility and learning in decision-making. It acknowledges that ecosystems are complex and constantly changing, requiring ongoing monitoring and adjustment of management strategies based on new knowledge and understanding.

3. **Precautionary approach:** EBM takes a cautious stance when faced with uncertainty and potential risks. It advocates for proactive measures to prevent or minimize harm to ecosystems, even in the absence of complete scientific evidence.

4. **Ecosystem resilience:** EBM aims to enhance the resilience of ecosystems, enabling them to withstand and recover from disturbances and stresses. This involves maintaining and restoring ecological processes, diversity, and functions that contribute to the overall health and productivity of ecosystems.

5. **Collaboration and stakeholder engagement:** EBM recognizes the importance of involving all stakeholders, including local communities, indigenous peoples, scientists, policymakers, and resource users, in the decision-making process. It promotes inclusive and participatory approaches to ensure that diverse perspectives and knowledge are considered.

Implementing Ecosystem-based Management

Implementing EBM requires a combination of scientific knowledge, policy frameworks, and management practices. Here are some key steps involved:

1. **Assessing and understanding ecosystems:** The first step is to conduct comprehensive assessments of ecosystems, including their structure, composition, and functioning. This involves gathering data on biodiversity, ecosystem services, and human activities that impact the ecosystems.

2. **Identifying management goals:** Based on the assessment, clear management goals and objectives should be established. These goals should be science-based, consistent with conservation targets, and consider the social and economic needs of the local communities.

3. **Designing management strategies:** EBM requires developing strategies that balance conservation priorities with sustainable use of resources. This may involve implementing zoning plans, regulating fishing practices, restoring degraded habitats, or reducing pollution inputs.

4. **Monitoring and evaluation:** Regular monitoring of ecosystem health, including biodiversity, water quality, and ecosystem services, is crucial to track the effectiveness of management strategies. Evaluation should be used to learn from successes and failures, allowing for adaptive management.

5. **Building partnerships:** Collaboration is vital in EBM, as it involves multiple stakeholders and sectors. Partnerships between government agencies, NGOs, local communities, and industry can facilitate knowledge sharing, resource pooling, and coordinated action.

Case Study: Ecosystem-based Fisheries Management

One example of EBM in practice is ecosystem-based fisheries management. Fisheries worldwide face numerous challenges, including overfishing, habitat degradation, and climate change impacts. EBM provides a framework for addressing these challenges and ensuring the long-term sustainability of marine resources.

In this approach, fisheries management is guided by an understanding of the ecosystem dynamics and the interactions between target species, other species, and the physical environment. It considers fish populations in the context of their habitat, food webs, and ecological roles. Building on this understanding, EBM aims to achieve sustainable harvest levels that maintain healthy fish populations while preserving the overall ecosystem's integrity and functioning.

Key elements of ecosystem-based fisheries management include:

1. **Protecting sensitive habitats:** EBM recognizes the importance of protecting critical habitats, such as coral reefs, seagrass beds, and mangroves, which serve as essential nurseries for fish species. It involves implementing measures to reduce or prevent habitat destruction from fishing practices.

2. **Setting harvest limits:** EBM considers the population dynamics of fish species, taking into account their reproductive rates, growth rates, and mortality rates. Harvest limits are set to ensure that fishing pressure does not exceed the capacity of fish populations to replenish themselves.

3. **Managing bycatch and discards:** EBM seeks to minimize the unintended capture of non-target species (bycatch) and the discard of unwanted catch. Measures such as selective fishing gear, area closures, and improved fishing practices can help reduce bycatch and protect vulnerable species.

4. **Monitoring and enforcement:** Regular monitoring of fish stocks, fishing effort, and compliance with regulations is essential for effective fisheries management. Adequate enforcement measures are necessary to deter illegal fishing and ensure compliance with management measures.

5. **Engaging stakeholders:** EBM involves consultation and collaboration with fishers, scientists, communities, and other stakeholders affected by fisheries management decisions. Their knowledge, experiences, and perspectives are valuable in developing and implementing effective management strategies.

By adopting an ecosystem-based approach, fisheries management can move beyond traditional single-species management and promote the conservation of marine biodiversity while supporting the livelihoods of fishing communities.

Conclusion

Ecosystem-based management provides a holistic and adaptive framework for addressing complex environmental challenges. By considering the interconnectedness of ecological, social, and economic systems, EBM strives to achieve sustainable development and preserve the integrity and resilience of ecosystems. Its implementation requires collaboration, stakeholder engagement, and science-based decision-making. Through examples like ecosystem-based fisheries management, EBM offers practical solutions for balancing conservation and human well-being in a rapidly changing world.

Adaptive Governance

Adaptive governance is a concept that emphasizes the need for flexible and responsive decision-making processes in order to address complex environmental challenges. It recognizes that the dynamics of social-ecological systems are constantly changing and unpredictable, requiring institutions and policies that can adapt and learn from new information and experiences.

Principles of Adaptive Governance

Adaptive governance is guided by several key principles:

1.   **Understanding Complexity:** Adaptive governance recognizes that social-ecological systems are highly complex and interconnected. It acknowledges the existence of multiple stakeholders with diverse interests and values. Decision-making processes must take into account this complexity and consider the multiple dimensions of sustainability.

2.   **Learning and Experimentation:** Adaptive governance encourages continuous learning and experimentation. It emphasizes the importance of monitoring and feedback mechanisms to assess the effectiveness of policies and interventions. This enables decision-makers to adapt their strategies based on new information and knowledge gained from real-world experiences.

3.   **Collaboration and Participation:** Adaptive governance promotes collaborative and participatory decision-making processes. It recognizes that diverse perspectives and local knowledge are valuable resources for generating innovative and context-specific solutions. Stakeholders are actively involved in planning, implementation, and evaluation processes, thus enhancing the legitimacy and effectiveness of decisions.

4.   **Resilience and Adaptability:** Adaptive governance aims to enhance the resilience and adaptability of social-ecological systems. It recognizes that change is inevitable and that systems must be able to absorb disturbances and reorganize themselves. Institutions and policies should be designed to foster resilience and the capacity to adapt to changing conditions.

5.   **Scale and Connectivity:** Adaptive governance recognizes that environmental challenges often transcend administrative boundaries.   It emphasizes the need for coordination and cooperation across different scales and levels of governance. This includes collaboration between different sectors, regions, and organizations to address issues that require collective action.

Challenges and Solutions

Implementing adaptive governance can face various challenges. These may include resistance to change, lack of institutional capacity, power imbalances, and limited resources. However, there are several strategies that can help overcome these challenges:

1. **Building Capacity for Adaptive Governance**: Capacity building is crucial for enabling adaptive governance. This involves providing training and support to individuals and organizations involved in decision-making processes. It includes fostering skills such as systems thinking, collaboration, and adaptive management.

2. **Promoting Learning Networks**: Creating networks and platforms for knowledge sharing and collaboration can facilitate learning and innovation. These networks can bring together diverse stakeholders and create spaces for dialogue, exchange of experiences, and joint problem-solving.

3. **Using Adaptive Management Approaches**: Adaptive management is an iterative process that involves setting clear objectives, implementing management strategies, monitoring outcomes, and adjusting actions based on new information. Adopting adaptive management approaches can enhance the ability to learn from experiences and make informed decisions.

4. **Inclusive Decision-Making Processes**: Inclusiveness and diversity in decision-making processes are crucial for achieving adaptive governance. This involves involving marginalized groups, indigenous communities, and other stakeholders who may be disproportionately affected by environmental challenges. Their knowledge, perspectives, and values should be fully recognized and integrated into decision-making processes.

Example: Adaptive Governance in Water Management

To illustrate the concept of adaptive governance, let's consider the case of water management in a river basin facing increasing drought conditions due to climate change. Traditional water management approaches may rely on fixed water allocation systems and predetermined operating rules.

However, adaptive governance would involve a more flexible and dynamic approach. Stakeholders would engage in collaborative processes to develop adaptive management plans that can respond to changing hydrological conditions. This could include the establishment of real-time monitoring systems, the integration of climate projections into decision-making, and the implementation of water-sharing agreements that consider the needs and vulnerabilities of different sectors and communities.

Adaptive governance would also involve ongoing monitoring and evaluation of the effectiveness of management strategies, as well as the regular review and adaptation of policies to ensure their continued relevance and effectiveness. This approach enables stakeholders to learn from their experiences, build resilience in the face of changing conditions, and promote more sustainable and equitable water management practices.

Resources

- Folke, C., Hahn, T., Olsson, P., & Norberg, J. (2005). Adaptive Governance of Social-Ecological Systems. Annu. Rev. Environ. Resour., 30, 441-473.

- Bixler, R. P., Kliskey, A. D., & Hall, T. E. (2016). Adaptive governance in the face of climate change: insights from drought management in pastoral systems in the American West. Ecology and Society, 21(2), 31.

- Pahl-Wostl, C. (2009). A conceptual framework for analysing adaptive capacity and multi-level learning processes in resource governance regimes. Global Environmental Change, 19(3), 354-365.

- Armitage, D., Plummer, R., Berkes, F., Arthur, R. I., Charles, A. T., Davidson-Hunt, I. J., ... & Weber, M. (2009). Adaptive co-management for social-ecological complexity. Frontiers in Ecology and the Environment, 7(2), 95-102.

- Westley, F. R., Olsson, P., Folke, C., Homer-Dixon, T., Vredenburg, H., Loorbach, D., ... & Thompson, M. (2011). Tipping toward sustainability: emerging pathways of transformation. AMBIO: A Journal of the Human Environment, 40(7), 762-780.

Exercises

1. Identify a local environmental challenge and discuss how adaptive governance principles could be applied to address it effectively.

2. Research a case study where adaptive governance has been successfully implemented. Analyze the key factors that contributed to its success and the lessons that can be learned from it.

3. Imagine you are part of a stakeholder group involved in a decision-making process related to environmental management. Describe how you would incorporate principles of adaptive governance to ensure a more inclusive, resilient, and adaptive outcome.

4. Discuss the role of technology in enabling adaptive governance. Provide examples of specific technologies that can support adaptive decision-making and enhance the resilience of social-ecological systems.

Remember, adaptive governance requires continuous learning and adjustment. It is a dynamic process that requires collaborative efforts and adaptive management practices to effectively address complex environmental challenges.

Multi-level Governance

Multi-level governance refers to the distribution of power and decision-making authority across different levels of government, including local, regional, national, and international levels, as well as non-governmental organizations (NGOs) and civil society. It recognizes that environmental issues are complex and interconnected, requiring collaboration and coordination among various actors at different levels.

Principles of Multi-level Governance

Multi-level governance is guided by several key principles, which are crucial for effective environmental management:

1. **Subsidiarity:** Decisions should be made at the most appropriate level of government, taking into account the specific context and needs of the issue at hand. It recognizes that local authorities often have a better understanding of local environmental conditions and can tailor solutions accordingly.

2. **Participation and inclusiveness:** Stakeholders, including local communities, NGOs, and indigenous groups, should be actively involved in decision-making processes. Their knowledge, perspectives, and experiences play a critical role in shaping policies and actions that are locally relevant and inclusive.

3. **Coordination and cooperation:** Effective multi-level governance requires strong coordination and cooperation mechanisms among different levels of government and stakeholders. This includes sharing information, resources, and expertise, and jointly addressing common challenges and goals.

4. **Transparency and accountability:** Decision-making processes should be transparent, ensuring that information is accessible to all relevant actors.

Additionally, mechanisms should be in place to hold government officials and other stakeholders accountable for their actions or lack thereof.

5. **Adaptive management**: Multi-level governance emphasizes the need for flexible and adaptive approaches to environmental management. It recognizes that environmental issues are dynamic and constantly evolving and requires the ability to respond and adapt to changing circumstances.

Challenges and Solutions

Implementing effective multi-level governance faces several challenges, including:

+ **Fragmentation**: The division of responsibilities and decision-making authority across different levels of government can lead to fragmentation and lack of coherence in environmental policies and actions. This can hinder effective environmental management.

+ **Power dynamics**: Power imbalances between different levels of government and stakeholders can influence decision-making processes and undermine the effectiveness of multi-level governance. It is essential to address these power dynamics and ensure equitable representation and participation.

+ **Lack of capacity and resources**: Some levels of government, particularly at the local level, may lack the capacity and resources to effectively participate in multi-level governance processes. Capacity-building programs and resource-sharing mechanisms can help address these gaps.

+ **Lack of awareness and understanding**: Another challenge is the limited understanding and awareness of the importance and benefits of multi-level governance among stakeholders. Education and awareness campaigns can help foster a culture of collaboration and cooperation.

To address these challenges, the following solutions can be implemented:

+ **Coordinating mechanisms**: Establishing formal and informal coordinating mechanisms, such as intergovernmental committees, task forces, and partnerships, can facilitate collaboration and coordination among different levels of government and stakeholders.

+ **Capacity-building**: Providing training and capacity-building programs for local authorities and other stakeholders can enhance their understanding of multi-level governance processes and equip them with the necessary skills and knowledge.

+ **Information sharing**: Developing platforms for sharing information and best practices can improve access to relevant data and experiences, allowing stakeholders to make more informed decisions.

+ **Incentives and support**: Providing incentives and support, such as financial grants and technical assistance, can encourage local authorities and other stakeholders to actively participate in multi-level governance processes.

Case Study: The European Union

The European Union (EU) provides an example of successful multi-level governance in environmental management. The EU's environmental policies and regulations are developed through a collaborative process involving the European Commission, the European Parliament, and the Council of the European Union, as well as consultations with member states and stakeholders.

The EU's multi-level governance approach recognizes the importance of subsidiarity, allowing member states to implement policies that are adapted to their specific circumstances while ensuring compliance with common environmental standards. The EU provides financial and technical support to member states, particularly to those with limited resources, to effectively implement and enforce environmental laws.

The EU also emphasizes the importance of stakeholder participation in decision-making processes. Civil society organizations, industry representatives, and other stakeholders are consulted during the development and implementation of environmental policies, ensuring diverse perspectives and experiences are considered.

Additionally, the EU promotes transparency and accountability through regular reporting on environmental performance, monitoring mechanisms, and sanctions for non-compliance. The EU's multi-level governance approach has led to significant environmental improvements, including cleaner air and water, reduced carbon emissions, and increased conservation efforts.

Exercises

1. Discuss the key principles of multi-level governance and their significance in environmental management.

2. Describe the challenges faced in implementing effective multi-level governance and propose solutions for overcoming them.

3. Choose a local or regional environmental issue and analyze how multi-level governance could be applied to address the problem effectively.

4. Research and compare the multi-level governance approaches of two different countries or regions. Highlight their similarities, differences, and lessons that can be learned from each.

Additional Resources

+ Bache, I., & Flinders, M. (Eds.). (2004). *Multi-level governance.* Oxford University Press.

+ Hooghe, L., & Marks, G. (2001). Multi-level governance and European integration. *Rowman & Littlefield.*

+ Ostrom, E. (2010). *Governing the commons: The evolution of institutions for collective action.* Cambridge University Press.

+ Pahl-Wostl, C. (Ed.). (2009). *The challenge of complexity in environmental management.* Edward Elgar Publishing.

+ Sørensen, E., & Torfing, J. (Eds.). (2007). *Theories of democratic network governance.* Palgrave Macmillan.

Remember, multi-level governance is a collaborative and inclusive approach that recognizes the interdependencies and complexities associated with environmental management. It requires the active participation of governments, stakeholders, and civil society to achieve sustainable and effective outcomes.

Policy Diffusion and Transfer

Policy diffusion and transfer refers to the process through which policies are spread and adopted across different jurisdictions or regions. It involves the exchange of knowledge, experiences, and best practices among policymakers, with the aim of addressing common challenges and achieving sustainable development goals. This section explores the concept of policy diffusion and transfer, its significance in environmental governance, and the key mechanisms involved in facilitating this process.

Understanding Policy Diffusion

Policy diffusion is a dynamic and complex phenomenon that occurs when policymakers in one jurisdiction adopt and implement policies that have been successful in other jurisdictions. It is driven by the recognition of policy innovations, the pressure to respond to common challenges, and the desire to benefit from the experiences of others.

Policy transfer, on the other hand, refers to the process of adapting and implementing policies from one jurisdiction to another, taking into account the specific context and conditions of the receiving jurisdiction. It involves the translation of policy ideas, instruments, and approaches to fit local needs and realities.

Policy diffusion and transfer can occur horizontally, between jurisdictions at the same level of governance (e.g., between cities or states), or vertically, from higher to lower levels of governance (e.g., from international agreements to national policies). It can also occur in a networked fashion, where multiple jurisdictions interact and learn from each other.

Significance of Policy Diffusion and Transfer

Policy diffusion and transfer play a crucial role in environmental governance for several reasons:

1. Knowledge Sharing: Policy diffusion facilitates the exchange of knowledge and experiences among policymakers, enabling them to learn from both successful and unsuccessful policy initiatives. It helps to overcome information gaps and improve decision-making processes.

2. Policy Learning: By adopting policies that have been proven effective elsewhere, policymakers can learn from the experiences of others, avoiding pitfalls and accelerating progress towards sustainable development goals. Policy diffusion enables learning from best practices and innovation.

3. Policy Innovation: Policy diffusion encourages policymakers to explore and experiment with new policy ideas and approaches. It promotes innovation by showcasing successful policy models and encouraging adaptation to local contexts.

4. Scaling up Impact: By spreading successful policies across jurisdictions, policy diffusion and transfer help to amplify the impact of sustainable development initiatives. It enables the replication of successful policies on a larger scale, leading to positive environmental outcomes.

5. Collaboration and Networking: Policy diffusion fosters collaboration and networking among jurisdictions, creating opportunities for joint problem-solving

and shared learning. It builds partnerships and strengthens the collective capacity to address shared environmental challenges.

Mechanisms of Policy Diffusion and Transfer

Several mechanisms facilitate the process of policy diffusion and transfer:

1. Policy Networks: Policymakers and experts form networks to exchange information, share experiences, and collaborate on policy development. These networks may be formal (e.g., international organizations) or informal (e.g., communities of practice) and play a critical role in policy diffusion.

2. Policy Entrepreneurs: Individuals or organizations who actively promote policy innovations and facilitate their adoption by policymakers are known as policy entrepreneurs. They act as change agents, advocating for policy transfer and fostering collaboration among stakeholders.

3. Knowledge Brokers: Knowledge brokers play a crucial role in facilitating policy diffusion and transfer by translating complex research into actionable information for policymakers. They bridge the gap between scientific knowledge and policy implementation, making evidence accessible and understandable.

4. Policy Pilots and Demonstrations: Piloting policies in specific jurisdictions or conducting demonstration projects helps in showcasing their effectiveness and building evidence of their impact. These pilots serve as examples for policymakers considering policy diffusion and transfer.

5. Policy Evaluation and Monitoring: Monitoring and evaluating the outcomes of adopted policies provide valuable feedback for policymakers. It helps identify successful policies that can be considered for diffusion and transfer, as well as areas for improvement and adaptation.

Challenges and Solutions

Despite its potential benefits, policy diffusion and transfer face several challenges:

1. Contextual Relevance: Policies cannot be transferred directly from one jurisdiction to another without considering the specific contextual factors. The unique socio-economic, cultural, and political conditions of the receiving jurisdiction need to be taken into account for successful policy adaptation.

2. Policy Conflict: Conflicts may arise when policies that have worked well in one jurisdiction clash with local values, interests, or resources in another jurisdiction. Addressing these conflicts requires careful negotiation and adaptation of policies to suit local needs.

3. Capacity and Resources: Adopting and implementing policies requires adequate capacity and resources. Some jurisdictions may lack the necessary technical expertise, financial resources, or institutional frameworks to effectively adopt and adapt policies.

4. Policy Distortion: During the process of policy diffusion and transfer, there is a risk of policy distortion, where the original intent or design of the policy is altered. This can result in unintended consequences or ineffective policy implementation.

To overcome these challenges, policymakers should:

- Conduct thorough policy analysis and assessment to determine the feasibility and relevance of adopting a particular policy. - Foster collaboration and knowledge exchange among policymakers, experts, and other stakeholders to ensure effective policy diffusion. - Consider the participation of local communities and indigenous knowledge in the policy diffusion process to ensure cultural appropriateness and relevance. - Build the capacity of policymakers and staff in receiving jurisdictions through training programs, knowledge-sharing platforms, and technical assistance. - Monitor and evaluate policy implementation to assess its impact and address any emerging challenges or shortcomings.

Case Study: Carbon Pricing

To illustrate the concept of policy diffusion and transfer, let's consider the case of carbon pricing. Carbon pricing, through mechanisms like carbon taxes or cap-and-trade systems, has been widely recognized as an effective policy tool to tackle climate change.

The policy diffusion of carbon pricing can be observed globally, with various jurisdictions adopting similar mechanisms to reduce greenhouse gas emissions. For example, Sweden implemented a carbon tax in the 1990s, which was later adopted by several other European countries, including Finland, Denmark, and the Netherlands. This successful diffusion of carbon pricing policies continues to expand, with jurisdictions like Canada, China, and South Africa implementing their own versions.

Policy diffusion and transfer played a significant role in the adoption of carbon pricing policies. Policymakers in these jurisdictions recognized the success of carbon pricing elsewhere, learned from the experiences of early adopters, and adapted the policy to suit their specific contexts. International organizations, such as the World Bank or the United Nations, facilitated knowledge exchange and provided technical assistance to support policy diffusion.

The case of carbon pricing demonstrates how policy diffusion and transfer can contribute to global efforts in combating climate change. By sharing knowledge and learning from successful examples, policymakers can leverage the experiences of others to adopt effective and contextually relevant policies.

Summary

Policy diffusion and transfer are vital components of environmental governance and sustainability. By sharing knowledge, learning from best practices, and adapting policies to local contexts, policymakers can collectively address shared challenges and accelerate progress towards sustainable development goals. Mechanisms such as policy networks, policy entrepreneurs, and knowledge brokers facilitate the process of policy diffusion. However, challenges such as contextual relevance and policy conflict need to be addressed through careful consideration and stakeholder engagement. Overall, policy diffusion and transfer are key drivers of policy innovation and collaborative problem-solving in the field of eco science.

Science-policy interfaces

In today's complex world, addressing environmental challenges and achieving sustainable development goals requires a close collaboration between the scientific community and policymakers. Science-policy interfaces play a crucial role in bridging the gap between scientific knowledge and policy formulation, implementation, and evaluation. This section explores the importance of science-policy interfaces, their key elements, challenges, and strategies for effective collaboration.

The role and importance of science-policy interfaces

Science-policy interfaces serve as a platform for exchanging knowledge, evidence, and expertise between scientists and policymakers. They facilitate the integration of scientific information into policy decisions, ensuring that policies are based on the best available evidence. By promoting dialogue and collaboration, science-policy interfaces contribute to informed decision-making, effective policy implementation, and evaluation of policy outcomes.

Effective science-policy interfaces are essential for addressing complex environmental issues, such as climate change, biodiversity loss, and pollution. These challenges require interdisciplinary approaches that combine scientific expertise, socio-economic considerations, and stakeholder perspectives. Science-policy interfaces foster this interdisciplinary collaboration by providing a space for scientists, policymakers, and other stakeholders to share their knowledge, perspectives, and experiences.

Moreover, science-policy interfaces enhance the credibility and legitimacy of policy decisions. By promoting transparency and inclusivity, they ensure that policies are based on a broad range of perspectives and are supported by scientific consensus. This, in turn, increases public trust in policy processes and outcomes.

Key elements of science-policy interfaces

Effective science-policy interfaces are characterized by several key elements:

1. **Accessible communication:** Science-policy interfaces should promote clear, concise, and accessible communication between scientists and policymakers. Scientists need to effectively communicate their research findings, uncertainties, and limitations in a way that policymakers can understand and use for decision-making. Conversely, policymakers should clearly articulate their information needs and policy priorities to scientists.

2. **Co-production of knowledge:** Science-policy interfaces should foster collaboration and co-production of knowledge between scientists and policymakers. This approach recognizes that both communities have valuable knowledge and expertise to contribute. By involving policymakers in the research process and scientists in the policy formulation process, science-policy interfaces can generate more relevant and actionable knowledge.

3. **Policy relevance and timeliness:** Science-policy interfaces should ensure that scientific information is policy-relevant and timely. Policymakers often need rapid access to scientific evidence to address urgent environmental challenges. Scientists should provide timely information that is responsive to policy needs, while maintaining scientific rigor and integrity.

4. **Inclusivity and stakeholder engagement:** Science-policy interfaces should be inclusive and engage a wide range of stakeholders, including scientists, policymakers, civil society organizations, indigenous communities, and the private sector. By incorporating diverse perspectives and knowledge systems, science-policy interfaces can generate more robust and context-specific policy solutions.

5. **Policy-learning and adaptive management:** Science-policy interfaces should promote policy-learning and adaptive management. They should provide feedback mechanisms for policymakers to evaluate the effectiveness of policy interventions and revise them based on new scientific insights and societal changes.

Challenges in science-policy interfaces

Despite the importance of science-policy interfaces, several challenges can hinder their effectiveness:

1. **Communication barriers:** Differences in language, terminology, and communication styles between scientists and policymakers can hinder effective communication. Scientists often use technical jargon and present their research findings in complex formats, making it difficult for policymakers to understand and use the information.

2. **Time constraints:** Policymakers often face tight deadlines and have limited time to engage with scientific research. This can lead to a disconnect between the timing of scientific outputs and policy needs, making it challenging to incorporate the latest scientific knowledge into policy decisions.

3. **Scientific uncertainty:** Scientific knowledge is often characterized by uncertainties and limitations. Policymakers may find it challenging to navigate these uncertainties and make decisions based on incomplete or conflicting evidence. Effective science-policy interfaces should acknowledge and communicate uncertainties transparently to support evidence-based decision-making.

4. **Power dynamics and vested interests:** Science-policy interfaces can be influenced by power dynamics and vested interests, which may hinder the objective integration of scientific evidence into policy decisions. It is important to establish mechanisms that safeguard the independence and integrity of scientific advice in policy processes.

Strategies for effective science-policy interfaces

To overcome these challenges and enhance the effectiveness of science-policy interfaces, several strategies can be employed:

1. **Enhancing science communication skills:** Scientists should receive training and support to enhance their science communication skills. They should learn how to effectively communicate their research findings, uncertainties, and limitations to policymakers and the general public in a clear and accessible manner.

2. **Strengthening collaboration and engagement:** Science-policy interfaces should foster collaboration and engagement between scientists, policymakers, and other stakeholders. Platforms for dialogue, such as workshops, conferences, and science-policy networks, can facilitate interactions, build relationships, and foster trust between these communities.

3. **Co-designing research and policy processes:** Scientists and policymakers should co-design research and policy processes to ensure that they are responsive to

policy needs and generate actionable knowledge. This can involve joint problem framing, iterative feedback loops, and ongoing collaboration throughout the research-policy cycle.

4. **Promoting interdisciplinary research:** Science-policy interfaces should encourage interdisciplinary research that integrates natural sciences, social sciences, humanities, and indigenous knowledge. By bringing together different perspectives and knowledge systems, interdisciplinary research can offer more comprehensive and context-specific insights for policy decisions.

5. **Strengthening institutional support:** Institutions, such as government agencies, research organizations, and international bodies, should provide adequate support and resources for science-policy interfaces. This includes funding for collaborative research projects, capacity building programs, and knowledge translation activities.

6. **Fostering evidence-informed policymaking:** Science-policy interfaces should promote a culture of evidence-informed policymaking, where policy decisions are based on the best available scientific evidence, combined with other considerations, such as societal values, feasibility, and equity.

In conclusion, science-policy interfaces play a critical role in addressing complex environmental challenges and promoting sustainable development. By fostering collaboration, communication, and co-production of knowledge between scientists and policymakers, science-policy interfaces can improve the quality and effectiveness of policy decisions. However, they also face challenges related to communication, time constraints, scientific uncertainty, and power dynamics. By implementing strategies such as enhancing science communication skills, strengthening collaboration, and promoting interdisciplinary research, we can overcome these challenges and enhance the role of science in policy processes.

Knowledge co-production

Knowledge co-production is a collaborative process that involves the active participation of multiple stakeholders in the production and use of knowledge. It recognizes the value of different types of knowledge, including scientific knowledge, indigenous knowledge, local knowledge, and experiential knowledge. By bringing together diverse perspectives and expertise, knowledge co-production aims to generate context-specific and actionable knowledge that can inform decision-making processes and contribute to sustainable development.

Principles of knowledge co-production

Knowledge co-production is guided by several key principles:

1. **Participatory approach:** Knowledge co-production involves the meaningful engagement and inclusion of stakeholders throughout the entire research process. This includes problem identification, data collection, analysis, interpretation, and application of knowledge. By involving stakeholders, such as local communities, policymakers, and practitioners, in every step of the process, knowledge co-production ensures that the generated knowledge is relevant, credible, and applicable to real-world contexts.

2. **Knowledge integration:** Knowledge co-production seeks to integrate different forms of knowledge, recognizing that diverse perspectives and ways of knowing contribute to a more comprehensive understanding of complex environmental and social issues. This includes integrating scientific knowledge with indigenous and local knowledge systems, as well as incorporating experiential knowledge from practitioners and community members. By combining different knowledge sources, knowledge co-production can provide a more holistic and nuanced understanding of problems and solutions.

3. **Transdisciplinary collaboration:** Knowledge co-production encourages collaboration among researchers, practitioners, policymakers, and communities from different disciplines and sectors. It recognizes that addressing complex sustainability challenges requires expertise and perspectives from various fields, such as ecology, social sciences, economics, and engineering. Transdisciplinary collaboration promotes the co-creation of knowledge, fosters mutual learning, and enhances the potential for innovation and transformative change.

4. **Contextualization and customization:** Knowledge co-production emphasizes the need to tailor research and knowledge generation processes to specific local contexts. This involves understanding the unique social, cultural, economic, and ecological conditions of a given region or community. By contextualizing knowledge production, it ensures that the generated knowledge reflects the specific needs and values of the stakeholders involved.

Process of knowledge co-production

The process of knowledge co-production typically involves several stages:

1. **Problem identification and scoping:** This stage involves identifying the key sustainability challenges or research questions that need to be addressed. It also includes scoping the boundaries of the research, defining the stakeholders and their

roles, and establishing the objectives and expected outcomes of the co-production process.

2. **Participatory data collection:** In this stage, various methods are used to collect data, including field surveys, interviews, focus group discussions, and participatory mapping exercises. These data collection methods engage stakeholders directly, ensuring their meaningful involvement in generating and validating data.

3. **Knowledge synthesis and analysis:** The collected data is analyzed using both quantitative and qualitative methods. This stage involves bringing together different forms of knowledge, integrating data from multiple sources, and identifying patterns, trends, and relationships. The analysis should be conducted in a way that respects and recognizes the different types of knowledge involved.

4. **Interpretation and co-creation of knowledge:** The findings are interpreted and synthesized through facilitated workshops, deliberative dialogues, and other collaborative processes. Stakeholders are actively involved in interpreting the data, discussing the implications, and co-creating knowledge. This stage fosters mutual learning, consensus building, and the integration of different perspectives.

5. **Application and action:** The co-produced knowledge is applied to inform decision-making processes, policy development, and sustainable development initiatives. Stakeholders are actively engaged in identifying and prioritizing actions, designing interventions, and monitoring the outcomes. The goal is to ensure that the knowledge generated leads to tangible on-the-ground changes and improvements.

Benefits and challenges of knowledge co-production

Knowledge co-production offers several benefits:

1. **Enhanced relevance and applicability:** By involving stakeholders directly, knowledge co-production ensures that the generated knowledge is tailored to the specific needs and contexts of the users. This increases the relevance and applicability of the knowledge in decision-making processes and actions.

2. **Improved legitimacy and credibility:** The inclusion of diverse stakeholders and knowledge sources enhances the legitimacy and credibility of the generated knowledge. It increases trust among stakeholders and encourages the use of knowledge in policy and practice.

3. **Increased innovation and transformative change:** Through transdisciplinary collaboration and the integration of different knowledge sources, knowledge co-production stimulates innovation and creative problem-solving. It can lead to transformative changes in policies, practices, and behavior.

4.   **Empowerment and capacity building:**   Knowledge co-production empowers stakeholders by involving them in the research and decision-making processes.   It strengthens their capacity to understand and address complex sustainability challenges, fostering a sense of ownership and accountability.

Despite its benefits, knowledge co-production also faces several challenges:

1. **Power dynamics and inequalities:** Power imbalances, unequal access to resources, and differences in knowledge systems can affect the co-production process. It is essential to address these power dynamics and ensure the meaningful inclusion and participation of marginalized and vulnerable groups.

2. **Time and resource constraints:** Knowledge co-production requires time, resources, and sustained commitment from all stakeholders involved.   Limited funding, time constraints, and administrative barriers can pose challenges to the implementation of knowledge co-production processes.

3. **Conflict and divergence of interests:** Stakeholders may have conflicting interests, values, and priorities, which can pose challenges to consensus-building and decision-making. Transparent and inclusive processes are necessary to navigate these conflicts and find common ground.

Case study: Indigenous knowledge and watershed management

An example of knowledge co-production is the integration of indigenous knowledge and scientific knowledge in watershed management.   Indigenous communities have unique knowledge and practices related to the management and conservation of water resources.   By combining indigenous knowledge with scientific research and monitoring techniques, more comprehensive and context-specific solutions to watershed management can be developed.

In this case, the process of knowledge co-production would involve engaging indigenous communities in the identification of research questions, data collection, and analysis. Local elders and community members can provide insights into the ecological dynamics, hydrological patterns, and traditional water management practices in the watershed. Scientists and researchers can contribute their expertise in data collection and analysis techniques, as well as in identifying potential threats and pressures on the watershed.

Through collaborative workshops and dialogues, the indigenous communities and researchers can co-create knowledge that integrates indigenous knowledge and scientific knowledge. This can lead to the development of sustainable watershed management strategies that incorporate both traditional practices and modern scientific approaches. The co-produced knowledge can inform policies, land use planning, and community-based initiatives aimed at preserving and restoring the watershed's ecological integrity.

Conclusion

Knowledge co-production is a valuable approach to generate context-specific and actionable knowledge for sustainable development. By involving diverse stakeholders and integrating different types of knowledge, it recognizes the importance of local contexts, promotes mutual learning, and fosters innovation. However, knowledge co-production also faces challenges related to power dynamics, resource constraints, and conflicting interests. Overcoming these challenges requires transparent and inclusive processes that prioritize equity and address power imbalances. Ultimately, knowledge co-production can contribute to more effective and sustainable solutions to complex sustainability challenges.

Future Directions in Eco Science

Emerging Technologies

Artificial intelligence

Artificial intelligence (AI) is a branch of computer science that aims to create intelligent machines capable of performing tasks that typically require human intelligence. AI systems can learn from experience, adapt to new inputs, and perform tasks with varying degrees of autonomy. This section explores the principles and applications of AI in the context of eco science and sustainability.

Introduction to Artificial Intelligence

AI is based on the idea of creating machines that can simulate human intelligence. It involves the development of algorithms and models that enable computers to understand, reason, learn, and make decisions. AI has the potential to revolutionize various fields, including eco science, by providing innovative solutions to complex environmental problems.

Machine Learning

One of the key techniques used in AI is machine learning (ML). ML algorithms enable computers to learn and improve from experience without being explicitly programmed. ML algorithms can analyze large datasets, identify patterns, and make predictions or decisions based on the data. In the context of eco science, machine learning can be used to analyze environmental data, identify trends, and make predictions about future outcomes.

Applications of Artificial Intelligence in Eco Science

AI has a wide range of applications in the field of eco science. Some of the notable applications include:

Environmental Monitoring and Data Analysis   AI algorithms can be used to analyze large volumes of environmental data collected from various sources such as satellites, sensors, and citizen science initiatives. These algorithms can identify patterns, detect anomalies, and provide valuable insights into the state of the environment. For example, AI can be used to analyze satellite imagery to monitor deforestation, track changes in biodiversity, and detect the presence of pollutants in water bodies.

Species Identification and Conservation   AI can play a crucial role in the identification and conservation of species. ML algorithms can be trained on large datasets of species images and audio recordings to accurately identify different species. This can help in monitoring and managing biodiversity, especially in remote or inaccessible areas. AI can also aid in predicting the impact of environmental changes on species populations and developing strategies for conservation and habitat restoration.

Smart Agriculture and Precision Farming   AI-powered systems can optimize agricultural practices by analyzing soil data, weather conditions, and crop health indicators. By analyzing these factors, AI can generate recommendations for precise fertilizer and pesticide application, water management, and optimized planting strategies. This can lead to increased crop yields, reduced resource usage, and minimized environmental impact.

Natural Language Processing for Environmental Policy   AI techniques, such as natural language processing (NLP), can help analyze and interpret large volumes of text data related to environmental policies and regulations. By extracting key information from policy documents, NLP algorithms can aid in the assessment of policy effectiveness, identification of gaps or inconsistencies, and support evidence-based decision-making in environmental governance.

Ethical Considerations and Challenges

As AI becomes more prevalent in eco science and sustainability, it is essential to consider the ethical implications and address the challenges associated with its use.

Some of the key ethical considerations include:

Bias and Fairness AI algorithms are trained on historical data, which may contain biases and inequities. These biases can be inadvertently perpetuated, leading to unfair outcomes and reinforcing existing social and environmental inequalities. It is crucial to ensure that AI systems are designed and trained in a way that promotes fairness, equity, and inclusiveness.

Privacy and Security AI systems often rely on vast amounts of personal and sensitive data. This raises concerns about privacy and data security. It is essential to adopt robust data protection measures and ensure transparency in data collection, storage, and usage to safeguard individual privacy and prevent misuse of data.

Transparency and Accountability As AI systems become more complex, it becomes challenging to understand and explain their decision-making processes. Ensuring transparency and accountability in AI algorithms is critical to building trust in their applications. Efforts should be made to develop explainable AI models and establish guidelines for auditing and validating AI systems.

Long-term Socioeconomic Impact The widespread adoption of AI in eco science may have significant socioeconomic implications, including job displacement and inequality. It is crucial to consider the potential negative impacts and implement measures to mitigate any adverse effects. This may include retraining and upskilling programs, ensuring equitable access to AI technologies, and promoting inclusive economic growth.

Conclusion

Artificial intelligence has the potential to revolutionize eco science and sustainability. By leveraging AI technologies like machine learning and natural language processing, we can gain deeper insights into environmental issues, develop innovative solutions, and make informed decisions to foster a sustainable future. However, it is essential to address the ethical considerations and challenges associated with AI to ensure its responsible and equitable use in eco science. The future of AI in eco science holds immense promise, but it requires careful consideration and ethical implementation to maximize its potential for positive change.

Gene editing

Gene editing is a powerful tool that allows scientists to make precise changes to the DNA of an organism. It has revolutionized the field of genetics and has the potential to transform various industries, including agriculture, medicine, and environmental conservation. In this section, we will explore the principles behind gene editing, its applications, and some of the ethical considerations associated with its use.

Principles of gene editing

Gene editing relies on molecular tools that can specifically target and modify genes within an organism's genome. There are several methods of gene editing, but one of the most widely used techniques is called CRISPR-Cas9.

CRISPR-Cas9 is a system that uses a molecule called RNA as a guide to locate and bind to specific DNA sequences. Once the RNA molecule is bound to its target, an enzyme called Cas9 cuts the DNA at that location. Scientists can then introduce a desired genetic change by providing a DNA template that the organism's cellular repair machinery can use to fix the cut.

Applications of gene editing

Gene editing has immense potential in various fields and can be used to address a wide range of challenges. Here are some examples of its applications:

- **Agriculture:** Gene editing can be used to create crops that are more resistant to pests, diseases, and environmental stressors. It can also be used to improve crop yield and nutritional value.

- **Medicine:** Gene editing holds promise in the development of new treatments for genetic diseases. It can be used to correct disease-causing mutations in human cells and potentially cure inherited disorders. Gene editing is also being explored as a tool for cancer treatment and the development of personalized medicine.

- **Conservation:** Gene editing can aid in the conservation of endangered species by promoting genetic diversity and increasing their resilience to environmental changes. It can also help in the management of invasive species by altering their reproductive abilities or eliminating harmful traits.

Ethical considerations

While gene editing presents exciting possibilities, it also raises important ethical considerations. The ability to modify the DNA of living organisms raises questions about the boundaries of human intervention in nature and the potential consequences of genetic alterations. Here are some of the key ethical considerations:

+ **Safety**: Ensuring the safety of gene-edited organisms is crucial. Extensive research and rigorous testing are necessary to understand the potential risks and unintended consequences associated with genetic modifications.

+ **Equity**: There are concerns about equitable access to gene editing technologies. It is important to consider how gene editing can be used to address social inequality and ensure that the benefits are shared by all.

+ **Environmental impact**: Gene editing can have unintended consequences on ecosystems. It is essential to evaluate the potential ecological impacts of releasing genetically modified organisms into the environment.

+ **Human germline editing**: Germline editing refers to making genetic changes that can be passed on to future generations. The ethical implications of editing the human germline are particularly complex, raising questions about consent, eugenics, and the alteration of human traits.

Real-world example: Editing disease-causing mutations

To illustrate the potential of gene editing in medicine, let's consider a real-world example. Cystic fibrosis (CF) is a genetic disorder that affects the lungs, digestive system, and other organs. It is caused by mutations in the CFTR gene.

Using gene editing techniques, scientists have been able to target the CFTR gene and correct disease-causing mutations in human cells. This opens up the possibility of developing a cure for CF by editing the gene in affected individuals.

However, there are still significant challenges to overcome before gene editing can be used as a widespread treatment for genetic diseases. These include ensuring the efficiency and safety of the editing process, addressing ethical concerns, and navigating regulatory frameworks.

Further resources

If you're interested in learning more about gene editing, here are some resources to explore:

+ *CRISPR: A Revolutionary Gene Editing System* by Jennifer Doudna and Samuel Sternberg

+ *The Gene: An Intimate History* by Siddhartha Mukherjee

+ *The Ethics of Invention: Technology and the Human Future* by Sheila Jasanoff

Key takeaways

In this section, we explored the principles of gene editing, its applications in agriculture, medicine, and conservation, and some of the ethical considerations associated with its use. Gene editing has the potential to revolutionize various fields and address complex challenges, but it also raises important questions about safety, equity, and the environment. As we continue to advance in our understanding and capabilities in gene editing, it is essential to approach this technology with careful consideration of its implications and a commitment to responsible innovation.

Biodegradable Materials

Biodegradable materials play a crucial role in promoting sustainability and minimizing environmental impact. These materials have the ability to break down naturally into non-toxic substances by the action of microorganisms such as bacteria and fungi. In this section, we will explore the concept of biodegradable materials, their advantages and challenges, and their applications in various industries.

Understanding Biodegradability

Biodegradability refers to the ability of a material to decompose under specific environmental conditions by the action of microorganisms. It is influenced by several factors such as temperature, moisture, oxygen availability, and the presence of enzymes. Biodegradable materials can undergo both aerobic (in the presence of oxygen) and anaerobic (in the absence of oxygen) degradation processes.

There are two main types of biodegradable materials: natural and synthetic. Natural biodegradable materials are derived from renewable resources such as plant-based products (e.g., cellulose, starch) and animal-based products (e.g., gelatin). Synthetic biodegradable materials, on the other hand, are chemically engineered to have biodegradable properties, typically using polymers derived from petrochemicals or renewable resources.

Advantages of Biodegradable Materials

The use of biodegradable materials offers several advantages:

+ **Reduced environmental impact:** Biodegradable materials break down into harmless byproducts, reducing pollution and minimizing waste accumulation in landfills and water bodies.

+ **Conservation of resources:** Most biodegradable materials are derived from renewable resources, reducing the dependence on finite fossil fuels.

+ **Lower carbon footprint:** Biodegradable materials generally have lower greenhouse gas emissions during production compared to conventional materials.

+ **Biocompatibility:** Many biodegradable materials are compatible with biological systems, making them suitable for medical applications such as biodegradable implants and drug delivery systems.

Challenges and Limitations

Despite their many advantages, biodegradable materials also face challenges and limitations:

+ **Limited lifespan:** Biodegradable materials have a finite lifespan and may not be suitable for long-term applications where durability and strength are key requirements.

+ **Variable degradation rates:** The rate of biodegradation can vary depending on the specific material, environmental conditions, and microbial activity, making precise prediction of degradation times challenging.

+ **Processing constraints:** Biodegradable materials may require specialized processing techniques, which can limit their use in certain industries.

+ **Cost considerations:** Biodegradable materials can have higher production costs compared to traditional materials, which can present economic barriers to widespread adoption.

Applications of Biodegradable Materials

Biodegradable materials find applications across a range of industries, including:

* **Packaging:** Biodegradable packaging materials, such as bioplastics, offer a sustainable alternative to conventional plastic packaging. They can reduce waste and facilitate composting.

* **Agriculture:** Biodegradable mulch films and plant pots made from biodegradable materials can enhance soil health and reduce plastic waste in agriculture.

* **Medical field:** Biodegradable polymers are used in various medical applications, including sutures, implants, drug delivery systems, and tissue engineering scaffolds.

* **Textiles:** Biodegradable fibers and fabrics, derived from natural materials like bamboo, hemp, and silk, offer sustainable alternatives to synthetic textiles.

* **Personal care products:** Biodegradable materials are used in the production of eco-friendly personal care products such as biodegradable wipes and menstrual products.

Case Study: Biodegradable Plastics

Biodegradable plastics have gained significant attention as an alternative to conventional plastics, which are known for their long persistence in the environment. These biodegradable plastics are typically derived from renewable resources like cornstarch or plant oils.

One example is polylactic acid (PLA), a biodegradable polymer derived from corn starch. PLA can be used to produce packaging materials, disposable cutlery, and food containers. When exposed to the appropriate conditions, such as high temperatures and moisture, PLA breaks down over time into carbon dioxide and water.

However, it is important to note that biodegradable plastics require proper disposal methods to ensure effective degradation. If not disposed of properly, they can still contribute to pollution and littering.

Conclusion

Biodegradable materials offer sustainable alternatives in various industries, helping to reduce pollution, conserve resources, and minimize waste. Despite some challenges and limitations, ongoing research and development are paving the way for improved biodegradable materials with enhanced durability and performance. As we continue to embrace eco-friendly solutions, the use of biodegradable materials will play a vital role in achieving a more sustainable future.

Nanotechnology

Nanotechnology is a rapidly growing field that focuses on the manipulation and control of matter at the nanoscale, typically between 1 and 100 nanometers in size. At this scale, the properties of materials can be significantly different from their macroscopic counterparts, allowing for the development of novel materials, devices, and applications with unique characteristics.

Background

The concept of nanotechnology was first proposed by physicist Richard Feynman in 1959 during his famous lecture, "There's Plenty of Room at the Bottom." However, it was not until the 1980s that the term "nanotechnology" was coined by Professor K. Eric Drexler, who envisioned the possibility of building complex structures and machines atom by atom.

Nanotechnology draws on principles and knowledge from various disciplines, including physics, chemistry, materials science, biology, and engineering. It offers immense potential for advancements in several fields, including electronics, medicine, energy, environmental science, and materials science.

Principles of Nanotechnology

Nanotechnology encompasses two main approaches: top-down and bottom-up.

In the top-down approach, researchers start with a bulk material and use various techniques to shrink it down to the desired nanoscale dimensions. This method often includes processes like lithography, etching, and milling. The top-down approach is commonly used in semiconductor industry to manufacture microchips with nanoscale features.

On the other hand, the bottom-up approach involves the assembly of individual atoms, molecules, or nanoparticles to build structures and devices from the ground up. This approach relies on self-assembly, chemical synthesis, and molecular engineering. It allows for precise control over the arrangement and composition of nanoscale materials.

Applications of Nanotechnology

Nanotechnology has a wide range of applications across various sectors. Some notable ones include:

1. Electronics: Nanotechnology has revolutionized the electronics industry by enabling the development of smaller, faster, and more efficient devices. Nanoscale transistors, memory chips, and displays are examples of the applications of nanotechnology in electronics.

2. Medicine: Nanotechnology has the potential to transform healthcare by enabling targeted drug delivery, early detection and diagnosis of diseases, and regenerative medicine. Nanoparticles can be designed to deliver drugs directly to diseased cells, minimizing side effects and improving therapeutic efficacy.

3. Energy: Nanotechnology plays a vital role in the development of clean and renewable energy sources. Nanomaterials, such as quantum dots and nanowires, are used in solar cells to enhance light absorption and improve energy conversion efficiency. Moreover, nanotechnology-enabled batteries and supercapacitors are being developed for efficient energy storage.

4. Environmental Science: Nanotechnology offers innovative solutions for environmental challenges. For example, nanomaterials can be used for pollution detection and remediation, water purification, and air filtration. Nanosensors can detect and monitor pollutants in real-time, enabling timely interventions to mitigate environmental risks.

5. Materials Science: Nanotechnology has led to the development of advanced materials with enhanced properties. Nanocomposites, for instance, combine different materials at the nanoscale to create stronger, lighter, and more durable materials with unique properties. These materials find applications in aerospace, automotive, and construction industries.

Challenges and Ethical Considerations

While nanotechnology holds great promise, it also presents several challenges and ethical considerations. Some of the key challenges include:

1. Safety Concerns: The potential health and environmental risks associated with exposure to engineered nanomaterials are a major concern. Researchers and policymakers must ensure the safe handling, disposal, and regulation of nanomaterials to minimize any potential adverse effects.

2. Regulation and Standardization: The rapid pace of nanotechnology development poses challenges for regulatory agencies to keep up with the evolving

field. Standards and guidelines need to be established to ensure the responsible development, use, and commercialization of nanotechnology.

3. Societal Implications: Nanotechnology may lead to economic disruption and potential job displacement in certain industries. Additionally, ethical considerations surrounding privacy, security, and the equitable distribution of nanotechnology benefits need to be addressed.

4. Environmental Impact: The production and disposal of nanomaterials can have environmental implications. Efforts must be made to minimize the environmental footprint of nanotechnology and consider its lifecycle impacts.

In conclusion, nanotechnology has the potential to revolutionize various fields by enabling precise control and manipulation at the nanoscale. While there are challenges to address, proper regulation, risk assessment, and ethical considerations can ensure the responsible development and use of nanotechnology for the benefit of society.

Blockchain Technology

Blockchain technology is a revolutionary concept that has gained significant attention in recent years. It is a distributed ledger system that allows for secure and transparent transactions, making it an ideal solution for various industries. In this section, we will explore the principles, applications, and potential challenges of blockchain technology.

Principles of Blockchain

At its core, blockchain technology is built upon three fundamental principles: decentralization, immutability, and security.

Decentralization: Unlike traditional centralized systems, blockchain operates in a decentralized manner. This means that there is no single governing authority or central point of control. Instead, multiple participants, known as nodes, maintain and validate the blockchain network. This decentralized structure ensures transparency, accountability, and resilience against single points of failure.

Immutability: The immutability of the blockchain refers to the inability to modify or tamper with data once it has been recorded. Each transaction or data entry, known as a block, is linked to the previous one through cryptographic hashes, creating a chain of blocks. This cryptographic linkage ensures that any attempt to alter a block will be easily detectable, thus preserving the integrity of the data.

Security: Blockchain technology utilizes advanced cryptographic algorithms to secure transactions and data. Each transaction is verified by network participants using consensus mechanisms such as proof-of-work or proof-of-stake. Once validated, the transaction is added to the blockchain, making it practically impossible to alter or delete. Additionally, the decentralized nature of the blockchain network makes it highly resistant to hacking or manipulation.

Applications of Blockchain

Blockchain technology has the potential to disrupt numerous industries and enable new applications. Let's explore some of the key areas where blockchain is being applied.

Financial Services: One of the most well-known applications of blockchain is in the financial services industry. Blockchain's ability to facilitate secure and transparent transactions without intermediaries has the potential to revolutionize processes such as cross-border payments, remittances, and trade finance. It can also improve the speed and efficiency of transactions while reducing costs.

Supply Chain Management: Blockchain provides a decentralized and transparent platform for tracking and verifying goods throughout the supply chain. By recording each step of the supply chain process, from sourcing to delivery, on the blockchain, companies can enhance traceability, eliminate counterfeit products, and improve trust between stakeholders.

Healthcare: Blockchain technology can address major challenges in healthcare, including data security, interoperability, and patient privacy. By storing medical records and securing them with cryptography, blockchain can ensure the privacy and integrity of patient data, while also enabling seamless sharing of information between different healthcare providers.

Identity Management: Blockchain has the potential to revolutionize identity management by providing a secure and decentralized system for verifying and managing digital identities. It can eliminate the need for multiple usernames and passwords by enabling users to have control over their personal data and granting access on a need-to-know basis.

Smart Contracts: Smart contracts are self-executing agreements written in code and stored on the blockchain. They automatically execute and enforce the terms of an agreement when predefined conditions are met. Smart contracts have the potential to automate various processes, reduce the need for intermediaries, and ensure trust and transparency in transactions.

Challenges and Considerations

While blockchain technology holds great promise, there are several challenges and considerations that need to be addressed for widespread adoption.

Scalability: The scalability of blockchain networks is a significant challenge. Current blockchain platforms have limitations in terms of transaction speed and processing capacity. As the number of participants and transactions increases, the scalability issue becomes more critical. Various solutions, such as layer-two protocols and sharding, are being explored to address this challenge.

Energy Consumption: Blockchain networks that rely on proof-of-work consensus mechanisms consume a significant amount of energy. The process of validating transactions requires extensive computational power, which contributes to environmental concerns. However, emerging consensus mechanisms like proof-of-stake aim to reduce energy consumption and mitigate these issues.

Regulatory Environment: Blockchain technology operates across national borders, raising questions about the regulatory environment. Policymakers should strike a balance between supporting innovation and ensuring consumer protection, privacy, and financial stability. Clear and adaptable regulations are essential to foster blockchain's growth.

Privacy and Security: While blockchain provides a secure platform, privacy concerns can arise when dealing with sensitive data. It is important to strike a balance between maintaining transparency and protecting confidential information. Additionally, new security vulnerabilities may arise as blockchain technology evolves, demanding continuous efforts to stay ahead of potential threats.

Real-World Example: Blockchain in Supply Chain

To better understand the potential of blockchain technology in supply chain management, let's consider a real-world example.

Imagine a global seafood company that wants to provide transparency and traceability for its products. By storing relevant data, such as the origin, processing, and transportation of each seafood batch, on a blockchain network, the company can share this information with consumers, suppliers, and regulatory authorities. This not only enhances trust but also helps identify and address any issues related to quality control or sustainability.

Additionally, the blockchain can be programmed to automatically trigger actions or payments based on predefined conditions. For instance, if a temperature sensor detects that the product has been exposed to suboptimal conditions during transportation, a smart contract can automatically initiate an insurance claim or trigger an investigation.

By leveraging blockchain technology, supply chain stakeholders can streamline processes, reduce fraud, improve product quality, and ensure sustainability throughout the supply chain.

Resources and Further Reading

If you want to dive deeper into blockchain technology, here are some recommended resources:

+ *Mastering Blockchain: Distributed Ledger Technology, Decentralization, and Smart Contracts Explained* by Imran Bashir.

+ *Blockchain Basics: A Non-Technical Introduction in 25 Steps* by Daniel Drescher.

+ *Blockchain Revolution: How the Technology behind Bitcoin Is Changing Money, Business, and the World* by Don Tapscott and Alex Tapscott.

+ *Blockchain: Blueprint for a New Economy* by Melanie Swan.

These resources provide comprehensive insights into the technical and practical aspects of blockchain technology, enabling you to explore its potential applications further.

Conclusion

Blockchain technology has the potential to revolutionize various industries by providing decentralized, secure, and transparent solutions. Its principles of decentralization, immutability, and security lay the foundation for innovative applications.

While challenges such as scalability and energy consumption exist, ongoing research and development aim to address these issues. With the right regulatory frameworks and continuous advancements in technology, blockchain has the potential to transform industries, improve efficiency, and create new economic opportunities.

So, whether you're interested in finance, supply chain management, healthcare, or any other field, understanding blockchain technology is becoming increasingly valuable in today's evolving digital landscape. Embrace the opportunities it presents and explore the limitless potential of blockchain.

Autonomous Systems

Autonomous systems, also referred to as autonomous agents or robots, are machines or computer systems capable of performing tasks without human intervention. These systems are designed to operate independently, making decisions and taking actions based on their programming and the information they gather from their environment. Autonomous systems have gained significant attention and importance in various fields, including transportation, manufacturing, healthcare, and defense.

Overview

The concept of autonomous systems is rooted in the field of artificial intelligence (AI) and robotics. It combines computer science, control theory, and engineering to create intelligent machines that can sense, perceive, reason, and act upon the world around them. Autonomous systems are designed to mimic human behavior and cognitive processes, enabling them to perform complex tasks efficiently and accurately.

Components of Autonomous Systems

Autonomous systems consist of several key components that enable them to function effectively. These components include:

1. **Sensors:** Autonomous systems are equipped with various sensors, such as cameras, lidar, radar, and GPS, to perceive and gather information about their surroundings. These sensors enable the system to detect objects, measure distances, and identify obstacles.

2. **Perception and Processing:** The gathered sensor data is processed using advanced algorithms and machine learning techniques to extract meaningful information. This information is used to generate a model of the system's environment and to make informed decisions.

3. **Decision-Making:** Autonomous systems use decision-making algorithms to analyze the processed data and make decisions based on predefined rules and objectives. These algorithms often use probabilistic models and optimization techniques to evaluate different courses of action and select the most appropriate one.

4. **Actuation:** Once a decision is made, the autonomous system must act upon it. This involves controlling motors, actuators, or other physical mechanisms to perform the desired action. Actuation is crucial for tasks such as navigation, manipulation, or communication.

5. **Learning and Adaptation:** Autonomous systems can also incorporate learning capabilities to improve their performance over time. By analyzing the outcomes of their actions and receiving feedback, the system can update its knowledge, refine its decision-making processes, and adapt to new situations or changes in its environment.

Applications of Autonomous Systems

Autonomous systems have a wide range of applications across various domains. Some notable examples include:

1. **Autonomous Vehicles:** Autonomous cars, drones, and unmanned aerial vehicles (UAVs) are becoming increasingly popular. They can navigate through busy city traffic, deliver packages, conduct surveillance, and even assist in search and rescue operations.

2. **Manufacturing and Logistics:** Autonomous robots are employed in factories and warehouses for tasks like assembly, packaging, and material handling. These robots can work collaboratively with humans or autonomously to optimize efficiency and reduce labor costs.

3. **Healthcare and Rehabilitation:** Autonomous systems are used in healthcare settings to assist with tasks such as surgery, patient monitoring, and rehabilitation. Robotic surgical systems, for example, enhance the precision and safety of surgical procedures.

4. **Military and Defense:** Autonomous drones and unmanned ground vehicles (UGVs) are used for surveillance, reconnaissance, and combat missions. These systems can operate in hostile environments and perform dangerous tasks that would put human lives at risk.

Challenges and Considerations

While the potential applications of autonomous systems are vast, there are several challenges and considerations that need to be addressed:

1. **Safety and Ethical Concerns:** Safety is paramount when deploying autonomous systems, particularly in safety-critical domains like transportation and healthcare. Robust fail-safe mechanisms, ethical decision-making algorithms, and adherence to legal and ethical standards are vital to ensure the well-being of both humans and the autonomous systems themselves.

2. **Reliability and Trust:** Autonomous systems must be reliable and trustworthy. They should be able to operate robustly in various environments and

handle unexpected situations. Ensuring system integrity, security, and resilience against cyber-attacks is crucial.

3. **Regulations and Policies**: The rapid development and deployment of autonomous systems require appropriate regulations and policies to govern their use. These regulations should address privacy concerns, liability issues, and ensure compliance with legal frameworks.

4. **Human-Machine Interaction**: The interaction between humans and autonomous systems needs to be user-friendly and intuitive. Designing effective user interfaces and incorporating natural language processing and gesture recognition technologies can enhance collaboration between humans and autonomous systems.

5. **Job Displacement**: The adoption of autonomous systems may lead to job displacement in certain sectors. Adequate measures should be taken to address the social and economic impact, such as retraining programs and the creation of new job opportunities in emerging fields.

Case Study: Autonomous Delivery Drones

One of the prominent examples of autonomous systems is the use of delivery drones for last-mile logistics. Companies like Amazon and DHL are experimenting with autonomous drones to deliver packages directly to customers' doorsteps. This innovative approach offers several advantages, including faster delivery times, reduced traffic congestion, and lower carbon emissions.

The autonomous delivery drone system involves multiple components working together. The drones are equipped with sensors and cameras to navigate and avoid obstacles. Advanced machine learning algorithms analyze the sensor data and generate a 3D map of the surroundings. Based on this map, the drone plans the most efficient route to the destination.

To ensure safety and reliability, the delivery drone system includes redundancy mechanisms and failsafe procedures. For example, the drones have backup power sources and alternative landing strategies in case of emergencies. They are also equipped with collision detection and avoidance systems to prevent accidents.

Before deployment, the autonomous delivery drone system undergoes rigorous testing and validation to comply with aviation regulations and safety standards. This includes simulation studies, real-world testing, and collaboration with aviation authorities to ensure the system's safety and compliance.

The use of autonomous delivery drones presents several challenges and considerations. Airspace regulations, privacy concerns, and public acceptance are critical factors that need to be addressed for widespread adoption. Additionally, the logistics infrastructure, including landing sites and maintenance centers, must be established to support the autonomous delivery drone network.

Despite these challenges, autonomous delivery drones have the potential to revolutionize the logistics industry, providing fast and efficient delivery services while reducing the carbon footprint associated with traditional transportation methods.

Resources

1. R. Arkin. "Behavior-Based Robotics". MIT Press, 1998.

2. S. Russell and P. Norvig. "Artificial Intelligence: A Modern Approach". Pearson, 2016.

3. M. Thrun, W. Burgard, and D. Fox. "Probabilistic Robotics". MIT Press, 2005.

4. J. Anderson. "Autonomous Agents: From Self-Control to Autonomy". Oxford University Press, 2019.

Summary

Autonomous systems are complex machines or computer systems capable of performing tasks without human intervention. They incorporate sensors, perception and processing, decision-making, and actuation mechanisms to operate independently. Autonomous systems are widely used in applications such as autonomous vehicles, manufacturing, healthcare, and defense. However, challenges related to safety, reliability, regulations, human-machine interaction, and job displacement need to be addressed. The case study of autonomous delivery drones illustrates the potential of autonomous systems in transforming last-mile logistics.

3D Printing

3D printing, also known as additive manufacturing, is a revolutionary technology that allows the creation of three-dimensional objects from a digital file. It has gained significant attention and is transforming various industries, including manufacturing, healthcare, architecture, and even fashion. In this section, we will explore the principles, applications, and future prospects of 3D printing.

Principles of 3D Printing

The basic principle of 3D printing involves building an object layer by layer, unlike traditional subtractive methods that involve cutting or drilling from a solid block of material. The process begins with a digital 3D model designed using Computer-Aided Design (CAD) software. This model is then sent to a 3D printer, which interprets the design and builds the object layer by layer.

There are different types of 3D printing technologies, each utilizing different materials and methods. The most common ones include:

- Fused Deposition Modeling (FDM): This method uses a thermoplastic filament that is heated and extruded through a nozzle. The nozzle moves according to the design, depositing the material layer by layer to create the object.

- Stereolithography (SLA): SLA employs a liquid photopolymer that is selectively cured by a UV laser to solidify each layer of the object.

- Selective Laser Sintering (SLS): SLS uses a high-powered laser to selectively fuse powdered materials (such as plastics, metals, or ceramics) layer by layer to form the object.

- Digital Light Processing (DLP): DLP utilizes a projector screen to flash a patterned image of each layer onto a vat of liquid resin, which is then cured by UV light.

Applications of 3D Printing

The applications of 3D printing are vast and continue to expand as new materials and techniques are developed. Here are some notable areas where 3D printing has made a significant impact:

1. **Prototyping and Product Development:** 3D printing allows rapid prototyping, enabling designers and engineers to quickly iterate and test their ideas before investing in expensive manufacturing processes. It accelerates product development cycles and reduces costs.

2. **Medical and Healthcare:** 3D printing has revolutionized medical practices. It enables the production of patient-specific implants, prosthetics, and medical devices, improving patient outcomes. Preoperative planning models, anatomical replicas, and customized surgical tools can be 3D printed, enhancing surgical precision.

3. **Architecture and Construction:** Architects and construction professionals utilize 3D printing to create complex and intricate models, enabling

visualization and testing of designs before actual construction. Additionally, 3D printed concrete structures are being explored as a sustainable and cost-effective alternative to traditional construction methods.

4. **Automotive and Aerospace:** 3D printing has disrupted the automotive and aerospace industries by enabling the production of lightweight and complex components with improved performance. Aerospace companies utilize 3D printing to create intricate designs with reduced weight, leading to fuel efficiency and cost savings.

5. **Fashion and Design:** Designers and artists employ 3D printing to materialize innovative and avant-garde creations that were previously impossible to produce using traditional methods. 3D printed fashion garments, accessories, and jewelry are pushing the boundaries of creativity and customization.

Future Prospects of 3D Printing

The future of 3D printing is brimming with possibilities. Here are some exciting prospects and emerging trends:

1. **Advanced Materials:** Researchers are actively exploring new materials with enhanced properties for 3D printing. This includes biocompatible polymers for medical implants, high-strength metals for aerospace applications, and conductive materials for electronics.

2. **Multi-material and Multi-component Printing:** Advancements in 3D printing technology are enabling the simultaneous printing of multiple materials and components within a single object. This opens up possibilities for creating complex, functional objects with embedded electronics, sensors, and integrated functionalities.

3. **Sustainable and Recyclable Printing:** As sustainability becomes a top priority, efforts are being made to develop eco-friendly 3D printing methods. This includes the use of recyclable materials, biodegradable resins, and optimizing energy consumption throughout the process.

4. **Mass Production:** Traditional manufacturing industries are exploring the use of 3D printing for mass production. Continuous advancements in speed, scalability, and cost-effectiveness are making it increasingly viable to produce end-use parts at scale, challenging the conventional manufacturing processes.

5. **Personalized Medicine:** 3D printing is expected to play a vital role in personalized medicine, where patient-specific organs, tissues, and implants can be fabricated. This includes the development of functional organs-on-a-chip for drug testing and the potential for bioprinting complex tissues and organs for transplantation.

Challenges and Considerations

Despite the incredible potential of 3D printing, there are several challenges and considerations to address:

1. **Intellectual Property (IP) and Regulation:** The ease of replicating objects using 3D printing raises concerns over intellectual property rights and copyright infringement. Developing appropriate regulations and frameworks to protect IP is crucial.

2. **Quality Control and Certification:** Ensuring consistent and reliable quality control is essential for industries utilizing 3D printing. Developing standardized certification processes and quality assurance protocols will be vital to gain trust and confidence in 3D printed products.

3. **Safety and Environmental Impact:** The materials used in 3D printing may have an impact on human health and the environment. Proper handling of potentially hazardous materials, waste management, and sustainable material choices are necessary to mitigate any adverse effects.

4. **Technological Limitations:** Despite rapid advancements, there are still limitations to 3D printing technology. Factors such as printing resolution, size constraints, printing speed, and material limitations need to be further improved to unlock its full potential.

Conclusion

3D printing is a game-changing technology that is revolutionizing various industries. Its ability to enable rapid prototyping, customization, and the production of complex objects has profound implications for manufacturing, healthcare, design, and more. As the technology continues to evolve and address its challenges, the prospects for 3D printing are limitless. It holds the potential to empower innovation, sustainability, and personalization in the future.

Quantum Computing

Quantum computing is an exciting and rapidly evolving field that seeks to harness the unique properties of quantum mechanics to perform computations that are infeasible for classical computers. While classical computers store and process information in bits that can represent a 0 or a 1, quantum computers use quantum bits, or qubits, which can represent 0, 1, or a superposition of both states simultaneously. This inherent parallelism of qubits allows quantum computers to perform certain calculations exponentially faster than classical computers.

Principles of Quantum Computing

To understand quantum computing, we need to delve into the principles of quantum mechanics. Quantum mechanics describes the behavior of particles at the atomic and subatomic levels. Three key principles of quantum mechanics form the foundation of quantum computing:

Superposition: Unlike classical bits, qubits can exist in a superposition of states. This means that a qubit can simultaneously represent both 0 and 1. The superposition allows quantum computers to process multiple possibilities in parallel, providing a significant advantage over classical computers.

Entanglement: Entanglement is a phenomenon where two or more qubits become linked in such a way that the state of one qubit is intrinsically related to the state of another, regardless of the distance between them. This entanglement enables the manipulation of qubits in a way that their collective behavior is correlated.

Quantum gate operations: Quantum gate operations are the building blocks of quantum algorithms. These operations manipulate the state of qubits, analogous to how classical gate operations manipulate bits in classical computers. Examples of quantum gate operations include the Hadamard gate, CNOT gate, and Toffoli gate.

Quantum Algorithms

Quantum algorithms are specifically designed to exploit the power of quantum computers. They leverage the unique properties of quantum mechanics to solve problems faster and more efficiently than classical algorithms.

Grover's Algorithm: Grover's algorithm is a quantum algorithm that can speed up the search of an unstructured database. It provides a quadratic speedup compared to classical algorithms and has applications in database search, optimization problems, and cryptanalysis.

Shor's Algorithm: Shor's algorithm is a groundbreaking quantum algorithm for integer factorization, which is the process of finding prime numbers that divide a given composite number. This algorithm demonstrates exponential speedup compared to the best-known classical algorithms and has profound implications for cryptography.

Quantum Simulation: Quantum simulation is an important application of quantum computing, particularly for simulating complex quantum systems that are difficult to model on classical computers. Such simulations have applications in molecular dynamics, material science, and drug discovery.

Challenges and Considerations

While quantum computing holds great promise, there are several challenges and considerations that need to be addressed:

Quantum Error Correction: Qubits are highly sensitive to noise and decoherence, which can introduce errors in quantum computations. Quantum error correction techniques are essential for mitigating these errors and preserving the quantum state of the qubits.

Scalability: Building quantum computers with large numbers of qubits presents a significant challenge. As the number of qubits increases, so does the complexity of implementing quantum gates and minimizing the impact of noise.

Quantum Supremacy: Achieving quantum supremacy refers to the point at which a quantum computer can perform a calculation that is infeasible for classical computers. While there have been demonstrations of quantum supremacy for specific problems, achieving it in a more general sense remains a significant hurdle.

Ethical Considerations: As quantum computing advances, there are ethical considerations to address, particularly in the area of cryptography. Quantum computers have the potential to break commonly used encryption algorithms, raising concerns about data security and privacy.

Real-World Applications

Quantum computing has the potential to revolutionize various fields and address complex problems that are beyond the reach of classical computers. Some potential real-world applications of quantum computing include:

Optimization: Quantum algorithms can offer improved solutions for optimization problems, such as logistics planning, resource allocation, and portfolio optimization.

Machine Learning: Quantum machine learning algorithms hold the potential to enhance pattern recognition, data analysis, and optimization algorithms in the field of artificial intelligence.

Material Science: Quantum simulation can enable a better understanding of materials at the atomic and molecular levels, leading to the development of novel materials with specific properties.

Cryptography: While quantum computers have the potential to break current cryptographic algorithms, they can also pave the way for the development of quantum-resistant encryption methods.

Resources and Further Reading

For those interested in exploring the field of quantum computing further, here are some recommended resources:

- IBM Quantum Computing

- Nature Quantum Information

- Quantum Computing Stack Exchange

- Quantum Computing for Computer Scientists by Noson S. Yanofsky and Mirco A. Mannucci

In conclusion, quantum computing represents a paradigm shift in computational power and has the potential to revolutionize various fields. While there are challenges to overcome, the future of quantum computing holds tremendous promise for solving complex problems efficiently and advancing science and technology.

Internet of Things

The Internet of Things (IoT) is a concept that refers to the interconnection of various devices and objects through the internet. It enables these devices to collect and exchange data, allowing for seamless communication and automation. In the context of eco science, the IoT plays a significant role in enhancing sustainability efforts by providing real-time data, increasing operational efficiency, and enabling smarter decision-making.

Principles of the Internet of Things

The IoT operates on certain fundamental principles that govern its functioning. These principles include:

+ **Connectivity**: The IoT relies on connectivity to enable communication between devices. This connectivity can be established through wired or wireless networks, such as Wi-Fi, Bluetooth, or cellular networks.

+ **Sensors and Actuators**: Devices in the IoT are equipped with sensors to gather data from the environment and actuators to interact with the physical world. Sensors can measure various parameters such as temperature, humidity, air quality, and energy consumption, while actuators can perform actions based on the received data.

+ **Data Collection and Analysis**: The IoT collects vast amounts of data from interconnected devices. This data is then analyzed using advanced algorithms and techniques to derive meaningful insights and patterns. Data analysis is crucial in identifying trends, predicting outcomes, and making informed decisions.

+ **Interoperability**: In an IoT system, devices from different manufacturers and with different functionalities should be able to communicate and work together seamlessly. Interoperability ensures compatibility, enables data exchange, and promotes collaboration among various devices and systems.

+ **Security and Privacy**: As the IoT involves the exchange of sensitive data, ensuring the security and privacy of this data is of utmost importance. Robust security measures, such as encryption, authentication, and authorization protocols, are necessary to protect against unauthorized access and data breaches. Privacy concerns also need to be addressed to ensure that individuals' personal information is protected.

+ **Scalability**: The IoT should be scalable to accommodate a large number of devices and users. It should be able to handle the increasing volume of data generated and support the growing demand for connectivity and functionality.

Applications of the Internet of Things in Eco Science

The IoT has numerous applications in the field of eco science. Here are some notable examples:

+ **Environmental Monitoring:** The IoT enables real-time monitoring of environmental parameters such as air quality, water quality, and soil moisture. Sensors deployed in various locations can continuously collect data and transmit it to a central monitoring system. This data can then be used to assess the impact of human activities on the environment, identify pollution sources, and develop effective mitigation and conservation strategies.

+ **Energy Management:** Connected devices in buildings, factories, and homes can gather data on energy consumption patterns. This data can be used to optimize energy usage, detect energy wastage, and identify areas for energy efficiency improvements. For example, smart thermostats can regulate temperature settings based on occupancy and weather conditions, thus reducing energy consumption.

+ **Precision Agriculture:** IoT devices such as soil moisture sensors, weather stations, and drones equipped with cameras can be used in precision agriculture. These devices collect data on soil conditions, weather patterns, and crop health. Farmers can then make data-driven decisions on irrigation, fertilizer application, pest control, and harvesting, leading to optimized resource usage and increased crop yield.

+ **Waste Management:** IoT-based waste management systems can track the fill levels of trash cans and recycling bins in real-time. Data from these sensors can be used to optimize waste collection routes, reduce unnecessary pickups, and improve waste segregation practices. This helps in reducing fuel consumption, greenhouse gas emissions, and overall waste management costs.

+ **Smart Grids:** IoT devices can play a crucial role in optimizing energy distribution in smart grids. By collecting real-time data on energy demand and supply, smart grids can efficiently balance the load, incorporate renewable energy sources, and prioritize energy distribution during peak demand periods. This leads to a more reliable and resilient energy infrastructure.

+ **Biodiversity Monitoring:** The IoT can aid in monitoring biodiversity by using connected sensors and cameras to collect data on species presence, abundance, and behavior. This data can be used to identify trends, track migration patterns, and assess the impact of climate change and human activities on ecosystems. It can also help in designing effective conservation and management strategies.

Challenges and Considerations

While the IoT offers great potential for enhancing sustainability efforts, it also presents certain challenges and considerations. These include:

+ **Data Privacy and Security**: As the IoT involves the collection and transmission of massive amounts of data, ensuring data privacy and security is crucial. Unauthorized access to sensitive data can have severe consequences. Therefore, robust security measures must be implemented to protect against cyber threats and data breaches.

+ **Compatibility and Interoperability**: Due to the presence of numerous devices and platforms in the IoT ecosystem, ensuring compatibility and interoperability can be challenging. Standardization efforts are necessary to enable seamless communication and integration between devices from different manufacturers.

+ **Data Management and Analytics**: Managing and analyzing the vast amounts of data generated by IoT devices can be complex. Adequate infrastructure and advanced analytics tools are required to process, store, and derive meaningful insights from this data.

+ **Ethical Considerations**: The widespread adoption of IoT technology raises ethical concerns related to data ownership, consent, and surveillance. Clear guidelines and regulations need to be established to address these ethical considerations and protect the rights of individuals.

+ **Sustainability of IoT Devices**: The sustainability of IoT devices themselves is an important consideration. Proper disposal and recycling mechanisms should be in place to minimize the environmental impact of electronic waste. Moreover, energy-efficient design and manufacturing practices should be adopted to reduce the energy consumption of IoT devices.

Case Study: Smart City Initiatives

One of the most prominent applications of the IoT is in the development of smart cities. Smart cities leverage IoT technologies to enhance the efficiency of various urban systems, including transportation, energy, waste management, and public services. Let's consider the example of a smart city's transportation system.

In a smart city, transportation infrastructure is equipped with IoT devices such as sensors and cameras. These devices collect real-time data on traffic flow, occupancy rates of parking spaces, and public transportation usage. This data is then analyzed to optimize traffic management, reduce congestion, and improve overall transportation efficiency.

Using the collected data, traffic signals can be dynamically adjusted to optimize traffic flow based on current conditions. Smart parking systems can inform drivers about available parking spaces, reducing the time spent searching for parking and minimizing traffic congestion.

Public transportation can also be enhanced through IoT integration. Real-time tracking of buses and trains enables accurate arrival time predictions, allowing commuters to plan their journeys more effectively. This not only improves the overall reliability and efficiency of public transportation but also encourages its use, leading to reduced traffic congestion and lower carbon emissions.

Additionally, IoT-enabled waste management systems can optimize garbage collection routes based on the fill levels of trash bins. This reduces unnecessary pickups, saves fuel, and minimizes the environmental impact of waste management.

By integrating various urban systems through the IoT, smart cities can achieve greater sustainability, efficiency, and quality of life for their residents.

Conclusion

The Internet of Things plays a significant role in advancing eco science and promoting sustainability. By connecting devices, collecting real-time data, and enabling smart decision-making, the IoT enhances environmental monitoring, energy management, waste management, and biodiversity conservation efforts. However, challenges related to data privacy and security, compatibility, and ethical considerations must be addressed to fully realize the potential of the IoT in eco science. Smart city initiatives serve as notable examples of IoT applications, showcasing how interconnected systems can lead to sustainable urban development. As technology continues to evolve, the IoT will play an increasingly crucial role in shaping a sustainable future.

Resilience and Adaptation

Climate Change Adaptation

Climate change is one of the most significant challenges facing our planet today. The increase in greenhouse gas emissions has led to rising global temperatures, changing weather patterns, and an increase in the frequency and intensity of extreme weather events. These changes have profound implications for human societies and natural ecosystems. In order to mitigate the impacts of climate change and ensure the long-term resilience of our planet, it is crucial to develop effective strategies for climate change adaptation.

Understanding Climate Change

Before delving into climate change adaptation, it is important to have a clear understanding of the phenomenon itself. Climate change refers to long-term shifts in temperature, precipitation, wind patterns, and other aspects of the Earth's climate system. While natural climate variability has always existed, the current trend of rapid and unprecedented warming is primarily attributed to human activities, particularly the burning of fossil fuels and deforestation.

One of the key impacts of climate change is the rise in global average temperatures. This is leading to the melting of glaciers, the polar ice caps, and permafrost, resulting in rising sea levels. These rising sea levels threaten coastal areas and low-lying islands, increasing the risk of flooding and saltwater intrusion into freshwater sources.

Climate change also affects weather patterns, leading to more frequent and intense extreme weather events such as hurricanes, droughts, heatwaves, and heavy rainfall. These events can have devastating impacts on ecosystems, infrastructure, agriculture, and human lives.

Climate Change Adaptation Strategies

Climate change adaptation involves taking measures to manage and reduce the risks and impacts associated with climate change. These strategies can be implemented at various scales, from individual households to communities, regions, and nations. Adaptation strategies can be grouped into several categories:

1. **Ecosystem-based approaches:** Ecosystems provide a range of goods and services that are critical for human well-being. Protecting and restoring natural ecosystems can enhance their resilience and ability to cope with the impacts of climate change. Examples of ecosystem-based adaptation measures include

reforestation, restoring mangroves and wetlands, and creating protected areas for biodiversity conservation.

2. **Infrastructure and built environment:** Infrastructure plays a crucial role in facilitating human activities and supporting economic development. Adapting infrastructure to climate change involves considering the projected impacts, such as increased flood risk, and incorporating resilience measures into design, construction, and operation. This can include raising buildings, implementing flood-proofing measures, improving drainage systems, and using green infrastructure solutions like permeable pavements and green roofs.

3. **Water resource management:** Water is a vital resource for both humans and ecosystems, and climate change can significantly affect its availability and quality. Adaptation strategies for water resource management include water conservation and efficiency measures, developing alternative water sources such as desalination or wastewater reuse, and implementing flood and drought management plans.

4. **Agriculture and food security:** Climate change poses significant challenges to agriculture, as it affects crop yields, livestock productivity, and food supply chains. Adaptation strategies in this sector may include adopting climate-resilient crop varieties, improving irrigation efficiency, implementing agroforestry practices, and promoting sustainable agricultural practices that reduce greenhouse gas emissions.

5. **Health and social services:** Climate change can have direct and indirect impacts on human health, including increased heat stress, changes in disease patterns, and mental health issues arising from natural disasters. Adaptation measures in the health sector include early warning systems, improved healthcare infrastructure, and public education campaigns.

6. **Community engagement and participation:** Adaptation efforts are most effective when they include active participation from local communities. Engaging communities in decision-making processes, providing access to climate information and resources, and supporting grassroots initiatives can enhance the resilience and adaptive capacity of communities.

Case Study: The Netherlands and Flood Management

The Netherlands is a country renowned for its innovative approaches to flood management. With a significant portion of its landmass located below sea level, the country faces the constant threat of flooding.

In response to this challenge, the Dutch have implemented a comprehensive system of flood defenses, including seawalls, dikes, and storm surge barriers. These structures are continuously monitored and maintained to ensure their effectiveness.

Additionally, the Dutch have embraced a concept known as "room for the river," which involves creating more space for rivers to expand during flood events. This approach involves reshaping riverbanks, removing obstacles, and creating floodplains that can temporarily store excess water. By allowing rivers to take a more natural course during floods, the risk of catastrophic flooding is reduced.

The Netherlands has also invested in advanced flood forecasting and early warning systems, enabling authorities to issue timely alerts and evacuate vulnerable areas when necessary.

This case study demonstrates the importance of adopting a multifaceted approach to climate change adaptation, incorporating both structural and non-structural measures. It also highlights the need for long-term planning, extensive monitoring, and continuous evaluation of adaptation strategies.

Challenges and Opportunities

Implementing effective climate change adaptation strategies is not without its challenges. Limited financial resources, political barriers, and lack of awareness or understanding can hinder progress in adapting to climate change.

However, there are also significant opportunities associated with climate change adaptation. Investing in adaptation measures can generate co-benefits, such as improved ecosystem health, enhanced food security, and increased resilience in the face of other hazards such as natural disasters. Furthermore, adaptation can create new economic opportunities, such as innovation in clean technologies and the development of climate-resilient infrastructure.

Conclusion

Climate change adaptation is an essential component of building a sustainable and resilient future. By understanding the impacts of climate change, identifying vulnerable areas, and implementing appropriate adaptation strategies, we can minimize the risks and maximize the benefits associated with a changing climate.

Through ecosystem-based approaches, improved infrastructure, sustainable water resource management, resilient agriculture, and active community engagement, we can enhance our capacity to cope with the challenges of climate change. While adaptation efforts may be complex and require concerted action at all levels, they offer a pathway towards a more sustainable and climate-resilient future. It is up to us to embrace this challenge and take the necessary steps to adapt to a changing world.

Further Reading

1. IPCC Fifth Assessment Report: Climate Change 2014: Impacts, Adaptation, and Vulnerability - *https://www.ipcc.ch/report/ar5/wg2*

2. United Nations Framework Convention on Climate Change (UNFCCC) Adaptation Knowledge Portal - *https://www.adaptation-undp.org/*

3. Global Climate Adaptation Partnership - *https://climate-adapt.eea.europa.eu/*

4. The World Bank Climate Change Adaptation - *https://www.worldbank.org/en/topic/climatechange/brief/adaptation*

5. U.S. Climate Resilience Toolkit - *https://toolkit.climate.gov/topics/*

6. The Nature Conservancy: Climate Change Adaptation Principles - *https://www.nature.org/en-us/what-we-do/our-insights/perspectives/climate-change-adaptati*

Resilience thinking

Resilience thinking is an interdisciplinary approach that focuses on understanding and managing complex social-ecological systems in the face of uncertainty and change. It recognizes that ecosystems, societies, and economies are interconnected and interdependent, and that they can undergo gradual or abrupt changes due to disturbances and stresses. Resilience thinking aims to enhance the capacity of these systems to absorb and recover from shocks, while also adapting and transforming to new conditions.

At the core of resilience thinking is the concept of resilience, which refers to the ability of a system to persist and maintain its structure, function, and identity in the face of disturbances. Resilience is not about preventing change or returning to a previous state, but rather about navigating and adapting to change in a sustainable and desirable way.

Principles of resilience thinking

Resilience thinking is guided by a set of principles that help inform its approach to understanding and managing complex systems:

1. System thinking: Resilience thinking acknowledges that social-ecological systems are complex, adaptive, and interconnected. It emphasizes the need to understand system dynamics, feedback loops, and non-linear interactions between different components of the system.

2. Adaptive governance: Resilience thinking recognizes the importance of adaptive governance systems that can respond and adapt to changing conditions. This involves fostering participatory decision-making, learning, and

experimentation, and creating mechanisms for collaboration and knowledge exchange among different stakeholders.

3. Multiple scales and levels: Resilience thinking considers multiple scales and levels of analysis, recognizing that social-ecological systems are nested within larger systems. It emphasizes the need to integrate knowledge and actions across different scales to address complex sustainability challenges.

4. Diversity and redundancy: Resilience thinking emphasizes the importance of diversity and redundancy within systems. Diversity provides options and variations that can help buffer against disturbances, while redundancy ensures that multiple elements perform similar functions, enhancing system stability.

5. Transformability: Resilience thinking recognizes that there are limits to the capacity of systems to cope with change. In some cases, transformation may be necessary to navigate through thresholds and avoid undesirable regime shifts. Transformability involves the ability of actors to intentionally and adaptively shape the future of the system.

Applying resilience thinking

Resilience thinking has been applied in various fields, including ecosystem management, disaster risk reduction, climate change adaptation, and social-ecological systems research. It provides a framework for understanding and managing the complexities and uncertainties associated with these domains.

One example of resilience thinking in action is the management of coral reef ecosystems. Coral reefs are highly vulnerable to climate change and other stressors, such as overfishing and pollution. Resilience thinking recognizes the importance of maintaining the diversity of coral species, as well as the structural complexity of reef habitats. It emphasizes the need for adaptive management strategies that can promote the recovery of damaged reefs and reduce the impacts of future disturbances.

Another example is the application of resilience thinking in urban planning. Cities are complex systems that face multiple challenges, including urbanization, climate change, and social inequalities. Resilience thinking calls for approaches that promote adaptive capacity and transformability in cities, such as creating green infrastructure, promoting community engagement, and integrating climate change adaptation into urban planning processes.

Challenges and future directions

While resilience thinking offers a valuable framework for addressing sustainability challenges, there are also a number of challenges and unanswered questions that researchers and practitioners face.

One challenge is the assessment and measurement of resilience. How can we quantify and assess the resilience of complex systems? What are the indicators and metrics that can help us understand the capacity of a system to absorb and recover from disturbances?

Another challenge is integrating resilience thinking into policy and governance processes. How can resilience thinking be effectively translated into practice? What are the institutional arrangements and decision-making processes that can support adaptive governance and transformative change?

Additionally, there is a need to better understand the social and cultural dimensions of resilience. How do social norms, values, and power relations shape the resilience of social-ecological systems? How can social equity and justice be included in resilience thinking?

Looking ahead, future research in resilience thinking could explore the role of emerging technologies, such as artificial intelligence and blockchain, in enhancing the adaptive capacity and transformability of social-ecological systems. These technologies have the potential to facilitate data integration, decision-making, and knowledge exchange across different scales and levels.

In conclusion, resilience thinking offers a holistic and dynamic approach to understanding and managing complex systems. By embracing the principles of resilience, adaptive governance, and systems thinking, we can navigate uncertainties and build a more sustainable and resilient future.

Ecosystem Services and Resilience

Understanding Ecosystem Services

Ecosystem services are the benefits that humans derive from ecosystems. These services can be categorized into four main types:

1. Provisioning Services: These are the direct products obtained from ecosystems, such as food, water, timber, and medicines.

2. Regulating Services: These services involve the regulation of natural processes and include things like climate regulation, water purification, and pest control.

3. Cultural Services: These services are non-material benefits, such as recreational opportunities, cultural heritage, and spiritual values associated with ecosystems.

4. Supporting Services: These services are essential for the functioning of ecosystems and include nutrient cycling, soil formation, and biodiversity maintenance.

Ecosystem services are crucial for human well-being and the sustainability of societies. They play a significant role in maintaining ecological balance and supporting economic development. However, the capacity of ecosystems to provide these services is under threat due to various factors like habitat degradation, climate change, and pollution.

Ecosystem Resilience

Ecosystem resilience refers to the ability of an ecosystem to resist, recover, and adapt to disturbances while maintaining its essential structure, function, and feedbacks. Resilient ecosystems can absorb shocks and maintain their stability and productivity.

There are several key factors that contribute to ecosystem resilience:

1. Biodiversity: Ecosystems with high biodiversity tend to be more resilient as they have a greater range of functional traits, which enables them to adapt and recover from disturbances.

2. Redundancy: Redundancy refers to the presence of multiple species and functional groups performing similar ecological roles. This redundancy provides insurance against the loss of specific species or functions.

3. Connectivity: The connectivity between different patches of habitat within a landscape increases resilience by facilitating species movement and gene flow. This movement allows species to colonize new habitats and recolonize areas that have been disturbed.

4. Adaptive Capacity: The ability of an ecosystem to adapt to changing conditions is crucial for its resilience. Adaptive capacity can be enhanced by maintaining genetic diversity, promoting ecological connectivity, and supporting the resilience of social-ecological systems.

5. Feedback Mechanisms: Feedback mechanisms within ecosystems can either reinforce or dampen the impacts of disturbances. Positive feedback amplifies changes and can lead to rapid shifts in ecosystem states, while negative feedback stabilizes the system and maintains its equilibrium.

Enhancing Ecosystem Resilience

To enhance ecosystem resilience, it is important to implement strategies that promote the maintenance and restoration of ecosystem services. Here are some approaches that can be taken:

1. Protecting Biodiversity: Conservation efforts should focus on protecting and restoring biodiversity, as it underpins the resilience of ecosystems. This can be achieved through the establishment of protected areas, habitat restoration, and conservation of endangered species.

2. Sustainable Land Use: Implementing sustainable land use practices helps maintain ecosystem services and reduces the pressure on natural resources. This includes practices like sustainable agriculture, agroforestry, and responsible forestry management.

3. Ecosystem Restoration: Restoring degraded ecosystems can help enhance their resilience and the provision of ecosystem services. Restoration efforts can include reforestation, wetland restoration, and the rehabilitation of degraded habitats.

4. Building Ecological Connectivity: Enhancing ecological connectivity between fragmented habitats promotes species movement and genetic exchange. This can be achieved through the creation of wildlife corridors and the restoration of ecological networks.

5. Adaptive Management: Adopting adaptive management approaches allows for the flexibility and adjustment of conservation and management strategies in response to changing conditions. This involves monitoring the effectiveness of interventions and adjusting them as required.

6. Incorporating Indigenous Knowledge: Indigenous knowledge and practices have often proven to be effective in managing ecosystems and maintaining resilience. Collaborating with indigenous communities and integrating their traditional knowledge into decision-making processes can contribute to more effective conservation efforts.

By implementing these strategies, it is possible to enhance the resilience of ecosystems and ensure the continued provision of ecosystem services for future generations.

Case Study: The Great Barrier Reef

The Great Barrier Reef is a globally significant ecosystem that provides a wide range of ecosystem services, including tourism, fisheries, and coastline protection.

However, it is facing numerous threats, including climate change, pollution, and coastal development.

To enhance the resilience of the Great Barrier Reef, various strategies have been implemented:

1. Climate Adaptation: Efforts are being made to reduce greenhouse gas emissions and mitigate the impacts of climate change. This includes the promotion of renewable energy sources, improvements in energy efficiency, and the development of adaptive management strategies.

2. Water Quality Improvement: Actions are being taken to reduce pollution from agricultural runoff and other sources. This involves improving land management practices, implementing buffer zones, and reducing nutrient inputs into the reef ecosystem.

3. Restoration Initiatives: Restoration projects are underway to restore damaged areas of the reef, including the planting of coral nurseries and the removal of invasive species.

4. Indigenous Involvement: Indigenous communities are playing a significant role in the conservation and management of the Great Barrier Reef through their traditional knowledge and stewardship practices.

These strategies aim to enhance the resilience of the Great Barrier Reef and ensure its long-term survival as a valuable ecosystem and provider of essential ecosystem services.

Conclusion

Ecosystem services play a vital role in supporting human well-being and the sustainability of societies. Understanding the concepts of ecosystem services and resilience is critical for effective conservation and sustainable development.

By promoting biodiversity, ecological connectivity, and adaptive management, we can enhance the resilience of ecosystems and ensure the continued provision of ecosystem services. Additionally, incorporating indigenous knowledge and practices can contribute to more effective conservation efforts.

Through these efforts, we can strive towards a future in which ecosystems are resilient, and ecosystem services are preserved for generations to come.

Community-based Adaptation

In the face of climate change, communities around the world are experiencing the impacts of extreme weather events, rising sea levels, and changing precipitation patterns. These changes pose significant challenges for vulnerable populations, especially those living in low-income countries with limited resources and infrastructure. Community-based adaptation (CBA) offers a bottom-up approach to address these challenges, empowering communities to take action and build resilience to climate change.

Principles of Community-based Adaptation

CBA is based on the principles of participation, inclusivity, and local knowledge. It recognizes that communities have valuable insights and understanding of their local environment and can contribute to finding sustainable solutions. By involving local people in the decision-making process, CBA ensures that adaptation strategies are context-specific, responsive to local needs, and supported by the community.

Understanding Vulnerability

To effectively implement CBA, it is crucial to understand the vulnerabilities of the community and the factors that contribute to their vulnerability. Vulnerability assessment involves identifying the social, economic, and environmental factors that make a community more susceptible to climate change impacts. This assessment helps in identifying the areas of highest risk and prioritizing adaptation actions.

Building Adaptive Capacity

Adaptive capacity refers to a community's ability to adjust and respond to changing conditions. CBA focuses on enhancing adaptive capacity by strengthening the skills, knowledge, institutions, and resources available to the community. This can be achieved through capacity-building activities, such as training on climate-smart agricultural practices, disaster preparedness, and early warning systems.

Examples of Community-based Adaptation

1. Community-managed forests: In some regions, communities have established forest management systems as a response to climate change. These initiatives involve local people in sustainable forest management, leading to increased biodiversity, carbon sequestration, and improved livelihoods.

    2. Climate-resilient agriculture: Farmers in vulnerable regions have adopted climate-smart agricultural practices to mitigate the impacts of climate change on their crops. Techniques such as agroforestry, conservation farming, and crop

diversification help in conserving water, improving soil fertility, and enhancing resilience to extreme weather events.

3. Community-based disaster risk reduction: Many communities are implementing early warning systems, creating disaster response plans, and conducting drills to enhance their preparedness and resilience to climate-related disasters. By involving the community in these efforts, the effectiveness of disaster risk reduction measures is significantly increased.

Challenges and Opportunities

While community-based adaptation has shown promise, it faces several challenges. Limited financial resources, lack of access to technology and information, and inadequate institutional support are some of the barriers to effective implementation. Additionally, communities may face social and cultural constraints that impede their capacity to adapt.

However, there are opportunities to overcome these challenges. Building partnerships between communities, governments, NGOs, and other stakeholders can provide the necessary support and resources for successful adaptation. Engaging women, youth, and marginalized groups in the decision-making process can ensure inclusivity and promote social equity in adaptation efforts.

Case Study: Community-based Adaptation in Bangladesh

Bangladesh is a country highly vulnerable to climate change impacts, including sea-level rise, flooding, and cyclones. In response, community-based adaptation initiatives have been implemented to protect livelihoods and increase resilience.

One example is the building of elevated homesteads in flood-prone areas. By raising their homes above the floodwater level, communities are able to protect their assets and continue their daily activities during floods. This simple yet effective adaptation measure has significantly reduced the vulnerability of households to climate-related disasters.

Another example is the establishment of community-managed saline agriculture in coastal regions. Saltwater intrusion due to sea-level rise has rendered agricultural land unproductive. Through community-led efforts, saline-tolerant rice varieties have been introduced and innovative farming techniques have been adopted, allowing farmers to continue agriculture in these challenging conditions.

Conclusion

Community-based adaptation is a vital approach to address the impacts of climate change at the local level. By empowering communities, recognizing local knowledge, and building adaptive capacity, CBA can enhance resilience, protect livelihoods, and ensure sustainable development. However, overcoming challenges and fostering collaboration among stakeholders are essential for the successful implementation of community-based adaptation initiatives. It is through these collective efforts that we can achieve a sustainable and resilient future for all.

Transformative Adaptation

Transformative adaptation refers to a proactive and comprehensive approach to addressing the impacts of climate change by transforming social, economic, and ecological systems. It recognizes the need for radical changes in the way we think, plan, and act, in order to adapt to the new environmental realities we face. This section explores the principles, strategies, and challenges associated with transformative adaptation.

Understanding Transformative Adaptation

Climate change presents complex and interconnected challenges that cannot be effectively addressed through incremental or reactive approaches alone. Transformative adaptation calls for a fundamental shift in our mindset, actions, and governance systems to tackle the root causes of vulnerability and build resilience against climate-related risks.

At its core, transformative adaptation is about reimagining and redesigning our social and ecological systems to better cope with and thrive in a changing climate. It involves understanding and addressing deep-seated societal and structural inequities that exacerbate vulnerability to climate change impacts. By challenging the status quo and promoting equity and justice, transformative adaptation aims to create more inclusive and sustainable societies.

Principles of Transformative Adaptation

Transformative adaptation is guided by several key principles:

1. Equity and Social Justice: Transformative adaptation seeks to address inequalities and prioritize the needs and voices of marginalized and vulnerable communities. It recognizes the differential impacts of climate change on different

groups and aims to ensure fair distribution of resources, benefits, and decision-making power.

2. Participation and Collaboration: Transformative adaptation emphasizes the importance of inclusive and meaningful engagement of stakeholders, including local communities, indigenous peoples, women, and youth. It encourages collaborative decision-making processes that draw on diverse knowledge systems and expertise.

3. Resilience and Sustainability: Transformative adaptation aims to enhance resilience at multiple scales, from local to global, by promoting sustainable practices and systems. It recognizes the interconnectedness between social, ecological, and economic dimensions of resilience and seeks to build adaptive capacity that can withstand future uncertainties.

4. Learning and Adaptiveness: Transformative adaptation embraces a learning-oriented approach, recognizing that adaptation strategies must evolve and adapt over time. It encourages experimentation, innovation, and the integration of traditional and scientific knowledge to inform decision-making.

Strategies for Transformative Adaptation

Transformative adaptation involves a range of strategies and actions that can be tailored to specific contexts. Some key strategies to consider include:

1. Mainstreaming Climate Considerations: Integrating climate considerations into existing policies, plans, and decision-making processes is essential for transformative adaptation. This involves considering climate change across sectors, such as agriculture, water management, and urban planning, and ensuring coherence and coordination among different stakeholders.

2. Building Adaptive Capacity: Enhancing the capacity of individuals, communities, and institutions to anticipate, respond to, and recover from climate-related shocks and stresses is crucial for transformative adaptation. This can be achieved through capacity-building programs, knowledge sharing, and the development of early warning systems.

3. Transforming Institutions and Governance: Transformative adaptation requires a shift towards more inclusive, transparent, and collaborative governance systems. This may involve institutional reforms, the creation of new decision-making structures, and the recognition of indigenous governance systems and traditional knowledge.

4. Promoting Nature-based Solutions: Nature-based solutions, such as ecosystem restoration, sustainable agriculture, and green infrastructure, can play a critical role in transformative adaptation. By harnessing the power of nature, these

solutions contribute to climate resilience while providing additional co-benefits, such as biodiversity conservation and water security.

5. Enhancing Social Safety Nets: Transformative adaptation should prioritize addressing social vulnerabilities and building social safety nets. This involves strengthening social protection systems, ensuring access to basic services, and promoting income diversification strategies to reduce reliance on climate-sensitive livelihoods.

Challenges and Considerations

Implementing transformative adaptation faces several challenges, including:

1. Power Dynamics and Inequalities: Transformative adaptation requires addressing deep-rooted power imbalances and social inequalities. This involves recognizing and challenging existing systems of privilege, marginalization, and discrimination that hinder inclusive and equitable decision-making.

2. Uncertainty and Complexity: Climate change is characterized by uncertainties and complex interactions between human and natural systems. Transformative adaptation must navigate these uncertainties, balancing short-term needs with long-term sustainability, and avoiding maladaptive actions.

3. Trade-offs and Synergies: Transformative adaptation often involves trade-offs between different goals, interests, and approaches. Balancing economic development, social justice, and environmental sustainability requires careful evaluation of the potential synergies and trade-offs associated with different adaptation strategies.

4. Financing and Resource Mobilization: Funding transformative adaptation initiatives remains a challenge. Innovative financing mechanisms, such as climate finance, green bonds, and public-private partnerships, can help mobilize resources for transformative adaptation at the required scale.

5. Capacity and Knowledge Gaps: Transformative adaptation requires the capacity to understand and respond to complex challenges. Enhancing knowledge and building the capacity of individuals, communities, and institutions is crucial for effective and equitable transformative adaptation.

Case Study: Transformative Adaptation in Bangladesh

Bangladesh, one of the most climate-vulnerable countries in the world, has been at the forefront of transformative adaptation efforts. The country faces multiple climate change impacts, including sea-level rise, cyclones, and floods, which pose significant challenges to its population and economy.

To address these challenges, Bangladesh has implemented transformative adaptation measures that prioritize community participation, gender equality, and ecosystem-based approaches. For example, the Community-based Adaptation program has empowered local communities to identify and implement adaptation actions, such as constructing cyclone shelters and embankments, promoting climate-resilient agriculture, and establishing early warning systems.

The government of Bangladesh has also invested in building climate-resilient infrastructure, such as disaster-resilient housing, schools, and healthcare facilities. These efforts have not only increased the resilience of local communities but have also contributed to poverty reduction and sustainable development.

The case of Bangladesh demonstrates the importance of local knowledge, community participation, and innovative adaptation strategies in addressing climate change impacts at the grassroots level. It serves as an inspiration and learning opportunity for other countries grappling with similar challenges.

Conclusion

Transformative adaptation represents a paradigm shift in how we approach climate change adaptation. It is a holistic and proactive approach that seeks to transform social, economic, and ecological systems to build resilience and ensure sustainable development in the face of mounting climate risks.

By embracing principles of equity, participation, resilience, and learning, transformative adaptation offers a pathway to create more just, sustainable, and inclusive societies. However, achieving transformative adaptation requires overcoming various challenges, including power imbalances, uncertainties, and resource constraints.

Realizing the potential of transformative adaptation will require collective action, strong governance, and the integration of diverse knowledge and perspectives. Ultimately, it is through transformative adaptation that we can pave the way towards a sustainable and resilient future for generations to come.

Social-ecological systems resilience

The concept of social-ecological systems resilience refers to the ability of a system to persist, adapt, and transform in the face of disturbances and changes, while maintaining desirable ecosystem services and human well-being. It recognizes the interdependence of social and ecological components and emphasizes the need for integrated management approaches that address the multiple dimensions of resilience.

Understanding social-ecological systems

Social-ecological systems (SES) are characterized by the interactions between humans and the natural environment. They encompass a range of systems, from local communities dependent on natural resources to larger-scale systems like cities and watersheds. An SES can include various stakeholders, such as government agencies, businesses, and civil society organizations, who have diverse values, interests, and power dynamics.

Understanding the dynamics of social-ecological systems requires a holistic approach that incorporates ecological, social, and economic dimensions. It involves analyzing the feedbacks and interactions between these dimensions, as well as the drivers of change that affect the system's resilience.

Resilience thinking and adaptive capacity

Resilience thinking provides a framework to analyze and enhance the resilience of social-ecological systems. It focuses on the capacity of a system to absorb disturbances and reorganize while retaining its essential structure, function, and identity. Resilience is not just about bouncing back from disturbances but also adapting and transforming in response to changing conditions.

One crucial aspect of resilience is adaptive capacity, which refers to the ability of a system to adjust its behavior in the face of change. Enhancing adaptive capacity involves building the knowledge, skills, institutions, and networks necessary for learning, collaboration, and collective action. It also requires promoting inclusivity, equity, and social cohesion to ensure the well-being and participation of all stakeholders.

Maintaining ecosystem services

Ecosystem services are the benefits that humans obtain from ecosystems, including provisioning services (e.g., food, water), regulating services (e.g., climate regulation, water purification), supporting services (e.g., nutrient cycling, soil formation), and cultural services (e.g., recreation, spiritual and aesthetic values). Social-ecological systems resilience aims to maintain and enhance these ecosystem services, recognizing their importance for human well-being and the sustainability of the system.

To achieve this, it is necessary to identify and understand the interactions between social and ecological components that contribute to the production and maintenance of ecosystem services. For example, the conservation and restoration of habitats can support biodiversity and ecosystem functioning, which, in turn, promote the provision of services. Similarly, sustainable land use practices can help preserve soil fertility and water resources, ensuring the continuity of provisioning services.

Transforming for sustainability

In some cases, resilience may require more than just adapting to changing conditions. It may involve transformative changes that address the root causes of unsustainability and promote new ways of thinking, organizing, and governing social-ecological systems. Transformative resilience aims to shift systems towards more sustainable trajectories by challenging dominant social norms, values, and institutions.

Achieving transformative resilience involves engaging with stakeholders and fostering inclusive decision-making processes that challenge power imbalances and enable the co-production of knowledge. It also requires fostering innovation, experimentation, and learning to explore alternative pathways towards sustainability. Transformative resilience recognizes the importance of social innovation, social-ecological entrepreneurship, and community-led initiatives in driving sustainability transitions.

Case study: Resilience of coastal communities

Coastal communities are highly vulnerable to multiple stressors, including sea-level rise, extreme weather events, and population growth. The resilience of these communities depends on their ability to adapt to changing coastal dynamics while safeguarding the ecological integrity of coastal ecosystems.

One example of building social-ecological systems resilience in coastal communities is the concept of "living shorelines." Living shorelines involve the use of natural and nature-based features, such as marshes, reefs, and dunes, to protect the coastlines from erosion and flooding. These features not only provide coastal protection but also support biodiversity and enhance ecosystem services.

To implement living shorelines effectively, it is crucial to involve local communities, scientists, and decision-makers in the planning and implementation processes. This participatory approach ensures that the interventions align with the needs and aspirations of the community, while also considering the ecological and social dynamics of the coastal system.

Key principles for enhancing resilience

Enhancing social-ecological systems resilience requires a set of key principles and strategies. These include:

1. Fostering collaboration and cooperation among diverse stakeholders to integrate knowledge and perspectives. 2. Promoting adaptive governance frameworks that enable collective decision-making and flexible management approaches. 3. Investing in capacity building and education to enhance the understanding and awareness of social-ecological systems and resilience thinking. 4. Implementing adaptive management practices that involve iterative learning, monitoring, and adjustment of management strategies. 5. Considering equity, justice, and distributional aspects in decision-making processes to ensure fairness and inclusivity. 6. Supporting local and indigenous knowledge systems to incorporate traditional ecological knowledge and enhance adaptive capacity. 7. Strengthening the connectivity and cross-scale interactions between different components of social-ecological systems. 8. Fostering innovation, experimentation, and transformative approaches that challenge the status quo and drive sustainability transitions.

Conclusion

The resilience of social-ecological systems is critical for addressing the complex challenges of sustainability. By understanding the interactions between social and ecological components, adopting a resilience thinking perspective, and promoting adaptive capacity, we can build more sustainable and resilient social-ecological systems. Achieving this requires collaborative and inclusive approaches that foster transformative changes towards a more sustainable future. Through these efforts, we can ensure the provision of ecosystem services and the well-being of both current and future generations.

Resilient infrastructure

Resilient infrastructure plays a crucial role in ensuring the sustainability and adaptability of communities and cities in the face of increasing climate change impacts and other environmental challenges. In this section, we will explore the principles, strategies, and technologies that contribute to the development of resilient infrastructure.

Principles of resilient infrastructure

Resilient infrastructure is built on several key principles that aim to enhance its ability to withstand and recover from shocks and stresses. These principles include:

- **Redundancy**: Resilient infrastructure incorporates redundancy to minimize the impact of disruptions. This can involve redundant systems, alternative routes, multiple power sources, or backup facilities.

- **Flexibility**: Resilient infrastructure is designed to be flexible and adaptable, capable of accommodating changing needs and conditions. Flexible infrastructure can be easily modified or repurposed to address future challenges.

- **Interconnectivity**: Resilient infrastructure is interconnected to enhance its effectiveness. Interconnectivity ensures the smooth flow of resources, services, and information across different infrastructure systems.

- **Diversity**: Resilient infrastructure embraces diversity in design and materials. Diversity reduces vulnerability to single-point failures and enhances the system's ability to withstand different types of shocks and stresses.

- **Modularity**: Resilient infrastructure is modular, consisting of smaller components that can be assembled or disassembled easily. Modularity enables the efficient replacement or repair of damaged components.

Strategies for resilient infrastructure

Building resilient infrastructure requires the implementation of various strategies and measures. Some of the key strategies include:

- **Risk assessment and planning**: Conducting thorough risk assessments is crucial to identify potential hazards and vulnerabilities. Based on the assessments, infrastructure planning can incorporate measures to mitigate risks and enhance resilience.

+ **Climate-responsive design:** Resilient infrastructure takes into account climate change impacts and adopts designs that can withstand extreme weather events, such as floods, hurricanes, or heatwaves. This can involve elevating structures, designing flexible water management systems, or incorporating green infrastructure.

+ **Nature-based solutions:** Nature-based solutions, such as green roofs, urban forests, or wetland restoration, can be integrated into infrastructure design to enhance resilience. These solutions provide multiple benefits, including stormwater management, temperature regulation, and biodiversity conservation.

+ **Smart technology integration:** Resilient infrastructure leverages smart technologies, such as sensors, data analytics, and automation, to monitor and manage infrastructure systems effectively. This enables early detection of potential issues, efficient resource allocation, and timely response to disruptions.

+ **Community engagement:** Engaging local communities in the planning and decision-making processes is vital for developing infrastructure that is aligned with their needs and values. Community participation fosters ownership, social cohesion, and collaboration in building resilient infrastructure.

Case study: Resilient transportation infrastructure

Transportation infrastructure is a critical component of urban systems that requires resilience to ensure its continuous operation and functionality. Let's consider a case study of a city's transportation infrastructure and explore the challenges it faces and the strategies employed to enhance its resilience.

Challenges: The city is susceptible to frequent heavy rainfall and floods, which often disrupt the transportation system. The existing road network suffers from poor drainage, resulting in waterlogging and road closures during monsoon season. Moreover, the city experiences increasing traffic congestion and unreliable public transportation, impacting the mobility and accessibility of its residents.

Strategies for resilience:

+ **Climate-responsive design:** The city integrates climate-responsive design principles into its transportation infrastructure. This includes elevating vulnerable road sections, constructing flood-resilient bridges, and improving stormwater drainage systems.

+ **Diversification of transportation modes:** The city promotes a shift towards sustainable and diverse transportation modes, such as cycling infrastructure, pedestrian-friendly streets, and efficient public transportation networks. This reduces the reliance on private vehicles and enhances the overall reliability and accessibility of the transportation system.

+ **Smart traffic management:** The city employs smart traffic management systems that utilize real-time data and predictive analytics to optimize traffic flow, manage congestion, and provide accurate travel information to commuters. This ensures efficient mobility even during peak hours or unexpected events.

+ **Public-private partnerships:** The city establishes partnerships with private entities to enhance the resilience of its transportation infrastructure. This includes collaborations for the development of electric vehicle charging infrastructure, sustainable logistics systems, and last-mile connectivity solutions.

+ **Community engagement:** The city actively involves local communities and stakeholders in transportation planning and decision-making processes. This includes public consultations, feedback mechanisms, and community-led initiatives, fostering a sense of ownership and resilience within the community.

Resources and further reading

Building resilient infrastructure requires a multidisciplinary approach and the integration of various knowledge domains. The following resources provide further reading on resilient infrastructure:

+ International Federation of Consulting Engineers (FIDIC) - Guidance for resilient infrastructure development.

+ United Nations Office for Disaster Risk Reduction (UNDRR) - Global assessment report on disaster risk reduction.

+ The World Bank - Building urban resilience: Principles, tools, and practice.

+ Resilient Cities Network - Resilience in urban infrastructure: Strategies and case studies.

Building resilient infrastructure is not only about preparing for future challenges but also about creating sustainable and livable communities. By implementing the principles, strategies, and technologies discussed in this section, we can ensure the resilience and adaptability of our infrastructure systems, contributing to a more sustainable and resilient future.

Exercise: Identify a vulnerable infrastructure system in your local area and propose strategies to enhance its resilience. Consider the specific hazards or challenges it faces and the principles discussed in this section. Discuss your proposal with classmates or local stakeholders to gather feedback and refine your ideas.

Adaptive Capacity Building

Adaptive capacity building is a crucial aspect of sustainability and resilience in the face of environmental change. It refers to the ability of individuals, communities, and systems to adjust, learn, and respond effectively to changing conditions, reducing vulnerability and enhancing resilience. In this section, we will explore the importance of adaptive capacity building, its principles, and strategies for implementation.

Importance of Adaptive Capacity Building

As the world experiences rapid environmental changes, including climate change, biodiversity loss, and natural disasters, building adaptive capacity becomes essential. Adaptive capacity allows individuals and communities to anticipate, respond, and recover from disturbances, thereby minimizing the negative impacts on livelihoods and ecosystems.

Enhancing adaptive capacity is vital for several reasons:

1. Resilience: Adaptive capacity enables systems to bounce back and recover quickly after disturbances. By building resilience, communities can better withstand shocks and maintain their well-being.

2. Sustainable Development: Adaptive capacity is closely linked to sustainable development. By enhancing individuals' and communities' ability to adapt, we can reduce poverty, improve food security, and promote equitable development.

3. Risk Reduction: Adaptive capacity building reduces vulnerability to environmental risks and disasters. It helps communities and systems prepare for and respond effectively to these challenges.

4. Innovation and Learning: Adaptive capacity fosters innovation and learning. It encourages the development and implementation of new approaches, technologies, and practices that increase resilience and sustainability.

Concepts and Principles of Adaptive Capacity Building

Adaptive capacity building involves a complex mix of social, economic, and environmental factors. Several key concepts and principles underpin the process:

1. Understanding Context: Adaptive capacity building requires a thorough understanding of the socio-cultural, economic, and political context in which it takes place. It considers both local and global factors influencing vulnerability and resilience.

2. Participation and Collaboration: Building adaptive capacity involves collaboration among different stakeholders, including local communities, government agencies, NGOs, and scientists. The participation of diverse perspectives and co-production of knowledge are vital for effective decision-making and action.

3. Knowledge and Learning: Adaptive capacity building emphasizes the importance of knowledge and learning. It involves combining traditional and indigenous knowledge with scientific expertise to inform decision-making and foster innovation.

4. Flexibility and Adaptability: Adaptive capacity building recognizes the need for flexibility and adaptability in the face of uncertainty and change. It encourages the development of strategies that can be adjusted and revised based on new information and evolving circumstances.

5. Equity and Social Justice: Building adaptive capacity should prioritize equity and social justice. It seeks to address underlying inequalities and empower marginalized and vulnerable groups, ensuring that they have the resources and opportunities to adapt and thrive.

Strategies for Adaptive Capacity Building

Building adaptive capacity requires a comprehensive and integrated approach. Here are some strategies that can be employed:

1. Knowledge Sharing and Transfer: Facilitating knowledge sharing and exchange between different stakeholders is essential to build adaptive capacity. This can be achieved through workshops, seminars, community meetings, and online platforms, fostering collaboration and learning.

2. Capacity Development: Enhancing the skills, knowledge, and capabilities of individuals and communities is crucial for adaptive capacity building. This can be achieved through training programs, educational initiatives, and awareness campaigns that focus on topics such as climate change resilience and disaster preparedness.

3. Strengthening Institutions and Governance: Adaptive capacity is closely linked to effective governance and institutions. Strengthening governance frameworks, improving coordination between stakeholders, and integrating climate and environmental considerations into policies and planning processes can enhance adaptive capacity.

4. Integrated Planning and Decision-Making: Integrating climate change and sustainability considerations into planning and decision-making processes is vital for adaptive capacity building. This involves considering the long-term implications of decisions and incorporating climate-risk assessments into development plans and projects.

5. Diversifying Livelihoods: Building adaptive capacity requires diversifying livelihoods and reducing dependence on vulnerable sectors. Encouraging alternative income-generating activities, promoting sustainable agriculture practices, and supporting small-scale enterprises can enhance resilience and adaptive capacity.

6. Enhancing Social Networks and Community Cohesion: Social networks and community cohesion play a crucial role in adaptive capacity building. Strengthening social connections, fostering collective action, and empowering local communities can enhance resilience and adaptive capacity.

By employing these strategies, we can enhance adaptive capacity at various levels, from individual households to entire regions, fostering sustainability and resilience in the face of environmental change.

Case Study: Building Adaptive Capacity in Small Island Developing States

Small Island Developing States (SIDS) face unique challenges concerning climate change, biodiversity, and economic vulnerability. Building adaptive capacity in SIDS is crucial for their long-term sustainability and resilience.

One example of successful adaptive capacity building in SIDS is the Pacific Adaptation to Climate Change (PACC) program. The PACC program, funded by the Global Environment Facility (GEF), supports 14 Pacific Island countries in implementing climate change adaptation measures.

The program employs several strategies:

1. Capacity Building: The PACC program focuses on enhancing the technical and institutional capacity of Pacific Island countries. It provides training and support for climate-risk assessments, development of adaptation plans, and implementation of resilience-building projects.

2. Knowledge Exchange and Learning: The PACC program facilitates knowledge exchange and learning among Pacific Island countries. It creates platforms for sharing experiences, best practices, and lessons learned in climate change adaptation, fostering collaboration and innovation.

3. Mainstreaming Climate Change Adaptation: The PACC program aims to mainstream climate change adaptation into national development planning processes. It helps countries integrate climate considerations into policies, plans, and projects, ensuring long-term resilience.

4. Community-Based Adaptation: Recognizing the importance of local knowledge and community participation, the PACC program promotes community-based adaptation approaches. It supports communities in identifying and implementing resilience-building activities, such as sustainable agriculture, water management, and infrastructure improvements.

The PACC program exemplifies the principles of adaptive capacity building, emphasizing local ownership, knowledge sharing, and cooperation. It demonstrates the importance of tailored approaches that consider the specific context and challenges faced by small island nations.

Summary

Adaptive capacity building is a fundamental component of sustainable development and resilience-building. It enables individuals, communities, and systems to respond effectively to environmental change, reducing vulnerability and enhancing long-term well-being.

Key principles of adaptive capacity building include understanding context, participation and collaboration, knowledge and learning, flexibility and adaptability, and equity and social justice. Strategies for building adaptive capacity range from knowledge sharing and capacity development to strengthening institutions, integrating climate considerations into planning, and diversifying livelihoods.

By employing these strategies, we can enhance adaptive capacity at various levels, fostering sustainability and resilience in a rapidly changing world. The case study on building adaptive capacity in Small Island Developing States highlights the importance of tailored approaches that consider specific challenges and foster collaboration and innovation.

Indigenous Knowledge in Adaptation

Indigenous knowledge refers to the collective knowledge, practices, and beliefs of indigenous communities that have been developed over generations through a deep connection with their environment. This knowledge system is rooted in sustainability and harmony with nature, making it highly relevant in the context of adaptation to environmental changes and promoting resilience.

Understanding Indigenous Knowledge

Indigenous knowledge systems are holistic, taking into account the interconnections between ecosystems, human societies, and the spiritual dimension of life. This worldview recognizes the inseparable relationship between humans and their environment, acknowledging that the well-being of one is dependent on the other. Indigenous knowledge is often passed down orally, from one generation to the next, and is deeply embedded in cultural practices, rituals, and traditions.

Indigenous communities have a profound understanding of their local ecosystems, including weather patterns, flora, fauna, and natural resources. They possess specific knowledge about their surroundings, such as the signs that indicate changes in weather, the behavior of animals and plants in different seasons, and the medicinal properties of various plants. This knowledge has been gained through careful observation, experience, and interaction with the environment.

Applying Indigenous Knowledge in Adaptation

Indigenous communities have a long history of adapting to environmental changes and maintaining their resilience in the face of adversity. Their knowledge and practices offer valuable insights into sustainable adaptation strategies. Here are some examples of how indigenous knowledge can be applied in the context of adaptation:

1. Traditional ecological knowledge: Indigenous communities have developed sophisticated methods for managing their resources sustainably. By combining traditional knowledge with modern technologies, their practices can contribute to the conservation and restoration of ecosystems. For example, indigenous fire management techniques can be used to prevent destructive wildfires and promote the regeneration of plant species.

2. Local weather prediction: Indigenous communities possess detailed knowledge of local weather patterns and can forecast short-term weather changes. This knowledge can be utilized to support climate adaptation efforts, such as

providing early warnings for extreme weather events or guiding agricultural practices based on anticipated rainfall patterns.

3. Agroecology and sustainable agriculture: Indigenous farming systems often employ agroecological practices that promote biodiversity, soil fertility, and resilience to climate variability. These practices include intercropping, crop rotation, and the use of organic fertilizers. Incorporating indigenous agricultural techniques can enhance the adaptive capacity of farming communities and contribute to food security.

4. Traditional water management: Indigenous communities have knowledge of water resource management practices that are well-suited for water-scarce regions. These include rainwater harvesting, canal irrigation systems, and the construction of water storage structures. Implementing such practices can enhance water security and resilience in the face of changing precipitation patterns.

5. Indigenous governance systems: Indigenous communities have well-established governance systems that integrate social, cultural, and ecological factors. These systems promote collective decision-making, consensus-building, and the equitable distribution of resources. Integrating indigenous governance principles into adaptation planning can foster greater community resilience and ensure the inclusion of marginalized voices.

Challenges and Opportunities

While indigenous knowledge holds great potential for adaptation, several challenges need to be addressed for its effective integration into mainstream policies and practices:

1. Recognition and respect: Indigenous knowledge and practices have often been marginalized or dismissed by dominant societies. Recognition and respect for indigenous knowledge systems are crucial to ensure its inclusion in decision-making processes.

2. Intellectual property rights: Intellectual property rights frameworks should be designed to protect indigenous knowledge from exploitation and appropriation, allowing indigenous communities to benefit from the commercialization of their traditional knowledge.

3. Cultural sensitivity: It is essential to approach indigenous knowledge with cultural sensitivity and avoid appropriating or commodifying it. Collaboration should be built on mutual respect, trust, and the recognition of indigenous rights and sovereignty.

4. Knowledge transmission and documentation: Efforts should be made to support the transmission of indigenous knowledge across generations and to

document it for future reference. This includes documenting traditional practices, ecological calendars, and oral histories.

The integration of indigenous knowledge in adaptation practices offers an opportunity to enhance the effectiveness and sustainability of adaptation efforts. By fostering collaboration and knowledge exchange between indigenous communities, scientists, and policymakers, we can create a more inclusive and resilient approach to addressing environmental changes.

Case Study: Indigenous Adaptation in the Arctic

The Arctic region is witnessing rapid environmental changes, including melting ice caps, thawing permafrost, and shifting ecosystems. Indigenous communities in the Arctic have been adapting to these changes for centuries, relying on their traditional knowledge and practices.

One example is the Inuit communities of Canada, who have a deep understanding of sea-ice conditions and the behavior of marine mammals. They have observed changes in the timing and thickness of ice, impacting their hunting and travel patterns. In response, they have adapted their hunting techniques, altered travel routes, and diversified their livelihood strategies. Additionally, Inuit communities have been actively involved in monitoring and documenting environmental changes, contributing valuable information for scientific research.

In the face of climate change, indigenous communities in the Arctic are advocating for the recognition of their rights, the protection of their cultural heritage, and their active participation in decision-making processes. By incorporating indigenous knowledge into climate adaptation strategies, policymakers can develop more effective and locally appropriate solutions that strengthen the resilience of these communities and their ecosystems.

Summary

Indigenous knowledge systems encompass a wealth of insights and practices developed by indigenous communities over generations. This knowledge is vital for adaptation to environmental changes and promoting resilience. By integrating indigenous knowledge into mainstream policies and practices, we can enhance the effectiveness and sustainability of adaptation efforts. However, challenges such as recognition, respect, and intellectual property rights need to be addressed to ensure the equitable inclusion of indigenous knowledge. Through collaboration and knowledge exchange, we can build a more inclusive and resilient approach to environmental adaptation.

Sustainable Urban Planning

Smart cities

In today's rapidly urbanizing world, the concept of a smart city has gained significant momentum. A smart city is an urban area that uses advanced technologies and data-driven solutions to improve the quality of life for its residents, enhance sustainability, and optimize resource consumption. It leverages a combination of information and communication technology (ICT), Internet of Things (IoT), and data analytics to optimize various aspects of urban living, including transportation, energy, infrastructure, waste management, and governance.

Principles of Smart Cities

The development of smart cities is guided by several key principles:

Integration of technology and infrastructure: Smart cities rely on the seamless integration of various technological solutions, such as sensors, connectivity, and data analytics, into existing urban infrastructure. This allows for real-time monitoring, evaluation, and optimization of different systems.

Sustainability and resource efficiency: Smart cities aim to minimize resource consumption and reduce their ecological footprint. They promote energy efficiency, sustainable transportation, waste management, and water conservation through the use of intelligent technologies.

Citizen engagement and empowerment: Smart cities prioritize citizen engagement and participation in decision-making processes. They facilitate two-way communication between citizens and the city administration, enabling residents to actively contribute to the city's development.

Data-driven decision-making: Smart cities rely on an extensive network of sensors and data collection systems to gather information on various aspects of urban life. This data is analyzed to generate insights and inform evidence-based decision-making by city planners and policymakers.

Collaboration and partnerships: Smart cities foster collaboration between government authorities, businesses, academia, and citizens. Public-private partnerships play a crucial role in implementing smart city projects, ensuring sustainable funding, and facilitating innovation.

Core Components of Smart Cities

Smart cities consist of several interconnected components that work together to create an intelligent and sustainable urban environment. Some of the core components include:

Smart infrastructure: This includes the integration of advanced technology into physical infrastructure, such as smart grids, intelligent transportation systems, and sensor networks. Smart infrastructure enables efficient resource management, improved connectivity, and optimized service delivery.

Smart mobility: Smart mobility focuses on developing intelligent transportation systems that reduce traffic congestion, enhance public transportation, and promote sustainable modes of transport. It involves the use of technologies like real-time traffic monitoring, smart parking, and multi-modal transportation solutions.

Smart energy: Smart energy aims to enhance energy efficiency, increase the use of renewable energy sources, and enable effective energy management. It involves the deployment of smart grids, smart meters, and energy storage systems, as well as demand response mechanisms.

Smart buildings: Smart buildings incorporate advanced technologies to optimize energy consumption, improve occupant comfort, and enhance overall building performance. This includes the use of automated systems for lighting, heating, ventilation, and air conditioning (HVAC), as well as monitoring devices to track energy usage and occupancy.

Smart governance: Smart governance focuses on leveraging digital technologies to improve the efficiency, transparency, and effectiveness of public sector operations. It involves the use of e-governance platforms, open data initiatives, and digital services to enhance citizen participation, streamline administrative processes, and deliver better public services.

Challenges and Solutions

The implementation of smart city projects comes with its own set of challenges. Some of the key challenges include:

Data privacy and security: As smart cities rely heavily on data collection and analysis, ensuring the privacy and security of sensitive information becomes crucial. Striking a balance between data-driven decision-making and protecting the privacy rights of citizens is a constant challenge.

Interoperability and standardization: Integrating various systems and technologies from different vendors can be a complex task. Developing interoperability standards and ensuring seamless communication between different components is essential for the success of smart city initiatives.

Digital divide: The digital divide refers to the disparity in access to technology and digital services among different sections of society. To create inclusive smart cities, efforts must be made to bridge this divide and provide equal access to all citizens, regardless of socioeconomic status.

Sustainability and scalability: Smart city projects need to be designed with long-term sustainability and scalability in mind. Implementing solutions that are adaptable and can accommodate future growth and technological advancements is vital for ensuring the longevity of smart city initiatives.

Public acceptance and engagement: For smart city projects to succeed, it is essential to gain the trust and active participation of citizens. Engaging the public from the early stages of planning, addressing concerns, and demonstrating the benefits of smart city solutions are key to fostering public acceptance.

Example: Barcelona's Smart City Initiative

Barcelona, Spain, is often hailed as a pioneer in the field of smart cities. The city embarked on a comprehensive smart city initiative in 2011, known as the Barcelona Smart City Project. This initiative aimed to transform Barcelona into a sustainable, technologically advanced city that improves the quality of life for its citizens.

One of the key projects under the Barcelona Smart City Project is the implementation of a smart lighting system. Traditional streetlights were replaced with energy-efficient LED lights equipped with sensors and wireless connectivity. This allows for remote monitoring and control, saving energy by adapting to real-time conditions. Additionally, the lighting system also serves as a backbone for other smart city applications, such as parking management and waste collection.

Another notable project is the development of a smart bus network. Barcelona implemented a bus fleet equipped with sensors and GPS devices to provide real-time information to passengers. This enables citizens to access accurate bus arrival times through mobile apps, reducing waiting times and improving the overall efficiency of public transportation.

Barcelona's smart city initiative also focuses on citizen engagement and participation. The city has launched various platforms and initiatives to involve residents in decision-making processes and gather feedback. For example, the "Decidim Barcelona" platform allows citizens to propose and vote on projects, ensuring that the development of the city aligns with the needs and aspirations of its residents.

These examples from Barcelona demonstrate the transformative potential of smart cities in improving urban living conditions and promoting sustainability. However, it is important to note that each city has its unique set of challenges and opportunities, necessitating tailored approaches to smart city development.

Conclusion

Smart cities hold immense promise for creating sustainable, efficient, and livable urban environments. By leveraging technology and data, smart cities can optimize resource consumption, improve public services, and enhance citizen well-being. However, the successful implementation of smart city projects requires careful planning, collaboration, and a focus on addressing challenges such as data privacy, interoperability, and public acceptance. Ultimately, the future of urban living lies in the integration of smart city solutions that prioritize sustainability, inclusivity, and citizen participation.

Green building design

Green building design plays a crucial role in promoting sustainable urban planning. It emphasizes the use of eco-friendly materials, energy-efficient technologies, and environmentally conscious practices to minimize the negative impact of buildings on the environment. In this section, we will delve into the principles, strategies, and benefits of green building design.

Principles of green building design

Green building design follows several fundamental principles that guide the construction and operation of environmentally responsible buildings. These principles include:

- **Energy efficiency:** Green buildings are designed to minimize energy consumption by utilizing efficient building materials, insulation, and advanced technologies for heating, cooling, and lighting systems. This reduces the reliance on fossil fuels and decreases greenhouse gas emissions.

- **Water conservation:** Green building design incorporates strategies to reduce water consumption through the use of low-flow fixtures, rainwater harvesting systems, and water-efficient landscaping. This helps in conserving water resources and mitigating the strain on local water supplies.

+ **Materials selection:** Sustainable materials, such as recycled content, locally sourced materials, and renewable resources, are preferred in green building design. This reduces the extraction of raw materials, decreases waste generation, and promotes a healthier indoor environment.

+ **Waste reduction and recycling:** Green buildings focus on minimizing waste generation during construction and operation. Strategies such as effective waste management plans, recycling programs, and the use of prefabricated building components help in reducing the environmental footprint associated with waste disposal.

+ **Indoor environmental quality:** Green building design prioritizes the health and well-being of occupants by ensuring optimal indoor air quality, thermal comfort, and access to natural daylight. This is achieved through proper ventilation systems, low-emission materials, and thoughtful building orientation.

Strategies for green building design

To achieve the principles of green building design, several strategies can be implemented during the planning, design, and construction phases. These strategies include:

+ **Passive design:** Passive design techniques maximize the use of natural resources, such as sunlight and airflow, to reduce energy consumption. Features like efficient insulation, shading devices, orientation for solar gain, and natural ventilation ensure a comfortable indoor environment without heavy reliance on mechanical systems.

+ **Energy-efficient systems:** Incorporating energy-efficient technologies like LED lighting, high-efficiency HVAC (heating, ventilation, and air conditioning) systems, and advanced building automation systems can significantly reduce energy demand in green buildings. Additionally, the use of renewable energy sources, such as solar panels and wind turbines, can further enhance the sustainability of the building.

+ **Water management:** Green building design includes rainwater harvesting systems, graywater recycling, and technologies like low-flow fixtures and dual-flush toilets to minimize water consumption. The use of drought-tolerant landscaping and efficient irrigation systems also contributes to water conservation.

- **Green roofs and walls:** Integrated green roofs and walls provide numerous environmental benefits. They help in reducing urban heat island effect, improving air quality, mitigating stormwater runoff, and providing additional insulation for the building.

- **Life-cycle assessment:** Life-cycle assessment (LCA) is a tool used in green building design to evaluate the environmental impacts of a building throughout its entire life cycle, including construction, operation, and end-of-life disposal. LCA helps in identifying areas for improvement and making informed decisions regarding material selection, energy use, and waste management.

- **Smart building technologies:** Integrating smart building technologies, such as energy monitoring systems, real-time data analytics, and automation systems, can optimize energy performance and enhance occupant comfort. These technologies allow for the monitoring, control, and optimization of various building systems, ultimately reducing energy consumption and improving operational efficiency.

Benefits of green building design

Green building design offers numerous social, environmental, and economic benefits. Some of the key benefits include:

- **Environmental benefits:** Green buildings reduce carbon emissions, minimize water consumption, preserve natural resources, and contribute to overall environmental sustainability. These buildings help in combating climate change and protecting ecosystems.

- **Health and well-being:** Green buildings prioritize the health and well-being of occupants by providing better indoor air quality, thermal comfort, and access to natural light. This improves productivity, reduces illnesses, and enhances the overall quality of life.

- **Energy and cost savings:** Green buildings are more energy-efficient, resulting in significant cost savings over the building's lifetime. By utilizing renewable energy sources, implementing energy-saving technologies, and optimizing systems, green buildings reduce energy bills and operational expenses.

+ **Increased property value:** Green buildings usually have higher property values due to their environmental and energy performance. These buildings are in high demand as they offer lower operating costs and healthier living spaces, making them attractive to potential buyers and tenants.

+ **Job creation and economic growth:** The green building sector creates job opportunities in various fields, such as construction, manufacturing, engineering, and renewable energy. Investing in green building design stimulates local economies and promotes sustainable economic growth.

Case study: The Edge Building, Amsterdam

A prominent example of innovative green building design is The Edge Building in Amsterdam, Netherlands. It has been widely recognized as the world's most sustainable office building.

The Edge Building incorporates several green building features, including:

+ Efficient use of energy through smart LED lighting, occupancy sensors, and climate control systems.

+ Rooftop solar panels that generate renewable energy to power the building.

+ Advanced ventilation systems that provide a comfortable and healthy indoor environment for occupants.

+ Rainwater harvesting and greywater reuse system for reducing water consumption.

+ Smart building technologies, such as smartphone apps that allow employees to control their work environment and optimize energy usage.

+ High levels of insulation, triple-glazed windows, and building materials with low environmental impact.

As a result of its sustainable design, The Edge Building operates at significantly lower energy consumption levels compared to traditional office buildings. It has achieved the highest rating (98.36%) in the Building Research Establishment Environmental Assessment Method (BREEAM), a leading sustainability assessment method for buildings.

The Edge Building demonstrates the potential of green building design to create environmentally responsible and energy-efficient buildings without compromising functionality or comfort.

Key Takeaways

Green building design is a crucial component of sustainable urban planning, emphasizing energy efficiency, water conservation, materials selection, waste reduction, and indoor environmental quality. Strategies like passive design, energy-efficient systems, water management techniques, green roofs and walls, life-cycle assessment, and smart building technologies contribute to the sustainability and overall performance of green buildings. The benefits of green building design include reduced environmental impact, improved occupant health and well-being, energy and cost savings, increased property value, and job creation. Real-world examples like The Edge Building in Amsterdam showcase the success and potential of green building design in creating high-performance and sustainable structures.

Urban Agriculture

Urban agriculture refers to the practice of cultivating and producing food within urban areas. It involves growing, processing, and distributing food in or around cities, often using vacant lots, rooftops, balconies, or other urban spaces. Urban agriculture offers numerous benefits, including improved food security, increased access to fresh and nutritious produce, environmental sustainability, and community development. In this section, we will explore the principles, methods, and potential of urban agriculture.

Principles of Urban Agriculture

Urban agriculture is guided by several principles that help ensure its effectiveness and sustainability:

1. **Localized Food Production:** Urban agriculture emphasizes the production of food close to the consumers, reducing the distance and energy required for transportation. This principle helps improve food security and reduces the carbon footprint associated with food distribution.

2. **Maximizing Land Use:** Urban agriculture aims to optimize the use of urban spaces, including rooftops, vacant lots, and unused land. Through innovative design and planning, cities can transform underutilized spaces into productive agricultural areas.

3. **Community Engagement and Integration:** Urban agriculture promotes community involvement, encouraging residents to actively participate in food production and distribution. Community gardens and shared urban farms foster a sense of ownership, social cohesion, and self-sufficiency.

4. **Environmental Sustainability**: Urban agriculture promotes sustainable farming practices, such as organic farming, permaculture, and agroecology. These methods reduce the use of chemical inputs, protect biodiversity, and promote soil and water conservation.

5. **Integration of Technology and Innovation**: Urban agriculture embraces technological advancements to enhance productivity and efficiency. Hydroponics, aquaponics, vertical farming, and rooftop greenhouse systems are examples of innovative approaches to urban agriculture.

Methods of Urban Agriculture

There are several methods commonly employed in urban agriculture:

1. **Community Gardens**: Community gardens are spaces where residents come together to collectively cultivate plants and vegetables. These gardens are often located in public parks, vacant lots, or underutilized spaces. Community gardens provide an opportunity for people to learn about gardening, share resources, and build relationships within the community.

2. **Rooftop Farming**: Rooftop farming utilizes the available space on rooftops for agricultural purposes. With proper structural support and innovative farming techniques, rooftops can be transformed into productive agricultural spaces. Rooftop farming offers numerous benefits, including insulation, stormwater management, and improved air quality.

3. **Vertical Farming**: Vertical farming involves the cultivation of crops in vertically stacked layers, usually in indoor environments. This method utilizes artificial lighting, hydroponics, and aeroponics to grow crops efficiently in limited spaces. Vertical farming has the potential to maximize agricultural productivity without the need for large land areas.

4. **Hydroponics and Aquaponics**: Hydroponics is a soil-less farming technique that involves growing plants in nutrient-rich water. Aquaponics combines hydroponics with aquaculture, where fish waste provides nutrients for the plants. Both methods are ideal for urban agriculture as they require minimal space and optimize resource use.

5. **Permaculture and Agroecology**: Permaculture and agroecology are holistic approaches to urban agriculture that aim to create sustainable and self-sufficient systems. These methods emphasize the natural characteristics of ecosystems, promoting biodiversity, soil health, and ecological balance. Permaculture and agroecology provide a framework for designing urban agricultural systems that are resilient and regenerative.

Challenges and Solutions

While urban agriculture offers numerous benefits, it also faces several challenges that need to be addressed:

1. **Limited Space:** Urban areas often have limited available land for agriculture. This challenge can be overcome by utilizing vertical spaces, such as rooftops, walls, and balconies. Community gardens and shared spaces can also help maximize land use.

2. **Access to Water:** Urban agriculture requires a reliable water source, which can be a challenge in water-scarce areas. Rainwater harvesting, graywater recycling, and efficient irrigation systems can help mitigate water scarcity issues.

3. **Soil Quality:** Urban soils are often contaminated with pollutants, making them unsuitable for agriculture. Remediation techniques, such as phytoremediation and soil amendments, can help improve soil quality and make it suitable for farming.

4. **Zoning and Regulations:** Many cities have zoning regulations that restrict or discourage urban agriculture. Engaging with local policymakers and advocating for supportive regulations can help overcome these barriers.

5. **Community Participation:** Engaging and sustaining community involvement can be a challenge in urban agriculture. Educating and empowering residents, organizing workshops and training programs, and fostering a sense of ownership can help build a strong and vibrant urban agriculture community.

Examples and Case Studies

1. **Brooklyn Grange (New York City, USA):** Brooklyn Grange operates the world's largest rooftop soil farms, spanning over two acres. They grow a wide variety of vegetables and herbs, supply local restaurants and markets, and offer educational programs. Brooklyn Grange demonstrates the potential of utilizing urban rooftops for large-scale agriculture.

2. **Copenhagen, Denmark:** Copenhagen has pioneered the concept of rooftop agriculture, with numerous rooftop farms and gardens across the city. These initiatives have transformed unused spaces into productive urban farmland, enhancing food security and promoting sustainable living.

3. **CUBOCC Rooftop Farm (São Paulo, Brazil):** CUBOCC is a digital advertising agency that transformed its rooftop into a productive farm. They grow vegetables and herbs, which are used in the company's cafeteria and served to employees. This case demonstrates the potential of urban agriculture in corporate settings.

Resources

1. *Urban Agriculture: A Guide to Container Gardening* by Alice Mary Alvrez: This book provides practical guidelines for starting and maintaining container gardens in urban areas. It covers topics like selecting containers, choosing appropriate plants, and utilizing small spaces effectively.

2. *The Vertical Farm: Feeding the World in the 21st Century* by Dr. Dickson Despommier: This book explores the concept of vertical farming and its potential to revolutionize urban agriculture. It discusses the technical aspects of vertical farming and its implications for sustainable food production.

3. *Growing Cities* (Documentary): This documentary film examines the rise of urban agriculture in America and its impact on communities. It features stories of individuals and organizations who are transforming vacant lots and urban spaces into thriving gardens.

Trivia

Did you know that the world's largest vertical farm is located in Newark, New Jersey? AeroFarms operates a 70,000 square foot indoor farm, utilizing vertical farming techniques to grow leafy greens and herbs. This modern farm can produce up to 2 million pounds of food per year, using 95% less water compared to traditional farming methods.

Exercises

1. Conduct research on the urban agriculture initiatives in your city or region. Identify the challenges they face and propose solutions to overcome them.

2. Design a vertical farming system for a small urban apartment. Consider the space constraints and optimize resource use for sustainable food production.

3. Analyze the environmental benefits of urban agriculture compared to conventional agriculture. Consider factors such as carbon emissions, water usage, and land utilization.

4. Develop a community engagement plan for a new community garden in your neighborhood. Identify strategies to encourage participation and ensure the long-term sustainability of the project.

5. Investigate the economic viability of urban agriculture. Analyze the potential for urban farms to generate income and contribute to local economies.

Remember, urban agriculture is not only about growing food but also creating a sense of community, promoting sustainable practices, and improving urban resilience. Get your hands dirty and explore the exciting possibilities of urban agriculture in your local context.

Sustainable transportation infrastructure

Transportation plays a crucial role in our daily lives, enabling the movement of people and goods. However, traditional transportation systems heavily rely on fossil fuels, contributing to air pollution, traffic congestion, and greenhouse gas emissions. To address these challenges and promote sustainable development, it is essential to transition towards a more sustainable transportation infrastructure.

Importance of sustainable transportation

Sustainable transportation aims to minimize the negative environmental, social, and economic impacts of transportation systems while ensuring efficiency, accessibility, and safety. It is an integral part of sustainable urban planning and a key factor in achieving the United Nations Sustainable Development Goals.

By implementing sustainable transportation infrastructure, we can reduce carbon emissions, improve air quality, enhance public health, promote energy efficiency, and create more livable and inclusive cities. Additionally, sustainable transportation systems can support economic growth, reduce congestion, and increase mobility options for all individuals, including those with limited access to private vehicles.

Principles of sustainable transportation

To achieve sustainable transportation infrastructure, several principles should be considered:

1. **Multimodal connectivity**: Designing transportation systems that provide convenient and seamless connections between different modes of transport, such as walking, cycling, public transit, and shared mobility services. This encourages people to choose sustainable modes of transportation and reduces dependency on private vehicles.

2. **Efficient land use**: Planning transportation infrastructure in a way that reduces sprawl and promotes compact, mixed-use development. By locating different land uses (residential, commercial, recreational) in close proximity, people can easily access daily needs without long commuting distances.

3. **Active transportation**: Encouraging walking and cycling as viable modes of transportation through the provision of safe and dedicated infrastructure, such as sidewalks, bike lanes, and pathways. Active transportation promotes physical activity, reduces traffic congestion, and improves air quality.

4. **Public transit**: Enhancing the accessibility, reliability, and efficiency of public transportation systems, including buses, trams, light rail, and subways. This involves investing in high-quality infrastructure, optimizing routes, improving affordability, and integrating technology for real-time information and ticketing.

5. **Shared mobility**: Promoting the use of shared transportation options, such as carpooling, ride-hailing, and bike-sharing services. Shared mobility reduces the number of vehicles on the road, lowers carbon emissions, and provides cost-effective transportation solutions for individuals.

6. **Electrification and alternative fuels**: Transitioning to electric vehicles (EVs) and alternative fuel technologies to reduce reliance on fossil fuels and decrease greenhouse gas emissions. This requires the development of charging infrastructure, incentives for EV adoption, and investments in research and development of sustainable fuels.

7. **Smart transportation systems**: Leveraging advanced technological solutions, such as intelligent transportation systems, data analytics, and real-time information, to optimize traffic flow, improve safety, and enhance the overall efficiency of transportation networks.

Challenges and solutions

Implementing sustainable transportation infrastructure faces several challenges, including:

+ **Limited funding**: The high costs associated with developing and maintaining sustainable transportation infrastructure can be a significant barrier. Governments, private sector entities, and international organizations need to collaborate and invest in sustainable transportation projects through innovative financing mechanisms.

+ **Resistance to change**: Shifting from car-centric transportation systems to sustainable alternatives often faces resistance from individuals and stakeholders. Public awareness campaigns, education, and incentives can help overcome this resistance and promote behavioral changes towards sustainable modes of transport.

+ **Infrastructure constraints**:   Retrofitting existing transportation infrastructure to accommodate sustainable modes of transport can be challenging. Governments and urban planners need to design and adapt infrastructure to ensure the safety and convenience of pedestrians, cyclists, and public transit users.

+ **Planning and coordination**:   Developing sustainable transportation infrastructure requires effective planning, coordination, and collaboration between various government agencies, policymakers, urban planners, transportation providers, and community stakeholders.

To address these challenges, innovative solutions can be implemented, such as:

+ **Integrated transportation planning**:   Adopting integrated and holistic approaches to transportation planning, considering land use, urban design, public health, and environmental factors.   This helps ensure that transportation systems are well-integrated, efficient, and contribute to the overall sustainability of cities.

+ **Public-private partnerships**: Collaborating with private sector entities to leverage expertise, resources, and investments in sustainable transportation projects.   Public-private partnerships can support the development of infrastructure, service provision, and technological innovations.

+ **Incentives and regulations**:   Implementing a mix of incentives and regulations to encourage the adoption of sustainable transportation modes. This may include offering subsidies for EV purchases, introducing congestion pricing, providing tax benefits for green vehicles, and establishing strict emissions standards for vehicles.

+ **Community engagement**: Engaging communities in the decision-making process and actively involving them in the design, implementation, and evaluation of sustainable transportation projects. Community engagement fosters ownership, increases public support, and ensures that infrastructure meets the needs of the local population.

Case study: Curitiba, Brazil

Curitiba, the capital city of the Brazilian state of Paraná, is often hailed as a model for sustainable urban transportation. The city implemented a groundbreaking transportation system known as the Bus Rapid Transit (BRT) system, which has become an inspiration for numerous cities worldwide.

The BRT system in Curitiba offers fast, reliable, and cost-effective transit by segregating bus lanes, providing dedicated stations, offering prepaid fare systems, and ensuring high-frequency service. The strategic integration of the BRT system with other modes of transport, such as pedestrian-friendly streets and bike-sharing programs, has created a seamless and sustainable transportation network.

Curitiba's BRT system has not only reduced congestion and air pollution but has also improved the social equity of transportation. The efficient and accessible public transit system caters to all income groups, promoting equity and inclusivity within the city.

Conclusion

Sustainable transportation infrastructure is a key component of achieving a more sustainable future. By implementing principles such as multimodal connectivity, efficient land use, active transportation, public transit, shared mobility, electrification, alternative fuels, and smart transportation systems, we can create transportation networks that are environmentally friendly, socially equitable, and economically viable.

While challenges exist, collaboration, innovative solutions, and community engagement can overcome these barriers and pave the way for sustainable transportation systems. Through case studies like Curitiba, we can learn valuable lessons and draw inspiration for creating sustainable transportation infrastructure in cities around the world. By prioritizing sustainable transportation, we can build more livable and resilient communities for generations to come.

Urban resilience

Urban resilience refers to the ability of a city or urban area to withstand and recover from a variety of shocks and stresses, including natural disasters, climate change impacts, economic crises, and social disruptions. It is a critical aspect of sustainable urban planning and development, as it focuses on building cities that can adapt and thrive in the face of uncertainty and change.

1. The need for urban resilience

Cities across the world are increasingly facing complex challenges and risks. Rapid urbanization, population growth, and climate change are putting significant pressure on urban systems and infrastructure. Cities need to be prepared to handle both acute shocks, such as floods and earthquakes, as well as chronic stresses, such as poverty, inequality, and environmental degradation.

2. Principles of urban resilience

Urban resilience is based on several key principles:

2.1. Integrated planning: Urban resilience requires the integration of various sectors and stakeholders in urban planning and decision-making processes. It involves considering social, economic, and environmental factors holistically to identify and prioritize actions that enhance a city's resilience.

2.2. Redundancy and diversity: Resilient cities recognize the importance of redundancy and diversity in their infrastructure, systems, and social networks. This means having backup systems in place and fostering diverse economic and social activities to minimize disruption and enhance adaptability.

2.3. Adaptive capacity: Resilient cities have the ability to learn, adapt, and transform in response to new challenges and changing circumstances. This involves building human, social, and institutional capacity to anticipate and respond to shocks and stresses effectively.

2.4. Ecosystem-based approaches: Resilient cities recognize the value of natural ecosystems in providing critical services, such as flood protection and climate regulation. They integrate nature-based solutions into urban planning and design to enhance resilience and promote sustainability.

3. Building urban resilience

3.1. Risk assessment and planning: The first step in building urban resilience is conducting a comprehensive risk assessment to identify the main hazards, vulnerabilities, and capacities of a city. This assessment helps prioritize action areas and informs the development of an urban resilience strategy and action plan.

3.2. Infrastructure and system design: Resilient cities invest in robust and flexible infrastructure systems that can withstand shocks and stresses. This includes designing buildings, transportation networks, and utility systems to be resilient to climate change impacts and other hazards.

3.3. Social and community resilience: Resilient cities also focus on building social and community resilience. This involves engaging and empowering local communities in decision-making processes, providing education and training opportunities, and fostering social cohesion and partnerships.

3.4. Green infrastructure: Green infrastructure, such as parks, green roofs, and urban forests, plays a vital role in enhancing urban resilience. It helps manage stormwater, reduce urban heat island effects, improve air quality, and provide recreational spaces. Resilient cities prioritize the creation and preservation of green spaces as part of their urban planning strategies.

3.5. Disaster preparedness and response: Resilient cities have robust disaster preparedness and response systems in place. This includes early warning systems, emergency response plans, and coordination mechanisms among different agencies and stakeholders. Regular drills, training, and public awareness campaigns are also essential components of a city's resilience.

4. Case study: The Netherlands' approach to urban resilience

The Netherlands is widely recognized as a global leader in urban resilience. The country has a long history of dealing with flooding and has developed innovative solutions to manage water and enhance resilience. Some key strategies adopted by the Netherlands include:

4.1. Water management: The Netherlands has implemented an extensive system of dykes, flood barriers, and water management infrastructure to protect its low-lying areas from flooding. This integrated approach to water management has made Dutch cities more resilient to climate change impacts.

4.2. Spatial planning: The Netherlands has implemented innovative spatial planning strategies, such as the concept of "room for the river," which involves creating extra space along riverbanks to accommodate floodwaters during periods of high water levels. This approach allows for natural water retention and reduces the impact of floods on urban areas.

4.3. Community involvement: The Netherlands emphasizes community involvement in decision-making processes and encourages active participation in water management and flood preparedness activities. Local communities are involved in the design and maintenance of water-related infrastructure, ensuring their needs and concerns are considered.

4.4. Ecosystem-based solutions: The Netherlands recognizes the importance of green infrastructure in enhancing urban resilience. The creation of urban parks, wetlands, and green roofs not only improves water management but also provides recreational spaces and enhances biodiversity.

5. Challenges and future directions

Building urban resilience is a complex and ongoing process that requires ongoing collaboration, adaptation, and innovation. Some challenges to consider include:

5.1. Financing: Building resilience can be costly, and securing adequate funding for resilience projects remains a challenge for many cities. Governments and urban planners need to explore innovative financing mechanisms, such as public-private partnerships and green bonds, to fund resilience initiatives.

5.2. Equity and social justice: Resilience efforts should prioritize the most vulnerable and marginalized communities to ensure that the benefits of urban resilience are distributed equitably. This requires addressing underlying social and economic inequalities and promoting inclusive decision-making processes.

5.3.   Integration of nature-based solutions: While the importance of nature-based solutions is increasingly recognized, their integration into urban planning and development processes is still limited. There is a need to scale up the adoption of nature-based solutions and mainstream them into urban policies and practices.

In conclusion, urban resilience is crucial for cities to navigate the challenges of the 21st century.   It requires integrated planning, investment in resilient infrastructure, community engagement, and the integration of nature-based solutions. By building resilience, cities can ensure the well-being and prosperity of their residents in the face of uncertainty and change.

Compact cities

Compact cities are a sustainable urban planning strategy that aims to address the challenges posed by urbanization, population growth, and environmental degradation. This approach promotes the development of densely populated urban areas with mixed land use, efficient infrastructure, and sustainable transportation systems. In compact cities, buildings are designed to be energy-efficient, and there is a focus on public spaces, pedestrian-friendly streets, and green spaces.

The concept of compact cities is based on the principles of sustainable development and seeks to reduce sprawl and promote efficient land use.   By concentrating development in a limited area, compact cities minimize the need for long-distance commuting and promote the use of sustainable modes of transport, such as walking, cycling, and public transit. This can lead to reduced greenhouse gas emissions, improved air quality, and enhanced quality of life.

Benefits of compact cities

Compact cities offer a range of benefits, both for the environment and for residents. Some key advantages include:

+ **Reduced environmental footprint:** Compact cities reduce the demand for land, water, and energy. By minimizing land use for development, they help to protect natural habitats and biodiversity. They also reduce the need for motorized transportation, leading to lower emissions of greenhouse gases and improved air quality.

+ **Increased accessibility:** In compact cities, amenities and services are more easily accessible to residents. With mixed land use and proximity to workplaces, schools, and recreational facilities, residents can walk or use sustainable modes of transport to reach their destinations. This reduces the reliance on private vehicles, traffic congestion, and the associated environmental and social costs.

+ **Social interaction and community cohesion:** Compact cities promote social interaction and community cohesion. With a high concentration of people in close proximity, there are more opportunities for socializing and collaboration. Public spaces, such as parks and plazas, play an important role in facilitating social engagement and enhancing the sense of community.

+ **Improved public health and well-being:** Compact cities encourage physical activity by providing safe and walkable neighborhoods. Access to green spaces and recreational facilities promotes active lifestyles and improves public health. Additionally, compact cities offer better access to essential services, such as healthcare, education, and public transportation, which can enhance the overall well-being of residents.

Challenges and solutions

While compact cities offer numerous benefits, their implementation can encounter certain challenges. These challenges include:

+ **Limited available land:** Finding suitable land for compact city development can be challenging, especially in densely populated areas. However, creative solutions such as redevelopment of existing urban areas, brownfield revitalization, and vertical expansion can help overcome this limitation.

+ **Resistance to change:** Compact city development often faces resistance from stakeholders who are accustomed to traditional urban development patterns. Effective communication and community engagement are crucial to address concerns, build consensus, and ensure the successful implementation of compact city plans.

+ **Infrastructure requirements:** Compact cities require well-planned and efficient infrastructure systems to support the high-density development. This includes transportation networks, utility services, waste management systems, and social infrastructure. Integrated planning, collaboration between different sectors, and innovative design solutions can help overcome infrastructure challenges.

+ **Affordability and equity:** Compact cities need to address issues of affordability and equity to ensure that all residents can access housing, amenities, and opportunities. This requires a balanced approach that combines affordable housing policies, inclusionary zoning, and social infrastructure investments to create vibrant and inclusive communities.

+ **Urban heat island effect:** Compact cities with high building density can experience the urban heat island effect, leading to increased energy consumption for cooling and health risks. Design strategies such as green roofs, urban greening, and the use of reflective materials can help mitigate the urban heat island effect and improve thermal comfort.

Case study: Curitiba, Brazil

Curitiba, the capital of the Brazilian state of Paraná, is an exemplary case of compact city planning. The city's innovative urban development strategies have resulted in a sustainable and livable urban environment.

One of Curitiba's most notable features is its integrated transportation system. The city implemented a Bus Rapid Transit (BRT) system that efficiently connects different parts of the city. The BRT system features dedicated lanes, prepaid boarding, and convenient stations, making public transportation a viable option for a significant portion of the city's population. This has reduced congestion and air pollution, making Curitiba one of the world's most pedestrian-friendly and transit-oriented cities.

Curitiba also prioritizes green spaces and parks. The city has an extensive network of parks, including the famous Barigui Park, providing residents with recreational areas and preserving the natural environment. These green spaces contribute to the overall quality of life in the city and help mitigate the urban heat island effect.

In terms of land use, Curitiba follows a compact city model with mixed land use and high-density development. The city's zoning regulations encourage a mix of residential, commercial, and recreational activities, promoting walkability and reducing the need for long commutes.

To address the challenges of affordable housing, Curitiba has implemented social housing programs that provide low-income residents with access to quality housing and amenities. These programs ensure that the benefits of compact city planning are accessible to all segments of society.

Conclusion

Compact cities are a promising approach to sustainable urban development. By focusing on efficient land use, sustainable transportation systems, and community-oriented design, compact cities can contribute to a more livable and resilient future. While challenges exist, innovative solutions, community engagement, and integrated planning can help overcome these obstacles and create vibrant, inclusive, and environmentally friendly cities. The case study of Curitiba demonstrates the potential of compact city planning in achieving these goals. By embracing the principles of compact cities, we can build a more sustainable and equitable future for urban environments.

Mixed-use development

Mixed-use development is a planning and design strategy that combines multiple types of land uses within a single project or neighborhood. It integrates residential, commercial, industrial, and/or institutional land uses, creating a diverse and vibrant urban environment. This approach promotes walkability, reduces the need for long commutes, and encourages social interaction.

Principles of mixed-use development

Mixed-use development is guided by several principles that aim to create sustainable and livable communities. These principles include:

1. **Compact design**: Mixed-use developments promote compact design by bringing different land uses closer together. This reduces the amount of land required for development, minimizing urban sprawl and preserving open spaces. Compact design also encourages the use of sustainable transportation modes such as walking, cycling, and public transit.

2. **Diversity of land uses**: A key feature of mixed-use development is the integration of diverse land uses. This includes a mix of residential units, offices, retail shops, restaurants, entertainment venues, and community spaces. By offering a variety of amenities and services in close proximity, mixed-use developments create vibrant and convenient communities.

3. **Walkability and connectivity**: Mixed-use developments prioritize pedestrian-friendly designs, with well-connected street networks and infrastructure. Sidewalks, bike lanes, and public transit options are incorporated to make it easy for residents to navigate the neighborhood without relying on cars. This reduces traffic congestion and promotes a healthier and more sustainable lifestyle.

4. **Density and intensity**: Mixed-use developments aim to maximize land utilization by creating higher density areas. Higher densities lead to more efficient use of infrastructure, utilities, and public services. This allows for a more sustainable use of resources and supports the economic viability of the development.

5. **Public spaces and amenities**: Mixed-use developments prioritize the creation of public spaces and amenities that enhance the quality of life for residents. Parks, plazas, community centers, and other gathering spaces promote social interaction and contribute to a sense of community. Access to amenities such as schools, libraries, and healthcare facilities is also considered an important aspect of mixed-use development.

Benefits of mixed-use development

Mixed-use development offers numerous benefits for both residents and the environment. Some of these benefits include:

1. **Reduced commute times**: By providing a mix of residential, commercial, and recreational spaces in close proximity, mixed-use developments reduce the need for long commutes. This saves time and reduces dependence on private vehicles, leading to less traffic congestion and lower carbon emissions.

2. **Improved walkability and health**: Mixed-use developments prioritize pedestrian-friendly design, making it easier for residents to walk or cycle to their destinations. This promotes physical activity, reduces sedentary lifestyles, and improves overall health and well-being.

3. **Enhanced sense of community**: The integration of different land uses in mixed-use developments creates opportunities for social interaction and community engagement. Residents have access to shared public spaces, such as parks and plazas, fostering a sense of belonging and connectedness.

4. **Economic vitality**: Mixed-use developments can stimulate local economies by attracting businesses, creating job opportunities, and increasing property values. The presence of commercial spaces within residential areas provides convenience and supports local businesses.

5. **Environmental sustainability**: By promoting compact design, mixed-use developments help reduce urban sprawl and preserve natural habitats. They also encourage the use of sustainable transportation modes, contributing to lower greenhouse gas emissions.

Challenges and considerations

While mixed-use development offers many benefits, there are also challenges and considerations to address. Some of these include:

1. **Zoning and regulations**: Creating mixed-use developments often requires changes to existing zoning regulations, which can be a complex and time-consuming process. Local authorities need to ensure that the proposed development aligns with the community's vision and land use policies.

2. **Infrastructure and services**: Mixed-use developments may require upgrades to existing infrastructure, including transportation, utilities, and public services. Adequate provisions need to be made to support the increased population density and the demands of commercial activities.

3. **Parking and traffic management**: Balancing parking needs and managing traffic flow in mixed-use developments can be challenging. Adequate parking facilities, parking management strategies, and well-designed transportation networks are necessary to avoid congestion and meet the needs of residents and businesses.

4. **Affordability and equity**: Mixed-use developments can lead to increased property values, which may result in gentrification and displacement of lower-income residents. It is essential to incorporate affordable housing options and ensure equitable access to the amenities and benefits of the development.

5. **Design and aesthetics**: Successful mixed-use developments require careful attention to design and aesthetics. Architectural coherence, compatibility with the surrounding context, and the preservation of local heritage and character are vital considerations.

Examples of mixed-use development

Numerous successful examples of mixed-use development can be found around the world. Here are a few notable examples:

1. **Battery Park City, New York City, USA**: Located in Manhattan, Battery Park City is a prime example of a mixed-use development. It combines housing, offices, parks, retail spaces, and cultural institutions in a pedestrian-friendly environment.

2. **Tianzifang, Shanghai, China**: Tianzifang is a vibrant mixed-use neighborhood in Shanghai known for its narrow alleys, art galleries, boutiques, and cafes. It seamlessly blends traditional Chinese architecture with modern amenities.

3. **Vancouver, Canada**: The city of Vancouver has embraced mixed-use development as a planning strategy. The False Creek area, for example, features a

mix of residential towers, commercial spaces, and public amenities, creating a highly livable and sustainable neighborhood.

Summary

Mixed-use development is a planning and design approach that integrates different land uses within a single project or neighborhood. It promotes compact and walkable communities, reduces commute times, enhances social interaction, and supports economic vitality. However, challenges such as zoning regulations, infrastructure needs, affordability, and design considerations must be addressed for successful implementation. From Battery Park City in New York City to Vancouver's False Creek, real-world examples demonstrate the benefits of mixed-use development in creating vibrant and sustainable urban environments. By embracing the principles of mixed-use development, we can work towards more livable and resilient cities.

Active Transportation Systems

Active transportation systems play a crucial role in promoting sustainable and healthy modes of transportation within urban environments. These systems prioritize non-motorized modes of transportation, such as walking, cycling, and skating, to provide efficient and environmentally friendly alternatives to traditional motorized transportation methods. In this section, we will explore the principles, benefits, challenges, and strategies associated with active transportation systems.

Principles of Active Transportation

Active transportation systems are built on the principle of creating safe and accessible infrastructure for non-motorized modes of transportation. These systems aim to prioritize the needs of pedestrians and cyclists by providing dedicated lanes, paths, and facilities that accommodate active transportation modes. The main principles include:

- **Safety**: Active transportation systems prioritize the safety of users by integrating design elements that reduce conflicts with motorized traffic, such as separated bike lanes, pedestrian crossings, and traffic calming measures.

- **Accessibility**: The design and layout of active transportation systems should be accessible to users of all ages and abilities. This includes the provision of features like ramps, elevators, and well-maintained sidewalks and pathways.

+ **Connectivity:** Active transportation systems should provide well-connected networks that allow users to easily access key destinations such as residential areas, schools, workplaces, and recreational facilities. This connectivity promotes the efficiency and convenience of non-motorized transportation options.

Benefits of Active Transportation

Active transportation systems offer numerous benefits to individuals, communities, and the environment. Some of these benefits include:

Health and Well-being: Active transportation promotes physical activity, which is essential for maintaining good health. Regular walking or cycling can reduce the risk of chronic diseases, including heart disease, obesity, and type 2 diabetes. It also helps improve mental health and overall well-being.

Environmental Sustainability: Active transportation contributes to reducing greenhouse gas emissions and air pollution compared to motorized transportation. By encouraging walking and cycling, active transportation systems mitigate the negative impacts of fossil fuel consumption, promote sustainability, and help combat climate change.

Economic Benefits: Active transportation systems can have positive economic effects for communities. They can reduce healthcare costs associated with physical inactivity, improve property values near active transportation infrastructure, and boost local businesses by increasing foot traffic and accessibility.

Social Equity: By prioritizing non-motorized transportation modes, active transportation systems provide equal opportunities for people of all socio-economic backgrounds to access transportation options. This promotes social equity by reducing transportation-related barriers and bridging gaps in mobility.

Challenges and Solutions

Implementing active transportation systems can present various challenges, including:

Limited Infrastructure: Many cities lack the necessary infrastructure to support active transportation. This can be addressed through comprehensive planning that incorporates the development of pedestrian-friendly sidewalks, dedicated bike lanes, and interconnected pathways.

Safety Concerns: Safety concerns related to active transportation, such as cyclist and pedestrian accidents, can discourage individuals from adopting these modes of transportation. Implementing traffic calming measures, improving intersections, and raising awareness about the benefits of active transportation can address these concerns.

Resistance to Change: Resistance from stakeholders, including motorists and business owners, can present a challenge in implementing active transportation systems. Effective communication, public engagement, and demonstration projects can help change perceptions and build support for these initiatives.

Seasonal Limitations: In regions with extreme climate conditions, active transportation may be limited during certain seasons. Providing adequate infrastructure, such as covered bike lanes or heated walkways, can help overcome these limitations.

Strategies for Successful Active Transportation Systems

To ensure successful implementation and uptake of active transportation systems, the following strategies should be considered:

Integrated Planning: Incorporate active transportation infrastructure into urban planning processes, considering factors like land use, population density, and proximity to key destinations.

Multi-modal Integration: Enhance connectivity between active transportation and other modes of transportation, such as public transit systems. Providing secure bike parking, bike-sharing facilities, and seamless transfer options can encourage multi-modal travel.

Education and Awareness: Promote the benefits of active transportation through education and awareness campaigns. These efforts should focus on safety, health, environmental benefits, and addressing common misconceptions.

Community Engagement: Involve the community in the planning, design, and decision-making processes to ensure their needs and preferences are considered. Engaging stakeholders through workshops, surveys, and public meetings can help build support for active transportation systems.

Monitoring and Evaluation: Regularly monitor the usage and effectiveness of active transportation systems through data collection and evaluation. This information can be used to identify areas for improvement, measure the impact, and justify further investment in these systems.

Example: Vancouver's Active Transportation Initiatives

Vancouver, Canada, is well-known for its successful implementation of active transportation initiatives. The city has invested in an extensive network of dedicated bike lanes, pedestrian-friendly streets, and scenic waterfront paths, making it one of the most bike-friendly cities in North America.

Vancouver's active transportation system includes features such as separated bike lanes along busy corridors, bicycle traffic signals, and bike-friendly intersections to ensure the safety of cyclists. The city has also implemented traffic calming measures, including reduced speed limits and traffic diverters, to create a safer environment for pedestrians.

Through community engagement and collaboration, Vancouver has established a culture of active transportation. The city regularly conducts public consultations and seeks input from citizens to improve the design and connectivity of its active transportation networks.

The success of Vancouver's active transportation initiatives is evident in their impact on the community. More people in Vancouver are choosing to walk, bike, or use other non-motorized modes of transportation for their daily commute. This shift has not only improved the overall health and well-being of residents but has also contributed to reduced traffic congestion and improved air quality in the city.

Conclusion

Active transportation systems offer sustainable and healthy alternatives to conventional motorized transportation methods. By prioritizing non-motorized modes of transportation, these systems promote physical activity, reduce environmental impacts, and contribute to creating vibrant and livable cities. Overcoming challenges and implementing effective strategies are integral to harnessing the transformative potential of active transportation systems and building a sustainable future.

Urban Heat Island Mitigation

Urban heat island (UHI) refers to the phenomenon where urban areas experience significantly higher temperatures compared to their surrounding rural areas. This is primarily due to human activities, such as the construction of buildings and roads, which lead to the modification of the natural landscape and the creation of urban heat islands.

In this section, we will explore various methods and strategies to mitigate the urban heat island effect. These approaches aim to reduce the temperature difference between urban and rural areas, creating more comfortable living conditions and promoting sustainable urban development.

Green Spaces and Urban Vegetation

One effective way to mitigate the urban heat island effect is by increasing the presence of green spaces and urban vegetation. Trees, grass, and shrubs provide shade, reduce the amount of heat absorbed by surfaces, and facilitate evapotranspiration, which cools the surrounding air.

The strategic planting of trees in urban areas can create a canopy that blocks the sun's rays and provides shade for buildings, sidewalks, and streets. Choosing tree species with large, dense canopies and heat-tolerant characteristics is crucial in achieving maximum cooling benefits.

Moreover, green roofs and walls can also contribute to urban heat island mitigation. These structures utilize vegetation and soil to insulate buildings, absorb solar radiation, and release moisture through evapotranspiration. Green roofs, in particular, can reduce the heat transfer into buildings, leading to energy savings and improved indoor comfort.

Cool Roofs and Heat-Reflective Materials

Another effective approach to mitigate the urban heat island effect is through the use of cool roofs and heat-reflective materials. Traditional dark-colored roofs and pavements absorb a significant amount of solar radiation, contributing to increased surface and ambient temperatures.

Cool roofs, on the other hand, are designed to reflect more sunlight and absorb less heat. They are typically made of light-colored or specially coated materials that have high solar reflectance and thermal emittance. Cool roofs can significantly reduce the surface temperature of buildings, leading to energy savings in cooling requirements. They can also contribute to a more comfortable urban environment by reducing the heat island effect.

Heat-reflective materials can also be used for pavements, parking lots, and road surfaces to minimize the heat absorption and subsequent heat release. These materials, such as reflective coatings or porous pavements, can reduce surface temperatures and improve pedestrian comfort during hot weather conditions.

Urban Design and Planning

In order to effectively mitigate the urban heat island effect, urban design and planning play a crucial role. Sustainable urban planning strategies can optimize the layout and configuration of buildings, streets, and open spaces to reduce heat buildup and increase natural ventilation.

Promoting compact, mixed-use development can minimize heat island formation by reducing the amount of impervious surfaces and maximizing natural green spaces. Designing streets and sidewalks with wider tree-lined boulevards can provide shade and create a more walkable and comfortable urban environment.

Additionally, incorporating water bodies and water features into the urban landscape can contribute to heat island mitigation. Water has a high thermal capacity and can absorb and store significant amounts of heat energy. Fountains, ponds, and artificial lakes can provide cooling effects through evaporative cooling and can also serve as recreational areas for the local community.

Cooling Technologies

Advancements in cooling technologies offer additional opportunities to mitigate the urban heat island effect. These technologies focus on reducing the demand for energy-intensive air conditioning while providing localized cooling solutions.

One such technology is the use of cool pavements, which are designed to reflect solar radiation and reduce surface temperatures. Cool pavements can be made of light-colored materials or coated with reflective materials to minimize heat absorption.

Another innovative approach is the use of cool paints, which have high solar reflectance and can be applied to buildings, infrastructure, and outdoor surfaces. These paints reflect a larger proportion of sunlight, reducing the amount of heat absorbed and contributing to lower surface and ambient temperatures.

Furthermore, the use of active cooling systems, such as misting sprays and evaporative coolers, can provide immediate relief from high temperatures in outdoor spaces. These systems utilize the evaporative cooling principle to cool the surrounding air, making outdoor areas more comfortable during hot weather conditions.

Community Engagement and Education

Lastly, effective urban heat island mitigation requires community engagement and education. Raising awareness about the causes and impacts of the urban heat island effect can empower individuals and communities to take action and implement sustainable practices.

Educational programs and outreach initiatives can provide information on the importance of urban green spaces, the selection of heat-tolerant plant species, and the benefits of energy-efficient building materials. Community participation in tree planting campaigns, green roof projects, and sustainable urban development initiatives can foster a sense of ownership and pride in creating a cooler and more livable city.

In conclusion, mitigating the urban heat island effect is crucial for creating sustainable and livable urban environments. Through the implementation of strategies such as increasing green spaces, using cool roofs and heat-reflective materials, incorporating sustainable urban design and planning principles, adopting cooling technologies, and promoting community engagement and education, we can significantly reduce the heat island effect and create more comfortable and resilient cities for the future.

Sustainable Consumption and Production

Circular Economy

The concept of a circular economy is gaining increasing attention worldwide as a sustainable alternative to the traditional linear economy. In a linear economy, resources are extracted, processed into products, used, and then discarded as waste. This linear model is highly wasteful and contributes to environmental degradation and resource scarcity.

The circular economy, on the other hand, aims to minimize waste and maximize resource efficiency by keeping materials and products in use for as long as possible. It is based on the principles of reducing, reusing, recycling, and recovering materials to create a closed-loop system. This shift towards a circular economy requires rethinking the way we design, produce, consume, and dispose of products.

Principles of the Circular Economy

The circular economy is guided by several key principles:

1. **Design for Circular:** Products should be designed with the intent of easy repair, modularization, and disassembly to enhance reusability, recyclability, and durability. This principle encourages businesses to adopt cradle-to-cradle approaches, where materials are recovered, reused, and recycled at the end of their life cycle.

2. **Resource Efficiency:** The circular economy prioritizes the efficient use of resources throughout the entire value chain. This involves optimizing production processes, reducing waste generation, and minimizing raw material extraction. Resource efficiency also includes adopting technologies and practices that enable the recovery and reuse of valuable materials from waste streams.

3. **Closing the Loop:** The circular economy aims to close the loop on material flows by promoting recycling and the use of recycled materials in the production of new products. This reduces the need for virgin resources and minimizes environmental impacts associated with extraction and manufacturing.

4. **Collaboration and Stakeholder Engagement:** The transition to a circular economy requires collaboration between various stakeholders, including businesses, government agencies, consumers, and non-governmental organizations. Such collaboration can foster innovative solutions, drive policy change, and promote the adoption of circular practices across different sectors.

Benefits of the Circular Economy

The circular economy offers numerous benefits and opportunities:

1. **Resource Conservation:** By keeping resources in use for longer periods and promoting recycling, the circular economy helps conserve finite resources, such as minerals, metals, and fossil fuels. This reduces the pressure on natural ecosystems and minimizes the environmental impacts of resource extraction.

2. **Waste Reduction:** The circular economy aims to minimize waste generation by prioritizing the reuse, repair, and recycling of products and materials. This reduces the need for landfilling and incineration, leading to lower greenhouse gas emissions and less pollution.

3. **Economic Growth and Job Creation:** The transition to a circular economy can stimulate economic growth and create new job opportunities. It can foster innovation, drive technological advancements, and generate revenue from waste and resource recovery.

4. **Resilience to Supply Chain Disruptions:** The circular economy promotes diversification of supply chains by reducing dependency on scarce or geographically restricted resources. This enhances resilience to supply chain disruptions, such as price volatility or political instability.

Challenges and Considerations

While the circular economy presents numerous benefits, its implementation is not without challenges and considerations:

1. **Complexity and Interdependencies:** Implementing circular economy practices requires coordination and collaboration among multiple stakeholders. It involves complex systems thinking, as changes in one part of the value chain can have ripple effects throughout the entire system.

2. **Lack of Infrastructure and Expertise:** The transition to a circular economy may require significant investments in infrastructure, technology, and human capital. This includes establishing efficient waste management systems, recycling facilities, and skills development programs.

3. **Consumer Behavior and Awareness:** Shifting consumer behavior and preferences towards more sustainable consumption patterns is crucial for the success of the circular economy. Educating and raising awareness among consumers about the benefits of circular products and services is vital.

4. **Policy and Regulatory Frameworks:** Governments play a critical role in creating an enabling environment for the transition towards a circular economy. This includes implementing supportive policies, providing financial incentives, and enacting regulations that promote circular practices and discourage wasteful behavior.

Real-world Examples

Several real-world examples demonstrate the potential of the circular economy:

1. **Circular Fashion:** Some fashion brands are adopting circular approaches, such as offering clothing rental services, promoting clothing repair and resale, and using recycled materials in their production processes.

2. **Recycling and Upcycling:** Recycling initiatives, such as plastic bottle recycling, paper recycling, and upcycling projects, demonstrate how waste can be transformed into valuable resources.

3. **Product Life Extension:** Companies that offer repair services for electronic devices, such as smartphones and laptops, contribute to extending the lifespan of products and reducing electronic waste.

4. **Industrial Symbiosis:** Industrial symbiosis is a concept where one company's waste becomes another company's raw material. Such collaborations foster resource efficiency, waste reduction, and cost savings.

Conclusion

The circular economy represents a transformative approach to address environmental challenges and drive sustainable development. By prioritizing resource conservation, waste reduction, and collaboration, it offers opportunities for economic growth, job creation, and increased resilience. However, its successful implementation requires concerted efforts from stakeholders across sectors, including governments, businesses, and consumers.

Product life cycle assessment

Product life cycle assessment (LCA) is a key tool in the field of eco science for evaluating the environmental impacts of a product throughout its entire life cycle. It provides a comprehensive analysis of the various stages a product goes through, from raw material extraction to production, use, and disposal. By quantifying the environmental impacts associated with each stage, LCA enables decision-makers to identify opportunities for improvement and make more sustainable choices.

Principles of LCA

LCA follows a set of principles that guide the assessment process. These principles include:

1. **Goal and scope definition:** Clearly defining the objectives and boundaries of the assessment, including the functional unit (e.g., one kilogram of the product) and the system boundaries (e.g., from cradle to grave).

2. **Inventory analysis:** Collecting and quantifying data on the inputs (e.g., energy, materials) and outputs (e.g., emissions, waste) associated with each stage of the product's life cycle.

3. **Impact assessment:** Evaluating the potential environmental impacts of the product by analyzing and interpreting the inventory data using impact categories (e.g., global warming potential, ozone depletion potential).

4. **Interpretation:** Drawing conclusions from the impact assessment and communicating the results to stakeholders in a clear and transparent manner.

Life cycle stages

A typical product life cycle consists of the following stages:

1. **Raw material extraction:** This stage involves the extraction or harvesting of raw materials from natural resources, such as mining for metals or logging for wood. LCA assesses the environmental impacts associated with these extraction processes, considering factors such as resource depletion, habitat destruction, and energy use.

2. **Production:** The production stage involves the transformation of raw materials into the final product. LCA analyzes the energy consumption, emissions, and waste generation during manufacturing processes, along with the associated environmental impacts, such as air and water pollution, greenhouse gas emissions, and waste disposal.

3. **Use:** During the use stage, the product is operated or consumed by the end-user. LCA examines the energy consumption and emissions resulting from the use of the product, as well as any potential environmental impacts during this period. For example, in the case of an electronic device, LCA may consider the energy consumed during its operation and the emissions associated with electricity generation.

4. **End-of-life:** This stage involves the disposal or recycling of the product at the end of its useful life. LCA assesses the environmental impacts associated with different waste management options, such as landfilling, incineration, or recycling. It considers factors such as waste generation, energy consumption, and emissions during the disposal process.

Quantifying environmental impacts

One of the key challenges in conducting LCA is the quantification of environmental impacts. This requires the use of specific indicators and models to convert inventory data into impact scores. The choice of impact categories depends on the goals and scope of the assessment, as well as the availability of data and the specific context of the product being evaluated.

Some commonly used impact categories in LCA include:

+ Global warming potential: Measures the potential of a substance to contribute to climate change through the emission of greenhouse gases, such as carbon dioxide (CO_2) or methane (CH_4).

- Acidification potential: Quantifies the potential of a substance to cause acidification of ecosystems, typically measured in terms of equivalents of sulfur dioxide (SO2).

- Eutrophication potential: Measures the potential of a substance to cause excessive nutrient enrichment in water bodies, leading to harmful algal blooms and oxygen depletion.

- Human toxicity potential: Assesses the potential adverse effects of a substance on human health, considering factors such as carcinogenicity, mutagenicity, and acute or chronic toxicity.

- Ecotoxicity potential: Evaluates the potential adverse effects of a substance on ecosystems and the environment, considering factors such as aquatic toxicity and the persistence and bioaccumulation of the substance.

Challenges and limitations

While LCA is a valuable tool for assessing the environmental performance of products, it also faces certain challenges and limitations. These include:

- Data availability and quality: Conducting a comprehensive LCA requires reliable data on the inputs, outputs, and environmental impacts associated with each stage of a product's life cycle. However, data availability and quality can vary significantly, particularly for emerging technologies or complex supply chains.

- System boundaries: Defining the system boundaries of an LCA can be challenging, as it involves deciding which processes and activities to include and exclude. This decision can significantly impact the results and the comparability of different LCAs.

- Impact assessment methods: The choice of impact assessment methods and the weighting of different impact categories can introduce subjectivity into LCA results. Different methods and weighting factors can lead to different conclusions and may require expert judgment.

- Interpretation and communication: Presenting LCA results in a clear and understandable manner to different stakeholders can be challenging. Balancing scientific rigor with accessibility is essential to ensure that the results are properly understood and can inform decision-making.

Example: LCA of a smartphone

To illustrate the application of LCA, let's consider a hypothetical LCA of a smartphone. The assessment would involve collecting data on the raw materials used (e.g., metals, plastics), energy consumption during manufacturing, emissions from production processes, energy use during the use phase, and potential impacts of disposal or recycling.

The LCA could reveal that the extraction of rare earth metals for the smartphone's components has significant environmental impacts, including habitat destruction and resource depletion. The manufacturing process might contribute to air and water pollution, as well as greenhouse gas emissions. The use phase could account for the largest share of energy consumption and emissions, depending on factors such as energy efficiency and the power source used. Finally, the end-of-life stage could reveal opportunities for recycling or proper disposal to minimize waste generation and associated environmental impacts.

By quantifying these impacts, decision-makers can identify areas where improvements can be made, such as using more sustainable materials, optimizing manufacturing processes, or promoting recycling initiatives.

Resources for LCA

Several resources are available to support the practice of LCA:

+ **Software tools:** Various software tools, such as SimaPro, GaBi, and openLCA, facilitate the conduct of LCA by providing databases, impact assessment methods, and modeling capabilities.

+ **Standards and guidelines:** International standards and guidelines, such as ISO 14040 and ISO 14044, provide a framework for conducting LCA and ensuring the quality and transparency of the assessment.

+ **Life cycle databases:** Databases, such as ecoinvent and ELCD, offer comprehensive inventories of life cycle data for a wide range of materials, processes, and products, enabling practitioners to access reliable and up-to-date information.

+ **Case studies and best practices:** Case studies and best practice examples provide real-world applications of LCA and insights into successful sustainability initiatives.

Exercises

1. Choose a product of your choice (e.g., a food item, electronic device) and conduct a simplified life cycle assessment. Identify the major environmental impacts associated with each life cycle stage and propose potential improvements.

2. Research current trends in LCA methodology and identify emerging techniques or developments that could enhance the accuracy and scope of future assessments.

3. Investigate businesses or industries in your local area that have implemented LCA as part of their sustainability practices. Assess the effectiveness of their initiatives and identify any challenges they may have faced.

Conclusion

Product life cycle assessment is a vital tool for evaluating the environmental impacts of a product throughout its entire life cycle. By systematically assessing the environmental impacts associated with each stage, decision-makers can make informed choices to minimize the ecological footprint of products. Although LCA faces challenges and limitations, it provides a valuable framework for promoting sustainability and guiding future eco science advances.

Sustainable packaging

In today's world, packaging plays a vital role in our daily lives. It helps to protect products, extends their shelf life, and provides convenience. However, the environmental impact of packaging cannot be ignored. The excessive use of packaging materials, especially non-biodegradable plastics, has led to significant environmental pollution, including increased waste generation and plastic pollution in our oceans and landfills. This has prompted a growing need for sustainable packaging solutions.

Sustainable packaging refers to the use of packaging materials and designs that have a reduced environmental footprint throughout their lifecycle. It aims to minimize waste, conserve resources, and promote environmental responsibility. Sustainable packaging encompasses various principles, including:

+ **Materials selection:** The choice of packaging materials is crucial in sustainable packaging. It involves opting for renewable, recyclable, and biodegradable materials that have a lower impact on the environment. For example, using plant-based plastics instead of traditional petroleum-based plastics.

+ **Minimalism:** Sustainable packaging focuses on minimizing the amount of packaging material used while still effectively protecting the product. This approach aims to reduce waste generation and conserve resources.

+ **Life cycle assessment:** Assessing the environmental impact of packaging throughout its entire life cycle is an essential aspect of sustainable packaging. This includes evaluating factors such as raw material extraction, production, distribution, use, and end-of-life disposal. Life cycle assessment helps identify areas where improvements can be made to reduce environmental impacts.

+ **Recycling and reuse:** Encouraging the recycling and reuse of packaging materials is a key principle of sustainable packaging. This involves designing packaging that is easily recyclable and implementing effective recycling programs. Promoting the use of refillable containers and encouraging customers to reuse packaging can also contribute to sustainability.

+ **Innovation and design:** Sustainable packaging involves developing innovative designs that minimize waste and maximize resource efficiency. This includes incorporating features such as lightweighting, which reduces the amount of material used in packaging. Designing packaging that is easy to disassemble and separate for recycling also enhances sustainability.

+ **Consumer education:** Educating consumers about the importance of sustainable packaging and their role in waste reduction and recycling is crucial. Providing clear instructions on proper disposal and recycling of packaging can help promote responsible consumer behavior.

Challenges and solutions

While sustainable packaging offers numerous benefits, there are several challenges that need to be addressed:

+ **Cost considerations:** Sustainable packaging materials and designs can sometimes be more expensive than traditional packaging options. However, advancements in technology and increasing demand for sustainable solutions are driving down costs. Companies can overcome this challenge by investing in research and development, leveraging economies of scale, and emphasizing the long-term environmental and economic benefits of sustainable packaging.

+ **Supply chain complexities:** Implementing sustainable packaging practices requires collaboration and coordination among various stakeholders in the supply chain, including raw material suppliers, manufacturers, retailers, and consumers. Effective communication and partnerships can help overcome these complexities and create a more sustainable packaging ecosystem.

+ **Consumer behavior:** Encouraging consumers to adopt sustainable packaging habits can be challenging. Many customers are accustomed to the convenience of single-use packaging and may not be aware of the environmental consequences. Educating consumers, creating incentives for sustainable packaging choices, and improving recycling infrastructure can help drive the shift towards sustainable packaging.

+ **Infrastructure and regulations:** The lack of appropriate recycling infrastructure and inconsistent regulations can hinder the adoption of sustainable packaging practices. Governments and industry organizations need to work together to invest in recycling facilities, improve waste management systems, and establish clear regulations and standards for sustainable packaging.

Real-world examples

Many companies and industries have already taken significant steps towards implementing sustainable packaging solutions. Here are a few examples:

+ **Biodegradable packaging materials:** Companies like Eco-Enclose and Natur-Tec offer biodegradable packaging materials made from renewable resources such as cornstarch and sugarcane. These materials are compostable and break down naturally, reducing their impact on the environment.

+ **Reusable packaging systems:** Loop, a global reuse platform, partners with leading consumer goods companies to offer products in durable, reusable packaging. Customers can order products online, receive them in reusable containers, and return the empty containers for cleaning and reuse. This eliminates single-use packaging waste.

+ **Paper-based alternatives:** Many companies are transitioning from plastic to paper-based packaging alternatives. For example, Nestlé has committed to eliminating all non-recyclable or hard-to-recycle plastics by 2025. They are replacing plastic straws with paper straws and introducing paper-based packaging for some of their products.

+ **Bulk and refillable options**: Retailers like zero-waste grocery stores and refill stations promote the use of bulk and refillable packaging. Customers bring their own containers and fill them with the desired quantity of products, reducing the need for single-use packaging.

Tricks and caveats

When considering sustainable packaging options, it is important to keep in mind the following tricks and caveats:

+ **Trade-offs**: Sustainable packaging solutions often involve trade-offs. For example, while plant-based plastics are biodegradable and renewable, their production may require significant land and water resources. It is essential to conduct a comprehensive life cycle assessment to determine the overall environmental impact of packaging options.

+ **Bioplastics and composting**: Bioplastics, including PLA (polylactic acid), are popular sustainable packaging alternatives. However, their proper disposal requires specific composting facilities that can accommodate biodegradable materials. Without proper composting, bioplastics may not break down efficiently and could contribute to environmental pollution.

+ **Recycling infrastructure**: The success of sustainable packaging relies heavily on the availability of efficient recycling infrastructure. It is crucial to ensure that the packaging materials chosen can be easily recycled and that appropriate recycling facilities are accessible to consumers.

+ **Reducing packaging waste**: While sustainable packaging is essential, reducing overall packaging waste should be a priority. Companies should explore options to minimize packaging materials and unnecessary packaging components, such as excessive plastic wrapping or packaging inserts.

Exercises

Here are a few exercises to deepen your understanding of sustainable packaging:

1. Research a company that has successfully implemented sustainable packaging practices. Describe their approach and the benefits they have achieved.

2. Conduct a life cycle assessment of a commonly used packaging material (e.g., plastic, glass, or paper). Evaluate its environmental impacts in terms of resource consumption, energy use, greenhouse gas emissions, and waste generation.

3. Identify a product in your household that has excessive packaging. Propose sustainable packaging alternatives for that product, considering factors such as materials, design, recyclability, and environmental impact.

4. Visit a local retailer or grocery store and assess their packaging choices. Identify examples of sustainable packaging and areas where improvements can be made.

Conclusion

Sustainable packaging is a critical component of eco science and plays a significant role in reducing the environmental impact of consumer goods. By adopting sustainable packaging practices, we can minimize waste, conserve resources, and protect our planet for future generations. It requires a collaborative effort from stakeholders across the supply chain and a shift in consumer behavior. With continued innovation and awareness, sustainable packaging has the potential to transform the way products are packaged and consumed. Let's strive for a future where packaging not only serves its functional purpose but also contributes to a more sustainable and resilient world.

Conscious Consumerism

Conscious consumerism is a growing movement that encourages individuals to make purchasing decisions based on ethical, social, and environmental considerations. It emphasizes the idea that our choices as consumers have far-reaching consequences and encourages us to be mindful of the impact our consumption has on the planet and society.

Understanding Conscious Consumerism

Conscious consumerism goes beyond traditional notions of shopping and promotes a deeper understanding of the lifecycle of products, from production to disposal. It encourages us to ask questions such as: Where was the product made? Were the workers paid fair wages? Is the product made using sustainable materials? How will the product be disposed of at the end of its life?

To engage in conscious consumerism, it is important to be aware of the social, economic, and environmental implications of our purchases. This awareness can be gained through education, research, and staying informed about current issues. By understanding the impact of different products and brands, we can make more informed choices that align with our values and contribute to a more sustainable future.

The Benefits of Conscious Consumerism

Conscious consumerism offers several benefits, both individually and collectively. Here are a few key benefits:

1. **Environmental Impact:** By choosing products that are sustainably produced, packaged, and transported, we can minimize our carbon footprint and reduce our contribution to climate change. For example, opting for organic and locally sourced food reduces the use of synthetic pesticides and decreases the energy required for transportation.

2. **Social Responsibility:** Conscious consumerism supports fair trade practices, fair wages, and safe working conditions for workers worldwide. By purchasing products that are ethically sourced and manufactured, we can help improve the lives of workers and contribute to a more equitable global society.

3. **Human Health:** Conscious consumerism prioritizes products that are safe for human health, avoiding chemicals that may be harmful. For instance, choosing natural and organic personal care products reduces exposure to potentially harmful substances found in conventional products.

4. **Economic Sustainability:** Supporting sustainable and ethical businesses not only benefits the environment and society but also contributes to a more sustainable economy. By investing in businesses that prioritize responsible practices, we can encourage economic growth that is aligned with sustainability.

Principles of Conscious Consumerism

Practicing conscious consumerism involves adopting certain principles and guidelines to make informed purchasing decisions. Here are some key principles to consider:

1. **Research and Education:** Stay informed about the social and environmental impact of different products and brands. Look for certifications and labels that indicate sustainable and ethical practices.

2. Rethink Consumption Habits: Challenge the idea of constantly buying and consuming more. Consider reducing, reusing, and repairing items instead of always buying new ones. Embrace minimalism and prioritize quality over quantity.

3. Support Local and Sustainable Businesses: Choose products that are locally produced or sourced to minimize transportation emissions. Look for businesses that prioritize sustainability, fair trade, and ethical practices.

4. Reduce Packaging Waste: Opt for products with minimal packaging or packaging made from recycled or biodegradable materials. Avoid single-use items whenever possible.

5. Vote with Your Wallet: Use your purchasing power to support businesses that align with your values and avoid those that don't. This sends a clear message to companies about the demand for sustainable and ethical products.

Examples of Conscious Consumerism

Here are a few examples of conscious consumerism in action:

1. Ethical Fashion: Many consumers are now opting for clothing brands that prioritize fair wages, safe working conditions, and sustainable sourcing of materials. They choose brands that use organic or recycled fabrics, promote slow fashion, and provide transparent supply chains.

2. Sustainable Food Choices: Conscious consumers support local and organic food producers, prioritize plant-based diets, and reduce food waste. They choose products that are free from genetically modified organisms (GMOs), synthetic pesticides, and hormones.

3. Green Electronics: Consumers are increasingly demanding electronics that are made using recycled materials, are energy-efficient, and have extended product lifecycles. They opt for brands that offer repair services and take-back programs for electronic waste.

Challenges and Caveats

While conscious consumerism offers many benefits, it is important to recognize its limitations and challenges. Here are a few things to consider:

1. Accessibility and Affordability: Sustainable and ethical products can sometimes be more expensive and less accessible to everyone. This poses a challenge for individuals with limited financial resources. Efforts are needed to make these products more affordable and widely available.

2. Greenwashing: Some companies may engage in greenwashing, which involves making false or exaggerated claims about the sustainability of their products or practices. This makes it important for consumers to do thorough research and look for credible certifications and third-party endorsements.

3. Systemic Change: While individual actions are important, systemic change is necessary for widespread sustainability. Conscious consumerism should be complemented by government regulations, corporate responsibility, and collective action to address broader issues.

4. Balancing Trade-offs: Conscious consumerism involves making trade-offs and compromises. For example, choosing organic produce may have higher environmental benefits but also carry a higher price tag. Finding the right balance is essential.

Conclusion

Conscious consumerism allows us to align our values with our purchasing decisions, promoting a more sustainable, ethical, and equitable world. By considering the social, economic, and environmental impact of our consumption, we can contribute to positive change and influence the practices of businesses and industries. Through education, research, and mindful choices, we can harness the power of our consumer decisions to create a better future for all.

Sharing Economy

The concept of a sharing economy has gained significant attention in recent years as a sustainable alternative to traditional consumption patterns. The sharing economy is based on the idea of sharing resources, goods, and services among individuals and organizations to optimize their use and reduce waste. It is founded on principles of collaborative consumption, access over ownership, and peer-to-peer exchange.

Overview and Importance

The sharing economy disrupts the traditional linear model of production and consumption, which is characterized by excessive resource extraction, high energy consumption, and generation of waste. Instead, it promotes a more sustainable and efficient use of resources by maximizing their utilization and minimizing waste. By enabling individuals and businesses to share underutilized assets, the sharing economy can contribute to reducing environmental impacts, fostering social and economic inclusivity, and enhancing resource efficiency.

Key Characteristics

The sharing economy is characterized by several key features:

- Peer-to-peer exchange: Individuals can directly interact and exchange resources or services without the need for intermediaries.

- Access-based consumption: The focus is on accessing and utilizing resources as needed, rather than owning them outright.

- Collaborative consumption: Multiple individuals can collectively share the ownership, use, and cost of resources.

- Online platforms: Digital platforms facilitate the matchmaking and coordination of sharing transactions, enabling efficient resource allocation and trust-building among participants.

Examples of Sharing Economy

The sharing economy encompasses a wide range of sectors and activities. Here are some popular examples:

- Ride-sharing services: Platforms like Uber and Lyft enable individuals to share car rides rather than relying solely on private vehicles or traditional taxi services.

- Home-sharing platforms: Services such as Airbnb allow individuals to rent out spare rooms or entire properties, promoting the efficient use of available housing.

- Peer-to-peer lending: Online platforms like LendingClub connect borrowers with lenders directly, bypassing traditional financial institutions.

- Co-working spaces: Shared workspaces provide an affordable and flexible alternative to traditional office spaces, encouraging collaboration and resource sharing among entrepreneurs and freelancers.

- Tool libraries: Community-based tool lending libraries allow people to borrow tools they need for specific projects, reducing the need for individual ownership.

- Sharing of household goods: Websites like Freecycle and Buy Nothing Project facilitate the free exchange of unwanted items within local communities, reducing waste and promoting reuse.

Benefits of the Sharing Economy

The sharing economy offers several benefits, including:

- Resource efficiency: By maximizing the use of underutilized assets, the sharing economy reduces the need for excessive production, leading to lower resource consumption.

- Economic empowerment: The sharing economy can provide economic opportunities for individuals and communities, allowing them to generate income by sharing their resources or skills.

- Social connectedness: Sharing platforms often foster social interactions and community engagement, enhancing social cohesion and trust among participants.

- Environmental sustainability: The sharing economy reduces waste, carbon emissions, and energy consumption associated with traditional consumption patterns.

- Cost savings: By sharing resources, individuals can access goods and services at a lower cost than purchasing or owning them individually.

Challenges and Considerations

While the sharing economy holds great promise, it also faces several challenges and considerations:

- Regulatory concerns: The rapid growth of sharing economy platforms has raised legal and regulatory questions surrounding issues such as safety, taxation, and labor rights.

- Trust and reputation: The success of sharing economy models relies on trust between participants. Building and maintaining trust is crucial for the sustained growth of the sharing economy.

- Accessibility and inclusivity: Ensuring that sharing economy benefits are accessible to all individuals and communities is essential to avoid contributing to existing inequalities.

- Data privacy and security: Sharing economy platforms often gather large amounts of personal data, necessitating robust privacy and security measures to protect user information.

+ Overconsumption risk: While the sharing economy promotes resource efficiency, there is a risk of overconsumption if access to shared resources becomes too convenient or affordable.

Conclusion

The sharing economy has the potential to transform consumption patterns, promote sustainability, and foster social and economic inclusivity. By embracing the principles of collaborative consumption and enabling peer-to-peer exchange, the sharing economy offers opportunities for individuals, businesses, and communities to make more efficient use of resources and contribute to a more sustainable future.

Collaborative Consumption

Collaborative consumption, also known as the sharing economy, is an emerging trend that promotes the sharing, borrowing, and renting of resources rather than individual ownership. It is a sustainable and cost-effective alternative to traditional consumerist practices that emphasizes access over ownership. In this section, we will explore the principles, benefits, challenges, and examples of collaborative consumption.

Principles of Collaborative Consumption

Collaborative consumption is rooted in the principles of sustainability, efficiency, and community. It is based on the idea that sharing resources and utilizing them collectively can lead to a more sustainable and equitable society. The following principles underpin the concept of collaborative consumption:

+ **Access-based consumption:** Instead of owning goods and services, individuals access them on a temporary basis. This reduces the demand for new products and promotes the efficient use of existing resources.

+ **Peer-to-peer sharing:** Collaborative consumption relies on peer-to-peer platforms that connect individuals who want to share or rent out their resources with those who need them. These platforms facilitate the exchange of goods and services between individuals, cutting out the middleman and creating a sense of community.

+ **Trust and reputation systems:** Collaborative consumption platforms often incorporate trust and reputation systems that allow users to rate and review each other. These systems help build trust among participants and ensure a reliable and secure sharing experience.

+ **Sustainable practices:** Collaborative consumption promotes sustainable practices by reducing waste, lowering carbon emissions, and conserving resources. By sharing and reusing items, the need for excessive production and consumption is minimized.

+ **Financial and social benefits:** Collaborative consumption offers cost savings for individuals who can access goods and services at a fraction of the cost of ownership. It also fosters a sense of community by creating opportunities for social interactions and shared experiences.

Benefits and Challenges of Collaborative Consumption

Collaborative consumption has several benefits that make it an attractive option for both individuals and society as a whole. However, it also faces certain challenges that need to be addressed to ensure its successful implementation and growth.

Benefits:

+ **Environmental sustainability**: By promoting the sharing and reuse of resources, collaborative consumption reduces the environmental impact associated with the production, consumption, and disposal of goods. It contributes to the conservation of resources, reduction of waste, and lower carbon emissions.

+ **Cost savings**: Collaborative consumption allows individuals to access goods and services without the high upfront costs associated with ownership. This can lead to significant cost savings for consumers, especially for items that are only needed occasionally or for a short duration.

+ **Increased resource efficiency**: Collaborative consumption optimizes the utilization of existing resources by sharing them among a larger group of individuals. This reduces the need for new production and helps maximize the value extracted from available resources.

+ **Social connections and community building**: Collaborative consumption platforms facilitate interactions between individuals, creating opportunities for social connections and community building. It fosters a sense of trust, cooperation, and mutual support among participants.

+ **Diversified consumer choices**: Collaborative consumption widens the range of options available to consumers by providing access to a variety of goods and services. It allows individuals to experiment with different products and experiences without the commitment of ownership.

Challenges:

+ **Regulatory concerns**: The sharing economy is a relatively new concept, and existing regulations may not always adequately address its unique dynamics. Issues related to liability, safety, taxation, and fair competition need to be carefully considered and addressed in order to ensure the responsible operation of collaborative consumption platforms.

+ **Trust and reputation:** Trust is a crucial element of collaborative consumption. Establishing and maintaining trust among participants can be challenging, especially in an online environment. Reputation systems and robust user verification mechanisms are necessary to build trust and mitigate risks.

+ **Unequal access and digital divide:** Collaborative consumption heavily relies on access to digital platforms and technology. The digital divide, where certain individuals or communities lack access to the necessary tools and infrastructure, can create barriers to participation and result in exclusion.

+ **Overconsumption disguised as sharing:** While collaborative consumption aims to promote sustainable practices, there is a risk of overconsumption disguised as sharing. Sharing platforms that encourage excessive consumption or offer unlimited access to resources without considering their environmental impact may undermine the sustainability goals of collaborative consumption.

+ **Quality control and maintenance:** Maintaining the quality and safety of shared resources can be challenging. Adequate measures need to be in place to ensure the reliability and usability of shared items, as well as their proper maintenance and repair.

Examples of Collaborative Consumption

Collaborative consumption has been successfully implemented in various sectors and industries. Here are a few examples:

+ **Ride-sharing:** Platforms like Uber and Lyft allow individuals to share rides, reducing the number of private cars on the road and promoting efficient use of resources.

+ **Accommodation sharing:** Airbnb and similar platforms enable individuals to rent out their spare rooms or entire properties to travelers, providing an alternative to traditional hotels and increasing accommodation options.

+ **Goods sharing:** Platforms like Craigslist and Freecycle facilitate the exchange of used goods among individuals, reducing waste and promoting a circular economy.

- **Co-working spaces:** Shared workspaces like WeWork provide individuals with affordable and flexible office spaces, promoting collaboration and reducing the need for individual office spaces.

- **Tool libraries:** Tool libraries allow individuals to borrow tools for temporary use, reducing the need for each person to own their own set of tools.

- **Clothing rental:** Clothing rental services, such as Rent the Runway, allow individuals to rent designer clothes for special occasions, reducing the demand for new clothing purchases.

- **Food sharing:** Platforms like Olio and LeftoverSwap enable individuals and businesses to share surplus food with others, reducing food waste and addressing food insecurity.

Conclusion

Collaborative consumption is an innovative approach to resource utilization that promotes sustainability, efficiency, and community building. By embracing the principles of sharing and access-based consumption, collaborative consumption offers environmental, economic, and social benefits. However, its implementation also poses challenges that require careful consideration and proactive solutions. Through the examples provided in this section, we can see how collaborative consumption is transforming various sectors, contributing to a more sustainable and equitable future.

Sustainable fashion

In recent years, there has been a growing concern about the negative environmental and social impacts of the fashion industry. The production and consumption of clothing have been associated with various sustainability challenges, including resource depletion, pollution, and unethical labor practices. To address these issues, the concept of sustainable fashion has emerged, aiming to create a more environmentally and socially responsible approach to fashion.

Definition and principles

Sustainable fashion, also known as ethical or eco-fashion, refers to the design, production, and consumption of clothing that minimizes harm to the environment and promotes social justice. It involves adopting practices that consider the entire lifecycle of a garment, from the sourcing of materials to disposal.

The principles underlying sustainable fashion include:

+ **Ethical sourcing:** Sustainable fashion emphasizes the use of materials that are responsibly sourced, such as organic cotton, reclaimed or recycled fabrics, and low-impact dyes. It encourages transparency and fair trade practices throughout the supply chain.

+ **Minimizing waste:** Sustainable fashion aims to reduce waste by promoting practices like upcycling, recycling, and designing for longevity. It also discourages the overproduction of garments that result in excessive waste.

+ **Reducing carbon footprint:** Sustainable fashion seeks to minimize greenhouse gas emissions by promoting local production, using renewable energy, and reducing transportation-related emissions. It also supports the adoption of sustainable packaging and shipping practices.

+ **Labor rights and fair working conditions:** Sustainable fashion advocates for fair wages, safe working conditions, and the abolition of exploitative labor practices. It encourages brands to collaborate with suppliers and manufacturers who prioritize worker well-being.

+ **Consumer awareness and education:** Sustainable fashion promotes consumer awareness by providing information about the environmental and social impacts of clothing. It encourages consumers to make informed choices and embrace responsible consumption habits.

Challenges and solutions

Despite the growing interest in sustainable fashion, there are several challenges that need to be addressed to achieve a more sustainable and ethical industry. These challenges include:

+ **Fast fashion culture:** The fast fashion industry, characterized by rapid production cycles and low-priced clothing, encourages a culture of disposable fashion. This leads to increased waste and exploitation of resources and labor. To counteract this, sustainable fashion promotes slow fashion, which focuses on quality, durability, and timeless designs.

+ **Supply chain complexity:** The fashion supply chain is often complex and globally dispersed, making it challenging to ensure transparency and traceability. Sustainable fashion addresses this challenge by promoting supply chain transparency, fair trade practices, and certifications that verify ethical and sustainable practices.

+ **Consumer behavior:** Shifting consumer behavior towards more sustainable fashion choices can be challenging due to factors like price considerations and lack of awareness. Sustainable fashion advocates for education and awareness campaigns to inform consumers about the environmental and social impacts of their choices. It also involves making sustainable fashion more accessible and affordable.

+ **Innovation and technology:** The fashion industry needs to embrace innovation and technology to develop more sustainable materials, textile recycling methods, and production processes. This involves exploring alternatives to conventional synthetic fibers, such as bio-based materials and innovative dyeing techniques.

Examples and initiatives

There are various initiatives and examples of sustainable fashion practices that highlight the potential for positive change within the industry. Some of these include:

+ **Collaboration for sustainability:** The Sustainable Apparel Coalition (SAC) is an industry-wide partnership that aims to reduce the environmental and social impacts of the fashion industry. It has developed the Higg Index, a standardized measurement tool that assesses the sustainability performance of brands and retailers.

+ **Circular fashion:** Circular fashion promotes a closed-loop system where garments are designed, produced, worn, and then recycled or upcycled at the end of their life. This approach aims to minimize waste and resource consumption. Brands like Patagonia and Eileen Fisher have embraced circular fashion by offering take-back programs and incorporating recycled materials into their products.

+ **Fair trade and ethical fashion:** Fair trade certification ensures that producers and workers receive fair wages and work in safe conditions. Brands like People Tree and Mata Traders focus on fair trade practices, supporting artisans in developing countries and providing consumers with ethically-made clothing.

+ **Slow fashion movement:** The slow fashion movement encourages consumers to buy fewer, higher-quality garments and to value craftsmanship and timeless design. Brands like Everlane and Amour Vert embody the principles of slow fashion by producing durable, sustainable, and versatile clothing.

+ **Innovative materials:** Sustainable fashion explores the use of innovative materials that have a lower environmental impact. For example, companies like Bolt Threads and Piñatex have developed bio-based and biodegradable alternatives to conventional textiles, such as spider silk and pineapple leaf fibers, respectively.

Conclusion

Sustainable fashion presents an alternative approach to the conventional fashion industry, addressing its negative environmental and social impacts. By embracing ethical sourcing, waste reduction, carbon footprint reduction, fair labor practices, and consumer education, sustainable fashion offers a path towards a more responsible and sustainable future. However, achieving a truly sustainable fashion industry requires collaboration between all stakeholders, including brands, consumers, policymakers, and innovators, to drive systemic change and create a fashion system that respects both people and the planet.

Interactive Exercise

Imagine you are a fashion designer tasked with creating a sustainable clothing line. Choose three sustainable fabrics or materials that you would incorporate into your designs and explain why you selected them. Consider their environmental impact, availability, and durability.

Green manufacturing

Green manufacturing refers to the production of goods using environmentally friendly and sustainable methods. It aims to reduce the environmental impact of manufacturing processes, minimize waste generation, conserve resources, and promote the use of renewable energy. By adopting green manufacturing practices, companies can contribute to the overall goal of achieving sustainability and mitigating climate change.

Principles of Green Manufacturing

Green manufacturing is guided by several key principles that help in the development and implementation of sustainable manufacturing processes:

1. **Resource efficiency**: Green manufacturing focuses on optimizing resource utilization, minimizing raw material consumption, and reducing energy usage. This involves using energy-efficient machinery, implementing recycling and waste reduction programs, and adopting lean manufacturing techniques to minimize waste generation.

2. **Life cycle assessment**: A life cycle assessment (LCA) is conducted to evaluate the environmental impact of a product throughout its entire life cycle, from raw material extraction to end-of-life disposal. Green manufacturing incorporates the findings of LCA into the decision-making process to identify areas for improvement and implement sustainable practices.

3. **Cleaner production**: Cleaner production involves minimizing or eliminating the use of hazardous materials, chemicals, and pollutants in manufacturing processes. This can be achieved by adopting cleaner technologies, substituting toxic substances with safer alternatives, and implementing pollution prevention measures.

4. **Design for the environment:**   Green manufacturing emphasizes the integration of environmental considerations into the product design phase. This involves using eco-friendly materials, designing products for disassembly and recycling, and reducing the overall environmental impact of the product's life cycle.

5. **Renewable energy adoption:**   Green manufacturing promotes the use of renewable energy sources, such as solar, wind, and hydropower, to power manufacturing operations. This helps reduce reliance on fossil fuels and lowers greenhouse gas emissions.

6. **Collaborative approach:**   Green manufacturing encourages collaboration and partnerships between manufacturers, suppliers, government agencies, and other stakeholders. This facilitates the sharing of best practices, knowledge, and resources to drive sustainable manufacturing.

Benefits of Green Manufacturing

Implementing green manufacturing practices offers numerous benefits for both businesses and the environment:

+ **Environmental benefits:**   Green manufacturing helps reduce carbon emissions, minimize air and water pollution, conserve natural resources, and protect ecosystems. It contributes to the preservation of biodiversity and the overall health of the planet.

+ **Cost savings:**   By optimizing resource usage and minimizing waste generation, green manufacturing can lead to significant cost savings for businesses.   Energy-efficient processes and equipment reduce energy consumption, resulting in lower utility bills. Recycling and waste reduction initiatives can also lower material costs.

+ **Improved reputation:** Adopting green manufacturing practices enhances a company's reputation and strengthens its brand image.   Consumers are increasingly conscious of environmental issues and prefer to support sustainable businesses. Public perception of a company as environmentally responsible can lead to increased customer loyalty and competitive advantage.

+ **Regulatory compliance:**   Green manufacturing helps businesses stay compliant with environmental regulations and standards. By implementing sustainable practices, companies can avoid penalties, fines, and legal disputes associated with environmental non-compliance.

+ **Employee morale:** Green manufacturing initiatives create a positive working environment and boost employee morale. Employees are more likely to be motivated and satisfied when working for an environmentally responsible company. Involving employees in sustainability initiatives can also foster a sense of ownership and pride.

Examples of Green Manufacturing Practices

Several green manufacturing practices are being implemented by companies worldwide. Here are a few examples:

+ **Energy-efficient production lines:** Companies are investing in energy-efficient machinery and equipment to reduce energy consumption during the manufacturing process. This includes the use of advanced monitoring systems, smart sensors, and automated controls to optimize energy usage.

+ **Waste reduction and recycling:** Manufacturers are implementing waste management programs to minimize waste generation and promote recycling. This involves segregating waste streams, reusing materials, and partnering with recycling facilities to ensure proper disposal of waste.

+ **Water conservation:** Companies are implementing water-saving measures, such as using water-efficient equipment, reusing water in manufacturing processes, and treating wastewater before discharge. Water recycling systems and rainwater harvesting are also utilized to conserve water resources.

+ **Green supply chain management:** Manufacturers are collaborating with suppliers who follow sustainable practices, such as using eco-friendly materials, reducing packaging waste, and optimizing transportation logistics. This ensures the entire supply chain is aligned with green manufacturing principles.

+ **Product life cycle assessment:** Companies are conducting life cycle assessments to identify areas for improvement, such as reducing the environmental impact of raw material extraction, improving energy efficiency during manufacturing, and designing products for easier disassembly and recycling.

Challenges and Considerations

While green manufacturing practices offer numerous benefits, there are also challenges that need to be addressed:

- **Initial investment costs:** Implementing green manufacturing practices may require capital investments, such as purchasing energy-efficient equipment or renovating production facilities. However, these costs can often be offset by energy and resource savings over time.

- **Supply chain complexities:** Green manufacturing involves collaborating with suppliers who follow sustainable practices. Ensuring the entire supply chain adopts green principles can be challenging, especially when dealing with suppliers from different regions with varying levels of sustainability awareness.

- **Technological limitations:** Some industries face technological limitations when it comes to implementing green manufacturing practices. Researchers and engineers need to continually develop and improve technologies that allow for more sustainable processes in these industries.

- **Changing consumer demands:** Consumer demand for eco-friendly products is constantly evolving. Manufacturers need to stay updated with changing consumer preferences and adapt their green manufacturing strategies accordingly.

- **Regulatory changes:** Environmental regulations and standards are subject to change over time. Manufacturers need to stay informed about evolving regulations and ensure their practices remain compliant.

Overall, green manufacturing plays a crucial role in promoting sustainability and reducing the environmental impact of manufacturing processes. By adopting resource-efficient practices, minimizing waste generation, and utilizing renewable energy sources, companies can contribute to a greener and more sustainable future.

Upcycling and Repurposing

In the quest for sustainability, finding ways to reduce waste and consume resources more efficiently is crucial. Upcycling and repurposing are two innovative approaches that address this challenge by transforming waste materials into new products with added value. In this section, we will explore the principles of upcycling and repurposing, their benefits, and examples of how these practices can be implemented.

Principles of Upcycling

Upcycling is the process of taking discarded or used materials and creatively transforming them into something of higher quality or value. Unlike recycling, which typically breaks down materials to be used again in a similar form, upcycling involves repurposing items in a way that enhances their function or aesthetics. The aim is to extend the lifecycle of materials, reduce waste, and minimize the need for new resources.

The principles of upcycling include:

1. **Creativity and Innovation**: Upcycling requires thinking outside the box and finding new uses for materials that were originally intended for a different purpose. It encourages designers and individuals to explore unconventional approaches to transform waste into valuable resources.

2. **Resource Conservation**: By upcycling materials, fewer resources are needed to create new products. This helps to reduce the environmental impact associated with extracting, manufacturing, and transporting raw materials.

3. **Minimal Processing**: Unlike recycling, which often requires energy-intensive processes to break down materials, upcycling focuses on repurposing items without extensive processing. This minimizes the energy and water consumption associated with traditional recycling methods.

4. **Sustainable Design**: Upcycling promotes the use of sustainable design principles, such as durability, modularity, and adaptability. By considering the entire lifecycle of a product, including its potential for upcycling, designers can create items that are easier to transform or repair.

Benefits of Upcycling

Upcycling offers several advantages over traditional waste management practices and conventional manufacturing processes. Some key benefits include:

- **Waste Reduction**: Upcycling diverts waste from landfills and reduces the demand for new raw materials. By finding innovative ways to repurpose materials, upcycling helps to minimize the environmental impact of waste disposal.

+ **Resource Conservation:** Upcycling reduces the need for extracting and processing virgin resources, leading to a more sustainable use of materials. It helps to conserve natural resources, such as timber, metals, and petroleum, which are often used in the production of new goods.

+ **Energy Savings:** Traditional manufacturing processes can be energy-intensive. By upcycling, energy consumption associated with extracting, refining, and processing raw materials is significantly reduced, resulting in lower greenhouse gas emissions.

+ **Promotion of Creativity and Craftsmanship:** Upcycling encourages creativity and allows for the preservation of traditional craftsmanship. It provides opportunities for artisans and designers to showcase their skills by transforming discarded materials into unique, handcrafted products.

+ **Engagement in Circular Economy:** Upcycling is a key component of the circular economy, where resources are kept in use for as long as possible. By extending the lifespan of materials through upcycling, the circular economy is promoted, reducing waste generation and creating economic opportunities.

Examples of Upcycling and Repurposing

The possibilities for upcycling and repurposing are vast and varied. Here are some examples that demonstrate the potential of these practices:

1. **Upcycled Furniture:** Old wooden pallets can be transformed into unique and stylish furniture pieces, such as coffee tables or bookshelves. By sanding, painting, and repurposing the pallets, their functional and aesthetic value is increased.

2. **Repurposed Clothing:** Unwanted clothing items can be creatively repurposed into new fashion pieces. For example, an old pair of jeans can be transformed into a trendy denim skirt or a patchwork bag.

3. **Upcycled Home Decor:** Glass bottles can be repurposed into decorative vases or lampshades. By adding paint, embellishments, or modifications, these bottles become unique and eye-catching home decor items.

4. **Repurposed Electronics:** Old electronic devices can be disassembled, and their components can be used in DIY projects. For instance, computer circuit boards can be turned into striking wall art or jewelry.

5. **Upcycled Art:** Artists often use discarded materials, such as scrap metal, old car parts, or broken ceramics, to create captivating sculptures or installations. These artworks not only showcase creativity but also raise awareness about waste and consumption.

Challenges and Considerations

While upcycling and repurposing offer significant benefits, there are also challenges and considerations to keep in mind:

+ **Quality Control:** Ensuring the quality and safety of upcycled products can be a challenge. It is essential to ensure that any repurposed item meets the necessary standards and regulations to guarantee its functionality and durability.

+ **Availability of Materials:** The availability of suitable materials for upcycling can vary regionally. Some areas may have limited access to specific waste streams, which can restrict the potential for upcycling projects.

+ **Market Demand:** Establishing a market demand for upcycled products is crucial for their long-term viability. Educating consumers about the benefits of upcycling and creating a preference for sustainable alternatives can drive the market demand.

+ **Design Challenges:** Upcycling often requires creativity and innovative design thinking. Designers need to consider the limitations and possibilities of the materials they are working with to ensure the desired outcomes are achieved.

Conclusion

Upcycling and repurposing offer exciting opportunities to reduce waste, conserve resources, and promote sustainable consumption and production practices. By challenging our conventional notions of waste and embracing creativity, we can transform discarded materials into valuable resources. Incorporating upcycling into our daily lives not only benefits the environment but also contributes to a more sustainable and creative society.

Global Collaboration for Sustainability

International cooperation

International cooperation plays a crucial role in addressing global environmental challenges and promoting sustainability. It involves collaboration and coordination among nations to tackle issues that transcend national boundaries and require joint efforts for effective solutions. In this section, we will explore the importance of international cooperation, discuss key initiatives and frameworks, and highlight some success stories in global environmental governance.

The need for international cooperation

The interconnected nature of environmental issues necessitates international cooperation. Many environmental challenges, such as climate change, biodiversity loss, and air and water pollution, do not adhere to political or geographical boundaries. They require collective action and collaboration among countries to mitigate their impacts and find sustainable solutions.

Moreover, addressing these challenges often goes beyond the capabilities and resources of individual nations. Global cooperation allows sharing of knowledge, technology, and resources, enabling countries to jointly develop and implement strategies for sustainable development.

Key initiatives and frameworks

Several global initiatives and frameworks have been established to facilitate international cooperation and address environmental issues on a global scale. These initiatives provide a platform for member countries to exchange ideas, negotiate agreements, and implement measures for sustainable development. Some key initiatives include:

1. United Nations Framework Convention on Climate Change (UNFCCC): Established in 1992, the UNFCCC aims to combat climate change and promote stability in greenhouse gas concentrations. The annual Conference of the Parties (COP) serves as a platform for negotiations on climate-related issues and the implementation of the Paris Agreement.

2. Convention on Biological Diversity (CBD): Adopted in 1992, the CBD focuses on the conservation and sustainable use of biodiversity. It provides a framework for cooperation among member countries to protect and restore ecosystems, conserve biological resources, and ensure fair and equitable sharing of benefits from the use of genetic resources.

3. United Nations Convention to Combat Desertification (UNCCD): The UNCCD addresses desertification, land degradation, and drought. It promotes sustainable land management practices and supports affected countries in implementing measures to combat desertification and restore degraded lands.

4. International Maritime Organization (IMO): The IMO is a specialized agency of the United Nations responsible for regulating international shipping. It works to reduce the environmental impacts of shipping by developing and enforcing standards, such as the International Convention for the Prevention of Pollution from Ships (MARPOL).

5. Basel Convention on the Control of Transboundary Movements of Hazardous Wastes and Their Disposal: The Basel Convention aims to minimize the generation of hazardous waste and promote environmentally sound management of such waste. It regulates the international movement of hazardous wastes and promotes their reduction, recycling, and proper disposal.

6. Ramsar Convention on Wetlands: The Ramsar Convention is an intergovernmental treaty that promotes the conservation and wise use of wetlands. It designates wetlands of international importance and seeks to halt their loss and degradation.

Success stories in global environmental governance

International cooperation has led to several success stories in global environmental governance. Here are a few notable examples:

1. Montreal Protocol on Substances that Deplete the Ozone Layer: The Montreal Protocol, adopted in 1987, has been highly successful in phasing out the production and use of ozone-depleting substances (ODS). As a result, the ozone layer is slowly recovering, and the protocol has prevented millions of cases of skin cancer and eye cataracts.

2. Paris Agreement on Climate Change: The Paris Agreement, adopted in 2015, aims to limit global warming to well below 2 degrees Celsius above pre-industrial levels and pursue efforts to limit the temperature increase to 1.5 degrees Celsius. The agreement has mobilized global action on climate change, with countries committing to Nationally Determined Contributions (NDCs) to reduce greenhouse gas emissions.

3. International Whaling Commission (IWC): The IWC was established in 1946 to regulate whaling activities and conserve whale populations. Through its efforts, several whale species, such as the humpback whale and the gray whale, have recovered from the brink of extinction.

Challenges and the way forward

While international cooperation has achieved significant milestones, challenges persist in effectively addressing global environmental issues. These challenges include differing national priorities, limited resources, conflicting interests, and the slow pace of decision-making.

To overcome these challenges, it is crucial to strengthen international partnerships, enhance coordination, and foster a spirit of collaboration among nations. Investing in capacity-building programs, technology transfer, and financial support for developing countries can promote equitable participation and ensure the effective implementation of environmental initiatives.

Furthermore, integrating environmental goals with broader development agendas, such as poverty eradication and social equity, is essential for promoting sustainable development and securing buy-in from diverse stakeholders.

In conclusion, international cooperation is indispensable for addressing global environmental challenges and advancing sustainability. Through initiatives and frameworks, countries can collaborate, share knowledge and resources, negotiate agreements, and work together to find sustainable solutions. While challenges persist, success stories demonstrate the potential for collective action to make a positive impact on the environment. Moving forward, it is crucial to strengthen partnerships, enhance coordination, and prioritize sustainable development as a shared global goal.

Global environmental governance

Global environmental governance refers to the mechanisms and processes through which countries and international actors work together to address global environmental challenges. It involves the coordination, negotiation, and implementation of policies, regulations, and agreements at the global level to promote sustainable development and protect the environment.

The need for global environmental governance

The global environment is facing numerous challenges, such as climate change, biodiversity loss, deforestation, pollution, and resource depletion. These challenges cannot be effectively addressed by individual countries alone. They require coordinated efforts and collaboration among all nations to ensure the sustainable use of natural resources and the protection of the environment for future generations.

Global environmental governance is necessary to overcome collective action problems, where the pursuit of individual interests by countries can lead to suboptimal outcomes. It provides a framework for cooperation, negotiation, and the establishment of rules and norms that guide countries' actions to achieve collective goals related to environmental protection and sustainable development.

Principles of global environmental governance

1. **Sovereignty and equity**: Global environmental governance respects the sovereignty of countries while promoting equity and fairness in the distribution of rights, responsibilities, and benefits related to environmental resources and services.

2. **Precaution and prevention**: Global environmental governance embraces the precautionary principle, which states that in the face of uncertainty, actions should be taken to prevent or minimize potential environmental harm. This principle guides decision-making on issues such as the introduction of new technologies or the release of potentially harmful substances.

3. **Common but differentiated responsibilities**: Recognizing the historical and current disparities in economic development and environmental impact, global environmental governance acknowledges that countries have different responsibilities and capacities in addressing environmental challenges. It calls for developed countries to take the lead in addressing global environmental issues and providing support to developing countries in their sustainable development efforts.

4. **Interdisciplinary and holistic approach**: Global environmental governance recognizes the interconnectedness of environmental, social, and economic systems. It promotes the integration of knowledge from various disciplines, such as ecology, economics, and sociology, to develop comprehensive and inclusive policies and strategies.

Key actors in global environmental governance

1. **United Nations (UN)**: The UN plays a central role in global environmental governance through its specialized agencies, such as the United Nations Environment Programme (UNEP) and the Intergovernmental Panel on Climate Change (IPCC). It facilitates international cooperation, negotiation, and the development of multilateral environmental agreements.

2. **Multilateral environmental agreements (MEAs)**: MEAs are international legally binding agreements that aim to address specific environmental issues. Examples include the United Nations Framework Convention on Climate Change

(UNFCCC), the Convention on Biological Diversity (CBD), and the Montreal Protocol on Substances that Deplete the Ozone Layer. MEAs provide a framework for countries to cooperate and take collective action on environmental challenges.

3. **Non-governmental organizations (NGOs)**: NGOs play a vital role in global environmental governance by advocating for environmental protection, conducting research, and collaborating with governments and international organizations. They often provide expertise, raise awareness, and mobilize public support for environmental causes.

4. **Private sector:** The private sector, including corporations and businesses, is increasingly recognized as an important actor in global environmental governance. Through corporate sustainability initiatives, green investments, and environmentally friendly practices, the private sector can contribute to addressing global environmental challenges.

5. **Civil society**: Civil society, including indigenous peoples, grassroots organizations, and local communities, plays a crucial role in global environmental governance. They often provide valuable knowledge, perspectives, and solutions based on their local experiences and traditional practices.

Challenges and opportunities

Global environmental governance faces several challenges that hinder its effectiveness:

1. **Fragmentation and overlap:** The existence of multiple organizations, agencies, and agreements can lead to complexity, redundancy, and gaps in global environmental governance. Coordination and cooperation among different actors and institutions are essential for addressing these challenges.

2. **Power imbalances:** Power imbalances between developed and developing countries, as well as between different interest groups, can hinder the equitable implementation of global environmental policies. Efforts are needed to enhance the inclusivity and representation of all stakeholders in decision-making processes.

3. **Lack of compliance and enforcement:** Some countries may not fully comply with international environmental agreements and regulations due to factors such as weak governance, lack of resources, or conflicting priorities. Strengthening compliance mechanisms and enforcement measures is crucial for the effectiveness of global environmental governance.

4. **Financing and resource mobilization:** Adequate funding and resources are essential for implementing global environmental policies and initiatives. Mobilizing

financial resources from both public and private sources and ensuring their effective and equitable allocation are ongoing challenges.

Despite these challenges, global environmental governance also presents opportunities for transformative change:

1. **Knowledge sharing and capacity building**: Global environmental governance facilitates the exchange of knowledge, best practices, and innovative solutions among countries, organizations, and stakeholders. This promotes learning and capacity building at the individual, institutional, and societal levels.

2. **Partnerships and collaboration**: Global environmental governance provides a platform for partnerships and collaborations among governments, NGOs, businesses, and civil society. These partnerships can leverage diverse expertise, resources, and perspectives to address environmental challenges more effectively.

3. **Technological innovation**: Advances in technology, such as remote sensing, data analytics, and renewable energy, present opportunities for more efficient and sustainable environmental management. Global environmental governance can help foster the development and transfer of environmentally friendly technologies.

4. **Public awareness and engagement**: Global environmental governance plays a vital role in raising public awareness about environmental challenges and engaging citizens in decision-making processes. Increased public participation can contribute to more accountable and transparent environmental governance.

In conclusion, global environmental governance is crucial for addressing global environmental challenges and achieving sustainable development. It requires the collaboration and cooperation of all stakeholders, including countries, international organizations, NGOs, the private sector, and civil society. While it faces various challenges, it also presents opportunities for transformative change and the integration of environmental concerns into global decision-making processes. By working together, we can create a more sustainable and resilient planet for future generations.

Climate Change Negotiations

Climate change negotiations play a crucial role in addressing the global challenge of climate change and formulating effective measures to mitigate its impacts. These negotiations involve a complex process of international cooperation, policy development, and consensus-building among nations. In this section, we will explore the key aspects of climate change negotiations, including the goals, main actors, challenges, and potential solutions.

Goals of Climate Change Negotiations

The primary goal of climate change negotiations is to establish an international framework for limiting global greenhouse gas emissions and promoting climate resilience. This framework seeks to achieve the following objectives:

1.   Mitigation: Negotiations aim to set targets and actions to reduce greenhouse gas emissions and limit global warming to well below 2 degrees Celsius above pre-industrial levels, while pursuing efforts to keep it below 1.5 degrees Celsius. This is based on the scientific consensus that exceeding these thresholds would have severe impacts on ecosystems and human societies.

2. Adaptation: Negotiations also focus on enhancing the adaptive capacity and resilience of countries to cope with the impacts of climate change. This involves supporting vulnerable communities and ecosystems to adapt to changing climate conditions, such as rising sea levels, extreme weather events, and altered precipitation patterns.

3. Finance and Technology Transfer: Another crucial aspect is the provision of financial resources and technology transfer from developed to developing countries to support their climate actions. This aims to address the capacity constraints and financial limitations that hinder effective climate mitigation and adaptation efforts in developing nations.

Main Actors in Climate Change Negotiations

Climate change negotiations involve a wide range of actors, including governments, intergovernmental organizations, non-governmental organizations, and civil society. The main actors in these negotiations are:

1.   United Nations Framework Convention on Climate Change (UNFCCC): The UNFCCC serves as the main international forum for climate change negotiations. It provides a platform for countries to come together and negotiate agreements, share knowledge, and coordinate their actions. The Conference of the Parties (COP) is the highest decision-making body under the UNFCCC.

2.   Parties to the UNFCCC: Parties are the countries that have ratified or acceded to the UNFCCC. They participate in the negotiations and make commitments to address climate change based on their national circumstances and capabilities.   The parties are divided into different groups, such as developed countries (Annex I) and developing countries (Non-Annex I), which have different responsibilities and expectations.

3.   Intergovernmental Panel on Climate Change (IPCC): The IPCC provides scientific assessments and reports on the state of the climate system, as well as the

potential impacts of climate change and possible mitigation and adaptation options. Its reports serve as a basis for informed decision-making in climate change negotiations.

4. Civil Society Organizations: Non-governmental organizations, research institutions, indigenous groups, and other civil society organizations play a crucial role in climate change negotiations. They advocate for climate action, provide expertise, and hold governments accountable for their commitments.

Challenges in Climate Change Negotiations

Climate change negotiations face several challenges that can hinder effective progress. Some of the key challenges include:

1. Differing National Interests: Countries have differing national interests, priorities, and capacities, which can lead to conflicts and difficulties in reaching consensus. Developed and developing countries often have divergent perspectives on issues such as emissions reductions, financial support, and technology transfer.

2. Equity and Justice: Achieving a fair and just distribution of the costs and benefits of climate action is a complex challenge. Negotiations need to address historical responsibility, the principle of common but differentiated responsibilities, and the different capacities of countries to address climate change.

3. Political Will and Leadership: Climate change negotiations require strong political will and leadership at both national and international levels. Shifting political landscapes, changes in government, and competing policy priorities can impact the commitment to climate action.

4. Complexity and Technicality: Climate change is a complex issue that requires interdisciplinary knowledge and expertise. Negotiators need a deep understanding of scientific, economic, and social dimensions of climate change to make informed decisions.

Potential Solutions

To overcome the challenges in climate change negotiations, several potential solutions can be explored:

1. Enhanced Dialogue and Cooperation: Improved dialogue and cooperation among countries can foster a better understanding of shared challenges and promote collaboration on climate action. Platforms for knowledge exchange, capacity-building programs, and peer learning can support this process.

2. Long-term Strategies: Countries can develop and communicate long-term low-emission development strategies. These strategies provide a roadmap for

decarbonization and signal national commitments to sustainable development and climate resilience.

3. Financial Mechanisms: Strengthening financial mechanisms, such as the Green Climate Fund, can ensure adequate and predictable funding for climate change adaptation and mitigation projects in developing countries.

4. Technology Transfer: Facilitating technology transfer from developed to developing countries can help bridge the technological gap and support the deployment of clean and sustainable technologies globally.

5. Empowering Vulnerable Communities: Ensuring the meaningful participation of vulnerable communities, such as indigenous peoples and women, in climate change negotiations can help address their unique needs, perspectives, and knowledge in crafting effective climate policies.

6. Multi-level Governance: Building effective multi-level governance structures that involve all levels of government, civil society, and private sector actors can promote coordination, synergies, and shared responsibility in climate action.

In conclusion, climate change negotiations are a critical component of global efforts to address the challenges posed by climate change. By setting goals, facilitating dialogue, and promoting international cooperation, these negotiations provide a platform for nations to work together towards a sustainable and resilient future. However, overcoming the challenges and finding solutions requires collective action, political will, and a commitment to equity and justice.

Biodiversity Conservation Agreements

Biodiversity conservation agreements play a crucial role in protecting and preserving the rich variety of life on Earth. These agreements are international efforts aimed at addressing the threats to biodiversity and promoting sustainable development. In this section, we will explore the key concepts, goals, and examples of biodiversity conservation agreements.

The Importance of Biodiversity Conservation

Biodiversity refers to the variety of life forms, including plants, animals, microorganisms, and their ecological complexes. It is essential for the functioning of ecosystems, providing a range of ecosystem services such as clean water, air purification, climate regulation, and nutrient cycling. Biodiversity also contributes to the aesthetic, cultural, and spiritual aspects of human life.

However, biodiversity is under severe threat due to various factors, including habitat loss, pollution, climate change, overexploitation of resources, and the spread of invasive species. These threats can lead to the extinction of species, disruption of ecological processes, and negative impacts on human well-being.

Goals of Biodiversity Conservation Agreements

Biodiversity conservation agreements aim to address these threats and promote the sustainable use of natural resources. The primary goals of these agreements include:

1. Conservation of biodiversity: The agreements strive to protect diverse ecosystems and their components, including species, habitats, and genetic resources.

2. Sustainable use of biodiversity: The agreements promote the use of biodiversity in a way that ensures its long-term sustainability and benefits both present and future generations.

3. Equitable sharing of benefits: The agreements recognize the importance of fair and equitable sharing of the benefits arising from the use of genetic resources and traditional knowledge.

4. Integration of biodiversity into development planning: The agreements emphasize the need to mainstream biodiversity conservation into various sectors, including agriculture, forestry, fisheries, and tourism.

Examples of Biodiversity Conservation Agreements

Several international agreements and initiatives have been established to address biodiversity conservation. Some of the notable ones include:

1. Convention on Biological Diversity (CBD): The CBD is the most comprehensive global agreement on biodiversity conservation. It sets out principles for the sustainable management and conservation of biodiversity and promotes the fair and equitable sharing of benefits arising from the use of genetic resources. The CBD has three main objectives: conservation of biodiversity, sustainable use of its components, and the fair and equitable sharing of benefits.

2. Ramsar Convention on Wetlands: The Ramsar Convention is an international treaty aimed at conserving and promoting the wise use of wetlands. It recognizes the ecological functions of wetlands and their importance for biodiversity conservation and sustainable development. The convention provides a framework for the designation and management of wetlands of international importance ("Ramsar Sites").

3. Convention on International Trade in Endangered Species of Wild Fauna and Flora (CITES): CITES is an international agreement that regulates the trade in endangered species to ensure their survival. It aims to prevent unsustainable

trade and protect species from overexploitation. CITES places restrictions on the international trade of certain species through a system of permits and certificates.

4.   World Heritage Convention: The World Heritage Convention is a UNESCO initiative that aims to identify and protect natural and cultural sites of outstanding universal value. Many of these sites are biodiversity hotspots and play a crucial role in the conservation of globally significant ecosystems.

Challenges and Opportunities

Despite the existence of biodiversity conservation agreements, significant challenges remain in achieving their goals. These challenges include insufficient funding, inadequate enforcement, lack of political will, and conflicting priorities in development planning.

However, there are also opportunities for enhancing biodiversity conservation efforts. Integrating traditional knowledge and practices of indigenous communities can contribute to more effective conservation strategies. Furthermore, technological advancements, such as remote sensing, DNA sequencing, and data analytics, provide new tools for monitoring biodiversity and informing conservation actions.

Conclusion

Biodiversity conservation agreements are essential instruments for protecting and preserving the Earth's rich natural heritage. They aim to address the threats to biodiversity, promote sustainable development, and ensure the equitable sharing of benefits. By implementing these agreements and taking collective action, we can safeguard biodiversity for future generations and maintain the delicate balance of ecosystems that support life on Earth.

Global sustainability initiatives

Global sustainability initiatives aim to address pressing environmental and social challenges at a global scale. These initiatives recognize the interconnectedness of human activities and their impact on the planet, and seek to promote sustainable development for the benefit of current and future generations. They are designed to bring together governments, organizations, and individuals to collaborate and take action towards achieving the United Nations Sustainable Development Goals (SDGs).

One of the key global sustainability initiatives is the United Nations Framework Convention on Climate Change (UNFCCC), which was established in 1992 with the objective of stabilizing greenhouse gas concentrations in the atmosphere. Under the UNFCCC, the Kyoto Protocol was adopted in 1997, which set binding emissions reduction targets for developed countries. The Paris Agreement, signed in 2015, builds upon the Kyoto Protocol and aims to limit global warming to well below 2 degrees Celsius above pre-industrial levels.

The Paris Agreement has spurred various global sustainability initiatives, such as:

1. **Renewable Energy Transition:** Many countries and organizations have committed to transitioning from fossil fuels to renewable energy sources in order to reduce greenhouse gas emissions and mitigate climate change. Initiatives such as the International Renewable Energy Agency (IRENA) and the Renewable Energy Policy Network for the 21st Century (REN21) promote the adoption and scaling up of renewable energy technologies worldwide.

2. **Sustainable Cities:** Urban areas are responsible for a significant share of global energy consumption and greenhouse gas emissions. The C40 Cities Climate Leadership Group brings together cities around the world to share knowledge and collaborate on climate action. The Global Covenant of Mayors for Climate & Energy is another initiative that supports cities in their efforts to develop and implement sustainable and low-carbon policies.

3. **Circular Economy:** The concept of a circular economy emphasizes the need to move away from the traditional linear model of production and consumption, where resources are extracted, used, and discarded. Initiatives such as the Ellen MacArthur Foundation's Circular Economy 100 and the World Economic Forum's Circular Economy Initiative promote the adoption of sustainable practices that minimize waste generation and maximize resource efficiency.

4. **Sustainable Agriculture:** Agriculture is a major contributor to environmental degradation, including deforestation, water pollution, and greenhouse gas emissions. Initiatives like the Sustainable Agriculture Initiative

Platform (SAI Platform) and the Global Alliance for Climate-Smart Agriculture (GACSA) promote the adoption of sustainable agricultural practices, including agroecology, organic farming, and climate-smart agriculture, to ensure food security while minimizing environmental impacts.

5. **Biodiversity Conservation:** Protecting and restoring biodiversity is essential for the health and resilience of ecosystems. The Convention on Biological Diversity (CBD) is a global sustainability initiative that aims to promote the conservation and sustainable use of biodiversity. The CBD's Aichi Biodiversity Targets set specific goals for the protection and restoration of ecosystems and the reduction of biodiversity loss.

6. **Green Finance:** The financial sector plays a crucial role in driving sustainable development. Initiatives such as the Global Green Finance Index (GGFI) and the Principles for Responsible Investment (PRI) encourage investment in environmentally and socially responsible projects. Green bonds, which finance projects with environmental benefits, have gained popularity as a means to mobilize capital for sustainable initiatives.

7. **Education and Awareness:** Building a sustainable future requires raising awareness and educating individuals about the importance of sustainable development. Initiatives such as the Global Education for Sustainable Development (ESD) and the United Nations Environment Programme (UNEP) promote educational initiatives that foster a deeper understanding of sustainability issues and inspire action.

It is important to note that global sustainability initiatives face various challenges, including the need for political will, financial resources, technological innovation, and behavioral change. However, they provide a framework for collaboration and collective action to address the pressing environmental challenges we face. By fostering international cooperation and promoting sustainable practices, these initiatives contribute to building a more sustainable and resilient future for all.

United Nations Sustainable Development Goals

The United Nations Sustainable Development Goals (SDGs) are a set of 17 global goals that aim to address the most pressing economic, social, and environmental challenges facing the world today. Adopted by UN member states in 2015 as part of the 2030 Agenda for Sustainable Development, the SDGs provide a comprehensive framework for achieving sustainable development in a balanced and integrated manner.

Background

The SDGs build on the success of the Millennium Development Goals (MDGs), which were in effect from 2000 to 2015. While the MDGs focused primarily on poverty alleviation and basic human needs, the SDGs have a broader scope, encompassing a wide range of interconnected issues such as poverty, hunger, health, education, gender equality, clean water and sanitation, renewable energy, sustainable cities, climate action, and biodiversity conservation.

Principles of the SDGs

The SDGs are guided by a set of principles that underpin sustainable development. These principles include universality, leaving no one behind, integrated and indivisible nature, and the need for global partnership and cooperation. Universality means that the goals apply to all countries, regardless of their income level or development status. Leaving no one behind emphasizes the importance of ensuring that the benefits of development are enjoyed by all, particularly the most vulnerable and marginalized populations. The integrated and indivisible nature of the goals recognizes that progress in one goal area is interlinked with progress in others, and therefore, a holistic and integrated approach is necessary. Lastly, achieving the SDGs requires a multi-stakeholder approach, with collaboration among governments, civil society, the private sector, and international organizations.

Overview of the SDGs

The 17 SDGs cover a wide range of sustainable development issues. Each goal is supported by specific targets and indicators to measure progress. Let's briefly explore some of the key goals:

1. Goal 1: No Poverty - End poverty in all its forms everywhere.

2. Goal 2: Zero Hunger - End hunger, achieve food security and improved nutrition, and promote sustainable agriculture.

3. Goal 3: Good Health and Well-being - Ensure healthy lives and promote well-being for all at all ages.

4. Goal 4: Quality Education - Ensure inclusive and equitable quality education and promote lifelong learning opportunities for all.

5. Goal 5: Gender Equality - Achieve gender equality and empower all women and girls.

6. Goal 6: Clean Water and Sanitation - Ensure availability and sustainable management of water and sanitation for all.

7. Goal 7: Affordable and Clean Energy - Ensure access to affordable, reliable, sustainable, and modern energy for all.

8. Goal 8: Decent Work and Economic Growth - Promote sustained, inclusive, and sustainable economic growth, full and productive employment, and decent work for all.

9. Goal 9: Industry, Innovation and Infrastructure - Build resilient infrastructure, promote inclusive and sustainable industrialization, and foster innovation.

10. Goal 10: Reduced Inequalities - Reduce inequality within and among countries.

11. Goal 11: Sustainable Cities and Communities - Make cities and human settlements inclusive, safe, resilient, and sustainable.

12. Goal 12: Responsible Consumption and Production - Ensure sustainable consumption and production patterns.

13. Goal 13: Climate Action - Take urgent action to combat climate change and its impacts.

14. Goal 14: Life Below Water - Conserve and sustainably use the oceans, seas, and marine resources for sustainable development.

15. Goal 15: Life on Land - Protect, restore, and promote sustainable use of terrestrial ecosystems, sustainably manage forests, combat desertification, halt and reverse land degradation, and halt biodiversity loss.

16. Goal 16: Peace, Justice and Strong Institutions - Promote peaceful and inclusive societies for sustainable development, provide access to justice for all, and build effective, accountable, and inclusive institutions at all levels.

17. Goal 17: Partnerships for the Goals - Strengthen the means of implementation and revitalize the global partnership for sustainable development.

Achieving the SDGs

Achieving the SDGs requires collective action at all levels - from individuals and communities to governments and international organizations. Governments play a crucial role in implementing and monitoring the goals. They are responsible for aligning national policies, allocating resources, and developing strategies to address the specific challenges and priorities of their countries. Additionally, it is essential to engage civil society, businesses, academia, and other stakeholders to contribute their expertise, resources, and innovative solutions to achieve the goals.

Challenges and Opportunities

While the SDGs present an ambitious agenda, there are several challenges to their implementation. One of the key challenges is the lack of funding and resources. Achieving the goals requires significant investments in infrastructure, technology, education, healthcare, and renewable energy, among others. Developing countries, in particular, face financial constraints and need support from the international community to realize the goals.

However, addressing these challenges also presents opportunities. The implementation of the SDGs can lead to inclusive and sustainable economic growth, poverty reduction, improved social outcomes, and a healthier environment. It can drive innovation and the development of new technologies, creating new business opportunities and jobs. By working together and embracing a sustainable development agenda, we can build a more equitable, resilient, and prosperous future for all.

Conclusion

The United Nations Sustainable Development Goals provide a blueprint for a better and more sustainable world. They address the interconnected challenges of poverty, inequality, climate change, environmental degradation, and much more. Achieving these goals requires the concerted efforts of governments, civil society, businesses, and individuals. By embracing the principles of the SDGs and taking action at all levels, we can pave the way for a brighter and more sustainable future for generations to come. Let's work together to make the vision of the SDGs a reality.

Green economy initiatives

Green economy initiatives are important strategies that aim to promote sustainable development by integrating environmental considerations into economic policies and practices.   These initiatives recognize that economic development and environmental protection can go hand in hand, resulting in long-term benefits for both people and the planet. They seek to foster a more sustainable and low-carbon economy that is resilient, inclusive, and resource-efficient.

One of the key objectives of green economy initiatives is the decoupling of economic growth from resource consumption and environmental degradation. This can be achieved through various strategies, including the promotion of renewable energy, the adoption of circular economy principles, and the implementation of sustainable production and consumption patterns.   Green economy initiatives promote the transition to a more sustainable future by encouraging innovation, creating green jobs, and reducing greenhouse gas emissions.

Renewable Energy

The promotion of renewable energy sources is a crucial component of green economy initiatives. Renewable energy, such as solar, wind, hydro, and geothermal power, offers a sustainable alternative to traditional fossil fuel-based energy sources.   Investing in renewable energy can reduce greenhouse gas emissions, enhance energy security, and create new job opportunities.

Governments and organizations can support the expansion of renewable energy through various measures, such as providing subsidies and incentives, setting renewable energy targets, and promoting research and development in clean energy technologies.   Transitioning to a renewable energy system requires a comprehensive approach that includes policy support, technological advancements, and infrastructure development.

Circular Economy

The circular economy is another important concept within green economy initiatives.   It is based on the idea of replacing the traditional linear model of production and consumption, which follows the "take-make-dispose" pattern, with a circular model that aims to minimize waste, maximize resource efficiency, and promote recycling and reusing.

In a circular economy, products are designed to be durable, repairable, and recyclable. Resources and materials are kept in use for as long as possible through recycling, remanufacturing, and repurposing.   This approach not only reduces waste and resource depletion but also offers economic opportunities, such as the creation of new industries and employment in the recycling and remanufacturing sectors.

Sustainable Production and Consumption

Promoting sustainable production and consumption practices is another key aspect of green economy initiatives. This involves adopting sustainable business models, improving resource efficiency, and minimizing environmental impacts throughout the entire lifecycle of products and services.

Sustainable production practices can include adopting cleaner production technologies, reducing the use of hazardous substances, optimizing energy and water consumption, and minimizing waste generation. Sustainable consumption practices, on the other hand, involve making informed choices as consumers, such as buying eco-friendly and ethically produced products, reducing waste, and embracing a more minimalist lifestyle.

To encourage sustainable production and consumption, policies and regulations can be implemented, awareness campaigns can be launched, and consumers can be provided with information and tools to make sustainable choices.

Green Jobs and Skills Development

Green economy initiatives also focus on creating green jobs and developing the necessary skills and knowledge for a sustainable economy. Green jobs are employment opportunities that contribute directly to the preservation or restoration of the environment or meet the demand for goods and services with lower environmental impacts.

Examples of green jobs include professionals working in renewable energy, energy efficiency, waste management, sustainable agriculture, and environmental consulting. These jobs not only contribute to environmental sustainability but also provide economic benefits and career opportunities.

Skills development is crucial for the successful transition to a green economy. Training programs and educational initiatives can be implemented to equip individuals with the skills and knowledge needed for green jobs. This includes technical skills in renewable energy technologies, waste management, and sustainable agriculture, as well as skills in green entrepreneurship, policy-making, and project management.

Public-Private Partnerships

Collaboration between the public and private sectors is essential for the success of green economy initiatives. Public-private partnerships can bring together the resources, expertise, and perspectives of both sectors to drive sustainable development.

Governments can provide policy support, incentives, and funding to encourage private sector engagement in green economy initiatives. On the other hand, businesses can contribute by adopting sustainable practices, investing in green technologies, and developing innovative solutions to environmental challenges. Public-private partnerships can also facilitate knowledge exchange, capacity building, and the scaling up of sustainable solutions.

International Cooperation

Green economy initiatives require global cooperation and coordination to address transboundary environmental challenges and achieve sustainable development outcomes. International collaboration is critical for addressing issues such as climate change, biodiversity loss, and pollution.

International organizations, such as the United Nations Environment Programme (UNEP) and the International Renewable Energy Agency (IREA), play a crucial role in promoting green economy initiatives at the global level. They facilitate dialogue, knowledge sharing, and capacity building among countries and support the implementation of international agreements and commitments related to sustainable development.

In addition to governmental and intergovernmental collaborations, non-governmental organizations (NGOs), civil society groups, and the private sector also play important roles in advancing green economy initiatives through advocacy, research, and innovative projects.

Overall, green economy initiatives offer a pathway towards a more sustainable and inclusive future. By embracing renewable energy, circular economy principles, sustainable production and consumption, green job creation, and international collaboration, societies can pave the way for a greener and more resilient economy that prioritizes the well-being of both people and the planet.

Key Takeaways

- Green economy initiatives promote the integration of environmental considerations into economic policies and practices.

+ Renewable energy, circular economy principles, and sustainable production and consumption are key components of green economy initiatives.

+ Green jobs and skills development are important for a successful transition to a green economy.

+ Public-private partnerships and international cooperation are crucial for the implementation of green economy initiatives.

Exercises

1. Conduct research on a green economy initiative implemented in your country or region. Describe its objectives, key strategies, and outcomes.

2. Analyze the potential economic and environmental benefits of transitioning to renewable energy sources in your area.

3. Imagine you are a sustainability consultant. Develop a proposal for a circular economy project that addresses a specific environmental issue in your community.

4. Conduct a life cycle assessment of a product of your choice, considering its environmental impacts from raw material extraction to end-of-life disposal.

5. Research a green job in a field that interests you. Describe the skills and qualifications required for that job and explain how it contributes to a green economy.

Resources

+ United Nations Environment Programme (UNEP): https://www.unenvironment.org/

+ International Renewable Energy Agency (IREA): https://www.irena.org/

+ Ellen MacArthur Foundation: https://www.ellenmacarthurfoundation.org/

+ World Green Building Council: https://www.worldgbc.org/

Global education and capacity building programs

Global education and capacity building programs play a crucial role in advancing eco science and promoting sustainability on a global scale. These programs aim to equip individuals and communities with the knowledge, skills, and tools necessary to understand and address environmental challenges. By providing education and building capacity, these programs empower people to make informed decisions, participate in sustainable practices, and contribute to the conservation and management of natural resources.

Importance of global education

Global education is essential for fostering a sense of global citizenship and promoting sustainable development worldwide. It helps individuals develop a holistic understanding of the interconnectedness of social, economic, and environmental issues. Through global education, people can gain the knowledge and perspective necessary to analyze complex environmental challenges and contribute to their solutions.

Moreover, global education can cultivate a sense of empathy and solidarity towards marginalized communities facing environmental injustices. It raises awareness of the unequal distribution of environmental resources and enables individuals to advocate for fair and equitable access to these resources.

Goals of global education and capacity building programs

Global education and capacity building programs have several goals, including:

1. Increasing environmental literacy: These programs aim to enhance understanding of ecological principles, environmental problems, and sustainability concepts among individuals of all ages and backgrounds.

2. Promoting interdisciplinary thinking: Global education encourages the integration of knowledge from various disciplines to address complex environmental challenges. It emphasizes the need for collaboration among scientists, policymakers, communities, and other stakeholders.

3. Fostering critical thinking and problem-solving skills: Through hands-on learning experiences, individuals develop the ability to analyze environmental issues, identify solutions, and implement sustainable practices in their daily lives.

4. Cultivating a sense of responsibility and stewardship: Global education programs instill a sense of responsibility towards the environment and promote environmentally conscious behaviors. They encourage individuals to become

responsible stewards of natural resources and work towards sustainable development.

5. Building global partnerships: These programs facilitate collaboration and knowledge exchange among individuals, organizations, and countries. They promote cross-cultural understanding and cooperation in addressing shared environmental challenges.

Approaches in global education and capacity building

Global education and capacity building programs employ various approaches to achieve their goals. Some of the common approaches include:

1. Curriculum integration: Global education can be integrated into formal education systems by incorporating sustainability principles and concepts across different subjects. This ensures that sustainability becomes an integral part of the learning process.

2. Experiential learning: Hands-on experiences, such as field trips, workshops, and simulations, provide individuals with practical knowledge and skills. Experiential learning helps individuals understand environmental issues in a real-world context and encourages active engagement in finding solutions.

3. Professional development: Programs targeting professionals in the field of eco science offer training and development opportunities to enhance their knowledge and skills. Workshops, conferences, and online courses are common avenues for professional capacity building.

4. Community engagement: Global education programs often involve working closely with communities to address local environmental challenges. This approach promotes community ownership, participation, and empowerment in sustainable practices and decision-making processes.

5. Online platforms and virtual learning: The use of technology has revolutionized global education and capacity building efforts. Online platforms, webinars, and virtual classrooms enable individuals from diverse backgrounds and geographical locations to access educational resources and participate in knowledge-sharing activities.

Challenges and considerations

While global education and capacity building programs offer numerous benefits, several challenges must be considered:

1. Access and equity: Ensuring equal access to global education for individuals from marginalized communities, developing countries, and disadvantaged

backgrounds is a significant challenge. Efforts should be taken to bridge the digital divide, language barriers, and socioeconomic disparities.

2. Local context and cultural sensitivity: Global education programs must be tailored to local contexts and cultures to be effective. It is essential to consider indigenous knowledge, traditional practices, and community values when designing educational initiatives.

3. Monitoring and evaluation: Regular monitoring and evaluation are necessary to assess the effectiveness and impact of education and capacity building programs. Continuous improvement based on feedback and data is crucial for program success.

4. Sustainability and long-term engagement: Building capacity and promoting sustainable practices require long-term engagement. Global education programs should focus on creating lasting change rather than short-term interventions.

Case study: The Global Learning and Observations to Benefit the Environment (GLOBE) Program

One notable example of a global education and capacity building program is the Global Learning and Observations to Benefit the Environment (GLOBE) Program. GLOBE is an international science and education program that encourages students and teachers to engage in hands-on learning and data collection to contribute to scientific research.

Through GLOBE, students and teachers from over 120 countries collaborate to collect environmental measurements and share their findings with the global community. This program promotes scientific literacy, cross-cultural understanding, and environmental stewardship.

GLOBE provides resources, protocols, and training materials to support participants in conducting scientific investigations related to atmosphere, hydrosphere, biosphere, and soil. By actively participating in data collection and analysis, students develop critical thinking and problem-solving skills while contributing to scientific research efforts.

The GLOBE Program underscores the importance of global collaboration in environmental education and empowers young minds to become future leaders in eco science and sustainability.

Conclusion

Global education and capacity building programs play a vital role in advancing eco science and promoting sustainability worldwide. By fostering environmental literacy, critical thinking, and cross-cultural collaboration, these programs empower individuals and communities to address environmental challenges and work towards a more sustainable future. Through education and capacity building, we can ensure that future generations are equipped with the knowledge and skills necessary to tackle the complex environmental issues of our time.

Knowledge sharing and capacity building

Knowledge sharing and capacity building are essential components of eco science and sustainable development. In order to address complex environmental challenges, it is crucial to promote the exchange of knowledge and skills among individuals, communities, organizations, and countries. This section explores the importance of knowledge sharing and capacity building in the context of creating a sustainable future.

Importance of knowledge sharing

Knowledge sharing plays a vital role in advancing eco science and promoting sustainability. By sharing knowledge and information, we can build a collective understanding of environmental issues and their potential solutions. It enables us to learn from one another's experiences, successes, and failures in tackling environmental challenges.

One of the key benefits of knowledge sharing is the ability to disseminate best practices. By sharing successful strategies and approaches, we can accelerate progress towards sustainable development. For example, if a community has implemented a successful waste management program, sharing their experience can inspire and guide others facing similar challenges.

Furthermore, knowledge sharing fosters innovation and creativity. Through collaboration and the exchange of ideas, new technologies, methodologies, and approaches can be developed. This drives continuous improvement and allows us to find more efficient and effective solutions to environmental problems.

Capacity building for sustainability

Capacity building refers to the development of knowledge, skills, and abilities required to address environmental challenges and promote sustainability. It focuses on enhancing the capabilities of individuals, organizations, and communities to participate in decision-making processes, formulate effective policies, and implement sustainable practices.

In the context of eco science, capacity building is crucial for empowering individuals and communities to become active contributors to sustainable development. It involves providing training, education, and resources to enhance understanding and competence in areas such as environmental management, conservation, and renewable energy.

Capacity building can take various forms, including workshops, training programs, seminars, and mentorship initiatives. These activities aim to develop technical expertise, analytical skills, leadership capabilities, and the ability to collaborate effectively with diverse stakeholders.

Methods and approaches

There are several methods and approaches to facilitate knowledge sharing and capacity building in the field of eco science. Some of these include:

1. Collaborative platforms: Online platforms and networks can promote the exchange of knowledge, experiences, and ideas among researchers, practitioners, and policymakers. These platforms can facilitate virtual collaboration, access to resources, and the sharing of research findings and best practices.

2. Training and education: Training programs and educational initiatives play a critical role in capacity building. They can be tailored to address specific needs and target different audiences, ranging from local communities to government officials. These initiatives can cover a wide range of topics, including sustainable agriculture, renewable energy, conservation, and environmental policy.

3. South-South and North-South cooperation: International collaboration between developing and developed countries can foster knowledge sharing and capacity building. South-South cooperation involves sharing experiences, expertise, and technologies among developing countries, while North-South cooperation involves partnerships and support from developed countries to assist in capacity building efforts.

4. Community-based learning: Engaging local communities and indigenous peoples in knowledge sharing and capacity building initiatives is crucial. By recognizing their traditional knowledge and practices, we can build a more inclusive and sustainable approach to environmental management. Community-based learning methods, such as participatory mapping and citizen science, can empower communities to contribute to data collection, monitoring, and decision-making processes.

5. Mentorship and knowledge transfer: Establishing mentorship programs and facilitating the transfer of knowledge from experienced professionals to the younger generation can effectively build capacity and promote the continuity of expertise. Mentors can provide guidance, support, and practical insights to mentees, helping them develop the skills necessary for sustainable development.

Challenges and opportunities

While knowledge sharing and capacity building are essential, there are several challenges and opportunities that need to be considered:

1. Access to information: Ensuring equitable access to information and knowledge resources is crucial. Disparities in access, particularly in the context of developing countries and marginalized communities, can hinder capacity building efforts. Efforts should be made to address these disparities and promote inclusivity in knowledge sharing initiatives.

2. Language and cultural barriers: Language differences and cultural diversity can pose challenges to effective knowledge sharing and capacity building. While promoting multilingual platforms and culturally sensitive content can help overcome these barriers, addressing these challenges requires careful attention and respect for diverse perspectives.

3. Technology and infrastructure: Access to technology and reliable infrastructure is essential for effective knowledge sharing and capacity building. Efforts should be made to bridge the digital divide and ensure that communities have the necessary tools and resources to participate in knowledge exchange activities.

4. Collaboration and coordination: Effective knowledge sharing and capacity building require collaboration and coordination among various stakeholders, including researchers, policymakers, NGOs, and communities. Building partnerships and establishing networks can enhance the effectiveness of these efforts and promote holistic approaches to sustainability.

5. Continuous learning and adaptation: Knowledge sharing and capacity building initiatives should be dynamic and adaptive to evolving environmental challenges. Continuous learning, evaluation, and feedback mechanisms can help identify gaps and improve the effectiveness of capacity building efforts.

Case study: The Global Learning Network

The Global Learning Network (GLN) is an example of a knowledge sharing and capacity building initiative in the field of eco science and sustainability. The GLN is a virtual platform that connects individuals, organizations, and communities working towards sustainable development goals.

The GLN provides a space for sharing best practices, research findings, and innovative solutions to environmental challenges. It offers a range of resources, including online courses, webinars, and interactive forums, to facilitate learning and collaboration.

Through the GLN, individuals and organizations can connect with experts, access up-to-date information, and participate in capacity building programs. The platform promotes diverse perspectives and encourages the integration of local knowledge and practices into global sustainability efforts.

By harnessing the power of technology and collaboration, the GLN aims to accelerate knowledge sharing, capacity building, and the implementation of sustainable practices worldwide.

Conclusion

Knowledge sharing and capacity building are crucial for advancing eco science and promoting sustainability. By facilitating the exchange of knowledge, skills, and best practices, we can tackle environmental challenges and work towards a more sustainable future. Through training, collaboration, and the use of innovative platforms, we can empower individuals, communities, and organizations to become active contributors to sustainable development. Efforts should focus on promoting inclusivity, bridging disparities, and adapting to evolving environmental realities. By fostering a culture of continuous learning and collaboration, we can drive positive change and build a global community committed to sustainable development.

Bibliography

[1] Food and Agriculture Organization of the United Nations. (2017). *Sustainable agriculture*. Retrieved from `http://www.fao.org/sustainable-agriculture/en/`

[2] Altieri, M. A., Nicholls, C. I., Henao, A. R., Lana, M. A., Redlich, B., & Poveda, K. (2015). *Agroecology and the design of climate change-resilient farming systems*. Agronomy for Sustainable Development, 35(3), 869-890.

Index

-up, 463

A. Sheikhzadeh, 283
ability, 3, 11, 27, 58, 59, 62, 71, 88,
        89, 103–105, 113, 144,
        204, 212, 270, 282, 331,
        332, 350, 459, 460, 475,
        486, 487, 489, 492, 498,
        499, 501, 504, 525, 526,
        592, 595, 596
absence, 275, 276, 281, 284, 387
absenteeism, 326
absorption, 323, 464, 539
academia, 159, 587
acceptance, 58, 59, 204, 471, 514
access, 105, 127, 129–133, 147, 154,
        157, 181, 191, 217, 247,
        377–379, 386, 395, 421,
        422, 425, 430–432, 457,
        493, 496, 513, 518, 522,
        530, 554, 558, 561, 592,
        593, 596–598
accessibility, 235, 242, 425, 522
account, 116, 386, 387, 416, 443,
        444, 508, 546
accountability, 43, 130, 316, 347,
        348, 378, 379, 388, 417,
        421, 422, 426, 430, 432,

441, 457
accumulation, 29, 39, 270, 324, 327,
        332
accuracy, 267, 547
acetogenesis, 281
achievement, 337, 350, 365, 409,
        420
acid, 270, 278, 280, 281, 314
acidification, 112
acidogenesis, 281
act, 106, 113, 308, 444, 494
action, 4, 6, 31, 35, 39, 43, 112, 125,
        129, 134, 281, 309, 317,
        343, 344, 347, 348, 350,
        352, 353, 356, 386, 387,
        423, 460, 485, 487, 497,
        498, 505, 506, 526, 540,
        553, 572, 574, 575, 579,
        580, 582, 584, 587
activism, 394
activity, 221, 282, 537
actuation, 472
adaptability, 27, 68, 70, 302, 501,
        504, 505, 507, 526
adaptation, 2, 5, 6, 42, 43, 57, 59, 68,
        71, 124, 309, 347, 348,
        350, 356, 391, 438, 443,
        444, 483, 485, 487,

492–497, 507–510, 527, 578, 580, 597

addition, 98, 115, 171, 181, 190, 213, 215, 219, 221, 251, 281, 282, 311, 315, 332, 345

address, 1, 2, 5, 6, 15, 16, 18, 31, 35, 39, 45, 49, 52, 58, 67, 69, 78, 87, 104, 109, 110, 117, 118, 131, 134–136, 138, 140, 145, 151, 154, 159, 162–164, 175, 180, 191, 195, 213, 216, 219, 237, 257, 260, 264, 267, 287–289, 309, 311, 313, 315, 326, 333, 345–347, 349–352, 355, 356, 359, 364, 367, 371, 374, 376, 379, 382, 385–389, 400, 413, 420–422, 430, 432, 436, 438–440, 444–446, 453, 456–458, 460, 465, 467, 468, 475, 477, 487, 494, 497–499, 505, 522, 524, 528, 530, 533, 543, 561, 568, 572, 574, 578–582, 584, 587, 590, 592, 593, 595–597

addressing, 2, 4, 5, 11, 16, 18, 21, 35, 40, 43, 51, 54, 59, 78, 82, 88, 113, 119, 122, 129, 132, 133, 146, 158, 160, 203, 204, 233, 250, 258, 261, 279, 290, 305, 306, 316, 330, 331, 334, 343, 347, 350, 353, 355, 356, 359, 365, 367, 374, 377, 378, 386, 387, 389, 391, 392, 415, 421, 429,

433–435, 442, 446, 449, 459, 488, 494, 496, 497, 500, 510, 514, 528, 564, 572, 574, 577, 580, 587, 590, 593, 597

adjustment, 439, 490, 500

adopting, 6, 14, 43, 44, 137, 143, 150, 151, 154, 159, 175, 211, 216, 250, 264, 271, 309, 317, 330, 372, 399, 407, 412, 435, 443, 445, 485, 500, 540, 551, 552, 561, 565, 568, 589, 590

adoption, 5, 54, 110, 135–137, 172, 178, 180, 183, 185, 186, 197, 213, 215, 217, 218, 223, 224, 229, 233, 235, 239, 252, 265, 286, 315, 324, 349, 408, 410, 444, 445, 457, 467, 471, 528, 588

adsorbent, 323

adsorption, 323

advance, 235, 460

advancement, 129, 184, 186, 204, 337

advantage, 10, 182, 229, 276, 399, 402

adversity, 508

advocacy, 128, 134, 394, 417, 432

advocate, 430, 579, 592

aeration, 269, 270

aerodynamic, 220

aerospace, 464

aesthetic, 27, 124, 580

affordability, 534

afforestation, 349

Africa, 44, 422

agenda, 587

agreement, 43, 348–350, 355

agriculture, 1, 6, 11, 29, 31, 35, 42, 44, 45, 52, 55, 65, 85, 87–89, 94, 103, 106, 113, 119, 144–147, 150, 151, 154, 174, 309, 310, 314, 316, 317, 329, 338, 344, 356, 417, 420, 421, 431, 458, 460, 483, 485, 490, 492, 493, 495, 497, 506, 507, 509, 518–522, 581, 589, 596

agroecology, 146, 149–151

agroforestry, 44, 87, 490, 492

agronomy, 147

aid, 456

aim, 11, 24, 43, 56, 69, 89, 106, 147, 161, 164, 175, 191, 212, 291, 293, 305, 315, 316, 356, 367, 379, 395, 420, 442, 468, 491, 501, 531, 534, 538, 569, 578, 581, 582, 587, 592, 596

air, 11, 29, 41, 91, 128, 144, 158, 160, 168, 172, 178, 180, 186, 190, 216, 219–221, 235, 239, 250, 251, 258–261, 275, 278–280, 284, 298, 305, 314–318, 320, 321, 324–327, 331, 334, 351, 356, 366, 431, 441, 464, 522, 525, 526, 528, 537–539, 546, 572, 580

aircraft, 219, 220, 251, 252, 256–261

airflow, 269

airspace, 259, 261

algae, 253

allocation, 355, 437

alteration, 191

alternative, 78, 99, 136, 179, 204, 214, 216, 219, 235, 243, 251, 253, 255, 258, 323, 324, 334, 337, 462, 471, 499, 506, 525, 540, 554, 558, 564, 588

aluminum, 215, 266

ambition, 356

amendment, 269, 276, 339

amino, 281

ammonia, 282, 285

amount, 172, 183, 187, 190, 204, 261, 270, 275, 301, 308, 322, 326, 413, 415, 538, 539

Amsterdam, 210, 517, 518

analysis, 4, 46, 317, 321, 404, 418, 419, 445, 452, 487, 594

animal, 29, 35, 55, 56, 74, 113, 117, 119, 154, 253

anode, 196, 201, 202

anticancer, 26

anxiety, 217

apartment, 521

apparel, 394

appeal, 124

application, 18, 19, 45, 54, 74, 114, 127, 146, 164, 198, 216, 277, 279, 280, 299, 301, 323, 333, 338, 339, 404, 415, 425, 456, 487

appreciation, 67

approach, 4–6, 12, 14, 15, 18, 20–23, 25, 34, 35, 44, 55, 62, 65, 77, 83, 84, 89, 112, 116–118, 133, 138, 144, 146, 147, 149, 150, 154,

157–159, 175, 178, 213, 214, 224, 242, 273, 279, 280, 292, 293, 308, 330, 334, 337, 363, 376, 378, 391, 394, 396, 405, 416–418, 422, 429, 431, 434, 435, 437, 438, 441, 442, 453, 460, 463, 471, 485, 486, 488, 494, 495, 497, 498, 500, 503, 505, 509, 510, 527, 528, 531, 534, 538, 539, 543, 561, 564, 585, 588, 593, 596

appropriateness, 445
appropriation, 391, 509
aquaculture, 90
architecture, 175–178, 472
Arctic, 510
area, 9, 44, 49, 82, 99, 124, 158, 168, 169, 189, 215, 235, 323, 332, 525, 528, 547, 585
arrangement, 463
array, 27, 113, 115
arrival, 265, 482, 513
art, 27, 167, 258, 280
asbestos, 327
ash, 104, 278, 280
aspect, 52, 68, 82, 90, 107, 110, 135, 140, 172, 214, 251, 259, 264, 271, 293, 305, 326, 401, 425, 426, 432, 498, 504, 525, 578, 589
assembly, 463
assessment, 2, 95, 99, 102, 168, 309, 330, 363, 404, 405, 445, 465, 488, 492, 518, 526, 543, 544, 547
asset, 180, 419
assistance, 389, 445

assurance, 70, 150
asthma, 315, 324, 326, 431
asymmetry, 391
atmosphere, 41, 277, 278, 298, 304, 314, 315, 317, 321, 349, 354, 594
atom, 463
attention, 99, 149, 186, 236, 253, 317, 344, 431, 432, 462, 465, 469, 472, 540, 554, 597
attractiveness, 124
attribute, 390
audits, 213, 267
Australia, 46, 67
authenticity, 418
authority, 391
auto, 216
automation, 261
automobile, 215
availability, 20, 27, 29, 62, 109, 135, 154, 179, 180, 191, 200, 252, 285, 313, 363, 405, 412, 544, 565
avenue, 413
average, 318, 348, 483
aviation, 219, 220, 250–253, 258, 259, 261, 471
avoidance, 471
awareness, 4, 18, 27, 35, 57, 84, 89, 90, 112, 113, 125, 128, 134, 159, 174, 175, 220, 267, 274, 303, 308, 316, 330, 345, 367, 416, 432, 485, 500, 506, 527, 540, 551, 552, 589, 592

back, 41, 119, 184, 199, 201, 221, 233, 305, 498, 504

backbone, 513

backup, 137, 198, 471, 526

backyard, 268

bacteria, 270, 281, 282, 310, 331, 460

balance, 24, 28, 32, 36, 44, 52, 54, 57, 58, 66, 83, 88, 102, 115, 133, 144, 158, 191, 208, 308, 328, 365, 489, 582

balancing, 14, 94, 192, 199, 435, 496

Bangladesh, 493, 496, 497

bank, 117

bar, 399

Barcelona, 513, 514

bark, 26

barrier, 133, 180, 298, 301, 333

base, 394, 399

baseload, 182

basic, 129, 190, 496

basin, 371, 437

basis, 79, 371

batch, 467

battery, 137, 195–199, 217, 218, 221, 230–233, 235, 252

bay, 306, 309

bed, 284

bedding, 269

behavior, 14, 21, 22, 58, 78, 144, 274, 368, 371, 394, 416, 476, 498, 508, 510, 551

Beijing, 316

being, 3, 5, 24, 26, 28, 35, 36, 39, 40, 54, 65, 66, 74, 78, 91, 110, 115, 119, 125, 131, 133, 138, 140, 144–146, 154, 157, 158, 175, 178, 182, 190, 191, 194, 203, 206, 211, 216, 250, 264, 276, 277, 288, 302, 324, 326, 353, 356, 365, 379, 385, 387, 395, 399, 400, 405, 431, 435, 464, 466, 489, 491, 498, 500, 504, 507, 508, 514, 518, 528, 537, 544, 567, 581, 590

benefit, 11, 12, 27, 65, 112, 118, 123, 391, 419, 443, 465, 509

Bernoulli, 180

betterment, 7

bicycle, 537

bike, 221, 247, 249, 250, 315, 525, 537

bin, 269

bio, 215

bioavailability, 329

biochar, 276

biodegradability, 214, 273

biodesign, 168

biodiesel, 253

biodiversity, 1, 3, 5, 6, 11, 20, 24, 26–32, 35, 42–46, 50, 52, 54, 55, 57, 59, 64–67, 70, 74, 78–80, 82–91, 94, 99, 101, 102, 105–107, 110, 111, 114, 116–119, 122–125, 128, 129, 144, 146, 147, 150, 154, 157, 158, 160, 168, 173, 315, 328, 337, 344, 347, 351–353, 356, 359, 364, 386, 390, 391, 435, 446, 456, 482, 489–492, 496, 499, 500, 504, 509, 527, 572, 574, 580–582, 590

bioenergy, 283

bioethanol, 253

biogas, 253, 275, 277, 281, 282

biology, 9, 18, 71, 79–83, 105, 305,
        463
biomass, 135, 182, 183, 215, 276,
        284, 285, 323, 324
biomimicry, 178
bioreactor, 290
bioremediation, 330, 331, 333, 334
biosecurity, 89
biosphere, 594
bird, 30, 124
blockchain, 389, 465–468, 488
blueprint, 160, 587
board, 416
Bogotá, 221
boiling, 323
Bokashi, 270, 271
bokashi, 270, 271
bond, 415, 416
book, 1, 2
borrowing, 558
bottom, 12, 14, 178, 463
boundary, 24
bran, 270
branch, 9
brand, 394, 402
Brazil, 118, 131, 159, 163, 164
break, 130, 174, 268, 270, 276, 281,
        306, 329, 331, 332, 460
breakdown, 275, 281, 290, 332
breaking, 85, 133
breast, 27
breath, 315
breeding, 27, 45, 57, 70, 113
bridge, 18, 350, 444, 580, 594, 597
bridging, 446, 598
bronchitis, 315
brown, 30
brownfield, 302, 329
buffer, 487, 491

building, 4–6, 62, 118, 132, 134,
        137, 140, 151, 159, 160,
        175, 178, 208–212, 247,
        285, 298, 324, 326, 348,
        349, 352, 367, 372, 385,
        386, 389, 425, 426, 432,
        444, 457, 463, 485,
        492–500, 504–507, 509,
        514–518, 525, 526, 528,
        537, 540, 561, 574, 577,
        579, 584, 590, 592–598
buildup, 391, 539
bulk, 333, 463
burden, 124, 174, 268, 350
burn, 182, 279
burning, 29, 46, 391, 483
bus, 513, 525
business, 14, 35, 272, 394–399, 401,
        402, 405–407, 420, 425,
        587, 589
buy, 574
by, 3–6, 10, 12, 14, 15, 17, 18, 22,
        27, 29–32, 35, 41, 44,
        54–57, 59, 60, 62, 66, 79,
        83, 85, 88, 90, 94,
        102–107, 109, 111, 113,
        116–118, 124, 131, 134,
        135, 137, 138, 140, 144,
        147, 150, 151, 157, 158,
        161, 164, 165, 168, 175,
        181, 182, 184, 186, 187,
        190, 194, 198–201, 204,
        212–215, 221, 224, 231,
        232, 240, 243, 258, 266,
        271–273, 276, 279, 281,
        283, 285, 286, 307, 309,
        314–316, 321, 322, 328,
        329, 332, 333, 337, 344,
        345, 349, 350, 355, 356,

359, 363, 364, 366, 368,
374, 377, 385–387, 389,
395, 399, 400, 405, 417,
420–422, 426, 430, 431,
433, 434, 436, 439,
443–447, 450, 455, 456,
458–460, 463–465, 468,
486, 489, 492, 494–496,
499, 507, 509–511, 513,
518, 525, 527, 528, 531,
534, 538–540, 554, 555,
565, 567, 568, 574, 575,
580, 585, 587, 588, 590,
593

byproduct, 268, 285

California, 267, 289
call, 56
calming, 537
camping, 124
can, 3–7, 9–12, 14, 15, 17–21,
24–31, 35, 36, 39, 40,
42–47, 50–59, 61, 62,
65–68, 70, 71, 73, 82–85,
88–92, 98, 100, 102–107,
110–114, 117–120,
122–125, 129–137, 140,
144, 146, 147, 150, 151,
154, 157–160, 164, 166,
167, 169–175, 178–184,
187, 189–191, 198,
200–203, 205, 208, 209,
211–217, 219–221, 225,
229, 233, 235, 237, 239,
240, 242, 243, 246, 250,
251, 253, 256, 258–260,
264–267, 269–282, 284,
285, 290–292, 298, 299,
301, 302, 304–310,

314–318, 321, 322,
324–334, 338, 353, 355,
367, 370, 372, 374,
377–379, 382, 385, 386,
388–396, 398, 399,
402–405, 408, 412,
415–425, 429, 431–433,
435–440, 443–446, 448,
449, 452, 453, 455–459,
462–465, 467, 468, 476,
482, 483, 485–497, 499,
500, 504–510, 514, 515,
522, 524–529, 531,
533–535, 538–540, 546,
547, 551, 552, 554, 558,
561, 565, 568, 569, 571,
574, 575, 577, 579–582,
587–590, 592, 593,
595–598

Canada, 58, 101, 445, 510, 536
canal, 509
cancer, 314, 315, 431
canopy, 538
cap, 350, 445
capability, 193
capacity, 5, 91, 118, 137, 140, 151,
174, 189, 193, 194, 196,
198, 200, 201, 218, 252,
261, 301, 306, 348, 349,
352, 358, 372, 385, 386,
389, 425, 432, 437, 444,
445, 485–489, 492–496,
498, 500, 504–507, 509,
526, 539, 574, 578, 579,
590, 592–598

capital, 130, 132, 163, 193, 200,
285, 351, 395, 413, 415,
417, 420–422, 530

capping, 301, 302

captivity, 57
capture, 180, 182, 184, 214, 216,
         276, 295–298, 323, 372,
         374, 419
car, 221, 225, 235, 273
carbon, 4, 10, 35, 43, 44, 91, 111,
         113, 117–119, 124, 137,
         158, 180–182, 204, 213,
         215, 216, 219, 233, 243,
         247, 250, 255, 272, 275,
         276, 281, 284, 298, 314,
         323, 324, 331, 333, 344,
         349, 355, 401, 413, 415,
         416, 419, 422, 441, 445,
         471, 472, 482, 492, 521,
         522, 564, 588
cardboard, 269
career, 589
carp, 90
carrying, 42
case, 2, 14, 35, 77, 82, 88, 99, 101,
         102, 104, 105, 110, 164,
         198, 216, 221, 280, 331,
         362, 374, 378, 379, 385,
         404, 411, 429, 431, 437,
         438, 445, 452, 471, 472,
         485, 497, 502, 507, 525,
         530, 531
cash, 130, 131
catalyst, 322
cathode, 196, 201, 202
cause, 83, 88, 89, 180, 285, 308, 310,
         315, 324, 328
cell, 183, 186, 196, 202, 204
cement, 285
century, 41, 115, 157, 221, 289, 306,
         528
ceramic, 322
certainty, 387

certification, 154, 415
chain, 17, 29, 39, 273, 308, 396,
         399, 401, 402, 419, 467,
         468, 551
challenge, 20, 43, 52, 58, 90, 104,
         118, 134, 135, 200, 252,
         277, 286, 317, 350, 378,
         388–391, 438, 484, 485,
         488, 496, 499, 500, 527,
         568, 577, 579, 594
chamber, 277
change, 1–6, 11, 13, 14, 27, 29, 30,
         35, 41–44, 54, 59, 61, 62,
         68, 71, 81–83, 87, 90, 94,
         102, 104–107, 110, 112,
         113, 116–119, 122, 124,
         125, 129, 131, 135, 137,
         140, 144, 160, 172, 179,
         183, 186, 191, 195, 204,
         208, 213, 221, 233, 250,
         256, 282, 298, 314, 316,
         321, 344, 345, 347–351,
         355, 356, 359, 377, 378,
         386, 391, 399, 412, 413,
         415, 417, 419–422, 431,
         432, 434, 437, 444–446,
         457, 458, 483, 485–489,
         491–498, 501, 504–507,
         510, 525–528, 554,
         563–565, 572, 574,
         577–580, 584, 587, 590,
         594, 598
char, 284
charge, 194, 227
charging, 196, 217, 218, 220,
         226–229, 232, 235
chemical, 15, 29, 89, 103, 114, 118,
         137, 150, 157, 174, 182,
         192, 196, 201, 202,

214–216, 264, 276, 282, 286, 292, 293, 306, 310, 321–323, 329, 332–334, 338, 463

chemistry, 18, 214–216, 284, 305, 463

child, 131

childhood, 130

China, 109, 189, 345, 349, 445

choice, 82, 105, 120, 190, 544

circuit, 196

circularity, 408

citizen, 4, 82, 164, 456, 513, 514, 596

citizenship, 592

city, 128, 158–160, 163, 164, 168, 239, 243, 246, 279, 280, 316, 431, 482, 502, 512–514, 521, 525–527, 530, 531, 536, 537, 540

claim, 468

clarity, 310

cleaner, 40, 183, 189, 213, 216, 224, 235, 294, 302, 309, 315, 317, 323, 324, 441, 589

cleaning, 178, 214, 324, 326, 332

cleanup, 285, 336

climate, 1, 3–6, 10, 11, 14, 27, 29, 30, 35, 41–44, 54, 59, 61, 62, 68, 71, 82, 87, 90, 91, 94, 104, 107, 110, 112, 113, 116–119, 122, 124, 125, 129, 135, 137, 140, 160, 172, 179, 183, 186, 195, 204, 208, 213, 221, 233, 250, 282, 298, 314, 316, 321, 344, 345, 347–351, 354–356, 359, 386, 391, 413, 415, 420,

431, 434, 437, 445, 446, 483, 485, 487–489, 491–497, 501, 504, 506–510, 525–527, 565, 572, 574, 577–580, 587, 590

climax, 102

clogging, 285, 322

closure, 316

clothing, 14, 394, 399, 407, 561, 565

co, 391, 449–453, 485, 496, 499, 505

coagulation, 306, 313

coal, 276, 284, 314, 316

coastline, 490

coconut, 269

coffee, 269

cogeneration, 213

coherence, 387, 388, 495

cohesion, 125, 498, 506, 526

coir, 269

collaboration, 2, 6, 7, 13, 14, 18, 21, 24, 45, 57, 59, 67, 74, 84, 85, 90, 97, 106, 112, 113, 115, 116, 139, 144, 160, 229, 252, 258, 267, 268, 274, 275, 316, 330, 346, 347, 349, 350, 352, 374, 385, 386, 388, 389, 402, 408, 422, 426, 435, 443–446, 449, 471, 487, 494, 498, 500, 505, 507, 510, 514, 525, 527, 537, 543, 564, 572, 574, 577, 579, 584, 585, 590, 592–598

collaborative, 21, 50, 54, 70, 78, 119, 157, 253, 258, 261, 272, 309, 359, 391, 429, 437,

439, 442, 446, 449, 452,
495, 500, 551, 554, 557,
558, 561
collecting, 318, 404, 482
collection, 4, 70, 174, 262, 273, 317,
320, 323, 452, 457, 482,
513, 594, 596
collector, 318
collision, 471
colonization, 102, 105
colony, 70
color, 430
combat, 27, 42, 43, 316, 344
combination, 31, 35, 62, 89, 98, 118,
129, 142, 265, 267, 292,
303, 316, 333, 334, 434
combustion, 182, 222, 252, 275,
278, 280, 284, 314, 322,
324
comfort, 208, 211, 324, 326, 517,
538, 539
commercialization, 218, 465, 509
commitment, 14, 43, 58, 160, 189,
239, 316, 394, 416, 418,
419, 422, 460, 579, 580
commodification, 391
common, 36, 47, 55, 92, 105, 120,
129, 137, 183, 205, 223,
249, 252, 266, 292, 301,
307, 310, 325, 327, 330,
339, 345–347, 386, 398,
421, 441–443, 473, 579,
593
communication, 4, 21, 259, 261,
279, 449
community, 10, 45, 46, 55, 58,
62–64, 66, 78, 82, 84, 85,
87, 99, 102–104, 110, 112,
124, 125, 128, 140, 157,

160, 168, 174, 175, 246,
247, 281, 389, 416, 417,
421, 422, 425, 426, 429,
432, 446, 452, 453, 485,
487, 492–494, 497, 499,
500, 505–507, 509, 518,
521, 525–528, 531, 537,
539, 540, 558, 561, 587,
593–595, 598
commute, 534, 537
commuting, 128, 158, 528
compactness, 164
company, 394, 399, 401, 407,
415–419, 422, 467
compatibility, 70, 482
compensation, 416
competence, 596
competition, 58, 90, 182
competitiveness, 204
complex, 2, 3, 5, 6, 15, 18, 20–25,
27, 39, 56, 57, 76, 97, 102,
104, 112, 129, 146, 157,
178, 214, 274, 279, 281,
333, 371, 376, 387, 388,
395, 421, 435, 436, 439,
443, 444, 446, 449, 453,
455, 457, 460, 463, 472,
475, 477, 478, 485–488,
494, 496, 500, 505, 526,
527, 577, 579, 592, 595
complexity, 35, 372, 487
compliance, 280, 293, 305, 366–368,
385, 419, 441, 471
component, 70, 71, 74, 94, 99, 113,
119, 132, 233, 247, 268,
317, 351, 367, 402, 485,
502, 507, 518, 525, 551,
580, 588
composition, 9, 20, 29, 55, 103, 277,

279–282, 290, 463

compost, 150, 174, 268, 270, 271, 282, 290

composting, 4, 159, 160, 174, 175, 261, 265, 268–271, 279, 282

compound, 26

compromise, 266

computer, 469, 472

computing, 476–478

concentration, 305, 317, 318, 324, 332, 333, 431

concept, 1, 6, 11, 12, 14, 21, 41, 59, 105, 157, 168, 169, 204, 208, 271, 274, 275, 290, 356, 390, 394, 413, 430, 436, 437, 442, 445, 460, 463, 465, 485, 486, 498, 499, 527, 528, 540, 554, 558, 561, 588

concern, 136, 154, 220, 309, 321, 326, 356, 464, 561

conclusion, 4, 11, 31, 46, 78, 82, 107, 116, 122, 125, 137, 172, 183, 186, 191, 215, 253, 280, 286, 307, 313, 324, 389, 422, 449, 465, 478, 488, 528, 540, 574, 577, 580

condensation, 323

conditioning, 539

conduct, 168

cone, 284

configuration, 196, 539

conflict, 129, 131, 429, 446

congestion, 128, 219, 235, 239, 242, 243, 247, 250, 471, 482, 522, 525, 537

conjunction, 332, 333

connection, 24, 122, 508

connectivity, 56, 62, 87, 105–107, 110, 118, 168, 242, 489–491, 500, 513, 525, 537

consensus, 447, 509, 577–579

consent, 390, 391

conservation, 1, 3–5, 11, 24, 29, 31, 35, 36, 43–46, 50, 52, 54, 57–59, 62–74, 76–79, 81–85, 87, 88, 94, 99, 100, 102, 105–107, 110, 111, 113, 117, 118, 122, 123, 125, 128, 129, 146, 158, 159, 175, 204, 205, 211, 268, 308, 337, 344, 347, 352, 353, 356, 359, 364, 367, 386, 390, 391, 408, 429, 435, 441, 452, 456, 458, 460, 482, 490–492, 496, 499, 508, 518, 543, 573, 580–582, 592, 596

conserve, 14, 27, 70, 78, 91, 116, 118, 119, 175, 261, 265, 271, 275, 353, 356, 379, 462, 547, 551, 565, 571

consideration, 24, 56, 277, 279, 337, 446, 457, 460, 561

construction, 56, 90, 106, 123, 136, 138, 175, 180, 181, 183, 200, 278, 280, 307, 310, 326, 327, 330, 331, 362, 415, 464, 509, 514, 515

consultation, 390

consulting, 425, 589

consumer, 14, 150, 274, 401, 402, 551, 554, 564

consumerism, 551–554

consumerist, 558

consumption, 2, 6, 17, 28, 125, 137,
        140, 141, 143, 175, 204,
        208, 211, 213–215, 220,
        251, 258, 261, 272, 274,
        275, 308, 316, 328, 356,
        389, 399, 404, 468, 514,
        546, 551, 554, 557–561,
        571, 588–590
contact, 290, 328, 332
container, 269, 270
containment, 298, 329–332, 334
contamination, 29, 265–267, 301,
        302, 307, 327–331, 333,
        431
content, 20, 282, 285, 597
context, 22, 24, 104, 105, 166, 290,
        358, 386, 387, 425, 426,
        434, 443, 449, 452, 453,
        492, 505, 507, 508, 522,
        544, 593–595, 597
continuity, 66, 499, 596
contour, 109, 308
contract, 468
contrast, 14
contribution, 189, 191, 316, 391
contributor, 172, 219, 314
control, 2, 4, 26, 55, 56, 84, 89–91,
        109, 113, 115, 118, 123,
        136, 146, 150, 182, 190,
        251, 259–261, 270,
        276–280, 282, 285,
        290–295, 303, 305–307,
        314, 315, 317, 321,
        323–325, 327, 329, 333,
        363, 364, 366, 368, 391,
        463, 465, 467, 488, 513
convenience, 547
convention, 352, 377–379
conversion, 85, 103, 180, 183, 202,

        205, 221, 277, 281, 284,
        323, 464
cool, 178, 538–540
cooler, 540
cooling, 178, 181, 323, 538–540
cooperation, 6, 35, 70, 252, 316,
        343, 345–347, 349, 352,
        353, 359, 367, 371, 386,
        388, 389, 500, 507,
        572–575, 577, 579, 580,
        584, 585, 590, 593, 596
coordinating, 6
coordination, 260, 261, 274, 349,
        358, 363, 367, 385, 386,
        388, 495, 506, 527, 572,
        574, 580, 590, 597
copper, 266
core, 22, 158, 181, 394, 398, 465,
        486, 494, 512
cornstarch, 462
corridor, 88, 168
corruption, 130
cost, 17, 89, 117, 135, 137, 138, 180,
        186, 191, 200, 204, 207,
        208, 218, 232, 233, 235,
        242, 268, 274, 313, 319,
        329, 331, 334, 337, 395,
        399, 402, 405, 408, 518,
        525, 558
country, 54, 127, 128, 153, 168, 189,
        353, 356, 359, 379, 484,
        493, 496, 527
course, 423, 485
cover, 1, 62, 109, 111, 125,
        298–302, 349, 585, 596
coverage, 356
creation, 2, 55, 87, 88, 110, 111,
        137, 160, 189, 271, 366,
        423, 472, 490, 495, 518,

527, 543, 589, 590
creativity, 571, 595
credibility, 447
credit, 130, 133, 399
crisis, 353, 431
criticism, 157, 355, 356
crop, 26, 27, 42, 44, 54, 66, 150,
        154, 157, 315, 323, 328,
        338, 339, 456, 492, 509
cropping, 144
cultivation, 117
culture, 175, 537, 598
cure, 459
Curitiba, 159, 164
current, 27, 105, 158, 183, 235, 266,
        275, 349, 350, 385, 482,
        483, 500, 547, 552
customer, 394, 395, 399
customization, 475
cut, 104, 431, 458
cycle, 4, 117, 129, 130, 133, 181,
        190, 194, 196, 232, 273,
        285, 404, 518, 544, 547
cycling, 10, 26, 68, 91, 119, 146,
        150, 158, 168, 221, 328,
        351, 489, 528, 534, 580
cyclone, 497

dairy, 270, 271
dam, 136, 190, 362
damage, 42, 56, 89, 90, 94, 111, 315,
        417, 419
data, 4, 83, 112, 305, 316–318, 320,
        321, 363, 404, 405, 412,
        452, 456, 457, 467, 471,
        482, 488, 514, 544, 582,
        594, 596
date, 598
day, 198, 232, 267

DC, 229
death, 29
debris, 17, 310
decade, 41
decarbonization, 580
decay, 181
decentralization, 465, 468
decision, 2–5, 12, 45, 54, 65, 70, 78,
        118, 130, 132–134, 147,
        159, 251, 261, 356, 359,
        361, 363, 367, 374,
        376–379, 386, 387,
        389–392, 405, 412, 417,
        418, 422, 423, 425–432,
        435–439, 441, 443, 446,
        449, 457, 472, 482, 486,
        488, 490, 492, 493, 495,
        496, 499, 500, 505, 506,
        509, 510, 513, 526–528,
        546, 547, 574, 577, 593,
        595, 596
decline, 30, 31, 71, 73, 83, 84, 89,
        90, 94, 111, 113, 114, 180,
        308
decomposition, 10, 17, 268–270,
        276, 281, 284, 290
decoupling, 588
decrease, 89, 215, 251
defense, 469, 472
deforestation, 29, 35, 52, 55, 83, 91,
        117, 118, 128, 356, 456,
        483, 574
degradation, 24, 44, 45, 49, 52, 54,
        68, 95, 98, 99, 109–111,
        113, 114, 117–120, 138,
        140, 157, 158, 164, 200,
        250, 271, 273, 307, 365,
        367, 430, 434, 462, 489,
        526, 528, 540, 573, 587,

588

degree, 105

delivery, 227, 464, 471, 472

demand, 30, 84, 136, 158, 190–193, 198–200, 213–215, 231–233, 267, 362, 416, 539, 589

demonstration, 444

dengue, 42

Denmark, 445

density, 9, 194, 196, 200, 218, 252, 266, 530

dependence, 4, 137, 181, 186, 213, 229, 253, 276, 506

dependency, 135

depletion, 4, 6, 30, 135, 138, 140, 182, 212, 271, 273, 306, 308, 359, 400, 404, 546, 561, 574, 588

deployment, 183, 193, 195, 204, 228, 252, 471, 580

deposition, 302, 315, 328

depression, 58

deprivation, 129

desalination, 186

design, 3, 6, 14, 45, 106, 112, 123, 138, 140, 144, 159, 165, 167, 174, 175, 178, 196, 208, 209, 211, 212, 214, 219, 220, 243, 251, 256–258, 261, 272, 273, 302, 326, 425, 445, 475, 514–518, 526, 527, 531, 534, 537, 539, 540, 561

designer, 565

desire, 443

destination, 471

destruction, 1, 3, 27, 29, 32–36, 44, 54, 68, 83, 84, 91, 110,

123, 351, 356, 546

detail, 262

detection, 89, 90, 118, 329, 464, 471

deterioration, 314

determination, 390

development, 1, 3–6, 18, 21, 27, 29, 35, 36, 45, 50, 52, 54, 56, 67, 71, 78, 83, 85, 87–90, 94, 110, 118, 119, 125, 127–132, 135, 136, 138, 140, 149, 150, 154, 158–160, 162–165, 171, 172, 179, 181–183, 189, 191, 195, 204, 214–218, 221, 229, 232, 233, 235, 239, 244, 246, 250–252, 255, 264–266, 270, 271, 274, 279, 315, 324, 327, 337, 343–347, 353, 356, 359, 363–365, 367, 374, 378, 379, 383, 385, 386, 388–392, 405, 413, 415–417, 420–422, 425, 432, 435, 441–443, 446, 449, 452, 453, 455, 463–465, 468, 481, 482, 485, 489, 491, 494–497, 501, 504–507, 511, 513, 514, 518, 522, 525, 526, 528, 530–534, 538–540, 543, 565, 572, 574, 575, 577, 579–582, 585, 587–590, 592, 593, 595, 596, 598

diagnosis, 464

dialogue, 347, 350, 386, 390, 417, 432, 446, 579, 580

diesel, 314

difference, 180, 200, 201, 413, 420,

538

diffusion, 269, 318, 442–446

digestate, 275, 281

digestibility, 282

digestion, 214, 275, 277, 281–283

diligence, 418

dimension, 508

dioxide, 35, 44, 91, 117, 124, 182, 275, 281, 284, 314, 315, 324, 331, 349, 366

directive, 371

disadvantage, 430

disaster, 124, 487, 492, 493, 497, 506, 527

discharge, 29, 193, 194, 196, 198, 218, 302, 304, 310

discipline, 56, 78

discomfort, 324

discrimination, 129, 131, 132, 430, 496

disease, 27, 42, 57, 68, 71, 102, 103, 459

disinfection, 306

disk, 284

dispersal, 30, 68, 88, 105

displacement, 31, 136, 191, 425, 457, 465, 472

disposal, 17, 135, 137, 172, 198, 214–216, 235, 267, 268, 272, 273, 278, 286, 290, 307, 309, 327–329, 331, 404, 462, 464, 465, 546, 551, 561, 573

dispose, 6, 14, 174, 214, 271, 540, 588

disruption, 29, 32, 42, 315, 465, 526, 581

dissemination, 174

distance, 229, 528

distinctiveness, 70

distortion, 445

distribution, 9, 129, 131, 386, 387, 430, 465, 495, 509, 579, 592

disturbance, 102–105

diversification, 130, 136, 324, 493, 496

diversion, 45, 56, 97, 115, 160, 293

diversity, 3, 9, 26, 27, 31, 45, 56–59, 62, 67–71, 73, 74, 84, 111, 118, 308, 351, 387, 416, 487, 489, 526, 597

divide, 594, 597

diving, 124

document, 168, 510

documentation, 509

dog, 70

domain, 212

dominance, 31

down, 66, 174, 199, 232, 268, 270, 276, 281, 306, 329, 331, 332, 389, 390, 460, 463, 508, 569

DPFs, 322

drainage, 45, 94, 113, 115

drill, 181

drinking, 308, 310, 431

driving, 10, 134, 175, 183, 186, 196, 199, 224, 233, 235, 350, 388, 408, 419, 421, 499

drone, 471, 472

drop, 322

drought, 437

drug, 464

duplication, 388

durability, 14, 17, 463, 565

duration, 193

dust, 314, 328

dynamic, 5, 437, 439, 443, 488, 597

Earth, 3, 26, 28, 31, 35, 41, 67, 113,
    117, 181, 183, 271, 315,
    351, 483, 580, 582
earth, 546
eco, 1–6, 11, 21, 22, 24, 25, 43, 52,
    59, 65–68, 70, 74, 78, 82,
    90, 94, 107, 110, 113, 119,
    124, 135, 158, 176–178,
    212, 215, 216, 219, 220,
    256, 258, 268, 271, 281,
    282, 295, 337, 343, 367,
    386, 426, 432, 446,
    455–457, 463, 479, 482,
    514, 547, 551, 561, 589,
    592–596, 598
ecology, 1, 11, 44, 46, 54–57, 71, 83,
    102–105, 116, 123, 147
economic, 3, 12, 14, 18, 20, 21, 45,
    52, 54, 57, 61, 82, 89, 98,
    105, 109, 117, 123–125,
    127, 129–134, 137, 138,
    140, 143, 144, 146, 147,
    149, 151, 154, 158, 160,
    168, 181, 189, 191, 204,
    208, 250, 262, 265, 267,
    271, 274, 275, 277, 283,
    298, 333, 356, 358, 359,
    365, 366, 385, 387, 389,
    395, 405, 408, 423, 429,
    435, 444, 446, 457, 465,
    468, 485, 489, 492,
    494–498, 505, 516, 521,
    522, 525, 526, 528, 534,
    543, 552, 554, 557, 561,
    579, 587–589, 592
economy, 4, 6, 7, 14, 54, 124, 144,
    159, 212, 213, 215, 216,

    268, 271–276, 282, 285,
    290, 350, 356, 365, 394,
    395, 405–408, 412, 413,
    415, 416, 419, 496,
    540–543, 554–558,
    587–590
ecosystem, 2, 4, 10, 11, 20, 21, 23,
    24, 26, 27, 29, 30, 35, 36,
    44, 46, 55–59, 61, 62, 66,
    68, 74, 78, 79, 82–99,
    103–107, 109–111,
    114–119, 121–123, 125,
    138, 140, 144, 146, 147,
    150, 164, 181, 191, 309,
    328, 339, 351, 356, 391,
    434, 435, 485, 487,
    489–491, 495, 497–500,
    580
ecotourism, 57, 122
Eddy, 266
Eddy, 266
editing, 5, 458–460
educating, 326
education, 4, 14, 18, 35, 40, 113,
    122, 124, 127, 129–133,
    135, 174, 175, 267, 268,
    327, 330, 402, 420, 421,
    432, 500, 526, 540, 552,
    554, 564, 587, 592–596
effect, 41, 158, 183, 530, 538–540
effectiveness, 49, 55, 96, 111, 112,
    115, 244, 261, 293, 302,
    309, 317, 321, 331–333,
    345, 350, 363, 367, 377,
    379, 388, 420–422, 438,
    444, 448, 449, 484, 490,
    493, 510, 518, 547, 576,
    594, 597
efficacy, 464

efficiency, 4, 14, 42, 137, 150, 159, 172, 175, 178, 183, 187, 189, 191, 193, 196, 204–206, 208, 209, 211, 213–216, 220, 235, 240, 244, 251, 258, 261, 266, 267, 271, 273, 279–282, 284, 285, 316, 333, 356, 399, 402, 405, 408, 419, 420, 459, 464, 468, 481, 482, 491, 513, 518, 522, 540, 546, 554, 558, 561, 588, 589

effort, 14, 115, 399, 551

electricity, 127, 136, 137, 180–184, 186, 187, 189–191, 193, 194, 199–201, 204, 213, 227, 243, 275–277, 279–281, 285, 298, 323, 422

electrification, 525

electrochemistry, 201, 232, 233

electrode, 201

electrolyte, 196, 201, 202

electromagnetism, 221

elephant, 77, 78

elevation, 180, 200, 201

elk, 58, 101

emergency, 527

emission, 182, 201, 215, 216, 275, 277, 285, 315, 316, 321, 323, 324, 349, 350, 354, 355, 579

emittance, 538

empathy, 432, 592

emphasis, 147

employability, 130

employee, 395, 399, 416

employment, 123, 125, 129, 132, 133, 589

empowerment, 133, 135, 593

end, 2, 174, 215, 268, 272, 281, 307, 404, 546, 551

endangerment, 83, 84

endeavor, 116

energy, 1, 2, 4, 5, 10, 42, 105, 125, 127, 130, 134–137, 158–160, 172, 175, 178–196, 198–206, 208, 209, 211–216, 218, 220, 221, 229–233, 235, 252–256, 258, 261, 272–282, 285, 286, 298, 314–317, 323, 324, 356, 362, 366, 386, 399, 404, 411, 415–417, 419–422, 431, 463, 464, 468, 482, 491, 513, 514, 517, 518, 522, 528, 538–540, 546, 554, 565, 568, 587–590, 596

enforce, 133, 441

enforceability, 350

enforcement, 84, 315, 356, 358, 359, 367, 378, 385, 388, 432, 582

engagement, 2, 4, 5, 24, 56, 57, 94, 99, 110, 112, 124, 134, 136, 140, 157, 160, 164, 242, 246, 279, 316, 350, 385, 390, 395, 399, 416, 422–426, 430, 432, 435, 446, 485, 487, 495, 513, 521, 525, 528, 531, 537, 540, 590, 593, 594

engine, 220, 222, 251, 252, 258, 322

engineering, 18, 256, 290, 463

enhance, 5, 14, 46, 56, 61, 62, 68, 85,
        91, 94, 98, 101, 106, 107,
        111, 112, 114, 116, 118,
        119, 124, 154, 158, 159,
        165, 168, 191, 209, 215,
        216, 242, 251, 269, 279,
        282, 324, 332, 333, 338,
        344, 347, 350, 356, 363,
        374, 377, 379, 389, 401,
        402, 408, 430, 439,
        447–449, 464, 481, 485,
        486, 490, 491, 493–495,
        498, 500–502, 506, 507,
        509, 510, 514, 522, 526,
        527, 547, 574, 588, 592,
        593, 596, 597
enhancement, 123, 395, 399
enrichment, 27
enrollment, 131
entanglement, 17
entity, 66
entrepreneurship, 130, 421, 499, 589
entry, 90
environment, 1, 2, 4, 5, 9–12, 15, 18,
        21, 24, 30, 36, 37, 42–44,
        57, 58, 66, 79, 103, 104,
        117, 130, 135, 146, 154,
        165, 167, 169, 172, 175,
        178, 204, 208, 211, 212,
        214, 236, 269, 270, 273,
        275, 278, 282, 284, 290,
        291, 293, 294, 298, 299,
        301, 302, 305, 308, 309,
        314, 321, 326–330, 334,
        356, 363, 365, 367, 368,
        374, 378, 379, 387, 389,
        390, 394, 395, 397–399,
        414, 416, 430, 432–434,
        456, 460, 462, 469, 492,

        508, 512, 514, 528, 530,
        532, 535, 537–539, 561,
        566, 571, 574, 587, 589,
        592
enzyme, 458
equality, 1, 125, 132–135, 416, 497
equation, 321, 322
equilibrium, 489
equipment, 137, 213, 268, 270, 285
equity, 52, 144, 158, 208, 350, 385,
        389, 432, 453, 457, 460,
        488, 493, 494, 497, 498,
        500, 505, 507, 525, 574,
        580, 593
eradication, 1, 89, 90, 125, 129–132,
        574
erosion, 24, 41, 52, 55, 67, 91, 109,
        111, 117, 144, 150, 307,
        308, 329, 500
eruption, 102, 104
essential, 3–5, 10, 14, 18, 21, 26, 27,
        42, 44–46, 50, 54, 56–58,
        67, 68, 74, 84, 85, 90, 91,
        94, 96, 105, 110, 111, 113,
        118, 119, 129, 131, 134,
        137, 138, 143, 146, 159,
        169, 172, 174, 180, 183,
        191, 193, 195, 211, 214,
        216, 217, 247, 251, 252,
        261, 266, 268, 272, 279,
        280, 285, 286, 292, 293,
        298, 309, 313, 316, 317,
        320, 324, 325, 328–330,
        334, 337, 343, 347, 351,
        356, 359, 364, 367, 371,
        385, 386, 390, 392, 399,
        402, 405, 416, 421, 426,
        446, 456, 457, 460, 485,
        489, 491, 494, 495, 498,

504, 505, 509, 522, 527,
574, 580, 582, 587, 590,
592, 594, 595, 597
establishing, 1, 35, 45, 57, 70, 106,
252, 274, 343, 356, 365,
385, 497, 597
establishment, 77, 83, 84, 87,
102–104, 174, 344, 390,
437, 490, 493, 575
estate, 124
ethnicity, 430, 432
Europe, 200, 371, 377
eutrophication, 308
evaluation, 112, 181, 367, 385, 419,
425, 432, 438, 446, 485,
496, 594, 597
evapotranspiration, 538
Everglades, 45, 56, 94, 97, 115
evidence, 3, 41, 444, 446
evolution, 27, 68
ex, 331, 333
example, 3, 9, 10, 14, 20, 23, 26, 29,
30, 39, 44, 45, 49, 54, 56,
59, 66–68, 73, 82, 83, 87,
89, 90, 94, 103–105, 109,
110, 115, 123, 124, 131,
132, 134, 140, 147, 150,
157, 159, 163, 167, 178,
189, 198, 200, 210,
213–215, 246, 266, 267,
277, 279, 282, 289, 292,
301, 305, 316, 331–333,
350, 356, 366, 367, 371,
378, 383, 391, 394, 399,
404, 407, 411, 415, 417,
419, 421, 422, 425, 429,
432, 434, 445, 452, 456,
464, 467, 471, 482, 487,
493, 497, 499, 508, 510,

513, 517, 595
excavation, 307, 328, 330, 333, 334
exchange, 273, 352, 389, 413, 442,
443, 445, 487, 488, 490,
505, 507, 510, 554, 557,
572, 579, 590, 593,
595–598
exercise, 418
exhaust, 29, 277, 322, 323, 328
existence, 41, 74, 582
exiting, 178
expansion, 35, 85, 87, 105, 106, 172,
217, 588
expense, 31
experience, 41, 229, 391, 431, 508,
595
experiment, 443
experimentation, 487, 495, 499, 500
expertise, 18, 20, 21, 59, 67, 112,
123, 139, 159, 378, 386,
390, 423, 425, 445, 446,
449, 452, 495, 505, 579,
587, 590, 596
exploitation, 364, 390, 509
exploration, 2, 21, 181
exposure, 308, 315, 324, 431, 464
expression, 68, 432
extent, 42, 95, 99, 103, 111, 120,
328, 333, 418
extinction, 30, 32, 35, 57, 68, 71, 82,
89, 581
extraction, 14, 214, 272, 274, 329,
330, 356, 546, 554

face, 4–6, 41, 43, 46, 49, 50, 58, 59,
68, 70, 84, 104, 105, 113,
116, 133, 135, 147, 158,
164, 183, 202, 204, 223,
224, 244, 247, 249, 254,

279, 316, 346, 347, 370,
377, 378, 387, 388, 399,
431, 434, 437, 438, 444,
449, 461, 485–488, 493,
494, 497, 498, 501,
504–506, 508–510, 521,
525, 528, 579, 584, 587
facilitating, 4, 50, 55, 89, 105, 389,
442, 444, 489, 580, 596,
598
facilitation, 103
facility, 200, 267, 278, 279, 331
factor, 83, 183, 187, 522
factory, 304
failure, 99, 181
fair, 130, 159, 386, 387, 399, 425,
430, 495, 551, 564, 579,
581, 592
fairness, 350, 457, 500
fall, 182
fare, 525
farm, 147, 189, 415
farmer, 147
farming, 54, 67, 144, 146, 147,
150–154, 315, 329, 492,
493, 509, 521
fashion, 14, 472, 561–565
fatigue, 324
fatty, 281
fauna, 508
favor, 417, 418
favoring, 31
feasibility, 131, 252, 298, 333, 402,
445
feed, 111
feedback, 22, 24, 25, 267, 444, 486,
489, 513, 594, 597
feeding, 113, 308

feedstock, 264, 276, 281, 282,
284–286
femicide, 133
fermentation, 270
fertility, 44, 66, 91, 99, 117, 154,
174, 276, 328, 338, 493,
499, 509
fertilizer, 275, 281, 456
fever, 42
fiber, 215
field, 2, 5, 15, 18, 21, 43, 45, 46, 54,
66, 71, 79, 83, 147, 160,
199, 229, 266, 281, 305,
332–334, 337, 343, 419,
421, 446, 456, 458, 463,
465, 468, 476, 478, 479,
513, 593, 596
file, 472
fill, 482
film, 186, 432
filter, 90, 111, 322
filtration, 306, 310, 313, 327, 464
finance, 348, 408–413, 415,
420–422, 468, 496
financing, 140, 168, 350, 415, 496,
527
finding, 58, 135, 192, 492, 568, 580,
593
fine, 284, 322
Finland, 445
fire, 46, 104, 207, 391, 508
firestick, 67
fish, 29, 30, 58, 67, 90, 101, 110,
111, 113, 136, 181, 191,
434
fishing, 30, 67, 124, 435
flavor, 154
fleet, 513

flexibility, 112, 193, 284, 355, 490, 505, 507

flight, 220, 251

Flint, 431

flood, 56, 113, 123, 136, 190, 484, 485, 493, 526, 527

flooding, 41, 483–485, 493, 500, 527

floodwater, 493

flora, 508

Florida, 45, 73, 97, 115

flow, 2, 10, 56, 68, 69, 105, 106, 115, 136, 183, 190, 191, 198, 199, 259, 261, 284, 323, 333, 430, 482, 489

flowering, 26, 30

flowing, 136, 180, 189–191, 199

flue, 278, 280, 322

fluid, 181, 284

flywheel, 194

focus, 3, 6, 11, 38, 56, 111, 112, 124, 130, 138, 215, 272, 324, 356, 411, 421, 490, 506, 514, 526, 528, 539, 578, 589, 594, 598

follow, 399

following, 22, 41, 46, 169, 175, 255, 256, 261, 265, 267, 291, 292, 301, 310, 313, 321, 322, 356, 360, 362, 402, 440, 503, 536, 544, 550, 558, 578

food, 9, 10, 17, 26, 29, 30, 39, 42, 58, 89, 144, 146, 150, 151, 154–157, 174, 182, 256, 270, 271, 275, 282, 308, 421, 431, 434, 485, 488, 504, 509, 518, 521

footprint, 14, 175, 213, 215, 216,

272, 280, 394, 401, 416, 425, 465, 472, 547, 564

force, 10, 180, 190

forecasting, 485

forefront, 496

forest, 10, 20, 23, 24, 35, 65, 102, 104, 106, 116–118, 123, 314, 349, 429, 492

forestry, 52, 83, 89, 181, 309, 344, 490, 581

form, 29, 79, 137, 147, 173, 180, 186, 199, 205, 323, 364, 413, 419, 476, 569

formation, 102, 215, 291, 314, 328, 489, 539

formula, 201

formulation, 376, 378, 379, 386, 388, 446

fossil, 4, 29, 135, 169, 179, 181, 182, 186, 190, 213, 215, 216, 229, 230, 253, 255, 272, 276, 285, 298, 314, 323, 417, 483, 522, 588

foster, 67, 158, 174, 183, 421, 432, 446, 457, 500, 505, 507, 509, 540, 557, 574, 579, 588, 596

foundation, 1, 5, 11, 26, 68, 79, 82, 140, 147, 173, 364, 377, 468, 476

fragmentation, 3, 30, 55, 83, 85–88, 106, 107, 118, 119, 378

framework, 129, 343, 345–349, 351–353, 356, 359, 364, 367, 371, 374, 377, 378, 418, 422, 434, 435, 487, 488, 498, 547, 575, 578, 584

fraud, 355, 468

freight, 167

frequency, 103, 194, 483, 525

freshwater, 29, 115, 182, 483

friction, 194

fruit, 269

fuel, 181, 190, 201–204, 213, 215,
220, 251, 258, 260, 284,
285, 298, 322, 391, 482,
588

function, 54, 85, 88, 98, 204, 324,
326, 469, 486, 489, 498,
569

functionality, 91, 107, 502, 517

functioning, 3, 9, 26, 57, 68, 70, 86,
90, 94, 105, 107, 119, 328,
434, 479, 489, 499, 580

fund, 413, 421, 422, 527

funding, 111, 112, 168, 239, 252,
358, 527, 580, 582, 587,
590

furniture, 324

future, 2–7, 11, 14, 15, 18, 21, 27,
28, 31, 35, 36, 40, 43, 46,
51, 54, 57, 67, 68, 70, 74,
82, 90, 95, 98, 107, 110,
113, 119, 125, 129, 134,
137, 138, 140, 144, 146,
151, 157–160, 172, 175,
178, 183, 186, 189–191,
195, 198, 203, 204, 208,
211, 212, 216, 220, 224,
232, 233, 235, 240, 243,
245, 247, 250, 253, 256,
261, 264, 265, 271, 275,
277, 282, 286, 290, 293,
294, 298, 302, 304, 306,
309, 313, 317, 321, 324,
327, 330, 334, 337, 344,
347, 353, 359, 367, 368,

374, 377, 379, 385, 387,
389, 392, 398–400, 402,
408, 412, 415, 416, 419,
422, 430, 433, 457, 463,
472, 474, 475, 478, 482,
485, 487, 488, 490, 491,
494, 495, 497, 500, 504,
510, 514, 525, 527, 531,
537, 540, 547, 551, 552,
554, 557, 561, 564, 568,
574, 577, 580–582, 584,
587, 588, 590, 594, 595,
598

G. Lettinga, 283

gain, 1, 2, 9, 10, 18, 221, 402, 419,
423, 425, 457, 592

game, 475

Gansu Province, 189

gap, 4, 134, 349, 350, 388, 444, 446,
580

garbage, 482

garden, 521

gardening, 269

garment, 561

gas, 42, 124, 135, 137, 138, 158,
172, 180, 181, 183, 189,
190, 204, 211–214, 216,
219, 221, 224, 235, 253,
255, 275–277, 282, 284,
285, 295, 297, 298, 314,
316, 322, 344, 345,
348–350, 356, 404, 415,
445, 483, 491, 522, 528,
546, 578, 588

gasification, 276, 284–286

gasifier, 284

gasoline, 225, 314

gathering, 67

gear, 399, 407

gender, 1, 125, 129, 130, 132–135, 497

Gene, 5, 68, 70, 458, 460

gene, 5, 68–70, 105, 106, 458–460, 489

generating, 12, 65, 181, 190, 201, 275, 409, 506

generation, 3, 4, 6, 11, 14, 66, 136, 137, 144, 175, 180, 181, 188, 190, 191, 199, 201, 204, 213–216, 232, 272, 273, 276, 278, 280, 281, 284, 286, 290, 291, 293, 298, 400, 404, 415, 508, 546, 547, 554, 565, 568, 573, 589, 596

generator, 136, 180, 186, 187, 190, 277

genetic, 26, 27, 45, 56–59, 62, 68–71, 73, 74, 83, 84, 102, 118, 351, 458, 459, 489, 490, 581

genome, 458

geology, 305

glacier, 102

glimpse, 2

global, 2, 3, 6, 7, 29, 41–43, 70, 154, 155, 164, 175, 189, 212, 213, 233, 250, 252, 316, 324, 343, 344, 346–353, 359, 367, 374, 385–389, 399, 412, 421, 445, 467, 483, 495, 505, 527, 572–574, 577, 578, 580, 583–585, 590, 592–594, 598

go, 420, 588

goal, 14, 117, 125, 129, 132, 265,

404, 422, 565, 574, 578, 585

good, 131, 324, 325, 371, 417

governance, 2, 6, 14, 62, 125, 130, 131, 343, 345–347, 358, 367, 374–379, 381–392, 416–419, 422, 425, 426, 429, 436–443, 446, 486, 488, 494, 495, 497, 500, 506, 509, 572–577, 580

governing, 499

government, 40, 50, 112, 127, 128, 154, 246, 253, 277, 283, 316, 367, 422, 425, 426, 432, 497, 505, 579, 580, 596

grass, 538

grassland, 119, 122

grease, 307

green, 1, 124, 128, 140, 158–160, 164–169, 175, 208, 211, 214–216, 356, 365, 366, 413–416, 420, 431, 487, 495, 496, 514, 515, 517, 518, 526–528, 530, 538–540, 565–568, 588–590

greenhouse, 41, 42, 124, 135, 137, 138, 158, 172, 179–183, 189, 190, 204, 211–213, 216, 219, 221, 224, 235, 253, 255, 275–277, 282, 285, 295, 297, 298, 314, 316, 344, 345, 348–350, 356, 404, 415, 445, 483, 491, 522, 528, 546, 578, 588

greenwashing, 395

grid, 135, 137, 159, 184, 191–195,

198–200, 227, 232, 233, 280, 422

ground, 251, 463

groundwater, 291, 301, 302, 308, 309, 328, 331–334, 371

group, 438

growth, 6, 9, 28, 29, 52, 54, 67, 89, 103, 104, 111, 118, 123, 130, 132, 134, 153, 154, 182–184, 186, 208, 233, 269, 274, 328, 338, 365, 366, 395, 401, 421, 457, 499, 522, 526, 528, 543, 559, 587, 588

GSCM, 400–402

Guam, 30

guidance, 330, 596

guide, 55, 63, 83, 91, 94, 107, 115, 243, 273, 351, 354, 367, 376, 386, 392, 416, 458, 514, 543, 575, 595

habitat, 1, 3, 4, 11, 24, 27, 29, 30, 32, 34–36, 44, 45, 49, 54–59, 62, 68, 77, 82–85, 88, 91, 94–99, 102, 106, 107, 110, 113, 115, 117–119, 123, 168, 351, 356, 434, 456, 489, 490, 546, 580

halt, 573

hand, 102, 103, 136, 184, 275, 302, 310, 322–324, 417, 443, 463, 538, 540, 588–590

handicraft, 65

handling, 214, 327, 464

Harare, 178

harassment, 133

harm, 88, 173, 305, 308, 315, 378, 387, 430, 561

harmony, 66, 67, 147, 178, 508

harvest, 434

harvesting, 30, 159, 509

health, 10, 17, 20, 24, 28, 29, 36, 39, 42, 59, 66, 70, 71, 73, 74, 82, 83, 89–91, 94, 96, 99, 107, 111, 116–118, 125, 127, 130, 131, 133, 144, 146, 147, 150, 154, 172, 173, 175, 214, 220, 221, 250, 276, 279, 286, 290, 302–305, 308, 309, 314, 315, 320, 324, 326–330, 366, 387, 400, 431, 432, 456, 464, 485, 518, 522, 537

healthcare, 42, 129–132, 420, 421, 464, 468, 469, 472, 475, 497, 587

heart, 11

heat, 6, 42, 158, 181, 183, 184, 205, 213, 268, 275–277, 279–281, 284, 298, 323, 526, 530, 538–540

heating, 181, 184, 215, 284, 285

help, 2–5, 21, 43, 44, 46, 53, 55–58, 62, 75, 83, 111, 112, 130, 135, 154, 164, 169, 174, 235, 239, 243, 251, 261, 266, 275–277, 285, 308, 309, 313, 315, 316, 321, 333, 334, 356, 367, 371, 386, 412, 417, 418, 420, 421, 425, 437, 443, 456, 486–488, 490, 493, 496, 499, 518, 530, 531, 565, 580, 597

heritage, 65, 128, 160, 489, 510, 582

hiking, 124

history, 430, 508, 527
home, 44, 49, 56
homelessness, 421
homogenization, 31
honeycomb, 322
horn, 84
hotspot, 118, 353
hour, 225
household, 269
housing, 35, 130, 159, 417, 420,
        497, 530
human, 3–5, 11, 15, 17, 23–31,
        35–37, 39, 42–44, 49,
        54–56, 66, 71, 74, 76–79,
        83, 85, 87–91, 94, 102,
        103, 106, 108, 110, 111,
        113, 118, 123, 130, 132,
        154, 172, 173, 204, 215,
        290, 302–305, 307, 308,
        314, 324, 326–328, 330,
        351, 356, 367, 368, 371,
        387, 416, 435, 455, 459,
        469, 472, 483, 489, 491,
        496, 498, 508, 526, 578,
        580, 581
humanity, 157
hunger, 157
hunting, 30, 65, 67, 510
hybridization, 74
hydro, 137, 193, 199–201, 588
hydroelectricity, 191
hydrogen, 204, 276, 281, 284, 332
hydrology, 111, 116
hydrolysis, 281
hydropower, 362
hydrosphere, 594
hyperspectral, 267

ice, 41, 483, 510

idea, 455, 551, 554, 558, 588
identification, 45, 267, 452, 456
identity, 125, 486, 498
image, 402
imagery, 456
imaging, 267
imbalance, 391
immobilization, 332
immutability, 465, 468
impact, 1, 2, 12, 14, 15, 24, 26, 32,
        34, 35, 41, 58, 59, 71, 79,
        89, 103, 104, 124, 125,
        131, 135, 150, 158, 175,
        178–180, 183, 191, 198,
        204, 208, 211, 212,
        214–216, 220, 233, 235,
        250, 253, 261, 264, 267,
        268, 272, 273, 276,
        279–281, 284–287, 289,
        290, 308, 324, 328, 343,
        363, 366, 367, 378, 394,
        395, 399, 401, 402, 404,
        416–422, 425, 443–445,
        456, 460, 473, 482, 514,
        518, 527, 537, 544, 547,
        551, 552, 554, 565, 568,
        574, 579, 594
implementation, 3, 4, 58, 63, 89, 91,
        94, 109, 112, 116, 119,
        139, 164, 202, 213, 215,
        220, 242, 249, 250, 252,
        261, 271, 273–275, 277,
        285, 289, 298, 309, 324,
        344, 351, 352, 359,
        363–367, 371, 374, 376,
        378, 379, 381, 386, 388,
        390, 429, 432, 435, 437,
        441, 444–446, 457, 493,
        494, 500, 501, 504, 505,

507, 512–514, 529, 534, 536, 540, 542, 543, 559, 561, 565, 574, 587, 588, 598

importance, 1–4, 6, 18, 23, 26–28, 40, 43, 46, 51, 58, 62, 65, 67, 68, 73, 84, 93, 94, 98, 99, 102, 104, 110, 112, 125, 128, 135, 136, 157, 160, 174, 204, 216, 228, 229, 239, 267, 280, 289, 309, 330, 345–348, 363, 367, 370, 374, 379, 386, 389, 391, 402, 425, 429, 441, 446, 448, 453, 469, 485–487, 495, 497, 499, 504, 505, 507, 526–528, 540, 572, 573, 581, 585, 594, 595

improvement, 45, 55, 197, 276, 309, 315, 367, 404, 405, 444, 594, 595

in, 1–6, 9–14, 16–18, 21–24, 26–30, 35, 39–46, 49, 50, 54–59, 62, 63, 65–68, 70, 71, 73–80, 82–85, 87, 89, 90, 93–95, 97–99, 101–107, 109–119, 122–125, 127–140, 144, 146, 147, 150, 151, 153, 154, 157–160, 163–165, 167–169, 172–176, 178–184, 186, 187, 189–192, 194–196, 198–204, 208, 210–217, 220, 221, 224, 225, 227, 229, 231–233, 235, 239, 243, 245–247, 249–253, 255–259, 261, 262, 264–282, 284–286, 289–293, 295, 298–302, 305–310, 312, 313, 315–319, 321, 323, 324, 326–335, 337, 338, 340, 343–345, 347–350, 352–356, 359, 362–364, 366, 367, 370, 374, 377–379, 383, 385–392, 394, 395, 399, 400, 402, 404–410, 413, 415–419, 421–423, 425, 426, 429–432, 434–439, 441–447, 449, 452, 456–460, 462–465, 467–469, 471, 472, 475, 476, 478, 479, 481–483, 485–501, 504–510, 513, 514, 517–519, 521, 522, 525–528, 530, 531, 534, 536–540, 544, 547, 550–554, 558, 560–562, 565, 568, 569, 571–574, 577–582, 585, 587–590, 592–598

inbreeding, 45, 58, 68, 73

incineration, 214, 275, 277–280, 282

incinerator, 278–280

inclusion, 350, 390, 509, 510

inclusiveness, 387, 457

inclusivity, 157, 377, 379, 389, 426, 447, 492, 493, 498, 500, 514, 525, 554, 557, 597, 598

income, 65, 117, 124, 129, 131, 159, 164, 413, 420, 430, 432, 496, 506, 521, 525, 530, 585 incorporation, 211

increase, 41, 70, 84, 102, 124, 134, 150, 187, 215, 232, 348, 349, 408, 483, 493, 505, 522, 539

independence, 232, 235, 416

India, 153, 154, 429

individual, 9, 22, 30, 34, 40, 175, 205, 229, 235, 386, 457, 463, 483, 506, 558, 572, 574, 575

indoor, 178, 211, 324, 325, 327, 518, 538

industrialization, 28, 52

industry, 14, 134, 135, 215, 216, 219, 253, 258, 259, 261, 267, 273, 276, 317, 367, 394, 419, 422, 441, 463, 464, 472, 561–564

inequality, 129, 131–134, 158, 159, 164, 416, 457, 526, 587

infiltration, 150

inflammation, 315

influence, 9, 10, 21, 103, 105, 416, 417, 554

information, 9, 71, 174, 208, 273, 318, 352, 377, 378, 388, 430, 436, 443, 444, 446, 467, 469, 476, 493, 505, 510, 513, 540, 589, 595, 597, 598

infrastructure, 5, 29, 31, 35, 42, 85, 87–89, 106, 124, 128, 131, 138–140, 158–160, 164–169, 200, 204, 216, 217, 220, 226–229, 235, 239, 243, 247, 250, 267, 274, 315, 356, 415, 425, 472, 482, 483, 485, 487, 495, 497, 501–504, 507, 522, 523, 525–528, 534, 539, 587, 588, 597

ingestion, 17

inhalation, 328

inhibition, 103

initiative, 14, 157, 258, 513, 582

injustice, 431

innovation, 13, 14, 98, 132, 134, 137, 154, 159, 160, 186, 189, 258, 274, 275, 280, 286, 313, 366, 377, 378, 388, 402, 408, 421, 443, 446, 453, 460, 475, 485, 495, 499, 500, 505, 507, 527, 551, 584, 587, 588, 595

inoculant, 270

input, 204, 205, 425, 537

insect, 150

insecurity, 431

insertion, 332

inspiration, 164, 168, 497, 525

installation, 180, 217, 302, 331, 415

instance, 3–5, 10, 14, 46, 67, 123, 282, 331, 333, 415, 419, 464, 467

instrument, 413, 415

insulation, 137, 159

insurance, 130, 468, 489

integration, 5, 18, 24, 54, 58, 146, 151, 167, 180, 186, 211, 224, 227, 229, 233, 235, 239, 256, 261, 277, 282, 333, 334, 344, 358, 365, 367, 378, 385, 387, 389, 391, 405, 418, 419, 425, 437, 446, 452, 482, 488, 495, 497, 503, 509, 510,

514, 525, 526, 528, 577,
   592, 593, 598
integrity, 52, 79, 90, 110, 118, 416,
   434, 435, 453, 499
intelligence, 5, 389, 455, 457, 488
intensification, 54
intensity, 9, 103, 112, 205, 483
intent, 445
intention, 420
interaction, 10, 472, 508, 534
interconnectedness, 21, 66, 125, 390,
   435, 495, 592
intercropping, 509
interdependence, 498
interest, 2, 410, 413, 415, 419, 422,
   562
interface, 385
interference, 44, 83, 354
intermittency, 135, 186, 189, 193
interoperability, 229, 514
interpretation, 321, 404
intervention, 55, 459, 469, 472
introduction, 27, 36, 68, 88, 89, 308,
   316, 327, 330
intrusion, 483, 493
inventory, 404, 544
investigation, 468
investing, 151, 158, 217, 229, 315,
   415–422, 590
investment, 189, 191, 200, 207, 242,
   247, 285, 408–412,
   416–422, 528
involvement, 56, 58, 98, 118, 272,
   309, 316, 348, 367, 378,
   386, 527
ion, 137, 198, 218, 232, 252
iron, 266, 333
irrigation, 136, 190, 509

island, 30, 102, 158, 507, 526, 530,
   538–540
issuance, 415
issue, 17, 36, 39, 67, 129, 135, 154,
   157, 213, 302, 307, 309,
   314, 316, 327, 330, 387,
   390, 415, 485, 579
issuer, 413
it, 1, 5, 12, 13, 15, 16, 22, 30, 35, 41,
   54, 57, 64, 66–68, 71, 74,
   75, 82, 83, 85, 94, 98, 100,
   103, 105, 106, 113, 128,
   136, 138, 144, 146, 149,
   150, 153, 154, 168, 172,
   180, 181, 186, 187,
   189–191, 198–200, 205,
   212, 214, 215, 221, 225,
   232, 233, 236, 239, 250,
   261, 263, 268–271, 277,
   278, 280, 281, 284, 285,
   291, 301, 302, 304, 305,
   307–309, 321, 323, 324,
   329, 331–333, 347, 350,
   353, 355, 361, 364, 365,
   378, 379, 386, 390, 392,
   393, 403, 405, 420–423,
   428, 438, 453, 456, 457,
   459, 460, 462–465, 468,
   481, 483, 490–493, 497,
   499, 500, 502, 505,
   508–510, 514, 520, 522,
   525, 537, 543, 545, 547,
   550, 552–554, 556, 559,
   574, 577, 578, 587, 595
ivory, 83

Japan, 270
job, 137, 189, 274, 457, 465, 472,
   518, 543, 588, 590

journey, 7
jurisdiction, 443, 444
justice, 130, 132, 134, 377–379,
    385, 387, 430–433, 488,
    494, 496, 500, 505, 507,
    527, 561, 580

K. Eric Drexler, 463
key, 1, 3, 5, 13, 16, 19, 21, 22, 24, 29,
    31–34, 43, 48, 49, 51, 55,
    56, 59, 63, 65, 69, 74–77,
    79, 80, 82, 85–87, 93, 95,
    100, 103, 107, 108, 117,
    119, 128, 132, 133, 138,
    140, 141, 144, 147–149,
    151, 154, 157, 158,
    160–162, 165, 173, 175,
    183, 186, 204, 208, 213,
    215, 216, 221–224, 227,
    231, 232, 236, 237,
    243–245, 247, 251, 256,
    259, 264, 271, 272, 282,
    284, 285, 288, 299, 303,
    305, 308, 317, 319–321,
    328, 329, 343–345, 350,
    354, 356, 364, 367, 371,
    375, 379, 381, 386, 387,
    389, 392, 396, 397, 399,
    400, 405, 413, 420–422,
    425, 426, 433, 434, 436,
    438, 439, 442, 446, 447,
    450, 457, 459, 464, 466,
    469, 476, 483, 489, 494,
    495, 500, 501, 505,
    511–513, 516, 522,
    525–528, 540, 544, 552,
    555, 565, 569, 572, 577,
    579, 580, 585, 587–589,
    595

keystone, 55, 77
kick, 104
knowledge, 1–6, 11, 18, 20, 24, 45,
    46, 59, 62, 66, 67, 70, 74,
    78, 84, 125, 133, 146, 151,
    154, 157, 346, 347, 349,
    359, 386, 388–392, 421,
    423, 429, 430, 432, 434,
    442–446, 449, 450, 452,
    453, 463, 487, 488,
    490–492, 494–500, 503,
    505–510, 572, 574,
    579–582, 589, 590,
    592–598
kudzu, 30

labeling, 273
labor, 131, 132, 150, 399, 416, 561,
    564
laboratory, 318
lack, 30, 127, 129, 208, 345, 356,
    371, 374, 388, 419, 431,
    437, 445, 485, 493, 582,
    587
lake, 190
land, 3, 6, 31, 35, 42, 44, 46, 52–54,
    66, 83, 87, 103, 104, 106,
    109, 110, 117–119, 128,
    133, 144, 150, 153, 158,
    160, 164, 172, 182, 189,
    256, 276, 290, 309, 344,
    356, 364, 383, 391, 425,
    452, 490, 491, 493, 499,
    521, 525, 528, 530, 531,
    534
landfill, 182, 214, 278, 280,
    285–290, 298, 301, 302
landfilling, 282
landing, 471, 472

landmark, 43, 350

landmass, 484

landscape, 20, 35, 87, 104–110, 118, 123, 165, 194, 468, 489, 539

language, 457, 594

larval, 112

lava, 104

law, 84, 130, 204

layer, 104, 270, 298

layout, 539

leachate, 290–295

lead, 29, 30, 42, 46, 52, 67, 68, 71, 89, 90, 110, 117, 123, 136, 191, 215, 229, 307–309, 315, 324, 327, 366, 388, 390, 391, 429, 431, 452, 456, 465, 482, 489, 528, 558, 575, 579, 581, 587

leader, 527

leadership, 132–134, 164, 385, 417, 579, 596

leak, 308, 329

leakage, 218

learning, 55, 178, 250, 267, 326, 385, 439, 443, 444, 446, 453, 457, 459, 471, 486, 495, 497–500, 505, 507, 579, 592, 593, 596–598

leave, 133

lecture, 463

led, 30, 35, 39, 84, 99, 109, 114, 134, 157, 168, 316, 344, 390, 430, 432, 441, 464, 483, 493, 499, 547, 573

ledger, 465

legislation, 133, 363–367, 371, 374, 385

legitimacy, 423, 429, 447

level, 9, 41, 101, 190, 198, 269, 346, 350, 354, 379, 381–385, 387, 388, 399, 429, 431, 439–442, 484, 493, 494, 496, 497, 499, 574, 580, 585

leverage, 66, 159, 388, 417, 446, 476, 481

lie, 139

life, 3, 4, 10, 14, 26–28, 31, 35, 39, 42, 99, 102, 113, 125, 128, 133, 154, 159, 160, 174, 181, 194, 196, 215, 221, 272, 273, 305, 351, 404, 431, 482, 508, 513, 518, 528, 530, 544, 546, 547, 551, 580, 582

lifecycle, 140, 143, 272, 273, 465, 547, 551, 561, 569, 589

lifespan, 190, 200

lifestyle, 589

lift, 421

light, 9, 183, 207, 243–247, 464, 538, 539

lighting, 137, 213, 513

lightning, 103

lime, 322

limestone, 322

limit, 42, 43, 133, 150, 193, 316, 348, 350, 578

line, 12, 14, 167, 246, 349, 350, 565

liner, 293

liquid, 218, 284, 290, 323, 329

list, 168

literacy, 592, 594

literature, 27, 432

lithium, 137, 198, 218, 232, 252

lithography, 463

littering, 462

livability, 158

livelihood, 78, 510

livestock, 27, 54, 66, 146, 298

living, 2, 4, 9–11, 36, 66–68, 175,
        309, 351, 459, 499, 500,
        514, 538

location, 106, 180, 331, 458

logging, 24, 30, 35, 65, 85, 87,
        102–104, 117

longevity, 6, 213, 218

look, 5, 37, 297

loop, 14, 213, 267, 271, 405, 540

Los Angeles County, 289

loss, 4, 6, 26, 27, 29–32, 35, 42, 44,
        45, 52, 55, 57, 59, 67–69,
        77, 82, 83, 87, 88, 91, 94,
        97, 99, 113, 114, 117, 128,
        136, 157, 191, 308, 315,
        328, 386, 446, 489, 504,
        572–574, 580, 590

loyalty, 395

lung, 27, 315

luxury, 30

machine, 267, 457, 471, 472

machinery, 458

mainstream, 67, 389, 391, 421, 507,
        509, 510, 528, 581

maintenance, 70, 183, 207, 302, 326,
        329, 472, 489, 490, 499,
        527

majority, 314, 327, 431

making, 2–6, 45, 52, 54, 65, 70, 75,
        78, 83, 103, 107, 118, 130,
        132–134, 136, 147, 159,
        179, 180, 190, 200, 217,
        225, 251, 261, 269, 279,
        307, 308, 321, 323, 329,
        332, 356, 359, 361, 363,
        367, 374, 376–379, 386,
        387, 389–392, 412, 417,
        418, 422, 423, 425–430,
        432, 435–439, 441, 443,
        444, 446, 449, 457, 465,
        469, 472, 482, 486, 488,
        490, 492, 493, 495, 496,
        499, 500, 505, 506,
        508–510, 513, 526–528,
        537, 539, 574, 577, 589,
        593, 595, 596

malaria, 42

management, 1, 3–6, 11, 18, 23, 24,
        29, 35, 42, 44–52, 54–58,
        62, 66–68, 70, 72–78, 80,
        83–85, 88–90, 94, 96, 98,
        102–104, 109, 110, 112,
        113, 115, 117, 118, 123,
        137, 140, 146, 150, 159,
        160, 172–175, 178, 190,
        191, 198, 213–216, 219,
        220, 251, 256, 258–261,
        264, 265, 267–274,
        276–282, 285–291, 293,
        295, 298, 299, 301, 304,
        306, 309, 316, 321, 326,
        329, 330, 337, 364, 365,
        371, 383, 385, 386, 389,
        391, 396, 402, 404, 416,
        417, 422, 429, 434, 435,
        437–439, 442, 452, 456,
        467, 468, 482, 484, 485,
        487, 490–492, 495, 498,
        500, 507–509, 513, 518,
        527, 569, 573, 589, 592,
        595, 596

managing, 15, 71, 74, 75, 77, 79, 82,
        94, 102, 105, 144, 193,
        317, 320, 326, 356, 359,

371, 374, 456, 486–488, 490, 508
mangrove, 110, 111
manipulation, 463, 465
manner, 140, 278, 290, 353, 364, 386, 422, 425
manual, 118, 265, 267
manufacturing, 123, 135, 214–216, 232, 233, 258, 262, 274, 305, 327, 404, 469, 472, 475, 546, 565–569
manure, 282, 298
map, 471
mapping, 596
marginalization, 67, 129, 496
market, 132, 140, 147, 218, 223, 267, 355, 416, 421
marketing, 396
mass, 279
material, 71, 183, 214, 267–269, 274, 276, 284, 298, 299, 322, 323, 331, 333, 404, 419, 463, 489
matrix, 106
matter, 2, 104, 146, 150, 174, 181, 182, 253, 255, 278, 280, 281, 285, 306, 314–316, 366, 463
maturity, 413, 415
means, 26, 88, 168, 526, 585
measure, 12, 420, 493, 585
measurement, 317, 318, 348, 421, 422, 488
meat, 270, 271
mechanism, 350, 356, 378
medicine, 30, 458, 460, 463, 464
medium, 202, 318
meeting, 3, 11, 355, 389, 430
melting, 41, 483, 510

member, 441, 572
membrane, 313
memory, 464
mentorship, 134, 596
metal, 322
methane, 172, 275, 281, 282, 296–298
methanogenesis, 281
methanol, 285
method, 95, 120, 199, 213, 268, 270, 271, 278–280, 297, 322, 323, 328, 463
methodology, 547
Michigan, 431
microfinance, 134, 420, 421
migration, 30, 44, 49, 68, 106, 136, 181, 291, 299, 329, 330, 333
mile, 167, 471, 472
milling, 463
million, 201, 349
mind, 200, 340, 550, 571
mindset, 14, 494
mini, 127, 168
minimization, 290
mining, 55, 290, 307, 327
misappropriation, 130
mission, 80
Mississippi River, 90
misuse, 457
Mitchell Joachim, 168
mitigation, 6, 38, 40, 42–44, 94, 110, 113, 124, 307, 309, 345, 347, 350, 395, 538–540, 578, 580
mix, 115, 136, 269, 324, 505, 530
mixing, 282, 284
mixture, 269, 270, 275

mobility, 169, 172, 224, 329, 522, 525
mode, 172, 247
model, 6, 14, 50, 94, 113, 140, 229, 239, 267, 271, 275, 394, 408, 530, 540, 554, 588
modeling, 5
moisture, 269, 270, 538
molecule, 458
momentum, 6, 154, 180, 221
money, 413
monitor, 20, 456, 464
monitoring, 5, 43, 57–59, 70, 74, 82, 84, 90, 91, 94, 96, 98, 102, 111, 112, 123, 270, 280, 282, 290, 298, 305, 307, 309, 316–321, 326, 356, 359, 367, 371, 374, 385, 388, 390, 425, 429, 437, 438, 441, 452, 456, 482, 485, 490, 500, 510, 513, 582, 587, 594, 596
monoxide, 276, 284, 314
mortality, 191
motion, 180, 182, 194
motor, 221, 366
Mount St. Helens, 104
mountain, 200
movement, 56, 105, 106, 118, 216, 259, 299, 331, 333, 334, 430, 489, 490, 522, 551, 573
moving, 186
multilateralism, 386
mural, 432
mussel, 111

nanoscale, 463–465
nanotechnology, 5, 463–465

nation, 385
nature, 5, 43, 66, 97, 124, 132, 178, 180, 330, 349, 390, 459, 495, 499, 508, 526, 528, 572, 585
navigation, 251, 259, 261
nectar, 10
need, 5, 35, 52, 57, 58, 72, 74, 88, 93, 99, 100, 104, 107, 116, 121, 135, 137, 139, 150, 153, 154, 158, 161, 174, 175, 178, 182, 185, 186, 208, 213–215, 217, 218, 221, 228, 231–233, 249, 252, 254, 256, 267, 272–274, 277, 278, 280, 282, 285, 286, 288, 293, 297, 298, 320, 322, 336, 347–351, 358, 359, 361, 363, 367, 385, 387, 395, 400, 412–414, 418, 429, 432, 436, 444, 446, 465, 467, 470–472, 476, 477, 485–488, 494, 498, 505, 509, 510, 520, 525–528, 530, 547, 548, 559, 562, 568, 569, 579, 581, 584, 585, 587, 592, 597
negotiation, 350, 386, 444, 574, 575
neighborhood, 521, 534
neighboring, 103
net, 130
Netherlands, 54, 210, 445, 484, 485, 517, 527
network, 106, 158, 164, 217, 221, 226, 229, 316, 467, 472, 513, 525, 530, 536
networking, 134, 443
New York City, 140, 167, 534

newspaper, 269
night, 198
nitrogen, 10, 44, 103, 280, 285, 308, 314, 315
noise, 136, 180, 235, 252, 258
non, 88, 103, 112, 213, 215, 221, 265, 266, 286, 308, 309, 331, 348, 350, 419, 422, 432, 441, 460, 485, 486, 489, 534, 537, 547, 578
nonpoint, 302, 307–310
Norman Borlaug, 157
North America, 67, 90, 267, 537
note, 150, 261, 462, 514, 584
number, 57, 219, 243, 259, 488
nursery, 111
nutrient, 10, 20, 26, 54, 68, 91, 119, 146, 150, 174, 268, 281, 282, 306, 308, 309, 328, 329, 351, 489, 491, 580

objective, 265, 354
observation, 508
occupancy, 482
occupant, 211, 518
ocean, 112, 182, 183
off, 191, 200, 322, 418, 422
offer, 19, 44, 67, 124, 137, 158, 194, 201, 202, 214, 218, 222, 223, 229, 235, 244, 247, 248, 250, 254, 265, 276, 298, 323, 339, 340, 405, 412, 414, 462, 485, 508, 528, 529, 535, 537, 539, 568, 571, 590, 593
offering, 27, 140, 230, 255, 389, 525
office, 210, 517
oil, 276, 302, 307, 310, 331

one, 26, 29, 32, 35, 66, 85, 94, 97, 103, 113, 118, 119, 125, 136, 183, 189, 192, 200, 205, 215, 267, 273, 275, 289, 443, 444, 458, 483, 496, 508, 537, 585, 595
operating, 190, 235, 437
operation, 138, 179, 180, 192, 228, 243, 252, 258, 261, 268, 271, 277, 318, 323, 502, 514
opportunity, 416, 417, 419, 497, 510
opposition, 279
optimization, 213, 214, 220, 261, 274
option, 254, 279, 285, 329, 559
order, 12, 16, 71, 93, 172, 374, 436, 483, 494, 539, 595
Oregon, 246
organic, 65, 104, 146, 150, 153, 154, 174, 181, 214, 253, 255, 268, 270, 271, 275–277, 281, 282, 290, 302, 306, 310, 315, 323, 329, 332, 333, 509
organism, 458
organization, 9, 12, 128, 389
origin, 46, 467
other, 9, 12, 26, 29, 50, 56–58, 66, 90, 101–103, 111, 119, 123, 129, 136, 137, 164, 179, 184, 200, 204, 207, 229, 233, 239, 261, 267, 272, 276, 278, 282, 285, 290, 302, 305, 308–310, 322–324, 328, 331–333, 345, 347, 366, 368, 386, 387, 399, 415, 417, 419, 420, 430, 431, 434, 441,

443, 445, 446, 463, 468,
483, 485, 487, 491, 493,
497, 501, 508, 513, 518,
525, 526, 537, 538, 540,
579, 587, 589, 590, 592
outcome, 438
outdoor, 124, 316, 324, 394, 399,
407, 539
output, 187, 204, 205, 208, 217
outreach, 4, 174, 540
overconsumption, 144
overexploitation, 68, 580
overfishing, 30, 61, 110, 434, 487
overgrazing, 58
overharvesting, 351
ownership, 4, 124, 274, 390, 425,
429, 507, 540, 554, 558,
593
oxidant, 201
oxidation, 196, 284, 313, 323, 332,
333
oxygen, 117, 268, 269, 275, 276,
281, 284, 306, 308
oyster, 111
ozone, 316, 332

pace, 464, 574
Pacific, 26
Pacific Island, 507
packaging, 174, 547–551
pain, 26
panel, 404
panther, 73
paper, 214, 270, 283
paradigm, 478, 497
parallelism, 476
Paraná, 163, 530
park, 44, 49, 58, 140, 167, 168
parking, 482, 513, 539

part, 25, 66, 169, 243, 279, 298,
351, 394, 438, 522, 527,
547, 593
participation, 4, 21, 62, 118, 130,
132–134, 164, 316, 363,
371, 374, 377–379, 385,
387, 389–391, 425, 430,
432, 441, 442, 445, 449,
492, 497, 498, 505, 507,
510, 513, 514, 521, 527,
540, 574, 580, 593
particulate, 182, 278, 280, 285,
314–316, 366
partnership, 385, 585
pass, 182, 322
Patagonia, 394, 395, 399, 407, 408
path, 146, 333, 564
pathway, 151, 275, 412, 485, 497,
590
pattern, 588
pay, 133
payload, 252
peace, 131
peak, 190, 191, 198, 200, 233
pedestrian, 140, 168, 315, 525, 528,
537, 539
peer, 273, 554, 557, 579
penstock, 190
people, 12, 45, 66, 154, 216, 247,
309, 421, 423, 426, 430,
492, 522, 537, 564, 588,
590, 592
perception, 239, 280, 290, 472
performance, 12, 196, 218, 225, 230,
266, 279, 324, 326, 416,
418–420, 422, 441, 463,
518, 545
period, 318
permafrost, 483, 510

permanganate, 332

permeability, 106, 299

perovskite, 186

peroxide, 332

persistence, 105, 133, 462

person, 273

personalization, 475

perspective, 5, 6, 24, 66, 150, 405, 500, 592

pest, 26, 68, 89, 91, 146, 150, 329, 488

pesticide, 456

petroleum, 327, 328, 332

pharmaceutical, 214

phase, 95, 316, 404, 546

phenomenon, 443, 483

phosphorus, 10, 308

photosynthesis, 10, 315

photovoltaic, 183, 323

physicist, 463

physics, 18, 305, 463

phytoremediation, 290, 329, 333, 334, 336, 337

pile, 269, 271

pilot, 131

pioneer, 102–104, 513

pipe, 190

place, 105, 160, 196, 235, 270, 273, 274, 279, 291, 363, 505, 526, 527

plan, 20, 82, 95, 99, 115, 147, 401, 482, 494, 521, 526

planet, 7, 12, 18, 26, 36, 40, 41, 51, 57, 90, 91, 94, 98, 137, 169, 183, 216, 351, 359, 367, 395, 483, 551, 564, 577, 588, 590

plankton, 90

planning, 2, 6, 7, 11, 56–59, 71, 83, 88, 91, 94, 95, 98, 102, 106, 128, 136, 158–161, 163–165, 172, 191, 239, 242, 245, 246, 261, 279, 364, 385, 425, 429, 453, 485, 487, 495, 500, 506, 507, 509, 514, 515, 518, 522, 525–528, 530, 531, 534, 539, 540, 581, 582

plant, 26, 29, 30, 35, 55, 56, 68, 103, 113, 115, 117, 119, 190, 191, 277, 279, 280, 304, 315, 328, 332, 337, 338, 390, 391, 462, 508, 540

planting, 44, 66, 117, 118, 123, 124, 456, 491, 538, 540

plasma, 276

plastic, 17, 18, 39, 83, 264, 276, 359, 547

platform, 347, 446, 513, 572, 580, 598

player, 186

playing, 232, 379, 491

plowing, 308

poaching, 49, 77, 82, 84

point, 302, 304–307, 310, 323

policy, 2, 5, 14, 18, 31, 35, 67, 125, 135, 139, 157, 273, 286, 298, 316, 334, 343, 345, 347, 353, 376, 378, 379, 385, 387, 388, 390, 392, 394, 434, 442–449, 488, 577, 579, 588–590, 596

pollen, 10

pollination, 11, 26, 30, 68, 91, 105, 150, 351

pollutant, 318

polluter, 367

pollution, 1, 2, 4, 6, 17, 18, 27, 29,
36–40, 44, 54, 55, 59, 61,
68, 82, 83, 91, 94, 97, 110,
113, 123, 135, 143, 144,
173, 180, 212, 216,
219–221, 235, 239, 250,
258, 261, 265, 272, 273,
275, 278–280, 291, 298,
302–310, 314–317, 320,
321, 324, 325, 330, 334,
351, 356, 359, 364, 366,
367, 386, 400, 431, 446,
462, 464, 487, 489, 491,
522, 525, 546, 547, 561,
572, 574, 580, 590
pool, 134, 421
pooling, 432
popularity, 150
population, 9, 10, 26, 28, 30, 31, 45,
52, 54, 57, 58, 70, 71, 73,
74, 84, 101, 105, 157, 158,
172, 496, 499, 526, 528
portion, 41, 213, 394, 484
Portland, 246
Portland Streetcar, 246
possibility, 459, 463
potassium, 332
potential, 1, 2, 5, 18, 19, 24, 26, 27,
35, 36, 55, 58, 62, 68, 82,
88, 96, 99, 102, 105, 110,
116, 123–125, 134, 137,
151, 168, 172, 179–183,
185, 186, 189–191, 193,
199–201, 216, 217, 220,
221, 224, 229, 233, 235,
246, 247, 252, 253, 255,
268, 275–277, 279, 282,
286, 289–291, 329, 333,
336, 355, 360, 363, 367,

368, 378, 387, 391, 392,
412, 417–423, 425, 428,
432, 444, 452, 455,
457–460, 463–468, 470,
472, 475, 477, 478, 481,
482, 488, 496, 497, 509,
514, 517, 518, 521, 531,
537, 542, 551, 557, 563,
570, 574, 577, 579, 595
poverty, 1, 14, 109, 110, 125, 127,
129–132, 157, 421, 431,
497, 504, 526, 574, 587
power, 4, 127, 133, 135–137,
180–184, 186–191,
194–196, 198–201, 204,
213, 217, 227, 229, 276,
282, 284, 285, 304, 306,
315, 318, 323, 355, 378,
391, 429, 437, 449, 453,
471, 476, 478, 488,
495–497, 499, 546, 554,
588, 598
practice, 24, 44, 67, 150, 176, 370,
391, 434, 488, 518, 546
pre, 43, 104, 270, 271, 278, 280,
348, 399, 578
precipitation, 29, 41, 42, 83, 483,
509, 578
precision, 54, 316, 329
precursor, 281
predation, 58
predator, 10, 101, 105
prediction, 508
premium, 154
preparedness, 42, 492, 493, 506, 527
presence, 58, 103–105, 268, 269,
326, 456, 489, 538
present, 3, 11, 14, 15, 125, 140, 144,
157, 168, 198, 268, 310,

317, 330, 331, 368,
377–379, 387, 430, 433,
535, 581, 587

preservation, 11, 35, 46, 59, 62, 66,
68, 74, 83, 85, 88, 160,
363, 367, 389, 395, 527,
589

pressure, 31, 52, 101, 180, 284, 322,
443, 490, 526

preventing, 44, 102, 117, 143, 329,
330, 333, 367, 486

prevention, 38, 55, 89, 214, 216,
304, 307, 309, 327, 329,
330, 391

prey, 10, 57, 101, 105

price, 180, 355

pricing, 349, 445

pride, 125, 540

principle, 132, 134, 180, 182, 183,
186, 190, 199, 204, 247,
281, 298, 345, 367, 386,
387, 405, 429, 534, 539,
579

printing, 472–475

privacy, 457, 465, 471, 482, 514

privilege, 496

problem, 443, 446, 592, 594

process, 20, 32, 55, 57, 83, 85, 94,
95, 99, 102–104, 107, 112,
116, 159, 174, 182, 196,
199, 213, 214, 261, 262,
265–271, 275–278,
280–282, 284, 292, 293,
297, 301, 317, 322, 323,
330–333, 350, 360, 362,
363, 404, 415, 418,
425–427, 432, 438, 439,
442–446, 449, 450, 452,
459, 476, 492, 493, 505,

527, 543, 546, 569, 577,
579, 593

processing, 262, 266, 307, 457, 467,
472, 518

procurement, 396

produce, 26, 144, 146, 154,
180–182, 190, 271,
275–277, 280, 285, 323,
518, 540

producer, 144, 175, 274

product, 6, 27, 123, 144, 174, 214,
268, 272, 273, 281, 322,
416, 467, 468, 544, 547,
551

production, 2, 6, 24, 26, 42, 44, 54,
65, 117, 125, 135, 137,
140, 141, 143, 144, 150,
179, 182, 193, 198, 201,
213–216, 235, 252, 253,
255, 256, 272, 274–277,
281–286, 314, 315, 317,
339, 356, 389, 396, 404,
449–453, 465, 475, 499,
505, 521, 551, 554, 561,
565, 571, 588–590

productivity, 29, 42, 44, 52, 68, 109,
117, 132, 144, 150, 154,
208, 324, 327, 328, 330,
489

professional, 593

profile, 418, 419

profiling, 46

profit, 112, 309, 422

profitability, 147

program, 70, 101, 112, 118, 131,
134, 160, 258, 309, 349,
394, 497, 506, 507, 594,
595

programming, 140, 469

progress, 20, 43, 55, 96, 112, 125,
131–134, 232, 252, 306,
309, 316, 352, 358, 383,
388, 417, 443, 446, 485,
579, 585, 595
project, 56, 82, 94, 95, 97–99, 109,
110, 115, 120, 140, 168,
189, 360, 362, 411, 415,
425, 513, 521, 534, 589
prominence, 114, 419
promise, 191, 204, 252, 274, 457,
464, 467, 477, 478, 493,
514, 556
promote, 4, 6, 11, 14, 18, 24, 43, 46,
57, 66, 67, 69, 70, 80, 104,
107, 116, 123, 124, 128,
133, 134, 137, 158, 163,
164, 183, 206, 208, 214,
216, 250, 285, 290, 309,
326, 351–353, 356, 367,
368, 374, 379, 382, 385,
387, 389–391, 407, 413,
417–419, 426, 429, 432,
435, 438, 444, 487, 490,
493, 499, 504, 508, 509,
522, 526, 528, 537, 547,
557, 565, 571, 573, 574,
579–582, 587, 588, 592,
593, 595–597
promotion, 62, 89, 94, 158, 172,
315, 316, 415, 491, 588
propagation, 111
property, 42, 124, 125, 168, 390,
509, 510, 518
proportion, 539
propose, 3, 82, 99, 513, 521
propulsion, 252, 258
prosperity, 158, 275, 528
protecting, 46, 50, 68, 77, 82, 84, 85,
106, 107, 112, 154, 302,
309, 359, 490, 580, 582
protection, 62, 67, 82, 84, 85, 110,
111, 129, 130, 132, 158,
208, 293, 301, 346, 347,
356, 359, 363–365, 367,
371, 374, 385, 430, 457,
490, 496, 500, 510, 526,
574, 575, 588
protocol, 280, 345, 354–356
prototyping, 475
provider, 491
provision, 94, 123, 490, 491, 499,
500, 578
provisioning, 499
proximity, 273
public, 4, 6, 18, 57, 84, 89, 90, 102,
112, 113, 128, 140,
158–160, 164, 167, 172,
204, 214, 219–221,
235–237, 239, 240, 242,
247, 266–268, 280, 290,
303, 308, 309, 315, 316,
330, 363, 366, 367, 371,
374, 377–379, 385, 422,
425, 430–432, 447, 471,
482, 496, 513, 514, 522,
525, 527, 528, 537, 590
Puente Hills Landfill, 289
pump, 191, 199, 200
purchase, 213, 415
purchasing, 551, 552, 554
purification, 56, 113, 351, 464, 488,
580
purpose, 368, 551
pursuit, 179, 195, 575
pyrolysis, 284

quality, 11, 20, 45, 55, 56, 58, 91,

111–113, 123, 125, 128, 130, 133, 136, 154, 158–160, 168, 178, 180, 181, 207, 221, 242, 266, 276, 278, 285, 292, 294, 305–307, 309, 313–321, 324–328, 338, 366, 371, 423, 425, 429, 431, 449, 467, 468, 482, 513, 518, 522, 526, 528, 530, 537, 569

quantification, 544

quantum, 464, 476–478

quarantine, 89

quest, 74, 138, 192, 568

quo, 494, 500

race, 430, 432

radiation, 41, 538, 539

radical, 494

rail, 167, 243–247

railway, 140, 168

rain, 314, 315

rainfall, 24, 41, 42, 483, 509

rainforest, 29, 35, 87, 88, 106

rainwater, 117, 159, 509

raising, 4, 18, 27, 90, 134, 316, 322, 345, 432, 493

range, 26, 30, 44, 57, 68, 76, 80, 90, 91, 99, 102, 103, 105, 112, 113, 117, 119, 123, 125, 136, 148, 158, 217, 218, 235, 252, 267, 270, 271, 275, 276, 285, 310, 314, 331, 332, 334, 339, 350, 378, 379, 422, 423, 426, 447, 456, 458, 462, 464, 470, 489, 490, 495, 507, 528, 555, 578, 580, 585, 596, 597

ranging, 43, 200, 328, 335, 596

rate, 41, 218, 266, 421

ratio, 205

re, 41, 45, 112, 459, 468

reach, 421, 477

reaction, 201, 202, 284, 321, 322, 332

reading, 313, 503

reality, 587

reason, 455

reception, 280

rechargeable, 198

recharging, 217

recognition, 5, 99, 389, 391, 413, 419, 443, 495, 509, 510

recovery, 4, 11, 44, 45, 54–56, 58, 94, 95, 107, 111, 112, 121–123, 213, 266, 268, 274, 277–280, 282, 286, 290, 306, 487

recyclability, 213, 273

recyclable, 265, 267, 268, 272, 273, 277, 278, 280, 588

recycle, 173–175

recycling, 4, 6, 14, 18, 159, 160, 174, 175, 213–216, 261–267, 271–274, 279, 329, 399, 404, 540, 546, 569, 573, 588, 589

reduce, 4, 6, 14, 18, 46, 106, 112, 123, 124, 128, 137, 144, 158, 173, 175, 186, 189, 191, 198, 204, 208, 213–216, 219, 229, 232, 235, 242, 243, 250–253, 258, 261, 265, 267, 268, 271, 272, 275–278, 280,

282, 285, 291, 305, 306,
308, 309, 315, 316, 321,
324, 329, 344, 345, 349,
391, 399, 401, 407, 408,
445, 462, 468, 482, 483,
487, 491, 496, 504, 522,
526, 528, 537–540, 554,
565, 568, 569, 571, 578,
588

reducing, 4, 18, 29, 30, 42, 58, 89,
101, 123, 124, 135, 137,
150, 158, 172, 174, 178,
181–183, 189, 191, 204,
208, 213, 215, 217, 220,
221, 224, 235, 239, 247,
251, 258, 260, 261, 268,
272, 273, 276, 285, 291,
295, 297, 298, 308, 309,
316, 321, 324, 328, 329,
332, 333, 345, 348, 354,
404, 421, 422, 472, 482,
491, 504, 506, 507, 513,
530, 538–540, 551, 554,
568, 588, 589

reduction, 4, 18, 38, 124, 127, 158,
159, 174, 175, 196, 211,
214–216, 218, 258, 271,
272, 278–280, 284, 350,
354–356, 366, 399, 415,
487, 493, 497, 518, 543,
564, 573, 587

redundancy, 26, 471, 487, 489, 526
reef, 29, 62, 111, 112, 487, 491
reestablishment, 55, 56, 117
reference, 55, 510
refining, 379
reflectance, 538, 539
reflection, 432
reforestation, 20, 55, 109, 128, 349,

490
refueling, 204
regeneration, 6, 46, 55, 117, 271,
274, 322, 391, 508
regime, 487
region, 35, 44, 87, 94, 99, 106, 109,
118, 123, 147, 168, 189,
243, 362, 510, 521
regrowth, 117
regulation, 11, 91, 124, 194, 270,
305, 307, 351, 386, 464,
465, 488, 526, 580
rehabilitation, 490
reinforcing, 457
reintroduction, 55, 57–59, 83, 95,
98–102
relatedness, 45
relationship, 1, 3, 5, 10, 11, 508
release, 58, 59, 88, 99, 179, 181, 182,
199, 275, 277–279, 282,
284, 298, 301, 304, 307,
310, 314, 316, 324, 327,
329, 538, 539
relevance, 405, 438, 445, 446
reliability, 189, 191, 405, 471, 472,
482
reliance, 157, 172, 198, 215, 232,
239, 272, 496
relief, 539
remanufacturing, 213, 588, 589
remediation, 38, 298, 299, 301, 302,
327–334, 337, 464
reminder, 347
remnant, 105
removal, 56, 62, 89, 90, 117, 118,
276, 310, 328, 331–334,
491
renewable, 42, 127, 134–137, 158,
159, 178–180, 182, 183,

186, 189–193, 195, 199,
213–216, 220, 229, 230,
233, 255, 272, 274–277,
281, 298, 315, 316, 323,
356, 404, 411, 415–417,
419, 420, 431, 462, 464,
491, 565, 568, 587–590,
596

renewal, 54

renovation, 326

rental, 14

renting, 273, 558

repair, 394, 399, 458

repairability, 14

repairing, 44, 110, 123, 399

replication, 443

report, 352, 420

reporting, 43, 280, 316, 348, 421,
441

representation, 130, 134

representative, 58

reproduction, 9, 30, 83

reputation, 394, 395, 399

request, 378

resale, 399

rescue, 45, 73

research, 4, 5, 11, 14, 18, 57, 74, 90,
98, 104, 115, 122, 151,
154, 186, 189, 195, 199,
203, 204, 218, 252, 253,
258, 266, 279, 280, 286,
313, 319, 337, 390, 418,
444, 449, 452, 463, 468,
487, 488, 510, 521, 552,
554, 579, 588, 594, 596,
597

reselling, 399

reservoir, 136, 137, 180, 190, 191,
193, 199–201

residence, 276

residue, 278

resilience, 2, 5, 6, 11, 14, 24, 26, 42,
46, 56, 59–62, 68, 70, 71,
84, 91, 96, 104, 105, 107,
111–113, 116, 118, 122,
144, 146, 147, 150, 169,
195, 233, 344, 348, 356,
391, 405, 408, 435, 438,
439, 483, 485–491,
493–500, 502, 504–510,
522, 525–528, 543, 578,
580

resistance, 27, 68, 437

resource, 3–6, 11, 14, 46, 54, 62, 66,
67, 70, 136, 138, 140, 143,
150, 175, 178, 180, 182,
212, 213, 216, 220, 256,
268, 271–274, 282, 286,
288, 290, 309, 313, 323,
356, 385, 388, 389, 391,
395, 400, 404, 405, 408,
416, 429, 432, 453, 456,
485, 497, 509, 514, 521,
540, 543, 546, 554, 561,
568, 574, 588, 589

respect, 66, 391, 395, 509, 510, 597

response, 89, 90, 102, 118, 194, 198,
200, 413, 431, 484, 490,
492, 493, 498, 510, 526,
527

responsibility, 4, 12, 27, 40, 144,
174, 175, 208, 274, 392,
394, 408, 416, 417, 419,
547, 579, 580, 592

rest, 280

restoration, 20, 35, 44–46, 55–57,
62, 74, 82, 83, 88–99, 103,
107–125, 309, 337, 356,

371, 456, 490, 495, 499, 508, 589
restoring, 11, 44, 45, 56, 57, 59, 83, 91, 96, 99, 102, 104, 106, 107, 115, 116, 118, 119, 123–125, 334, 453, 490
result, 27, 41, 42, 55, 65, 68, 71, 85, 89, 181, 307, 308, 324, 326, 327, 330, 385, 400, 418, 445
retention, 44, 109, 174, 281, 527
rethinking, 275, 540
return, 44, 58, 413, 415
reuse, 6, 173–175, 213, 271, 329, 394
reusing, 399, 540, 588
revegetation, 110, 115
reverence, 66
review, 349, 378, 438
revision, 349
revitalization, 168
revolution, 233
rhino, 84
rhinoceros, 84, 85
rhizosphere, 332
rice, 493
Richard Feynman, 463
richness, 29
right, 20, 27, 125, 160, 430, 468
rise, 30, 41, 43, 182, 348, 350, 431, 483, 493, 496, 499
risk, 46, 57, 68, 74, 82, 123, 124, 181, 218, 290, 301, 308, 329, 331, 391, 395, 418, 419, 445, 465, 483, 485, 487, 492, 493, 506, 507, 526
river, 107, 123, 190, 191, 371, 431, 437, 485, 527

road, 219, 539
roadmap, 6, 129, 579
Robert Anderson, 221
role, 1–4, 6, 11, 18, 26, 27, 35, 43, 45, 46, 54, 57, 66, 68, 74, 75, 82–84, 89, 90, 94, 98, 101, 103, 105, 107, 110, 112, 113, 116, 117, 119, 122, 124, 129, 130, 136–138, 144, 147, 157, 160, 164, 165, 175, 178, 183, 189, 192, 195, 199, 200, 203, 204, 208, 211, 212, 214, 216, 220, 224, 229, 232, 233, 235, 239, 247, 250, 253, 256, 261, 265, 268, 271, 274, 278, 282, 286, 295, 302, 305, 309, 313, 316, 317, 321, 324, 334, 338, 343, 345, 347, 350, 353, 356, 359, 363, 367, 374, 378, 379, 385, 386, 389, 395, 399, 402, 405, 408, 413, 415, 419, 422, 425, 432, 439, 443–446, 449, 456, 460, 463, 464, 482, 488, 489, 491, 495, 501, 506, 514, 522, 526, 534, 539, 547, 551, 568, 572, 577, 579, 580, 582, 587, 592, 594–596
roof, 540
room, 485, 527
root, 4, 129, 494, 499
rotation, 66, 186, 509
rotor, 180
route, 425, 471
routing, 251

rule, 130

run, 191, 200

runoff, 29, 159, 302, 306–310, 329,
        491

safety, 130, 133, 134, 196, 218, 227,
        250, 261, 416, 459, 460,
        471, 472, 496, 522, 537

sale, 65

saline, 493

salmon, 67

saltwater, 483

sampling, 317, 318

San Francisco, 267

sanitation, 129, 130

satellite, 456

savanna, 10

savannahs, 44, 49

saving, 70, 137, 213, 260, 513

scalability, 136, 150, 284, 468

scale, 56, 94, 102, 103, 106, 109,
        110, 113, 115, 118, 136,
        147, 154, 155, 179, 184,
        189–191, 193, 199, 200,
        206, 216, 252, 268–271,
        316, 331, 422, 443, 463,
        496, 500, 506, 528, 572,
        592

scaling, 590

scarcity, 14, 109, 290, 356, 540

scheme, 355

school, 131, 326

science, 1–6, 11, 15–19, 21, 22, 24,
        25, 43, 52, 59, 66–68, 70,
        74, 78, 82, 90, 94, 107,
        110, 113, 119, 135, 232,
        233, 256, 268, 271, 281,
        295, 305, 317, 337, 343,
        367, 385, 386, 426, 432,

        435, 446–449, 455–457,
        463, 478, 479, 482, 547,
        551, 592–596, 598

scope, 544, 547

screening, 417

scrubbing, 322

scuba, 124

sea, 41, 112, 431, 483, 484, 493,
        496, 499, 510, 578

seafood, 17, 467

seagrass, 110, 111

seating, 167

seawater, 41

second, 204

secrete, 281

section, 1, 2, 12, 26, 29, 32, 35, 36,
        43, 46, 59, 62, 68, 74, 79,
        82, 85, 107, 119, 123, 129,
        135, 140, 154, 157, 165,
        169, 172, 179, 183, 186,
        190, 192, 195, 208, 212,
        216, 230, 235, 243, 247,
        250, 253, 256, 261, 268,
        271, 275, 278, 286, 289,
        298, 302, 307, 310, 314,
        317, 321, 324, 327, 330,
        338, 343, 356, 363, 367,
        374, 379, 386, 389, 395,
        408, 419, 426, 442, 446,
        458, 460, 465, 472, 494,
        501, 504, 514, 518, 534,
        538, 558, 561, 568, 572,
        577, 580, 595

sector, 54, 134, 188, 216, 220, 224,
        253, 315, 347, 366, 386,
        577, 580, 585, 590

security, 42, 137, 144, 154–157,
        182, 183, 189, 204, 255,
        256, 324, 457, 465, 468,

482, 485, 496, 504, 509, 518, 588

sediment, 123, 307–309, 331

sedimentation, 306, 309, 310, 313

seed, 55, 68, 70, 105, 117, 154

segregation, 173, 290

selection, 58, 71, 117, 118, 323, 333, 337, 518, 540

selectivity, 215

self, 178, 218, 390, 463

semiconductor, 463

sense, 4, 57, 124, 174, 425, 521, 540, 592

sensing, 82, 305, 582

sensitivity, 509, 594

sensor, 267, 467, 471

separation, 173–175, 260, 267, 279, 323, 328

sequencing, 582

sequester, 91, 124, 276, 349

sequestration, 111, 113, 118, 119, 276, 492

series, 266, 281, 360, 431

service, 243, 272, 420, 525

set, 75, 91, 136, 183, 204, 266, 305, 356, 359, 368, 374, 378, 391, 395, 399, 421, 486, 500, 512, 514, 543, 578, 585

setting, 6, 67, 95, 355, 364, 366, 371, 580, 588

settlement, 111

severity, 103

sewage, 275, 282, 302, 304, 305, 310

shade, 44, 117, 538, 539

shape, 5, 15, 374, 422, 487, 488

share, 6, 137, 159, 195, 346, 432, 446, 467, 546, 554, 574, 594

shareholder, 417, 418

sharing, 66, 67, 118, 144, 221, 247–250, 252, 272, 273, 347, 349, 386, 388, 389, 391, 421, 432, 437, 445, 446, 495, 505, 507, 525, 554–558, 561, 572, 581, 582, 593, 595–598

sheet, 331

shelf, 547

shellfish, 111

shelter, 10

shift, 6, 14, 172, 274, 275, 419, 478, 494, 495, 497, 499, 537, 540, 551

shopping, 551

shortness, 315

side, 464

signal, 580

significance, 2, 68, 119, 194, 442

Sikkim, 153

silicon, 183

silver, 90

simulation, 471

Singapore, 383–385

site, 95, 99, 104, 117, 200, 213, 299, 301, 331, 333, 334, 337

siting, 180

situ, 331, 333

size, 9, 70, 74, 187, 200, 266, 463

skating, 534

sludge, 275, 282, 306

slurry, 322, 329, 331

smartphone, 546

smog, 314

snake, 30

snorkeling, 124

society, 1, 2, 4, 5, 11, 12, 37, 43, 128, 131, 132, 144, 159, 165,

169, 216, 235, 344, 347, 349, 374, 378, 379, 386, 388, 394, 395, 399, 416, 420, 441, 442, 465, 530, 551, 558, 559, 571, 577–580, 585, 587

socio, 24, 57, 61, 98, 109, 123, 125, 358, 429, 444, 446, 505

soil, 10, 20, 23, 24, 29, 44, 52, 55, 66, 91, 99, 103, 109–111, 117, 118, 144, 146, 147, 150, 154, 173, 174, 269, 271, 276, 291, 298–302, 307, 308, 327–334, 338–340, 356, 456, 489, 493, 499, 509, 538, 594

solar, 4, 127, 135, 136, 179, 180, 183–186, 195, 198, 199, 211, 213, 232, 315, 323, 404, 464, 538, 539, 588

solid, 1, 198, 218, 252, 275–280, 284–286, 306, 322, 329

solidarity, 592

solution, 6, 201, 233, 235, 242, 250, 277, 281, 282, 285, 297, 465

solvent, 323

solving, 443, 446, 478, 592, 594

sorbent, 322

sorting, 262, 265–268, 279, 280

source, 26, 58, 135, 136, 173, 179, 181–183, 186, 189–191, 214, 255, 275–277, 281, 298, 304–310, 323, 324, 327, 331, 333, 390, 431, 546

sourcing, 273, 394, 561, 564

South Africa, 445

Southern Florida, 94

sovereignty, 509

space, 30, 103, 140, 168, 184, 200, 268, 269, 271, 446, 485, 521, 527, 597

Spain, 513

span, 106, 420

spawning, 67

species, 2, 3, 5, 9–11, 20, 26–33, 35, 36, 42–46, 49, 50, 55–59, 62, 68, 70, 71, 73, 74, 77, 82–85, 88–91, 94–96, 98–107, 110–113, 115, 117–119, 183, 308, 332, 353, 356, 391, 434, 435, 456, 487, 489–491, 508, 538, 540, 581

speed, 187, 194, 268, 537

spill, 329, 330

spirit, 574

sprawl, 158, 528

spread, 42, 88–90, 117, 118, 298, 324, 330, 331, 442, 580

stability, 10, 26, 27, 68, 74, 78, 131, 154, 180, 192, 199, 218, 282, 487, 489

stabilization, 329, 330

staff, 326, 445

stage, 281, 378, 404, 546, 547

stakeholder, 2, 24, 94, 98, 110, 112, 136, 140, 242, 385, 395, 402, 422–426, 435, 438, 441, 446, 585

standardization, 252, 412

starting, 104, 109

state, 44, 83, 95, 104, 105, 107, 115, 116, 119, 153, 163, 198, 218, 252, 258, 280, 309, 348, 350, 378, 456, 486, 530

status, 283, 430, 432, 494, 500, 585
steam, 181, 182, 184, 277, 280, 284
steel, 266, 285
step, 315, 344, 355, 526
stewardship, 4, 18, 46, 57, 124, 133, 134, 275, 330, 401, 491, 592, 594
stone, 56
storage, 56, 111, 135–137, 180, 190–195, 198–201, 216, 230–233, 266, 327, 329, 331, 457, 464, 509
store, 113, 137, 190–192, 194, 195, 199, 200, 232, 233, 399, 476, 485, 539
storm, 112, 484
stormwater, 140, 159, 302, 304, 306, 526
story, 160
storytelling, 432
strategy, 44, 57, 62, 84, 89, 100, 213, 221, 298, 315, 323, 394, 401, 419, 526, 528
stream, 266, 284, 322
street, 164
streetcar, 243, 246
strengthening, 62, 347, 348, 367, 385, 388, 449, 492, 496, 507
strike, 183
stroke, 322
structure, 10, 45, 54, 55, 59, 62, 102, 104, 117, 174, 322, 328, 374, 378, 416, 486, 489, 498
study, 9–11, 14, 15, 20, 35, 71, 77, 79, 82, 83, 88, 102, 104–106, 110, 164, 198, 216, 280, 305, 362, 374,

379, 385, 404, 411, 431, 438, 472, 485, 502, 507, 527, 531
subset, 103
subsidiarity, 387, 441
substrate, 102, 111
subsurface, 333
success, 12, 21, 30, 48, 50, 54–56, 58, 59, 68, 71, 94, 100, 110, 112, 113, 115, 118, 160, 164, 168, 172, 174, 189, 226, 229, 233, 246, 285, 337, 394, 399, 408, 417, 420, 438, 445, 518, 537, 572–574, 590, 594
succession, 55, 102–105
suit, 399, 444, 445
suitability, 102, 106, 268, 271
sulfur, 285, 314, 315, 366
summary, 6, 14
sun, 10, 135, 136, 183, 186, 538
sunlight, 10, 41, 135, 183, 184, 323, 538, 539
superposition, 476
supply, 180, 182, 192–195, 199, 200, 233, 235, 273, 277, 285, 310, 395, 396, 399–402, 419, 467, 468, 551
support, 20, 29, 31, 41, 43, 46, 50, 56, 58, 59, 62, 91, 95, 110, 112, 117, 118, 130, 134, 144, 151, 154, 158, 217, 235, 246, 247, 260, 274, 286, 298, 348, 350, 352, 355, 356, 387, 389, 390, 394, 409, 413, 415, 417–419, 423, 425, 432, 439, 441, 445, 472, 488,

493, 499, 500, 507–509,
522, 546, 574, 578–580,
582, 587, 588, 590, 594,
596
surface, 41, 180, 183, 200, 291, 293,
315, 323, 328, 371, 538,
539
surge, 484
surrounding, 104, 106, 124, 140,
168, 290, 291, 298, 299,
301, 465, 538, 539
survival, 3, 9, 28, 30, 43, 45, 57, 58,
61, 62, 68, 70, 71, 83, 84,
105–107, 111, 118, 328,
491
susceptibility, 71
sustainability, 1, 2, 4–6, 11–14,
18–20, 24, 49, 56, 57, 59,
66, 74, 80, 90, 91, 94, 98,
107, 116, 118, 119, 138,
140, 143, 146, 147, 149,
151, 157, 158, 164, 168,
169, 189, 191, 204, 208,
209, 211, 212, 215, 216,
220, 232, 252, 256, 258,
261, 265, 271, 276, 280,
282, 286, 288, 290, 293,
301, 302, 309, 324, 334,
338, 346, 356, 359, 367,
371, 374, 379, 381, 383,
385, 386, 388, 389, 394,
395, 398, 399, 401, 402,
405, 407, 408, 411, 413,
416–419, 426, 432, 434,
446, 453, 456, 457, 460,
467, 468, 475, 481, 482,
487–489, 491, 496,
499–501, 504–508, 510,
514, 518, 521, 526, 547,

557, 558, 561, 565, 568,
572, 574, 581, 583, 584,
589, 592–595, 597, 598
sustenance, 66
Sweden, 445
symbiosis, 273
synchrony, 30
synergy, 273
syngas, 276, 284, 285
synthesis, 276, 284, 463
system, 14, 21–24, 90, 115, 117,
137, 147, 154, 164, 172,
175, 178, 189, 198, 200,
201, 205, 224, 232, 239,
240, 246, 267, 269, 271,
277, 280, 302, 350, 354,
374, 425, 458, 465, 471,
482–484, 486–489, 498,
500, 508, 513, 521,
525–527, 537, 540, 564,
588

take, 4, 6, 14, 37, 112, 147, 166, 174,
196, 235, 260, 270, 271,
297, 344, 348, 387, 390,
420, 485, 540, 588, 596
talent, 134
Tanzania, 44, 49, 50
tar, 284, 285
target, 348, 434, 458, 459, 596
task, 56, 204
tax, 445
tea, 269
team, 20
technique, 117, 174, 289, 298, 301,
328–333
techno, 283
technology, 1, 5, 6, 43, 82, 133, 135,
137, 159, 180, 189, 191,

193, 204, 217, 218, 220, 221, 224, 229–233, 235, 239, 245, 252, 258, 268, 274–276, 279, 281, 284–286, 321, 323, 348, 349, 352, 389, 421, 439, 460, 465–468, 472, 475, 478, 482, 493, 514, 539, 572, 574, 578–580, 587, 593, 597, 598

temperature, 41–43, 112, 184, 269, 270, 276, 281, 290, 322, 323, 348–350, 467, 483, 538

template, 458

terawatt, 189

term, 20, 24, 28, 40, 41, 43, 44, 49, 52, 54, 56–59, 61, 66–68, 70, 71, 73, 74, 78, 84, 90, 94, 98, 102, 107, 110, 112, 113, 115, 118, 140, 144, 146, 151, 157, 160, 164, 191, 207, 247, 288, 301, 302, 309, 314, 315, 318, 319, 326, 334, 337, 349, 350, 395, 399, 400, 408, 417–419, 422, 426, 434, 463, 483, 485, 491, 496, 506–508, 521, 579, 581, 588, 594

termism, 412

termite, 178

terracing, 109, 308

test, 131

testing, 471

textbook, 295

texture, 268

The Chesapeake Bay, 306, 309

the Chesapeake Bay's, 306

the East Coast, 306

the Flint River, 431

the Florida Everglades, 56

the Great Lakes, 90

The Loess Plateau, 109

the United States, 30, 90, 289, 306, 345, 366, 430

theater, 432

theft, 250

thickness, 510

thinking, 5, 18, 21–25, 140, 164, 486–488, 498–500, 592, 594

thinning, 117

threat, 29, 30, 32, 42, 68, 83, 84, 88, 90, 106, 111, 113, 304, 484, 489, 580

tillage, 308

timber, 24, 117, 488

time, 2, 5, 55, 102–104, 194, 200, 217, 227, 235, 251, 260, 268, 276, 281, 290, 305, 316, 318, 326, 327, 421, 437, 449, 464, 482, 495, 513, 595

timing, 30, 510

tobacco, 417

today, 2, 4, 223, 446, 468, 483, 547

tolerance, 27, 103

tomorrow, 129

tool, 57, 59, 74, 102, 174, 208, 320, 359, 361, 363, 367, 403, 405, 426, 445, 458, 545, 547

toolkit, 150

top, 58, 102, 178, 463

topic, 2, 255

topography, 191

total, 101, 189, 201

tourism, 29, 42, 65, 124, 490, 581
tower, 269
toxicity, 214, 215
trace, 181
traceability, 467
track, 280, 305, 456
tracking, 186, 482
traction, 255
trade, 18, 30, 46, 83, 84, 89, 130,
    350, 418, 419, 445, 496
trading, 349, 355, 356
tradition, 153
traffic, 128, 219, 220, 235, 243, 247,
    250, 251, 259–261, 471,
    482, 522, 537
trafficking, 46
training, 130, 134, 445, 492, 506,
    507, 526, 527, 593, 594,
    596, 598
trajectory, 5, 103, 105
tram, 243
transfer, 10, 43, 151, 180, 183, 196,
    284, 348, 349, 352, 389,
    442–446, 538, 574,
    578–580, 596
transformability, 487, 488
transformation, 408, 487
transit, 164, 242, 525, 528
transition, 4, 43, 144, 179, 183, 186,
    189, 203, 213, 229, 233,
    250, 268, 273, 274, 276,
    282, 344, 356, 365, 395,
    399, 408, 413, 415, 416,
    419, 522, 588, 589
translation, 443
translocation, 45
transmission, 66, 89, 415, 509
transparency, 130, 316, 347, 348,
    363, 372, 378, 379, 394,
    395, 399, 416, 421, 422,
    426, 432, 441, 447, 457,
    467
transplantation, 62, 111
transplanting, 111
transport, 158, 159, 239, 525, 528
transportation, 1, 3, 4, 6, 42, 88,
    128, 158, 164, 169–172,
    181, 186, 204, 205, 216,
    218–221, 224, 229, 232,
    235–237, 239, 240, 242,
    243, 246, 247, 250, 256,
    259, 266, 314, 315, 317,
    425, 467–469, 472, 482,
    502, 513, 522, 523, 525,
    526, 528, 531, 534–537
trap, 113
trapping, 322
trash, 482
travel, 89, 202, 220, 229, 242, 247,
    258, 261, 510
treatment, 27, 89, 159, 273, 276,
    278–280, 282, 291–293,
    302, 304–306, 309–313,
    329, 331, 334, 430, 459
treaty, 345, 573
tree, 20, 26, 30, 102, 117, 118, 123,
    160, 538–540
trend, 5, 419, 483, 558
trigger, 102, 467, 468
trip, 193
trust, 423, 426, 447, 457, 467, 509
turbine, 136, 180, 184, 186, 187,
    189–191, 277, 280
turbofan, 251
turn, 58, 124, 190, 282, 447, 499
turnaround, 251
type, 111, 168, 183, 218, 268, 270,
    271, 279, 328, 333, 339,

413

UK, 200
uncertainty, 387, 449, 486, 505, 525, 528
underperformance, 418
understanding, 1–4, 9–11, 15, 18, 21, 24, 25, 28, 31, 36, 39, 46, 51, 58, 66, 71, 79, 82, 88, 98, 103, 107, 119, 122, 125, 147, 150, 199, 284, 295, 313, 317, 320, 327, 374, 389, 390, 419, 423, 433, 434, 460, 468, 483, 485–488, 492, 494, 500, 505, 507, 508, 510, 550–552, 579, 592–596
undertaking, 102
unemployment, 123
United States, 309
universality, 585
unpredictability, 104
unsustainability, 499
up, 136, 172, 184, 261, 267, 268, 275, 286, 290, 307, 330, 332, 334, 421, 443, 459, 463, 464, 485, 528, 590, 598
upcycling, 569–571
upgrade, 306
upskilling, 457
uptake, 10, 328, 332, 536
urbanization, 29, 31, 52, 55, 83, 88, 106, 113, 159, 160, 164, 172, 359, 487, 526, 528
USA, 45, 94, 97, 104, 115
usage, 144, 235, 273, 416, 456, 457, 482, 521

use, 3, 6, 11, 14, 18, 35, 39, 42, 44, 46, 52–54, 58, 62, 66, 67, 74, 78, 83, 87, 89, 90, 106, 111, 113, 115, 117, 128, 135, 137, 140, 144, 146, 150, 158, 160, 164, 172, 182, 192, 198, 205, 213–217, 220, 221, 232, 247, 256, 264, 270–274, 284, 298, 306, 307, 309, 315, 318, 322, 323, 326, 327, 329, 330, 332, 344, 351–353, 356, 364, 371, 390, 391, 404, 420, 425, 449, 452, 456–458, 460, 461, 463, 465, 471, 476, 482, 490, 499, 509, 514, 521, 525, 528, 530–534, 537–540, 544, 546, 547, 554, 557, 565, 573, 574, 581, 588, 589, 593, 598
user, 227
utility, 179, 184, 526
utilization, 6, 154, 178, 181, 183, 261, 272, 286, 323, 521, 554, 561

vacuum, 329
validation, 471
value, 6, 27, 28, 30, 35, 74, 106, 124, 154, 265, 266, 274, 390, 398, 418, 449, 518, 526, 568, 569, 582
Vancouver, 534, 536, 537
vapor, 329, 330
variability, 45, 68, 483, 509
variety, 3, 26, 28, 58, 60, 66, 71, 92, 94, 108, 119, 120, 144,

182, 318, 322, 351, 418, 420, 525, 580
vegetable, 269
vegetation, 29, 30, 44, 46, 58, 101, 102, 104, 105, 109, 117, 118, 168, 538
vehicle, 29, 172, 215, 221, 227, 229, 315, 328
velocity, 284
ventilation, 178, 324, 326, 327, 539
verification, 348, 415, 416
vermicompost, 269
vermicomposting, 269, 271
versatility, 285, 302
vessel, 270, 271, 284
viability, 52, 54, 58, 67, 71, 73, 74, 78, 106, 117, 144, 154, 208, 267, 277, 298, 395, 400, 521
view, 419
village, 429
vine, 30
violence, 129, 133, 134
vision, 587
vitality, 534
voice, 430, 432
voltage, 196, 198
volume, 275–278, 280, 281, 292
voting, 417
vulnerability, 87, 124, 348, 492–494, 504, 505, 507

waiting, 513
Wales, 200
walkability, 128, 158, 530
walking, 158, 528, 534
warming, 3, 43, 483, 578
warning, 42, 485, 492, 493, 495, 497, 527

washing, 328, 330
Washington, 104
waste, 1, 4, 6, 14, 18, 29, 39, 83, 137, 144, 159, 160, 172–175, 181, 182, 204, 205, 208, 211, 213–216, 253, 261, 264–282, 284–286, 289, 290, 293, 295, 298, 299, 301, 307, 309, 316, 323, 327, 329, 330, 364, 399–401, 404, 405, 407, 408, 416, 430, 431, 462, 482, 513, 518, 540, 543, 546, 547, 551, 554, 564, 565, 568, 569, 571, 573, 588, 589, 595
wasteland, 104
wastewater, 159, 302, 304–306, 308, 309
watching, 124
water, 10, 11, 20, 29, 39, 41, 42, 44, 45, 54–56, 83, 91, 94, 97, 109–113, 115, 117, 123, 124, 129, 130, 136, 137, 144, 146, 150, 158, 159, 173, 174, 178, 180, 181, 184, 189–191, 193, 199–201, 211, 277, 290, 291, 293, 294, 301–313, 315, 322, 323, 328, 329, 331, 333, 334, 351, 356, 371, 386, 401, 404, 431, 437, 438, 441, 452, 456, 464, 485, 488, 493, 495, 496, 499, 507, 509, 518, 521, 527, 539, 546, 572, 580, 589
waterborne, 310, 313
waterfront, 537

watershed, 309, 452, 453

waterway, 199

way, 1, 7, 14, 21, 29, 44, 102–104, 136, 146, 172, 175, 178, 186, 208, 220, 229, 233, 253, 258, 290, 377, 390, 457, 463, 486, 494, 497, 525, 538, 540, 551, 569, 581, 587, 590

wealth, 46, 70, 129, 389, 510

weather, 29, 41, 42, 66, 83, 124, 136, 180, 270, 431, 456, 483, 493, 499, 508, 509, 539, 578

weathering, 327

web, 2, 9, 27, 129

webs, 10, 30, 434

website, 399

weed, 268

weight, 215, 251, 252

welfare, 154, 399

well, 3, 5, 9, 10, 12, 15, 24, 26, 28, 32, 35, 36, 39, 40, 43, 62, 65, 66, 74, 78, 79, 85, 90, 91, 102, 110, 125, 133, 138, 140, 144, 146, 154, 157–159, 164, 172, 175, 178, 186, 196, 211, 214, 216, 221, 240, 250, 271, 274, 275, 302, 315, 324, 326, 327, 333, 348, 350, 353, 356, 365, 367, 372, 374, 378, 379, 385, 387, 395, 399, 400, 431, 435, 438, 444, 452, 487, 489, 491, 498, 500, 504, 507–509, 514, 518, 526, 528, 536, 537, 544, 546, 578, 581, 589, 590

wet, 322

wetland, 45, 56, 97, 113–116, 490

wheezing, 315

whole, 22, 36, 123, 314, 559

wild, 57, 70, 74

wildfire, 102, 391

wildlife, 17, 23, 24, 30, 31, 43, 44, 46, 49, 74–78, 83–85, 94, 106, 115, 118, 123, 128, 144, 168, 490

will, 1, 2, 5, 12, 26, 29, 32, 36, 43, 46, 59, 62, 68, 74, 79, 82, 85, 107, 119, 123, 129, 132, 135, 140, 154, 157, 164, 169, 172, 179, 183, 186, 189, 190, 192, 195, 204, 208, 212, 230, 232, 233, 235, 243, 247, 250, 253, 256, 261, 266, 268, 271, 275, 278, 286, 294, 298, 302, 307, 310, 314, 317, 321, 327, 330, 337, 338, 343, 356, 363, 367, 385–387, 389, 395, 402, 405, 419, 426, 458, 460, 463, 465, 472, 482, 497, 501, 504, 514, 518, 534, 538, 551, 558, 568, 572, 577, 579, 580, 582, 584

wind, 4, 134–137, 180, 183, 186–189, 195, 199, 213, 315, 323, 411, 415, 483, 588

windthrow, 103

wisdom, 66, 67

withdrawal, 345

wolf, 58, 59, 101, 102

wood, 323

work, 6, 11, 14, 28, 35, 36, 40, 43,

46, 67, 70, 90, 110, 113,
122, 133, 144, 146, 147,
157, 183, 216, 253, 267,
274, 281, 304, 346, 347,
368, 385, 386, 391, 392,
512, 534, 574, 580, 587,
593, 595, 598

working, 59, 118, 129, 159, 175,
399, 417, 430, 433, 471,
577, 587, 589, 593

world, 2, 6, 11, 12, 15, 25, 43, 50,
59, 62, 66, 67, 82, 87, 94,
97, 104, 118, 122, 129,
132, 135, 143, 153, 154,
157, 172, 189, 190, 192,
194, 198, 200, 210, 220,
224, 225, 229, 233, 242,
247, 349, 356, 372, 399,

415, 421, 425, 430, 432,
433, 435, 446, 467, 471,
477, 485, 496, 504, 507,
517, 518, 525, 526, 533,
534, 542, 547, 551, 554,
587, 593
worldview, 508
worm, 269

yard, 174
year, 168, 270
yeast, 270
Yellowstone, 59, 101
yew, 26
yield, 154, 276
youth, 493, 495

Zimbabwe, 178
zoning, 160, 530, 534